Lecture Notes in Computer Science 11189

Commenced Publication in 1973
Founding and Former Series Editors:
Gerhard Goos, Juris Hartmanis, and Jan van Leeuwen

More information about this series at http://www.springer.com/series/7407

Geno Nikolov · Natalia Kolkovska
Krassimir Georgiev (Eds.)

Numerical Methods and Applications

9th International Conference, NMA 2018
Borovets, Bulgaria, August 20–24, 2018
Revised Selected Papers

 Springer

Editors
Geno Nikolov (iD)
Sofia University "St. Kliment Ohridski"
Sofia, Bulgaria

Krassimir Georgiev (iD)
Bulgarian Academy of Sciences
Sofia, Bulgaria

Natalia Kolkovska
Bulgarian Academy of Sciences
Sofia, Bulgaria

ISSN 0302-9743   ISSN 1611-3349   (electronic)
Lecture Notes in Computer Science
ISBN 978-3-030-10691-1   ISBN 978-3-030-10692-8   (eBook)
https://doi.org/10.1007/978-3-030-10692-8

Library of Congress Control Number: 2018965924

LNCS Sublibrary: SL1 – Theoretical Computer Science and General Issues

This Springer imprint is published by the registered company Springer Nature Switzerland AG
The registered company address is: Gewerbestrasse 11, 6330 Cham, Switzerland

Preface

The ninth issue of the series of international conferences on Numerical Methods and Applications (NMA 2018) held in Bulgaria took place in the beautiful resort Borovets during the period August 20–24, 2018. The conference was organized by the Faculty of Mathematics and Informatics of Sofia University St. Kliment Ohridski, in co-operation with two institutes of the Bulgarian Academy of Sciences: the Institute of Mathematics and Informatics and the Institute of Information and Communication Technologies.

In total, 112 participants from 23 countries all over the world attended the conference. The nice weather and the fresh air of Rila mountain highly contributed to the creative atmosphere of the conference, providing an opportunity for researchers to present their recent achievements, share ideas, continue existing or start new fruitful scientific cooperations.

A wide range of problems concerning both recent theoretical advances in numerical methods and the application of numerical methods in mathematical modeling were discussed at NMA 2018. In total, 92 talks, including four plenary lectures, were delivered at the conference. Five special sessions featured in the scientific program: Numerical Search and Optimization, Problem-driven Numerical Methods, Numerical Methods for Fractional Diffusion Problems, Orthogonal Polynomials and Numerical Quadratures, and Monte Carlo and Quasi-Monte Carlo Methods, along with a stream of talks that formally do not fall into these sessions.

This volume contains 56 papers, based on the talks of the participants at NMA 2018, including the plenary lectures of Jean-Claude Latché (France) and Francisco Gaspar (The Netherlands). The abstracts of the other two plenary lectures, delivered by Jun Hu (China) and Rafael Kruse (Germany), are also presented here. Each of the papers in this volume has passed a single-blind review procedure. We thank all the authors who contributed to the volume.

The success of NMA 2018 would not have been possible without the joint efforts and hard work of many colleagues from various institutions and organizations. We are grateful to all members of the Organizing and Scientific Committees, to the organizers of the special sessions, and to all reviewers. We are also thankful to the local staff for the excellent service.

The conference was partially supported by the Sofia University Research Fund through Grant 80-10-231/2018.

November 2018

Geno Nikolov
Natalia Kolkovska
Krassimir Georgiev

Organization

Organizing Committee Chair

Stefka Dimova   Sofia University St. Kliment Ohridski, Bulgaria

Managing Editors

Stefka Dimova   Sofia University St. Kliment Ohridski, Bulgaria
Rumen Uluchev   Sofia University St. Kliment Ohridski, Bulgaria

International Scientific Committee

Andrey Andreev   Technical University of Gabrovo, Bulgaria
Ján Buša   Technical University of Košice, Slovakia
Tatiana Chernogorova   Sofia University St. Kliment Ohridski, Bulgaria
Boris Chetverushkin   Keldysh IAM and Moscow State University, Russia
Pasqua D'Ambra   Institute for Applied Computing - CNR, Italy
Ivan Dimov   Institute of Information and Communication
                                Technologies, BAS, Bulgaria
István Faragó   Eötvös Loránd University, Hungary
Stefka Fidanova   Institute of Information and Communication
                                Technologies, BAS, Bulgaria
Francisco Gaspar   Centrum Wiskunde & Informatica, The Netherlands
Krassimir Georgiev   Institute of Information and Communication
                                Technologies, BAS, Bulgaria
Snezhana Gocheva-Ilieva   University of Plovdiv Paisii Hilendarski, Bulgaria
Jean-Luc Guermond   Texas A&M University, USA
Stefan Heinrich   University of Kaiserslautern, Germany
Raphaele Herbin   Aix-Marseille Université, France
Jun Hu   Peking University, China
Oleg Iliev   Fraunhofer Institute for Industrial Mathematics ITWM,
                                Germany
Aneta Karaivanova   Institute of Information and Communication
                                Technologies, BAS, Bulgaria
Tzanio Kolev   Lawrence Livermore National Laboratory, USA
Natalia Kolkovska   Institute of Mathematics and Informatics, BAS,
                                Bulgaria
Johannes Kraus   Radon Institute for Computational and Applied
                                Mathematics, Austria
Raphael Kruse   Technische Universität Berlin, Germany
Jean-Claude Latché   Institut de Radioprotection et de Sûreté Nucléaire,
                                France

Raytcho Lazarov	Texas A&M University, USA
Svetozar Margenov	Institute of Information and Communication Technologies, BAS, Bulgaria
Svetoslav Markov	Institute of Mathematics and Informatics, BAS, Bulgaria
Piotr Matus	Institute of Mathematics, Belarus
Gradimir Milovanović	Mathematical Institute of SASA, Serbia
Peter Minev	University of Alberta, Canada
Geno Nikolov	Sofia University St. Kliment Ohridski, Bulgaria
Kalin Penev	Solent University, UK
Bojan Popov	Texas A&M University, USA
Igor Puzynin	Joint Institute for Nuclear Research, Russia
Stefan Radev	Institute of Mechanics, BAS, Bulgaria
Pedro Ribeiro	University of Porto, Portugal
Wil Schilders	TU Eindhoven, The Netherlands
Blagovest Sendov	Institute of Information and Communication Technologies, BAS, Bulgaria
Vidar Thomée	Chalmers University of Technology, Sweden
Michail Todorov	Technical University of Sofia, Bulgaria
Petr Vabishchevich	Nuclear Safety Institute of RAS, Russia
Maria Vasilyeva	M. K. Ammosov North-Eastern Federal University, Russia
Yuri Vassilevski	Institute of Numerical Mathematics, RAS, Russia
Ludmil Zikatanov	The Pennsylvania State University, USA

Organizing Committee

Stefka Dimova	Sofia University St. Kliment Ohridski, Bulgaria
Ana Avdzhieva	Sofia University St. Kliment Ohridski, Bulgaria
Tatiana Chernogorova	Sofia University St. Kliment Ohridski, Bulgaria
Ivan Georgiev	Institute of Information and Communication Technologies, BAS, Bulgaria
Vesselin Gushev	Sofia University St. Kliment Ohridski, Bulgaria
Ivan Hristov	Sofia University St. Kliment Ohridski, Bulgaria
Tihomir Ivanov	Sofia University St. Kliment Ohridski, Bulgaria
Galina Lyutskanova-Zhekova	Sofia University St. Kliment Ohridski, Bulgaria
Nikola Naidenov	Sofia University St. Kliment Ohridski, Bulgaria
Geno Nikolov	Sofia University St. Kliment Ohridski, Bulgaria
Vaia Rakidzi	Sofia University St. Kliment Ohridski, Bulgaria
Rumen Uluchev	Sofia University St. Kliment Ohridski, Bulgaria

Abstract of Invited Talks

Adaptive and Multilevel Mixed Finite Element Methods

Jun Hu

Peking University, Beijing, China
hujun@math.pku.edu.cn

The problems that are most frequently solved in scientific and engineering computing may probably be the elasticity equations. The finite element method (FEM) was invented in analyzing the stress of the elastic structures in the 1950s. The mixed FEM within the Hellinger–Reissner (H-R) principle for elasticity yields a direct stress approximation since it takes both the stress and displacement as an independent variable. The mixed FEM can be free of locking for nearly incompressible materials, and be applied to plastic materials, and approximate both the equilibrium and traction boundary conditions more accurate. However, the symmetry of the stress plus the stability conditions make the design of the mixed FEM for elasticity surprisingly hard. In fact, "Four decades of searching for mixed finite elements for elasticity beginning in the 1960s did not yield any stable elements with polynomial shape functions" [D. N. Arnold, Proceedings of the ICM, Vol. I: Plenary Lectures and Ceremonies (2002)]. Since the 1960s, many mathematicians have worked on this problem but compromised to weakly symmetric elements, or composite elements. In 2002, using the elasticity complexes, Arnold and Winther designed the first family of symmetric mixed elements with polynomial shape functions on triangular grids in 2D.

The first part of the talk presents a new framework to design and analyze the mixed FEM of elasticity problems, which yields optimal symmetric mixed FEMs. In addition, those elements are very easy to implement since their basis functions, based on those of the scalar Lagrange elements, can been explicitly written down by hand. The main ingredients of this framework are a structure of the discrete stress space on both simplicial and product grids, two basic algebraic results, and a two-step stability analysis method.

The second part of the talk gives a unified analysis of both convergence and optimality of adaptive mixed finite element methods for a class of problems when the finite element spaces and corresponding a posteriori error estimates satisfy five hypotheses. We prove that these five conditions are sufficient for convergence and optimality of the adaptive algorithms under consideration. The main ingredient for the analysis is a new method to analyze both discrete reliability and quasi-orthogonality. As applications, we prove these five hypotheses for the Raviart–Thomas and Brezzi–Douglas–Marini elements of the Poisson and Stokes problems in both two and three dimensions. To extend the above result to linear elasticity problems, we propose a reliable and efficient a posteriori error estimator for the symmetric mixed finite element methods for linear elasticity problems. In addition, we construct nested mixed finite elements by relaxing C^0 continuity of the existing mixed elements in the literature.

The third part of the talk constructs a block diagonal preconditioner with the minimal residual method and a block triangular preconditioner with the generalized minimal residual method for the symmetric mixed finite element methods of linear elasticity. A fast auxiliary space preconditioner based on the H^1 conforming linear element of the linear elasticity problem is then designed for solving the Schur complement. For both diagonal and triangular preconditioners, it is proved that the conditioning numbers of the preconditioned systems are bounded above by a constant independent of both the crucial Lamé constant and the mesh-size.

Error Analysis of Randomized Time-Stepping Methods for Non-autonomous Evolution Equations with Time-Irregular Coefficients

Raphael Kruse

Technische Universität Berlin, Straße des 17. Juni 136, 10623 Berlin, Germany
kruse@math.tu-berlin.de

In this talk, we consider the numerical approximation of Carathéodory-type differential equations of the form

$$u'(t) = f(t, u(t)), \quad t \in (0, T), \quad u(0) = u_0,$$

and of nonlinear and non-autonomous evolution equations of the form

$$u'(t) + \mathcal{A}(t)u(t) = f(t), \quad t \in (0, T), \quad u(0) = u_0,$$

where f and $\mathcal{A}$ may be discontinuous with respect to the time variable. In this non-smooth situation, it is notoriously difficult to construct numerical algorithms with a positive convergence rate. In fact, it can be shown that any deterministic algorithm depending only on point evaluations may fail to converge if, for instance, $\mathcal{A}$ and f only satisfy an L^2-integrability condition with respect to t.

Instead, we propose to apply randomized Runge–Kutta methods to such time-irregular evolution equations as, for instance, a randomized version of the backward Euler method. We obtain positive convergence rates with respect to the mean-square norm under considerably relaxed temporal regularity conditions. An important ingredient in the error analysis consists of a well-known variance reduction technique for Monte Carlo methods, the stratified sampling. We demonstrate the practicability of the new algorithm in the case of a fully discrete approximation of a more explicit parabolic PDE.

This talk is based on joint works [1, 2] with Monika Eisenmann (Technische Universität Berlin), Mihály Kovács and Stig Larsson (both Chalmers University of Technology) as well as Yue Wu (University of Edinburgh).

References

1. Eisenmann, M., Kovács, M., Kruse, R., Larsson, S.: On a randomized backward Euler method for nonlinear evolution equations with time-irregular coefficients (2017, Preprint). arXiv: 1709.01018
2. Kruse, R., Wu, Y.: Error analysis of randomized Runge-Kutta methods for differential equations with time-irregular coefficients. Comput. Methods Appl. Math. **17**(3), 479–498 (2017). https://doi.org/10.1515/cmam-2016-0048

Contents

Problem-Driven Numerical Method: Motivation and Application

Contributed Talks

Invited Papers

New Stabilized Discretizations
for Poroelasticity Equations

Francisco J. Gaspar[1]([✉]) [iD], Carmen Rodrigo[2] [iD], Xiaozhe Hu[3] [iD],
Peter Ohm[3], James Adler[3] [iD], and Ludmil Zikatanov[4] [iD]

[1] Centrum Wiskunde & Informatica (CWI), Science Park 123,
1090 Amsterdam, The Netherlands
[2] IUMA and Department of Applied Mathematics,
University of Zaragoza, Zaragoza, Spain
F.J.Gaspar@cwi.nl
[3] Department of Mathematics, Tufts University, Medford, MA 02155, USA
[4] Department of Mathematics, The Pennsylvania State University,
University Park, PA 16802, USA

Abstract. In this work, we consider two discretizations of the three-field formulation of Biot's consolidation problem. They employ the lowest-order mixed finite elements for the flow (Raviart-Thomas-Nédélec elements for the Darcy velocity and piecewise constants for the pressure) and are stable with respect to the physical parameters. The difference is in the mechanics: one of the discretizations uses Crouzeix-Raviart nonconforming linear elements; the other is based on piecewise linear elements stabilized by using face bubbles, which are subsequently eliminated. The numerical solutions obtained from these discretizations satisfy mass conservation: the former directly and the latter after a simple postprocessing.

Keywords: Stable finite elements · Poroelasticity equations
Mass conservation

1 Introduction

The interaction between the deformation and fluid flow in a fluid-saturated porous medium is the object of study in poroelasticity theory. Such coupling was already modeled in the early one-dimensional work of Terzaghi [1]. A more general three-dimensional mathematical formulation was later established by Maurice Biot in several pioneering publications [2,3]. Biot's models are widely used nowadays in the modeling of many applications in different fields, ranging from geomechanics and petroleum engineering, to biomechanics. The existence and uniqueness of the solution for these problems have been investigated by Showalter in [4] and by Ženíšek in [5]. Regarding the numerical simulation of the poroelasticity equations, there have been numerous contributions using finite-difference schemes [6,7] and finite-volume methods (see [8,9] for recent developments). Finite-element methods, which are the subject of this work, have also

© Springer Nature Switzerland AG 2019
G. Nikolov et al. (Eds.): NMA 2018, LNCS 11189, pp. 3–14, 2019.
https://doi.org/10.1007/978-3-030-10692-8_1

been considered (see for example the monograph by Lewis and Schrefler [10] and the references therein).

For the three-field formulation, which includes as unknowns, the displacements, the pressure, and the Darcy velocity, several conforming and nonconforming discretizations involving Stokes-stable finite-element spaces were proposed in recent years. For instance, a stable finite-element method based on non-conforming Crouzeix-Raviart finite elements for the displacements, lowest order Raviart-Thomas-Nédélec elements for the Darcy velocity, and piecewise constants for the pressure was proposed in [11]. In [12], a family of *parameter-robust* three-field finite-element schemes were proposed and analyzed and a general theory for the error analysis was introduced. Additionally, a novel three-field formulation based on displacement, pressure, and total pressure was proposed in [13] with error estimates independent of the Lamé constants, yielding a locking-free approach. Furthermore, in [14], one finds a parameter-robust error analysis and optimal preconditioning techniques for several discretizations of three-field formulations for Biot's model.

2   Preliminaries: Model Problem and Notation

We consider the quasi-static Biot model for soil consolidation in a linearly elastic, homogeneous, and isotropic porous medium saturated by a Newtonian fluid. The weak form of Biot's three-field consolidation model is given as: For each $t \in (0, T]$, find $(\boldsymbol{u}(t), \boldsymbol{w}(t), p(t)) \in \boldsymbol{V} \times \boldsymbol{W} \times Q$ such that

$$a(\boldsymbol{u}, \boldsymbol{v}) - (\alpha p, \operatorname{div} \boldsymbol{v}) = (\rho \boldsymbol{g}, \boldsymbol{v}), \quad \forall \, \boldsymbol{v} \in \boldsymbol{V}, \tag{1}$$

$$(\boldsymbol{K}^{-1} \mu_f \boldsymbol{w}, \boldsymbol{r}) - (p, \operatorname{div} \boldsymbol{r}) = (\rho_f \boldsymbol{g}, \boldsymbol{r}), \quad \forall \, \boldsymbol{r} \in \boldsymbol{W}, \tag{2}$$

$$\left(\frac{1}{M} \frac{\partial p}{\partial t}, q \right) + \left(\alpha \operatorname{div} \frac{\partial \boldsymbol{u}}{\partial t}, q \right) + (\operatorname{div} \boldsymbol{w}, q) = (f, q), \quad \forall \, q \in Q, \tag{3}$$

where,

$$a(\boldsymbol{u}, \boldsymbol{v}) = 2\mu \int_\Omega \boldsymbol{\varepsilon}(\boldsymbol{u}) : \boldsymbol{\varepsilon}(\boldsymbol{v}) + \lambda \int_\Omega \operatorname{div} \boldsymbol{u} \operatorname{div} \boldsymbol{v}. \tag{4}$$

The initial condition at $t = 0$ is $p + M\alpha \operatorname{div} \boldsymbol{u} = 0$ for $\boldsymbol{x} \in \Omega$. Further, λ and μ are the Lamé coefficients, M is the Biot modulus, α is the Biot-Willis constant, $\boldsymbol{K}$ stands for the absolute permeability tensor, and μ_f is the viscosity of the fluid. The unknown functions are the displacement vector, $\boldsymbol{u}$, the pore pressure, p, and the Darcy velocity, $\boldsymbol{w}$. The function spaces used in the variational form are

$$\boldsymbol{V} = \{ \boldsymbol{u} \in \boldsymbol{H}^1(\Omega) \mid \boldsymbol{u}|_{\overline{\Gamma}_c} = \boldsymbol{0} \},$$
$$\boldsymbol{W} = \{ \boldsymbol{w} \in \boldsymbol{H}(\operatorname{div}, \Omega) \mid (\boldsymbol{w} \cdot \boldsymbol{n})|_{\Gamma_c} = 0 \},$$
$$Q = L^2(\Omega).$$

Notice that these function spaces incorporate, as usual, the considered boundary conditions, which here are the following:

$$p = 0, \quad \text{for} \quad x \in \overline{\Gamma}_t, \quad \boldsymbol{\sigma}' \boldsymbol{n} = \boldsymbol{0}, \quad \text{for} \quad x \in \Gamma_t, \tag{5}$$

$$\boldsymbol{u} = \boldsymbol{0}, \quad \text{for} \quad x \in \overline{\Gamma}_c, \quad \frac{\partial p}{\partial \boldsymbol{n}} = 0, \quad \text{for} \quad x \in \Gamma_c, \tag{6}$$

with $\overline{\Gamma} = \overline{\Gamma}_t \cup \overline{\Gamma}_c$ and Γ_t and Γ_c open (with respect to Γ) subsets of Γ with nonzero measure. The well-posedness of the continuous problem was established by Showalter [4]. Next, we focus on the discretizations of Biot's model.

2.1 Discretizations

We partition Ω into shape regular (bounded ratio of the diameter of the simplex and the radius of the inscribed ball) n-dimensional simplices, so that we have a valid triangulation, that is, the mesh is a n-homogenous simplicial complex and $\overline{\Omega} = \cup_{T \in \mathcal{T}_h} \overline{T}$. We denote the partition with $\mathcal{T}_h$, and we associate a triple of piecewise polynomial, finite-dimensional spaces,

$$\boldsymbol{V}_h \subset \boldsymbol{V}, \quad \boldsymbol{W}_h \subset \boldsymbol{W}, \quad Q_h \subset Q. \tag{7}$$

While we specify two choices of the spaces $\boldsymbol{V}_h$ later, we fix $\boldsymbol{W}_h$ and Q_h as follows,

$$\boldsymbol{W}_h = \{\boldsymbol{w}_h \in \boldsymbol{W} \mid \boldsymbol{w}_h|_T = \boldsymbol{a} + \eta \mathbf{x}, \ \boldsymbol{a} \in \mathbb{R}^d, \ \eta \in \mathbb{R}, \ \forall T \in \mathcal{T}_h\},$$

$$Q_h = \{q_h \in Q \mid q_h|_T \in \mathbb{P}_0(T), \ \forall T \in \mathcal{T}_h\},$$

where $\mathbb{P}_0(T)$ is the one-dimensional space of constant functions on T. We note that the inclusions listed in (7) imply that the elements of $\boldsymbol{V}_h$ are continuous on Ω, the functions in $\boldsymbol{W}_h$ have continuous normal components across element boundaries, and that the functions in Q_h are in $L^2(\Omega)$. This choice of the pair $(\boldsymbol{W}_h, Q_h)$ is the standard lowest order Raviart-Thomas-Nédélec space and the piecewise constant space (P0) (see [15–17]). For time discretization, we use a backward Euler scheme with constant time-step size τ. The discrete scheme corresponding to the three-field formulation (1)–(3) reads:

Find $(\boldsymbol{u}_h^m, \boldsymbol{w}_h^m, p_h^m) \in \boldsymbol{V}_h \times \boldsymbol{W}_h \times Q_h$ such that

$$a(\boldsymbol{u}_h^m, \boldsymbol{v}_h) - (\alpha p_h^m, \operatorname{div} \boldsymbol{v}_h) = (\rho \boldsymbol{g}, \boldsymbol{v}_h), \quad \forall \, \boldsymbol{v}_h \in \boldsymbol{V}_h, \tag{8}$$

$$\tau(\boldsymbol{K}^{-1} \mu_f \boldsymbol{w}_h^m, \boldsymbol{r}_h) - \tau(p_h^m, \operatorname{div} \boldsymbol{r}_h) = \tau(\rho_f \boldsymbol{g}, \boldsymbol{r}_h), \quad \forall \, \boldsymbol{r}_h \in \boldsymbol{W}_h, \tag{9}$$

$$\left(\frac{1}{M} p_h^m, q_h\right) + (\alpha \operatorname{div} \boldsymbol{u}_h^m, q_h) + \tau(\operatorname{div} \boldsymbol{w}_h^m, q_h) = (\widetilde{f}, q_h), \quad \forall \, q_h \in Q_h, \tag{10}$$

where $(\widetilde{f}, q_h) = \tau(f, q_h) + \left(\frac{1}{M} p_h^{m-1}, q_h\right) + \left(\alpha \operatorname{div} \boldsymbol{u}_h^{m-1}, q_h\right)$, and,

$$(\boldsymbol{u}_h^m, \boldsymbol{w}_h^m, p_h^m) \approx (\boldsymbol{u}(\cdot, t_m), \boldsymbol{w}(\cdot, t_m), p(\cdot, t_m)), \quad t_m = m\tau, \ m = 1, 2, \dots$$

2.2   Some Additional Notation

We consider the set of $(d-1)$ dimensional faces from $\mathcal{T}_h$ and denote this set by $\mathcal{E} = \mathcal{E}^o \cup \mathcal{E}^\partial$, where $\mathcal{E}^o$ is the set of interior faces (shared by two elements) and $\mathcal{E}^\partial$ is the set of faces on the boundary. In addition, $\mathcal{E}^{\Gamma_t}$ is the set of faces on the boundary Γ_t and $\mathcal{E}^{o,t} = \mathcal{E}^o \cup \mathcal{E}^{\Gamma_t}$. Note, if $\Gamma_t = \partial\Omega$ (pure traction boundary condition), then $\mathcal{E}^{\Gamma_t} = \mathcal{E}^\partial$ and $\mathcal{E}^{o,t} = \mathcal{E}$. For any face $e \in \mathcal{E}^o$, such that $e \in \partial T$, and $T \in \mathcal{T}_h$, let $\boldsymbol{n}_{e,T}$ be the outward (with respect to T) unit normal vector to e. With every face $e \in \mathcal{E}^o$, we also associate a unit vector $\boldsymbol{n}_e$ which is orthogonal to it. Clearly, if $e \in \partial T$ we have $\boldsymbol{n}_e = \pm\boldsymbol{n}_{e,T}$. For the boundary faces $e \in \mathcal{E}^\partial$, we always set $\boldsymbol{n}_e = \boldsymbol{n}_{e,T}$, where T is the <u>unique</u> element for which we have $e \subset \partial T$. For the interior faces, the particular direction of $\boldsymbol{n}_e$ is not important, although it is important that this direction is fixed. More precisely,

$$\boldsymbol{n}_e = \boldsymbol{n}_{e,T^+} = -\boldsymbol{n}_{e,T^-} \quad \text{if} \quad e = T^+ \cap T^-, \quad \text{and} \quad T^\pm \in \mathcal{T}_h. \tag{11}$$

Further, with every face $e \in \mathcal{E}$, $e = T^+ \cap T^-$, we associate a vector-valued function $\boldsymbol{\Phi}_e$,

$$\boldsymbol{\Phi}_e = \varphi_e \boldsymbol{n}_e, \quad \text{with} \quad \varphi_e\Big|_{T^\pm} = \varphi_{e,T^\pm}, \quad \text{and} \quad \varphi_{e,T^\pm} = \prod_{k=1,k\neq j^\pm}^{d+1} \lambda_{k,T^\pm}, \tag{12}$$

where $\lambda_{k,T^\pm}$, $k = 1,\dots,(d+1)$ are barycentric coordinates on $T^\pm$ and $j^\pm$ is the vertex opposite to the face e in $T^\pm$. We note that $\boldsymbol{\Phi}_e \in V$ is a continuous piecewise polynomial function of degree d.

3   Conforming Choice of Displacement Space

We first introduce a well-known stabilization technique based on enrichment of the piecewise linear continuous finite-element space, $\boldsymbol{V}_{h,1}$, with edge/face (2D/3D) bubble functions (see [18, pp. 145–149]). The discretization described below is based on a Stokes-stable pair of spaces $(\boldsymbol{V}_h, Q_h)$ with $\boldsymbol{V}_h \supset \boldsymbol{V}_{h,1}$ and follows [18]. The stabilized finite-element space $\boldsymbol{V}_h$ is defined as

$$\boldsymbol{V}_h = \boldsymbol{V}_{h,1} \oplus \boldsymbol{V}_b, \quad \boldsymbol{V}_b = \text{span}\{\boldsymbol{\Phi}_e\}_{e \in \mathcal{E}^{o,t}}. \tag{13}$$

The degrees of freedom associated with $\boldsymbol{V}_h$ are the values at the vertices of $\mathcal{T}_h$ and the total flux through $e \in \mathcal{E}^{o,t}$ of $(I - \Pi_1)\boldsymbol{v}_h$, where Π_1 is the standard piecewise linear interpolant, $\Pi_1 : C(\overline{\Omega}) \mapsto \boldsymbol{V}_{h,1}$. Then, the canonical interpolant, $\Pi : C(\overline{\Omega}) \mapsto \boldsymbol{V}_h$, is defined as:

$$\Pi\boldsymbol{v} = \Pi_1\boldsymbol{v} + \sum_{e \in \mathcal{E}^{o,t}} v_e \boldsymbol{\Phi}_e, \quad v_e = \frac{1}{|e|}\int_e (I - \Pi_1)\boldsymbol{v}.$$

With this choice of $\boldsymbol{V}_h$, the variational form, (8)–(10), remains the same and we have the following block form of the discrete problem:

$$\mathcal{A}\begin{pmatrix} \boldsymbol{U}_b \\ \boldsymbol{U}_l \\ \boldsymbol{W} \\ \boldsymbol{P} \end{pmatrix} = \boldsymbol{b}, \quad \text{with} \quad \mathcal{A} = \begin{pmatrix} A_{bb} & A_{bl} & 0 & G_b \\ A_{bl}^T & A_{ll} & 0 & G_l \\ 0 & 0 & \tau M_w & \tau G \\ G_b^T & G_l^T & \tau G^T & -M_p \end{pmatrix}, \tag{14}$$

where U_b, U_l, W and P are the unknown vectors for the bubble components of the displacement, the piecewise linear components of the displacement, the Darcy velocity, and the pressure, respectively. The blocks in the definition of $\mathcal{A}$ correspond to the following bilinear forms:

$$a(u_h^b, v_h^b) \to A_{bb}, \quad a(u_h^l, v_h^b) \to A_{bl}, \quad a(u_h^l, v_h^l) \to A_{ll},$$
$$-(\alpha p_h, \operatorname{div} v_h^b) \to G_b, \quad -(\alpha p_h, \operatorname{div} v_h^l) \to G_l, \quad -(p_h, \operatorname{div} r_h) \to G,$$
$$(K^{-1}\mu_f w_h, r_h) \to M_w, \quad \left(\frac{1}{M}p_h, q_h\right) \to M_p,$$

where $u_h = u_h^l + u_h^b$, $u_h^l \in V_{h,1}$, $u_h^b \in V_b$, and an analogous decomposition for v_h. As shown in [19] the block A_{bb} can be replaced by a diagonal matrix and then all bubbles can be eliminated by static condensation.

Following the parameter robust analysis in [12] we introduce a norm on $V_h \times W_h \times Q_h$:

$$|||(u_h, w_h, p_h)||| := \left[\|u_h\|_A + \tau \|w_h\|_{K^{-1}\mu_f}^2 + \tau^2 \xi^{-1} \|\operatorname{div} w_h\|^2 + \xi \|p_h\|^2 \right]^{1/2}. \tag{15}$$

Above, we have defined $\xi = \frac{\alpha^2}{\zeta^2} + \frac{1}{M}$ where $\zeta = \sqrt{\lambda + 2\mu/d}$, and $\|r\|_{K^{-1}\mu_f} := (K^{-1}\mu_f r, r)^{1/2}$. We introduce the composite bilinear form on the space $V_h \times W_h \times Q_h$,

$$B(u_h, w_h, p_h; v_h, r_h, q_h) := a^D(u_h, v_h) - (\alpha p_h, \operatorname{div} v_h) + \tau(K^{-1}\mu_f w_h, r_h)$$
$$-\tau(p_h, \operatorname{div} r_h) - \left(\frac{1}{M}p_h, q_h\right) - (\alpha \operatorname{div} u_h, q_h) - \tau(\operatorname{div} w_h, q_h).$$

Note that the bilinear form for the mechanics part of the model, $a(\cdot, \cdot)$, has been replaced by a bilinear form with a diagonal matrix, $a^D(\cdot, \cdot)$. Since these two forms are spectrally equivalent (see [19]) we have the following theorem.

Theorem 1. *If the triple (V_h, W_h, Q_h) is Stokes-Biot stable, then:*

$B(\cdot, \cdot, \cdot \; ; \; \cdot, \cdot, \cdot)$ is continuous with respect to $|||(\cdot, \cdot, \cdot)|||$; and the following inf-sup condition holds.

$$\sup_{(v_h, r_h, q_h) \in V_h \times W_h \times Q_h} \frac{B(u_h, w_h, p_h; v_h, r_h, q_h)}{|||(u_h, w_h, p_h)|||} \geq \gamma |||(v_h, r_h, q_h)|||, \tag{16}$$

with a constant $\gamma > 0$ independent of mesh size h, time step size τ, and the physical parameters.

For the definition of Stokes-Biot stability and the proofs of the spectral equivalence of $a(\cdot, \cdot)$ and $a^D(\cdot, \cdot)$ we refer to [19]. As a result, we have the following error estimates for the fully discrete problem.

Theorem 2. *Let u, w, and p be the solutions of (1)–(3) and u_h^m, w_h^m, and p_h^m be the solutions of the fully discrete Biot's system. If the following regularity assumptions hold,*

$$u(t) \in L^\infty\left((0,T], \mathbf{H}_0^1(\Omega) \cap \mathbf{H}^2(\Omega)\right),$$

$$\partial_t u \in L^1\left((0,T], \mathbf{H}^2(\Omega)\right), \ \partial_{tt} u \in L^1\left((0,T], \mathbf{H}^1(\Omega)\right),$$

$$w(t) \in L^\infty\left((0,T], H_0(\mathrm{div}, \Omega) \cap \mathbf{H}^1(\Omega)\right),$$

$$p \in L^\infty\left((0,T], H^1(\Omega)\right), \ \partial_t p \in L^1\left((0,T], H^1(\Omega)\right),$$

then,

$$\|(u(t_m) - u_h^m, w(t_m) - w_h^m, p(t_m) - p_h^m)\|_\tau \leq c \left\{ \|e_u^0\|_1 + \frac{1}{M}\|e_p^0\| + \tau \int_0^{t_m} \|\partial_{tt} u\|_1 \mathrm{d}t \right.$$

$$\left. + h\left[\|u\|_2 + \tau^{1/2}\|w\|_1 + \|w\|_1 + \|p\|_1 + \int_0^{t_m} (\|\partial_t u\|_2 + \|\partial_t p\|_1)\,\mathrm{d}t \right] \right\}, \qquad (17)$$

where $\|(u, w, p)\|_\tau^2 := \|u\|_1^2 + \tau\|w\|_{K^{-1}\mu_f}^2 + \left(\frac{1}{M} + 1\right)\|p\|^2.$

3.1   Implementation Issues

Since $a^D(\cdot, \cdot)$ has a diagonal matrix representation corresponding to the bubbles' space, we can eliminate these degrees of freedom obtaining the same degrees of freedom as in the original P1-RT0-P0 method for the three-field formulation. After eliminating such unknowns we obtain a (3×3) block discrete linear system:

$$\widehat{\mathcal{A}}^D = \begin{pmatrix} A_{ll} - A_{bl}^T D_{bb}^{-1} A_{bl} & 0 & G_l - A_{bl}^T D_{bb}^{-1} G_b \\ 0 & \tau M_w & \tau G \\ G_l^T - G_b^T D_{bb}^{-1} A_{bl} & \tau G^T & -M_p - G_b^T D_{bb}^{-1} G_b \end{pmatrix}. \qquad (18)$$

3.2   Mass Conservation

Finally, we briefly comment on an efficient post-processing step to ensure that the numerical solution obtained above preserves mass. Let $(u_h, p_h) \in V_h \times Q_h$, with $u_h = u_l + u_b$ be the numerical solution to Stokes' equation obtained in the following way: first, we solve System (18) for u_l; and then, we compute u_b. Note that the second step requires only the solution of systems with D_{bb}, which is a diagonal matrix. A mass-conserving approximation is then obtained by interpolating the numerical solution using the interpolant from the lowest-order BDM space (see Brezzi, Douglas and Marini [20], and Brezzi, Douglas, Duran and Fortin [21] for more details).

More specifically, let Π_h^{BDM} be the standard interpolation operator in the BDM space as defined in [22], [23, Sect. 5.4]. From the commuting diagram property of BDM elements (see, e.g. [24, Proposition 2.5.2]),

$$\mathrm{div}\, \Pi_h^{\mathrm{BDM}} v = \Pi_h^0\, \mathrm{div}\, v,$$

for all sufficiently smooth $v \in V$. Here, Π_h^0 is the $L^2(\Omega)$-orthogonal projection on the space of piecewise constants, Q_h. This implies that

$$\int_\Omega \operatorname{div} \Pi_h^{\mathrm{BDM}} u_h q_h = \int_\Omega \operatorname{div} u_h q_h = 0, \quad \text{for all} \quad q_h \in Q_h, \tag{19}$$

which shows that $\Pi_h^{\mathrm{BDM}} u_h$ is indeed mass conservative.

Furthermore, we show that $\Pi_h^{\mathrm{BDM}} u_h$ also approximates the solution, u, to Stokes' equation in the $L^2(\Omega)$-norm. We recall the following classical error estimate for the BDM interpolant (see, e.g. [24, Proposition 2.5.4], [23, Theorem 5.25]):

$$\|w - \Pi_h^{\mathrm{BDM}} w\| \lesssim h|w|_1. \tag{20}$$

As a consequence from (20),

$$\|\Pi_h^{\mathrm{BDM}} w\| \le \|w - \Pi_h^{\mathrm{BDM}} w\| + \|w\| \lesssim h|w|_1 + \|w\|. \tag{21}$$

Now, using estimates (20) and (21), we obtain the following a priori error estimate,

$$\begin{aligned}
\|u - \Pi_h^{\mathrm{BDM}} u_h\| &\le \|u - \Pi_h^{\mathrm{BDM}} u\| + \|\Pi_h^{\mathrm{BDM}}(u - u_h)\| \\
&\lesssim h|u|_1 + h|u - u_h|_1 + \|u - u_h\| \\
&\lesssim h|u|_1 + |u - u_h|_1 \lesssim h\|u\|_2.
\end{aligned}$$

Thus, (19) and the a priori estimate above guarantee that the BDM interpolant of the numerical solution, $\Pi_h^{\mathrm{BDM}} u_h$, is a mass-conserving approximation to u, which requires little extra cost to compute.

4 Nonconforming Choice of Displacement Space

In this section, we consider a spatial discretization using a nonconforming finite-element method. We again have the following finite-element discretization corresponding to the three-field formulation:

Find $(u_h, w_h, p_h) \in V_h \times W_h \times Q_h$ such that

$$a_h(u_h, v_h) - (\alpha p_h, \operatorname{div} v_h) = (\rho g, v_h), \quad \forall\, v_h \in V_h, \tag{22}$$

$$\left(K^{-1}\mu_f w_h, r_h\right)_h - (p_h, \operatorname{div} r_h) = (\rho_f g, r_h), \quad \forall\, r_h \in W_h, \tag{23}$$

$$\left(\frac{1}{M}\frac{\partial p_h}{\partial t}, q_h\right) - \left(\alpha \operatorname{div}\frac{\partial u_h}{\partial t}, q_h\right) - (\operatorname{div} w_h, q_h) = (f, q_h), \quad \forall\, q_h \in Q_h. \tag{24}$$

Here, V_h is the Crouzeix-Raviart finite-element space [25]. Note that we have $a_h(\cdot, \cdot)$ instead of $a(\cdot, \cdot)$ in (22) and $(\cdot, \cdot)_h$ instead of $(\cdot, \cdot)$ in (23). These are perturbations of the bilinear forms which target important issues: the former satisfies the discrete Korn inequality and the latter leads to monotone pressure approximations. More details on the definition of the bilinear forms $a_h(\cdot, \cdot)$ and $(\cdot, \cdot)_h$ are given below.

We begin by the definition of nonconforming $\boldsymbol{V}_h$. For a function u, its jump across an interior face $e \in \mathcal{E}^o$ is denoted by $[\![u]\!]_e$, and defined as

$$[\![u]\!]_e(x) = u_{T^+(e)}(x) - u_{T^-(e)}(x), \quad x \in e.$$

The Crouzeix-Raviart space $\boldsymbol{V}_h$ consists of vector valued functions which are linear on every element $T \in \mathcal{T}_h$ and satisfy the following continuity conditions

$$\boldsymbol{V}_h = \left\{ \boldsymbol{v}_h \in \boldsymbol{L}^2(\Omega) \ \Big| \int_e [\![\boldsymbol{v}_h]\!]_e = 0, \ \text{for all } e \in \mathcal{E}^o \right\}.$$

Equivalently, all functions from $\boldsymbol{V}_h$ are continuous at the barycenters of the faces in $\mathcal{E}^o$. For the boundary faces, the elements of $\boldsymbol{V}_h$ are zero in the barycenters of any face on the Dirichlet boundary.

Let us now consider the bilinear form $a_h(\cdot, \cdot) : \boldsymbol{V}_h \times \boldsymbol{V}_h \mapsto \mathbb{R}$. Before we write out the details, we have to assume that Γ_c is non-empty. If $\Gamma_c = \emptyset$, i.e., $\Gamma_t = \Gamma$ (the pure traction problem), $a(\cdot, \cdot)$ is a positive semidefinite form and the dimension of its null space equals the number of edges on the boundary (for both 2D and 3D). Therefore, Korn's inequality fails. Even if $\Gamma_c \neq \emptyset$, for some cases, Korn's inequality may fail for the standard discretization by Crouzeix-Raviart elements without additional stabilization. In another words, if we take $a_h(\cdot, \cdot) = a(\cdot, \cdot)$ then it does not satisfy the discrete Korn's inequality and, therefore, $a_h(\cdot, \cdot)$ is not coercive. Moreover, it is also possible that Korn's inequality holds, but the constant will approach infinity as the mesh size h approaches zero. If we use $a_h(\cdot, \cdot) = a(\cdot, \cdot)$, the coercivity constant blows up when h approaches zero. For discussions on nonconforming linear elements for elasticity problems and the discrete Korn's inequality, we refer to [26, 27] for more details.

One way to fix this potential problem is to add a stabilization. The following perturbation of the bilinear form which does satisfy the Korn's inequality was proposed by Hansbo and Larson [28]:

$$a_h(\boldsymbol{v}, \boldsymbol{w}) = a(\boldsymbol{v}, \boldsymbol{w}) + a_{\mathtt{J}}(\boldsymbol{v}, \boldsymbol{w}), \quad \text{where} \quad a_{\mathtt{J}}(\boldsymbol{v}, \boldsymbol{w}) = 2\mu\gamma_1 \sum_{e \in \mathcal{E}} h_e^{-1} \int_e [\![\boldsymbol{v}]\!]_e [\![\boldsymbol{w}]\!]_e.$$

Here, the constant $\gamma_1 > 0$ is a fixed real number away from 0 (i.e. $\gamma_1 = \frac{1}{2}$ is an acceptable choice). As shown in Hansbo and Larson [28] the bilinear form $a_h(\cdot, \cdot)$ is positive definite and the corresponding error is of optimal (first) order in the corresponding energy norm. Moreover, the resulting method is <u>locking free</u>. In [28], the jump term $a_{\mathtt{J}}(\cdot, \cdot)$ includes all the edges, i.e., the stabilization needs to be done on both interior and boundary edges. In [29], it has been shown that the jump stabilization only needs to be added to the interior edges and boundary edges with Neumann boundary conditions and the discrete Korn's inequality still holds. In fact, in [30], it is suggested that only the normal component of the jumps on the edges is needed for the stabilization in order to satisfy the discrete Korn's equality.

We next consider the bilinear form in (23), denoted by $(\cdot, \cdot)_h$. The first choice for such a form is just taking the usual $L^2(\Omega)$ inner product, i.e. $(\boldsymbol{w}, \boldsymbol{r})_h = (\boldsymbol{w}, \boldsymbol{r}) = \int_\Omega \boldsymbol{w} \cdot \boldsymbol{r}$. This is the standard choice and leads to a mass matrix in the Raviart-Thomas-Nédélec element when we write out the matrix form.

The second choice, which is the bilinear form we use here, is based on mass lumping in the Raviart-Thomas space, i.e.,

$$(\boldsymbol{r}, \boldsymbol{s})_h = \sum_T \sum_{e \subset \partial T} \omega_e \, e(\boldsymbol{r}) e(\boldsymbol{s}). \tag{25}$$

We refer to [31] and [32] for details on determining the weights ω_e, which are $\omega_e = \dfrac{|e| d_e}{d}$ with d_e being the signed distance between the Voronoi vertices adjacent to the face e. Roughly speaking, such weights, in the two-dimensional case, are chosen so that

$$(\boldsymbol{w}, \boldsymbol{r})_h = \int_\Omega \boldsymbol{w} \cdot \boldsymbol{r}, \quad \boldsymbol{w}, \boldsymbol{r} \in \boldsymbol{W}_h \text{ and } \boldsymbol{w}, \boldsymbol{r} \text{ are piecewise constants}, \tag{26}$$

which implies the equivalence between $(\boldsymbol{w}, \boldsymbol{r})_h$ and the standard L^2 inner product $(\boldsymbol{w}, \boldsymbol{r})$. The situation in the three-dimensional case is a little bit involved since (26) does not hold in general. Nevertheless, the equivalence between $(\boldsymbol{w}, \boldsymbol{r})_h$ and the standard L^2 inner product $(\boldsymbol{w}, \boldsymbol{r})$ can still be shown. Overall, such mass lumping, in both two- and three-dimensional cases, maintains the optimal convergence order, see [31] for details.

In practice, a lumped mass approximation results in a block diagonal matrix and, therefore, we can eliminate the Darcy velocity $\boldsymbol{w}$ and reduce the three-field formulation to a two-field formulation involving only the displacements $\boldsymbol{u}$ and pressure p. This elimination reduces the size of the linear system that needs to be solved at each time step and saves computational cost. Moreover, for Biot's model, as shown by the numerical experiments section in [11], the lumped mass approximation actually gives an oscillation-free approximation while maintaining the optimal error estimates.

4.1   Analysis of the Fully Discrete Scheme: Nonconforming Case

Next, we consider the fully discrete scheme for (1)–(3) at time $t_m = m\tau$, $m = 1, 2, \ldots$: Find $(\boldsymbol{u}_h^m, \boldsymbol{w}_h^m, p_h^m) \in \boldsymbol{V}_h \times \boldsymbol{W}_h \times Q_h$ such that

$$a_h(\boldsymbol{u}_h^m, \boldsymbol{v}_h) - \alpha(p_h^m, \operatorname{div} \boldsymbol{v}_h) = (\rho \boldsymbol{g}, \boldsymbol{v}_h), \quad \forall \, \boldsymbol{v}_h \in \boldsymbol{V}_h, \tag{27}$$

$$(K^{-1} \mu_f \boldsymbol{w}_h^m, \boldsymbol{r}_h)_h - (p_h^m, \operatorname{div} \boldsymbol{r}_h) = (\rho_f \boldsymbol{g}, \boldsymbol{r}_h), \quad \forall \, \boldsymbol{r}_h \in \boldsymbol{W}_h, \tag{28}$$

$$\left(\frac{1}{M} \bar{\partial}_t p_h^m, q_h \right) + \alpha(\operatorname{div} \bar{\partial}_t \boldsymbol{u}_h^m, q_h) + (\operatorname{div} \boldsymbol{w}_h^m, q_h) = (f, q_h), \quad \forall q_h \in Q_h, \tag{29}$$

where τ is the time step size and $\bar{\partial}_t \boldsymbol{u}_h^m := (\boldsymbol{u}_h^m - \boldsymbol{u}_h^{m-1})/\tau$. For the initial data $\boldsymbol{u}_h^0$, we use the discrete counterpart of the divergence free condition: $\operatorname{div} \boldsymbol{u}_h^0 = 0$.

The well-posedness of the linear system (27)–(29) at each time step t_m follows from considerations similar to the conforming case. Again, we use the composite bilinear form,

$$B(\boldsymbol{u}_h, \boldsymbol{w}_h, p_h; \boldsymbol{v}_h, \boldsymbol{r}_h, q_h) := a_h(\boldsymbol{u}_h, \boldsymbol{v}_h) - (p_h, \operatorname{div} \boldsymbol{v}_h) + \tau(K^{-1}\boldsymbol{w}_h, \boldsymbol{r}_h)_h$$
$$- \tau(p_h, \operatorname{div} \boldsymbol{r}_h) - (\operatorname{div} \boldsymbol{u}_h, q_h) - \tau(\operatorname{div} \boldsymbol{w}_h, q_h).$$

As the nonconforming spaces and the bilinear forms involved in the definition above are Stokes-Biot stable then we have the following theorem showing the solvability of the linear system corresponding to the discrete Biot's model.

Theorem 3. *The bilinear form* $B(\cdot, \cdot, \cdot; \cdot, \cdot, \cdot)$ *satisfies the following inf-sup condition,*

$$\sup_{(\boldsymbol{v}_h, \boldsymbol{r}_h, q_h) \in V_h \times W_h \times Q_h} \frac{B(\boldsymbol{u}_h, \boldsymbol{w}_h, p_h; \boldsymbol{v}_h, \boldsymbol{r}_h, q_h)}{\|\|(\boldsymbol{v}_h, \boldsymbol{r}_h, q_h)\|\|} \geq \gamma \|\|(\boldsymbol{u}_h, \boldsymbol{w}_h, p_h)\|\|, \qquad (30)$$

with a constant $\gamma > 0$ *independent of mesh size* h *and time step size* τ. *Moreover, the discrete three field formulation is well-posed.*

As a consequence, we also have the following theorem regarding the errors in the fully discrete scheme, (27)–(29). The proofs follow from the standard error analysis for time-dependent problems in Thomée [33].

Theorem 4. *Let* $\boldsymbol{u}$, $\boldsymbol{w}$, *and* p *be the solutions of* (1)–(3) *and* $\boldsymbol{u}_h^m$, $\boldsymbol{w}_h^m$, *and* p_h^m *be the solutions of* (27)–(29). *If the following regularity assumptions hold,*

$$\boldsymbol{u}(t) \in L^\infty\left((0,T], \mathbf{H}_0^1(\Omega) \cap \mathbf{H}^2(\Omega)\right),$$
$$\partial_t \boldsymbol{u} \in L^1\left((0,T], \mathbf{H}^2(\Omega)\right), \; \partial_{tt}\boldsymbol{u} \in L^1\left((0,T], \mathbf{H}^1(\Omega)\right),$$
$$\boldsymbol{w}(t) \in L^\infty\left((0,T], H_0(\operatorname{div}, \Omega) \cap \mathbf{H}^1(\Omega)\right),$$
$$p \in L^\infty\left((0,T], H^1(\Omega)\right), \; \partial_t p \in L^1\left((0,T], H^1(\Omega)\right),$$

then we have the error estimates

$$\|(\boldsymbol{u}(t_m) - \boldsymbol{u}_h^m, \boldsymbol{w}(t_m) - \boldsymbol{w}_h^m, p(t_m) - p_h^m)\|_\tau \leq c \left\{ \|e_{\boldsymbol{u}}^0\|_{a_h} + \tau \int_0^{t_m} \|\partial_{tt}\boldsymbol{u}\|_1 \mathrm{d}t \right.$$
$$\left. + h \left[\|\boldsymbol{u}\|_2 + \tau^{1/2}\|\boldsymbol{w}\|_1 + \|\boldsymbol{w}\|_1 + \|p\|_1 + \int_0^{t_m} \left(\|\partial_t \boldsymbol{u}\|_2 + \|\partial_t p\|_1\right) \mathrm{d}t \right] \right\}.$$
$$(31)$$

Remark 1. We require full regularity of the solution in space in Theorems 2 and 4 as we use standard approximation results in Sobolev spaces. With respect to the smoothness in time, we follow the standard theory in [33]. Assuming less regularity in space, requires approximation estimates in fractional order Sobolev spaces and allowing less regularity of the solution in time would require special energy estimates (see, e.g. [34]). While such extensions of the results given earlier are plausible, the analysis would be much more involved and falls beyond the scope of our presentation.

Acknowledgements. The work of F. J. Gaspar is supported by the European Union's Horizon 2020 research and innovation programme under the Marie Sklodowska-Curie grant agreement NO 705402, POROSOS. The research of C. Rodrigo is supported in part by the Spanish project FEDER /MCYT MTM2016-75139-R and the DGA (Grupo consolidado PDIE). The work of Zikatanov was partially supported by NSF grants DMS-1720114 and DMS-1819157. The work of Adler, Hu, and Ohm was partially supported by NSF grant DMS-1620063.

References

1. Terzaghi, K.: Theoretical Soil Mechanics. Wiley, New York (1943)
2. Biot, M.A.: General theory of three-dimensional consolidation. J. Appl. Phys. **2**(12), 155–164 (1941)
3. Biot, M.A.: Theory of elasticity and consolidation for a porous anisotropic solid. J. Appl. Phys. **2**(26), 182–185 (1955)
4. Showalter, R.E.: Diffusion in poro-elastic media. J. Math. Anal. Appl. **1**(251), 310–340 (2000)
5. Ženíšek, A.: The existence and uniqueness theorem in Biot's consolidation theory. Apl. Mat. **3**(29), 194–211 (1984)
6. Gaspar, F., Lisbona, F., Vabishchevich, P.: A finite difference analysis of Biot's consolidation model. Appl. Numer. Math. **4**(44), 487–506 (2003)
7. Gaspar, F., Lisbona, F., Vabishchevich, P.: Staggered grid discretizations for the quasi-static Biot's consolidation problem. Appl. Numer. Math. **6**(56), 888–898 (2006)
8. Nordbotten, J.M.: Cell-centered finite volume discretizations for deformable porous media. Int. J. Numer. Methods Eng. **6**(100), 399–418 (2014)
9. Nordbotten, J.M.: Stable cell-centered finite volume discretization for Biot equations. SIAM J. Numer. Anal. **2**(54), 942–968 (2016)
10. Lewis, R.W., Schrefler, B.A.: The Finite Element Method in the Static and Dynamic Deformation and Consolidation of Porous Media. Wiley, New York (1998)
11. Hu, X., Rodrigo, C., Gaspar, F., Zikatanov, L.: A nonconforming finite element method for the Biot's consolidation model in poroelasticity. J. Comput. Appl. Math. **310**, 143–154 (2017)
12. Hong, Q., Kraus, J.: Parameter-robust stability of classical three-field formulation of Biot's consolidation model. Elec. Transact. Numer. Anal. **48**, 202–226 (2018)
13. Oyarzúa, R., Ruiz-Baier, R.: Locking-free finite element methods for poroelasticity. SIAM J. Numer. Anal. **5**(54), 2951–2973 (2016)
14. Lee, J., Mardal, K.-A., Winther, R.: Parameter-robust discretization and preconditioning of Biot's consolidation model. SIAM J. Sci. Comput. **1**(39), A1–A24 (2017)
15. Raviart, P.A., Thomas, J.M.: A mixed finite element method for 2-nd order elliptic problems. In: Galligani, I., Magenes, E. (eds.) Mathematical Aspects of Finite Element Methods. LNM, vol. 606, pp. 292–315. Springer, Heidelberg (1977). https://doi.org/10.1007/BFb0064470
16. Nédélec, J.-C.: A new family of mixed finite elements in $\mathbf{R}^3$. Numerische Mathematik **1**(50), 57–81 (1986)
17. Nédélec, J.-C.: Mixed finite elements in $\mathbf{R}^3$. Numerische Mathematik **3**(35), 315–341 (1980)
18. Girault, V., Raviart, P.: Finite Element Methods for Navier-Stokes Equations. Springer Series in Computational Mathematics, Berlin (1986)

19. Rodrigo, C., Hu, X., Ohm, P., Adler, J.H., Gaspar, F.J., Zikatanov, L.T.: New stabilized discretizations for poroelasticity and the Stokes' equations. Comput. Methods Appl. Mech. Eng. **341**(1), 467–484 (2018)
20. Brezzi, F., Douglas Jr., J., Marini, L.D.: Recent results on mixed finite element methods for second order elliptic problems. In: Vistas in Applied Mathematics, Optimization Software, New York, pp. 25–43 (1986)
21. Brezzi, F., Douglas Jr., J., Durán, R., Fortin, M.: Mixed finite elements for second order elliptic problems in three variables. Numerische Mathematik **2**(51), 237–250 (1987)
22. Brezzi, F., Fortin, M.: Mixed and Hybrid Finite Element Methods. Springer Series in Computational Mathematics 15, New-York (1986)
23. Monk, P.: Finite Element Methods for Maxwell's Equations. Oxford University Press, New York (2003)
24. Boffi, D., Brezzi, F., Fortin, M.: Mixed Finite Element Methods and Applications. Springer, Heidelberg (2013)
25. Crouzeix, M., Raviart, P.-A.: Conforming and nonconforming finite element methods for solving the stationary Stokes equations I. Rev. Française Automat. Informat. Recherche Opérationnelle Sér. Rouge **R–3**(7), 33–75 (1973)
26. Falk, R.S.: Nonconforming finite element methods for the equations of linear elasticity. Math. Comput. **196**(57), 529–550 (1991)
27. Falk, R.S., Morley, M.E.: Equivalence of finite element methods for problems in elasticity. SIAM J. Numer. Anal. **6**(27), 1486–1505 (1990)
28. Hansbo, P., Larson, M.G.: Discontinuous galerkin and the crouzeix-raviart element: application to elasticity. M2AN. Math. Modeling Numer. Anal. **1**(37), 63–72 (2003)
29. Brenner, S.C.: Korn's inequalities for piecewise H^1 vector fields. Math. Comput. **247**(73), 1067–1087 (2004)
30. Mardal, K.-A., Winther, R.: An observation on Korn's inequality for nonconforming finite element methods. Math. Comput. **253**(75), 1–6 (2006)
31. Brezzi, F., Fortin, M., Marini, L.D.: Error analysis of piecewise constant pressure approximations of Darcy's law. Comput. Methods Appl. Mech. Eng. **13–16**(195), 1547–1559 (2006)
32. Baranger, J., Maitre, J.-F., Oudin, F.: Connection between finite volume and mixed finite element methods. RAIRO Modél. Math. Anal. Numér. **4**(30), 445–465 (1996)
33. Thomée, V.: Galerkin Finite Element Methods for Parabolic Problems. Springer, Berlin (2006)
34. Nochetto, R.H., Savare, G., Verdi, C.: A posteriori error estimates for variable time step discretizations of nonlinear evolution equations. Comm. Pure Appl. Math. **53**, 525–589 (2000)

A Class of Staggered Schemes
for the Compressible Euler Equations

Raphaele Herbin[1] and Jean-Claude Latché[2]([✉])

[1] Aix-Marseille Université, Marseille, France
raphaele.herbin@univ-amu.fr
[2] Institut de Radioprotection et de Sûreté Nucléaire (IRSN),
Cadarache, France
jean-claude.latche@irsn.fr

Abstract. We present a class of numerical schemes for the solution of the Euler equations; these schemes are based on staggered discretizations and work either on structured meshes or on general simplicial or tetrahedral/hexahedral meshes. The time discretization is performed by fractional-step algorithms, either based on semi-implicit pressure correction techniques or segregated in such a way that only explicit steps are involved (referred to hereafter as "explicit" variants). These schemes solve the internal energy balance, with corrective terms to ensure the correct capture of shocks, and, more generally, the consistency in the Lax-Wendroff sense. To keep the density, the internal energy and the pressure positive, positivity-preserving convection operators for the mass and internal energy balance equations are designed, using upwinding with respect of the material velocity only. The construction of the fluxes thus does not need any Riemann or approximate Riemann solver, and yields particularly efficient algorithms. The stability is obtained without restriction on the time step for the pressure correction time-stepping and under a CFL-like condition for explicit variants: the preservation of the integral of the total energy over the computational domain and the positivity of the density and of the internal energy are ensured, and entropy estimates are derived.

Keywords: Euler equations · Staggered schemes

1 Introduction

We address in this paper the solution of the Euler equations for an ideal gas, which read:

$$\partial_t \rho + \mathrm{div}(\rho\, \boldsymbol{u}) = 0, \tag{1a}$$

$$\partial_t(\rho\, \boldsymbol{u}) + \mathrm{div}(\rho\, \boldsymbol{u} \otimes \boldsymbol{u}) + \boldsymbol{\nabla} p = 0, \tag{1b}$$

$$\partial_t(\rho\, E) + \mathrm{div}(\rho\, E\, \boldsymbol{u}) + \mathrm{div}(p\, \boldsymbol{u}) = 0, \tag{1c}$$

$$p = (\gamma - 1)\, \rho\, e, \qquad E = \frac{1}{2}|\boldsymbol{u}|^2 + e, \tag{1d}$$

© Springer Nature Switzerland AG 2019
G. Nikolov et al. (Eds.): NMA 2018, LNCS 11189, pp. 15–26, 2019.
https://doi.org/10.1007/978-3-030-10692-8_2

where t stands for the time, ρ, $\boldsymbol{u}$, p, E and e are the density, velocity, pressure, total energy and internal energy respectively, and $\gamma > 1$ is a coefficient specific to the considered fluid. The problem is supposed to be posed over $\Omega \times (0, T)$, where Ω is an open bounded connected subset of $\mathbb{R}^d$, $1 \le d \le 3$, and $(0, T)$ is a finite time interval. System (1) is complemented by initial conditions for ρ, e and $\boldsymbol{u}$, let us say ρ_0, e_0 and $\boldsymbol{u}_0$ respectively, with $\rho_0 > 0$ and $e_0 > 0$, and by suitable boundary conditions (not specified for short).

Finite volume schemes for the solution of System (1), and, more generally speaking, of hyperbolic problems, generally use a collocated arrangement of the unknowns, all of them being associated to the cell centers, and apply a Godunov-like technique for the computation of the fluxes at the cells faces: the face is seen as a discontinuity line for the beginning-of-time-step numerical solution, supposed to be constant in the two adjacent cells; a solution, either exact or approximate, of the so-posed Riemann problem is constructed and the numerical solution is advanced in time by projection of this construction on piecewise constant functions (see *e.g.* [1,17] for the development of such solvers). Thanks to the properties of the projection, at least for exact Riemann solvers, application of this process to the Euler equations yields consistant schemes which preserve the non-negativity of the density and the internal energy and, for first-order variants, satisfy an entropy inequality. The price to pay is the computational cost of the evaluation of the fluxes, and the fact that this issue is intricate enough to put almost out of reach implicit-in-time formulations, which would allow to relax CFL time step constraints. In addition, preserving the accuracy for low Mach number flows is a difficult task (see *e.g.* [9] and references herein).

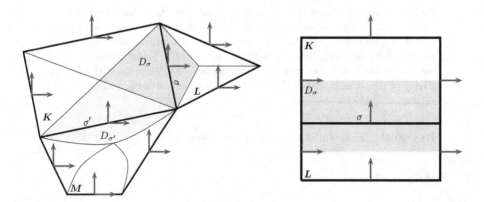

Fig. 1. Meshes and unknowns – Left: unstructured discretizations (the present sketch illustrates the possibility, implemented in our software CALIF^{3}S [2], of mixing simplicial and quadrangular cells); scalars variables are associated to the primal cells (here K, L and M) while velocity vectors are associated to the faces (here, σ and σ') or, equivalently, to dual cells (here, D_σ and $D_{\sigma'}$). – Right: MAC discretization; scalars variables are associated to the primal cells and each face is associated to the component of the velocity normal to the face.

The aim of this paper is to review recent developements performed to derive a class of schemes following a different route. The space discretization is staggered: scalar variables are associated to cell centers while the velocity is associated to the faces, or, equivalently, to staggered mesh(es). Two different space discretizations may be considered: either the so-called Marker-And-Cell (MAC) scheme for structured grids [11] or, for general meshes, a space discretization using degrees of freedom similar to low-order Rannacher-Turek [16] or Crouzeix-Raviart [3] finite elements (see Fig. 1). With this space discretization, the use of Riemann solvers seems difficult (scalar unknowns and velocities may still be considered as piecewise constant functions, but not associated to the same partition of the computational domain). The positivity of the internal energy is thus ensured by a non-standard argument: the internal energy balance is discretized instead of the actual (total) energy balance (1c) by a positivity-preserving scheme. This leads to consistency problems, which are the main difficulty faced here. We develop two time discretizations: a pressure correction technique and a fractional step scheme involving only explicit steps. We finally obtain a class of schemes which offer many interesting properties: both the density and internal energy positivity are preserved, unconditionnally for the pressure correction scheme and under CFL-like conditions for the (quasi) explicit variant, and the integral of the total energy on the computational domain is conserved (which yields a stability result); the construction of the fluxes is very simple (essentially based on standard upwinding techniques of the convection operators with respect to the material velocity); finally, the space approximation, the fluxes and the choice of the internal energy balance are consistent with usual discretizations of quasi-incompressible flows, so the pressure correction scheme is asymptotic preserving by construction in the limit of vanishing low Mach number flows. In addition, an entropy estimate is obtained for the pressure correction scheme, while only a weak entropy estimate seems to hold for the explicit variant. The development of this class of schemes started ten years ago, and we review here the essential arguments; details may be found in [8, 12, 13].

The use of staggered discretization for compressible flows began with the very first papers on the MAC scheme [10], and has been the subject of a wide litterature (see [18] for a textbook and references in [8, 12, 13]). However, the use of the internal energy equation associated to a consistency correction seems to be restricted to the context of Lagrangian approaches, up to a very recent work implementing a Lagrange-remap technique on staggered meshes [4].

2   A Pressure Correction Scheme

2.1   A Basic Lemma

Let ρ and $\boldsymbol{u}$ be regular scalar and vector-valued functions, respectively, such that

$$\partial_t \rho + \mathrm{div}(\rho \boldsymbol{u}) = 0.$$

Let z be a regular scalar function. Then we have:

$$\mathcal{C}(z) = \partial_t(\rho z) + \text{div}(\rho z \boldsymbol{u}) = \rho\big(\partial_t z + \boldsymbol{u} \cdot \boldsymbol{\nabla} z\big) + z\big(\partial_t \rho + \text{div}(\rho \boldsymbol{u})\big)$$
$$= \rho\big(\partial_t z + \boldsymbol{u} \cdot \boldsymbol{\nabla} z\big). \tag{2}$$

Let φ be a regular real function. Then:

$$\varphi'(z)\,\mathcal{C}(z) = \varphi'(z)\,\rho\left(\partial_t z + \boldsymbol{u} \cdot \boldsymbol{\nabla} z\right) = \rho\left(\partial_t \varphi(z) + \boldsymbol{u} \cdot \boldsymbol{\nabla}\varphi(z)\right).$$

Now, reversing the computation performed in Relation (2) with $\varphi(z)$ instead of z, we get:

$$\varphi'(z)\,\mathcal{C}(z) = \partial_t\big(\rho\varphi(z)\big) + \text{div}\big(\rho\varphi(z)\boldsymbol{u}\big). \tag{3}$$

The following lemma states a time semi-discrete version of this computation.

Lemma 1. *Let ρ^n, ρ^{n+1}, z^n and z^{n+1} be regular scalar functions, let $\boldsymbol{u}$ be a regular vector-valued function and let φ be a twice-differentiable real function. Let us suppose that*

$$\frac{1}{\delta t}\,(\rho^{n+1} - \rho^n) + \text{div}(\rho^{n+1}\boldsymbol{u}) = 0, \tag{4}$$

with δt a positive real number. Then

$$\varphi'(z^{n+1})\,\Big[\frac{1}{\delta t}\,(\rho^{n+1}z^{n+1} - \rho^n z^n) + \text{div}(\rho^{n+1}z^{n+1}\boldsymbol{u})\Big]$$
$$= \frac{1}{\delta t}\,\Big(\rho^{n+1}\varphi(z^{n+1}) - \rho^n\varphi(z^n)\Big) + \text{div}\big(\rho^{n+1}\varphi(z^{n+1})\boldsymbol{u}\big) + \mathcal{R}^n, \tag{5}$$

with

$$\mathcal{R}^n = \frac{1}{2\,\delta t}\rho^n\varphi''(\bar{z})\,(z^{n+1} - z^n)^2, \quad \bar{z} = \theta z^n + (1-\theta)z^{n+1}, \quad \theta \in [0,1].$$

Proof. We first begin by deriving a discrete analogue to Identity (2):

$$\frac{1}{\delta t}\,(\rho^{n+1}z^{n+1} - \rho^n z^n) + \text{div}(\rho^{n+1}z^{n+1}\boldsymbol{u})$$
$$= \frac{1}{\delta t}\,\rho^n\,(z^{n+1} - z^n) + \rho^{n+1}\boldsymbol{u} \cdot \boldsymbol{\nabla} z^{n+1} + z^{n+1}\Big[\frac{1}{\delta t}\,(\rho^{n+1} - \rho^n) + \text{div}(\rho^{n+1}\boldsymbol{u})\Big]$$
$$= \frac{1}{\delta t}\,\rho^n\,(z^{n+1} - z^n) + \rho^{n+1}\boldsymbol{u} \cdot \boldsymbol{\nabla} z^{n+1}. \tag{6}$$

Then the result follows by multiplying this relation by $\varphi'(z^{n+1})$, using a Taylor expansion for the first term and the same combination of partial derivative as in the continuous case for the second one, and finally, still as in the continuous cas, by performing this computation in the reverse sense with $\varphi(z^n)$ and $\varphi(z^{n+1})$ instead of z^n and z^{n+1}.

2.2   The Time Semi-discrete Scheme

We begin with a formal reformulation of the energy equation. Let us suppose that the solution is regular, and let E_k be the kinetic energy, defined by $E_k = \frac{1}{2}\,|\boldsymbol{u}|^2$. Taking the inner product of (1b) by $\boldsymbol{u}$ yields, after the formal compositions of partial derivatives described in the previous section:

$$\partial_t(\rho E_k) + \operatorname{div}\big(\rho\, E_k\, \boldsymbol{u}\big) + \boldsymbol{\nabla} p \cdot \boldsymbol{u} = 0. \tag{7}$$

This relation is referred to as the kinetic energy balance. Subtracting this relation to the total energy balance (1c), we obtain the so-called internal energy balance equation:

$$\partial_t(\rho e) + \operatorname{div}(\rho e \boldsymbol{u}) + p\operatorname{div}\boldsymbol{u} = 0. \tag{8}$$

Since,

- as seen in the previous section, thanks to the mass balance equation, the first two terms in the left-hand side of (8) may be recast as a transport operator,
- and, from the equation of state, the pressure vanishes when $e = 0$,

this equation implies that, if $e \geq 0$ at $t = 0$ and with suitable boundary conditions, then e remains non-negative at all time. The same result would hold if (8) featured a non-negative right-hand side, as for the compressible Navier-Stokes equations. Solving (8) instead of the total energy balance is thus appealing, to preserve this positivity property by construction of the scheme. In addition, it avoids introducing a space discretization for the total energy which, for a staggered discretization, combines cell-centered (the internal energy and the density) and face-centered (the velocity) variables. However, a raw discretization of a non-conservative equation derived from a conservative system (formally, *i.e.* supposing unrealistic regularity properties of the solution) may be non-consistent (and the numerical test presented in Sect. 4 shows that, for the problem at hand, a such a scheme would be unable to capture shock solutions). To deal with this problem, we implement the following strategy:

- First, we derive a discrete kinetic energy balance, by mimicking at the discrete level the computation used to obtain Eq. (7). This relation allows to identify the terms which are likely to lead to non-consistency: the numerical diffusion in the momentum balance equation yields dissipation terms in the kinetic energy balance which are observed to behave, when the space and time step tend to zero, as measure born by the shocks which modify the jump conditions.
- These terms are thus compensated in the internal energy balance.

At the fully discrete level, for staggered discretizations, the kinetic and internal energy balances are not posed on the same mesh (the dual and primal mesh respectively) and cannot be combined to provide a local discrete total energy balance, even though the dissipation and correction terms have opposite integrals over the computational domain, so that the integral of the total energy over the domain is conserved. However, we are able to show that the scheme is consistent, in the Lax-Wendroff sense, to the weak form of the total energy balance: indeed, for a given sequence of discrete solutions (obtained with a sequence of discretizations where the space and time steps tend to zero) controlled and converging to a limit in suitable norms (namely, controlled in BV norms and converging in L^p norms, for $p \in [1, +\infty)$), we show that the limit is a weak solution of the Euler equations. As far as the total energy balance is concerned, the key trick to this purpose is to use two interpolates of the test function, on the dual and primal mesh for the kinetic and energy balance respectively, and to pass to the limit in the equation obtained by summing the two corresponding relations.

We now derive the time semi-discrete formulation of a pressure correction scheme following these guidelines. This scheme takes the following general form:

$$\frac{1}{\delta t}(\rho^n\, \tilde{\boldsymbol{u}}^{n+1} - \rho^{n-1}\, \boldsymbol{u}^n) + \operatorname{div}(\rho^n\, \boldsymbol{u}^n \otimes \tilde{\boldsymbol{u}}^{n+1}) + \zeta^n \boldsymbol{\nabla} p^n = 0, \tag{9a}$$

$$\frac{1}{\delta t}\rho^n\, (\boldsymbol{u}^{n+1} - \tilde{\boldsymbol{u}}^{n+1}) + \boldsymbol{\nabla} p^{n+1} - \zeta^n \boldsymbol{\nabla} p^n = 0, \tag{9b}$$

$$\frac{1}{\delta t}(\rho^{n+1} - \rho^n) + \operatorname{div}(\rho^{n+1}\, \boldsymbol{u}^{n+1}) = 0, \tag{9c}$$

$$\frac{1}{\delta t}(\rho^{n+1}\, e^{n+1} - \rho^n\, e^n) + \operatorname{div}(\rho^{n+1}\, e^{n+1}\, \boldsymbol{u}^{n+1}) + p^{n+1}\operatorname{div}\boldsymbol{u}^{n+1} = S^{n+1}, \tag{9d}$$

$$p^{n+1} = (\gamma - 1)\,\rho^{n+1}\, e^{n+1}. \tag{9e}$$

The first equation allows for the computation of a tentative velocity $\tilde{\boldsymbol{u}}^{n+1}$; it is decoupled from the other equations of the system, and referred to as the velocity prediction step. Equations (9b)–(9e) constitute the correction step, and are solved simultaneously; note however that using the equation of state to recast $\rho^{n+1}\, e^{n+1}$ as a function of the pressure only in (9d) and eliminating $\boldsymbol{u}^{n+1}$ in this relation thanks to the divergence of (9b) divided by ρ^n yields a nonlinear and nonconservative elliptic problem for the pressure only. This process must be performed at the fully discrete level to preserve the properties of the scheme. The coefficient ζ^n in Eq. (9a) and the correction term S^{n+1} in (9d) are computed in the derivation of the scheme so as to ensure stability and consistency. The first step of this process is to obtain a discrete kinetic energy balance. To this purpose, let us multiply (9a) by $\tilde{\boldsymbol{u}}^{n+1}$ and apply Lemma 1 component by component, with $\varphi(s) = \frac{1}{2}s^2$.

We get:

$$\frac{1}{2\,\delta t}\left(\rho^n\, |\tilde{\boldsymbol{u}}^{n+1}|^2 - \rho^{n-1}\, |\boldsymbol{u}^n|^2\right) + \frac{1}{2}\operatorname{div}\left(\rho^n\, |\tilde{\boldsymbol{u}}^{n+1}|^2 \boldsymbol{u}^n\right) + \zeta^n \boldsymbol{\nabla} p^n \cdot \tilde{\boldsymbol{u}}^{n+1} + R_1^n = 0, \tag{10}$$

with

$$R_1^n = \frac{1}{2\,\delta t}|\tilde{\boldsymbol{u}}^{n+1} - \boldsymbol{u}^n|^2.$$

Note that the mass balance equation (9c), which is a fundamental assumption of Lemma 1, only holds at this stage of the algorithm with the previous time step values, hence the shift of the time level of the density in (9a). Let us now recast Eq. (9b) as

$$\alpha^n \boldsymbol{u}^{n+1} + \frac{1}{\alpha^n} \boldsymbol{\nabla} p^{n+1} = \alpha^n \tilde{\boldsymbol{u}}^{n+1} + \frac{\zeta^n}{\alpha^n} \boldsymbol{\nabla} p^n, \quad \alpha^n = [\frac{\rho^n}{\delta t}]^{1/2}$$

and square this relation, to get

$$\frac{1}{2\,\delta t} \rho^n |\boldsymbol{u}^{n+1}|^2 + \boldsymbol{\nabla} p^{n+1} \cdot \boldsymbol{u}^{n+1} + R_2^n = \frac{1}{2\,\delta t} \rho^n |\tilde{\boldsymbol{u}}^{n+1}|^2 + \zeta^n \boldsymbol{\nabla} p^n \cdot \tilde{\boldsymbol{u}}^{n+1}, \quad (11)$$

with

$$R_2^n = \frac{\delta t}{\rho^n} |\boldsymbol{\nabla} p^{n+1}|^2 - (\zeta^n)^2 \frac{\delta t}{\rho^n} |\boldsymbol{\nabla} p^n|^2.$$

Summing (10) and (11) yields the kinetic energy balance that we are seeking:

$$\frac{1}{2\,\delta t} \left(\rho^n |\boldsymbol{u}^{n+1}|^2 - \rho^{n-1} |\boldsymbol{u}^n|^2 \right) + \frac{1}{2} \mathrm{div}\left(\rho^n |\tilde{\boldsymbol{u}}^{n+1}|^2 \boldsymbol{u}^n \right) + \boldsymbol{\nabla} p^{n+1} \cdot \boldsymbol{u}^{n+1} + R_1^n + R_2^n = 0.$$

We now choose the coefficient ζ^n in such a way that the remainder term R_2^n becomes the difference of two consecutive time levels of the same quantity, which is realized by choosing

$$\zeta^n = [\frac{\rho^n}{\rho^{n-1}}]^{1/2}.$$

Supposing the control in $L^1(0, T, BV)$ of the pressure and in L^∞ of the pressure and of the inverse of the density, the term R_2^n may thus be seen to tend to zero with the discretization parameters in a distributional sense. We just need to compensate R_1^n in the internal energy balance, which is done by choosing $S^{n+1} = R_1^n$, which thus ensures $S^{n+1} \geq 0$. The definition of the time-discrete scheme is now complete.

2.3   The Fully Discrete Scheme

The fully discrete scheme is obtained from System (9) by applying the following guidelines:

- The mass and internal energy balances (*i.e.* Eqs. (9c) and (9d) respectively) are discretized on the primal mesh, while the velocity prediction (9a) and correction (9b) are discretized on the dual mesh(es). The equation of state only involves cell quantities, and its expression is obtained by writing (9e) for these latter.
- The space arrangement of the unknowns (density discretized at the cell and velocity at the faces) yields a natural expression of the mass fluxes in the mass balance, performed by a first-order upwind scheme (with respect to the velocity). By construction, the density is thus non-negative (in fact, positive, at the discrete level, if the initial density is positive). The discrete mass balance equation on the cell K of measure $|K|$ and faces $\sigma \in \mathcal{E}(K)$ takes the form:

$$\frac{|K|}{\delta t} (\rho_K^{n+1} - \rho_K^n) + \sum_{\sigma \in \mathcal{E}(K)} F_{K,\sigma}^{n+1} = 0, \quad (12)$$

where $F_{K,\sigma}$ is the mass flux across σ outward K.

- The form of the time-derivative and convection operator from the internal energy (let us denote $\mathcal{C}_K(e^{n+1})$ the sum of these two terms) in Eq. (9d) follows from this relation:

$$\mathcal{C}_K(e^{n+1}) = \frac{|K|}{\delta t}(\rho_K^{n+1}e_K^{n+1} - \rho_K^n e_K^n) + \sum_{\sigma \in \mathcal{E}(K)} F_{K,\sigma}e_\sigma^{n+1},$$

where e_σ^{n+1} is the upwind approximation of e^{n+1} at σ with respect to $F_{K,\sigma}^{n+1}$ (or, equivalently, since the density is positive, with respect to the velocity). This was shown in [14] to be a sufficient condition to obtain a positivity-preserving operator, and is also a necessary condition for a fully discrete version of Lemma 1 to hold; this is of course linked since both results rely on the possibility to recast $\mathcal{C}_K$ as a transport operator, and the positivity-preserving property of $\mathcal{C}_K$ may be proved by applying Lemma 1 with $\varphi(s) = \min(s,0)^2$. Once again, thanks to the arrangement of the unknowns, a natural discretization for $\mathrm{div}\,u^{n+1}$ is available. Since p^{n+1} is a function of e^{n+1} given by the equation of state and invoking the corrective term is non-negative, we are able to show that the discrete internal energy is kept positive by the scheme.

- For the derivation of a discrete kinetic energy balance, the same structure is needed for the time-derivative and convection operator in the velocity prediction step (9a). This raises a difficulty since this equation is posed on the dual mesh, and thus we need an analogue of the mass balance (12) to also hold on this mesh. The way to build the face density and the mass fluxes across the faces of the dual mesh for such a relation to hold, while still ensuring the scheme consistency, is a central ingredient of the scheme; it is detailed in [5] for the MAC discretization and in [15] for unstructured discretizations.

Once the face density is defined, the discretization of the coefficient ζ^n is straightforward. In order to combine the discrete equivalents of $u \cdot \nabla p$ (kinetic energy balance) and $p\,\mathrm{div}\,u$ (internal energy balance), the discrete gradient is defined as the transposed of the divergence operator with respect to the L^2 inner product (if $u \cdot \nabla p + p\,\mathrm{div}\,u = \mathrm{div}(p\,u)$, the integral of this quantity over the computational domain vanishes when the normal velocity is prescribed to zero at the boundary). Note that this definition is consistent with the usual treatment in the incompressible case, and is a key ingredient for the scheme to be asymptotic preserving in the limit of vanishing Mach number flows. As in the incompressible case, it also allows to control the L^2 norm of the pressure by a weak norm of its gradient, which is central for convergence studies; with this respect, a discrete *inf-sup* condition is required in some sense, which is true for staggered discretizations.

3   A "quasi-Explicit" Variant

A variant of the proposed scheme which consists only in explicit steps (in the sense that these steps do not require the solution of any linear or non-linear algebraic system) reads, in the time semi-discrete setting:

$$\frac{1}{\delta t}(\rho^{n+1} - \rho^n) + \mathrm{div}(\rho^n \, \boldsymbol{u}^n) = 0, \tag{13a}$$

$$\frac{1}{\delta t}(\rho^{n+1} \, e^{n+1} - \rho^n \, e^n) + \mathrm{div}(\rho^n \, e^n \, \boldsymbol{u}^n) + p^n \mathrm{div}\boldsymbol{u}^n = S^n, \tag{13b}$$

$$p^{n+1} = (\gamma - 1)\, \rho^{n+1} \, e^{n+1}, \tag{13c}$$

$$\frac{1}{\delta t}(\rho^{n+1} \, \boldsymbol{u}^{n+1} - \rho^n \, \boldsymbol{u}^n) + \mathrm{div}(\rho^n \, \boldsymbol{u}^n \otimes \boldsymbol{u}^n) + \boldsymbol{\nabla} p^{n+1} = 0. \tag{13d}$$

The update of the pressure before the solution of the momentum balance equation is crucial in our derivation of entropy estimates (see Sect. 5 below). This issue seems to be supported by numerical experiments: ommitting it, we observe the appearance of non-entropic discontinuities in rarefaction waves.

The space discretization differs from the pressure correction scheme in two points:

- the discretization of the convection operator in the momentum balance equation (13d) is performed by the first order upwind scheme (still with respect to the material velocity $\boldsymbol{u}^n$),
- the corrective term S^n is still obtained by deriving a kinetic energy balance multiplying Eq. (13d) by $\boldsymbol{u}^{n+1}$, but its expression is quite different, due to the time-level used in the convection operator. The time-discretization is now anti-diffusive but, as usual for explicit schemes, this anti-diffusion is counterbalanced by the diffusion in the approximation of the convection (hence the upwinding) and S^n is non-negative only under a CFL condition.

4   A Numerical Test

In this section, we assess the behaviour of the scheme on a one dimensional Riemann problem. We choose initial conditions such that the structure of the solution consists in two shock waves, separated by the contact discontinuity, with sufficiently strong shocks to allow an easy discrimination of correct numerical solutions. These initial conditions are those proposed in [17, Chap. 4], for the test referred to as Test 5. The computations are performed with the open-source software CALIF³S [2], and results shown here are obtained with the pressure correction scheme (9).

The density fields obtained with $h = 1/2000$ (or a number of cells $n = 2000$) at $t = 0.035$, with and without assembling the corrective source term in the internal energy balance, together with the analytical solution, are shown on Fig. 2. We observe that both schemes seem to converge, but the corrective term is necessary to obtain the right solution. Without a corrective term, one can check that the obtained solution is not a weak solution to the Euler system (Rankine-Hugoniot conditions are not verified). We also observe that the scheme is rather diffusive especially for contact discontinuities for which the beneficial compressive effect of the shocks does not apply; this may be cured in the explicit variant by implementing MUSCL-like algorithms [7].

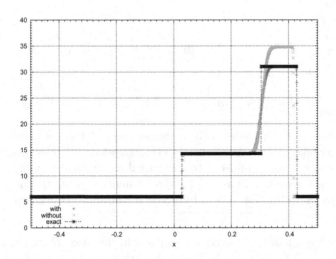

Fig. 2. Test 5 of [17, Chap. 4] - Density obtained with $n = 2000$ cells, with and without corrective source terms, and analytical solution.

5   Entropy Estimates

Let us consider the following subsystem of the Euler equations:

$$\partial_t \rho + \operatorname{div}(\rho\, \boldsymbol{u}) = 0, \tag{14a}$$

$$\partial_t(\rho\, e) + \operatorname{div}(\rho\, e\, \boldsymbol{u}) + p \operatorname{div}(\boldsymbol{u}) = \mathcal{R} \geq 0, \tag{14b}$$

$$p = (\gamma - 1)\, \rho\, e. \tag{14c}$$

The derivation of an entropy estimate for the continuous Euler system may be deduced from the subsystem (14) in the following way. We seek an entropy function η satisfying the entropy balance:

$$\partial_t \eta(\rho, e) + \operatorname{div}\big[\eta(\rho, e)\, \boldsymbol{u}\big] \leq 0. \tag{15}$$

To this end, let the convex functions φ_ρ and φ_e be defined as follows:

$$\varphi_\rho(z) = z \log(z), \quad \varphi_e(z) = \frac{-1}{\gamma - 1} \log(z), \quad \text{for } z > 0.$$

Let us multiply (14a) by $\varphi'_\rho(\rho)$, which yields:

$$\partial_t\big[\varphi_\rho(\rho)\big] + \operatorname{div}\big[\varphi_\rho(\rho)\, \boldsymbol{u}\big] + \big[\rho\varphi'_\rho(\rho) - \varphi_\rho(\rho)\big]\operatorname{div}(\boldsymbol{u}) = 0. \tag{16}$$

Then, multiplying (14b) by $\varphi_e'(e)$ yields, once again formally, since $\varphi_e'(z) < 0$ for $z > 0$:

$$\partial_t \left[\rho\, \varphi_e(e) \right] + \mathrm{div} \left[\rho\, \varphi_e(e)\, \boldsymbol{u} \right] + \varphi_e'(e)\, p\, \mathrm{div}(\boldsymbol{u}) \leq 0. \tag{17}$$

Summing (16) and (17) and noting that $\rho\varphi_\rho'(\rho) - \varphi_\rho(\rho) + \varphi_e'(e)\, p = 0$, we obtain (15) for $\eta(\rho, e) = \varphi_\rho(\rho) + \rho\varphi_e(e)$.

Depending on the time and space discretization, we obtain two types of results [6]:

- local entropy estimates, *i.e.* discrete analogues of (15), in which case the scheme is called entropy-stable,
- "weak local" entropy inequalities, *i.e.* results of the form:

$$\partial_t \eta(\rho, e) + \mathrm{div} \left[\eta(\rho, e)\, \boldsymbol{u} \right] + \mathcal{R} \leq 0,$$

with $\mathcal{R}$ tending to zero with respect to the space and time discretization steps, provided that the solution is controlled in reasonable norms (here, L^∞ and BV norms). Such an inequality readily yields a "Lax-consistency" property, in the sense that the limit of a convergent sequence of solutions, bounded in suitable norms, satisfies the following weak entropy inequality:

$$-\int_0^T \int_\Omega \eta(\rho, e)\, \partial_t \varphi + \eta(\rho, e)\, \boldsymbol{u} \cdot \boldsymbol{\nabla}\varphi \,\mathrm{d}\boldsymbol{x}\, \mathrm{d}t - \int_\Omega \eta(\rho, e)(\boldsymbol{x}, 0)\; \varphi(\boldsymbol{x}, 0)\, \mathrm{d}\boldsymbol{x} \leq 0,$$

$$\text{for any function } \varphi \in C_c^\infty \big([0, T) \times \bar{\Omega}\big), \varphi \geq 0.$$

The pressure correction scheme - In the pressure correction scheme, the correction step includes a fully implicit discretization of Subsystem (14). A fully discrete analogue of the relation (16) may be found in [12, Lemma A1], while the relation (17) is a direct application of Lemma 1. The pressure correction scheme is thus entropy-stable.

The explicit variant - In this variant, the time discretization of Subsystem (14) is explicit. An adaptation of Lemma 1 still holds under a CFL condition, but it does not seem to be the case for Relation (16). Consequently, we only obtain a "weak local" entropy inequality, under the restrictive assumption that the ratio of the time to the space step tends to zero for the sequence of discretization at hand. Nevertheless, we never observed in numerical experiment any phenomena likely to lead to thinking that the scheme could converge to a non-entropy weak solution.

References

1. Bouchut, F.: Nonlinear Stability of Finite Volume Methods for Hyperbolic Conservation Laws. Birkhauser, Basel (2004)
2. CALIF³S: a software components library for the computation of reactive turbulent flows. https://gforge.irsn.fr/gf/project/isis

3. Crouzeix, M., Raviart, P.: Conforming and nonconforming finite element methods for solving the stationary Stokes equations. RAIRO Série Rouge **7**, 33–75 (1973)

4. Dakin, G., Jourdren, H.: High-order accurate Lagrange-remap hydrodynamic schemes on staggered Cartesian grids. Comptes rendus Mathématique **34**, 211–217 (2016)

5. Gallouët, T., Herbin, R., Latché, J.C.: Kinetic energy control in explicit finite volume discretizations of the incompressible and compressible Navier-Stokes equations. Int. J. Finite Vol. **7**(2), 1–6 (2010)

6. Gallouët, T., Herbin, R., Latché, J.C., Therme, N.: Entropy estimates for a class of schemes for the Euler equations (2017). arXiv:1707.01297

7. Gastaldo, L., Herbin, R., Latché, J.C., Therme, N.: A MUSCL-type segregated - explicit staggered scheme for the Euler equations. Comput. Fluids **175**, 91–110 (2018)

8. Grapsas, D., Herbin, R., Kheriji, W., Latché, J.C.: An unconditionally stable staggered pressure correction scheme for the compressible Navier-Stokes equations. SMAI J. Comput. Math. **2**, 51–97 (2016)

9. Guillard, H.: Recent developments in the computation of compressible low Mach flows. Flow Turbul. Combust. **76**, 363–369 (2006)

10. Harlow, F., Amsden, A.: A numerical fluid dynamics calculation method for all flow speeds. J. Comput. Phys. **8**, 197–213 (1971)

11. Harlow, F., Welsh, J.: Numerical calculation of time-dependent viscous incompressible flow of fluid with free surface. Phys. Fluids **8**, 2182–2189 (1965)

12. Herbin, R., Kheriji, W., Latché, J.C.: On some implicit and semi-implicit staggered schemes for the shallow water and Euler equations. Math. Model. Numer. Anal. **48**, 1807–1857 (2014)

13. Herbin, R., Latché, J.C., Nguyen, T.: Consistent segregated staggered schemes with explicit steps for the isentropic and full Euler equations. Math. Model. Numer. Anal. **52**, 893–944 (2018)

14. Larrouturou, B.: How to preserve the mass fractions positivity when computing compressible multi-component flows. J. Comput. Phys. **95**, 59–84 (1991)

15. Latché, J.C., Saleh, K.: A convergent staggered scheme for variable density incompressible Navier-Stokes equations. Math. Comput. **87**, 581–632 (2018)

16. Rannacher, R., Turek, S.: Simple nonconforming quadrilateral Stokes element. Numer. Methods Partial Differ. Equ. **8**, 97–111 (1992)

17. Toro, E.: Riemann Solvers and Numerical Methods for Fluid Dynamics - A Practical Introduction, 3rd edn. Springer, Heidelberg (2009). https://doi.org/10.1007/b79761

18. Wesseling, P.: Principles of Computational Fluid Dynamics. Springer Series in Computational Mathematics, vol. 29. Springer, Heidelberg (2001). https://doi.org/10.1007/978-3-642-05146-3

Numerical Search and Optimization

An Effective Guided Fireworks Algorithm for Solving UCAV Path Planning Problem

Adis Alihodzic[✉], Damir Hasic, and Elmedin Selmanovic

Department of Mathematics, University of Sarajevo, Zmaja od Bosne 33-35,
71000 Sarajevo, Bosnia and Herzegovina
{adis.alihodzic,eselmanovic}@pmf.unsa.ba, damir.hasic@gmail.com

Abstract. The use of the unmanned aerial vehicles is rapidly growing in the ever more extensive range of applications where the military is the oldest ones. One of the fundamental problems in the unmanned combat aerial vehicles (UCAV) control is the path planning problem that refers to optimization of the flight route subject to various constraints inside the battlefield environments. Since the number of control points is high as well as the number of radars, the traditional methods could not produce acceptable results when tackling this problem. In this paper, we propose an adjustment of the recent guided fireworks algorithm from the class of swarm intelligence algorithms for locating the optimal path by unmanned combat aerial vehicle taking into consideration fuel consumption and safety degree. For experimental purposes, we compared it with eight different methods from the literature. Based on the experimental results, it can be concluded that our proposed approach is robust, exhibits better performance in almost all cases.

Keywords: Unmanned combat aerial vehicle · Path planning
Swarm intelligence · Metaheuristics · Fireworks algorithm

1 Introduction

Featured by high mobility and flexible deployment, unmanned combat aerial vehicles (UCAVs) have often become solutions applied to the civilian applications over the past few years, including traffic control, cargo delivery, surveillance, aerial inspection, rescue and search, video streaming, precision agriculture, and so forth [15]. The UCAV partially known as a drone is an aircraft which does not require a human pilot on the board. Also, unmanned aerial vehicles as a remotely piloted or self-piloted aircraft can carry a lot of various pieces of equipment such as cameras, sensors, and communications equipment [11]. Comparing to the aircraft, UCAVs have the advantage of low-cost, high-security, high survival ability, good maneuvering performance. Also, driven by the continuous cost reduction in UCAV manufacturing, as well as the recent government efforts in devising UCAV-related regulations in many countries, it

© Springer Nature Switzerland AG 2019
G. Nikolov et al. (Eds.): NMA 2018, LNCS 11189, pp. 29–38, 2019.
https://doi.org/10.1007/978-3-030-10692-8_3

is expected that they will be used for skyrocket soon. For example, the global UCAV market was valued at 18.14 billion U.S. dollars in 2017 and is projected to reach 52.30 billion U.S. dollars by 2025 [2]. One important design aspect of UCAVs is path planning [9]. In general, a path planning is one of the most critical parts for the UCAV, and it presents a large-scale multi-constrained optimization problem. On the one hand, these problems inherently involve an infinite number of variables due to the continuous UCAV's trajectory to be determined. On the other hand, the problems are also usually subject to a variety of practical constraints (e.g., connectivity, fuel limitation, collision, and terrain avoidance), many of which are time-varying in nature and are difficult to model accurately. Therefore, the UCAV path planning problem relays on the calculating of the sub-optimal route between an initial safe location and the desired hazardous destination considering several objectives and constraints. Since the UCAV path planning is a hard problem, there are no deterministic methods for its solving in a reasonable time. In the last two decades, nature-inspired algorithms, especially the swarm intelligence ones, have been widely and successfully used for rapid finding of approximative (acceptable) solutions. Some of these algorithms have exploited for tackling of this problem, such as: the stud genetic algorithm (SGA) [10], particle swarm optimization (PSO) [13], differential evolution (DE) [6], ant colony optimization (ACO) [1], modified firefly algorithm [8], bat algorithm (BA) [16], fireworks algorithm [3], elephant herding optimisation [4], brain storm optimisation (BSO) [7], and so forth. In this paper we adjusted the recent guided fireworks algorithm (GFWA) [12] for solving the UAV planning problem. To prove the feasibility and effectiveness of our GFWA method, we will make a comparative analysis between quality of the proposed method and other the state-of-the-art algorithms from [7]. The simulation experiments indicate that our approach can produce an acceptable, feasible sub-optimal trajectory for UCAV path planning problem as well it almost always gives better results compared to rest twelve algorithms from literature, considering both accuracy and, especially, stability of the algorithms. In subsequent sections, we foremost introduce ourselves to the mathematical definition of the UCAV path planning problem while in Sect. 3 our proposed method for solving the UCAV is presented. In Sect. 4 simulation results are reported. Finally, our conclusions are discussed in the last part of the paper.

Table 1. Information about known threatening areas

No.	Location (km)	Threat radius (km)	Threat level
1.	(45, 50)	10	2
2.	(12, 40)	10	10
3.	(32, 68)	8	1
4.	(36, 26)	12	2
5.	(55, 80)	9	3

2 Mathematical Problem Formulation for the UCAV

In this section, we propose the mathematical model for seeking a sub-optimal path along which a UCAV can freely fly from one point A to second point B due to some metrics. It is important to highlight here that the sub-optimal paths can be defined in numerous ways since different criteria can be taken into considerations (the shortest or the smoothest path, the minimal fuel consumption, the safest path, and so on). In this article, two different criteria were used, fuel consumption and safety degree. Also, consider that on the path between these points there are enemy-threat areas such as radars, artillery, missiles and so on. Information about these dangerous areas is shown in Table 1. In the model, the dangerous areas are being presented in the form of circles, inside of which UCAV will be sensitive to the danger with a certain probability proportional to the distance away from the threat centers, while outside of which it will not be stricken. Hence, each of the dangerous areas has a particular risk grade. To discover the optimal trajectory connecting the points $A(x_1, y_1)$ and $B(x_2, y_2)$, we suggest the model which maps the path planning problem into a d-dimensional optimization problem in the following way. Foremost, we transform the original Descartes's coordinate system **XOY** into a new rotated one **X'O'Y'**, where X'O'Y' is obtained by rotating XOY counter-clockwise for the angle θ as well as by translating it along the vector $\overrightarrow{OA}$, as is depicted in Fig. 1. The angle θ denotes the angle between the segment $\overline{AB}$ and the positive part of the x-axis, and it is given by:

$$\theta = \arctan\left(\frac{y_2 - y_1}{x_2 - x_1}\right) \tag{1}$$

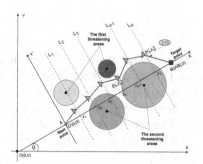

Fig. 1. The UCAV battle model in new rotated coordinate system

In the X'O'Y' system, the starting point O' has coordinates (0, 0), while the target point B has the coordinates $(|\overline{AB}|, 0)$. To formulate the path planning problem as a d-dimensional constrained optimization problem, we first divide the segment $\overline{AB}$ into $(d + 1)$ equals sub-segments. Then, in each of the endpoints $(x'_k, 0)$ of the generated sub-segments, we draw lines

$L_k (k = 1, 2, \cdots, d)$ which are orthogonal to the segment $\overline{AB}$. Further, by taking the values y'_k along the lines L_k, we obtain a discrete points collection $S = \{(x'_1, y'_1), (x'_2, y'_2), \cdots, (x'_d, y'_d)\}$. By extending of the set S with two additional points $O'(0, 0)$ and $B'(|\overline{AB}|, 0)$ added at beginning and end of the S, as a consequence we get the flight path. Hence, the path planning problem has formulated as a problem of optimizing the coordinates y'_k $(k = 1, 2, \cdots, d)$ to achieve a suboptimal path. Since the process of searching values, y'_k will be performed in the new coordinate system, the coordinates of the centers for the dangerous areas should be determined in the same coordinate system as well. To calculate the new coordinates for an arbitrary center $C(x_a, y_a)$ of the dangerous area, we first should translate the point C along the x-axis for $-x_1$ and also for $-y_1$ along the y-axis. Then, we rotate the new translated point $D(x_a - x_1, y_a - y_1)$ counterclockwise around the point $O(0, 0)$ by the angle θ. As a result, we obtain the point $C'(x'_a, y'_a)$ in the new coordinate system. As is depicted in Fig. 1, it is important to observe that coordinate x'_k of each point belonging to the trajectory can be calculated by a simple formula $x'_k = k|AB|/(d + 1)$. Therefore, the path will pass through the points of the set S.

2.1 Performance Indicators and Constraints

In this paper, two performance indicators will be taken into consideration as well as several constraints. The first performance indicator, fuel consumption F_C is proportional to the path length so it can be represented as a line integral of a function w_f (fuel cost) over a piecewise smooth curve L. For the implementation purposes, we assume that $w_f = 1$ so F_C is equal to the path length which is the sum of distances $l_i (i = 1, 2, \cdots, d + 1)$ between corresponding neighbor points in the path. Hence, the discrete model of total fuel consumption is defined by:

$$F_C = \sum_{i=1}^{d+1} l_i \tag{2}$$

The second performance indicator is safety. To determine the optimal route, a UCAV needs to escape all threats during the determination of the total threat cost. Similarly to the fuel consumption F_C, the continuous model of risk cost T_C can be calculated as a line integral of a function w_t (threat weight) over a piecewise smooth curve L. The total threat cost produced by N_t threats while a UCAV flies along the path segment L_{ij} can be calculated by [8]

$$w_{t, L_{ij}} = \int_{L_{ij}} \sum_{i=1}^{N_t} \frac{t_i}{\left((x - x_i)^2 + (y - y_i)^2\right)^2} \, dl \tag{3}$$

where t_i denotes the threat grade of the threat i with the center at point (x_i, y_i). To simplify the calculation of the T_C, this discrete model was used as in [16]:

$$T_C = \frac{1}{5} \sum_{i=1}^{d+1} l_i^5 \sum_{j=1}^{N_t} t_j \sum_{k \in K} \frac{1}{D_k^4(i, j)} \tag{4}$$

where d denotes the dimension of the problem, l_i is the length of the ith sub-segment, N_t is the overall number of the dangerous areas, t_j is the threat grade of the jth threat, $K = \{0.1, 0.3, 0.5, 0.7, 0.9\}$, and $D_k(i, j)$ is Euclidian distance between kth point belonging to the ith subsegment and the jth threat center. The 1/10 location (for $k = 0.1$) which is located in the last subpath is shown in Fig. 1. The overall cost C of a UCAV from the start position to the desired destination can be calculated from a weighted sum of the threat and fuel costs:

$$C = \lambda \cdot T_C + (1 - \lambda) \cdot F_C \tag{5}$$

where $\lambda \in [0, 1]$ is a variable which is used to establish the balance between the threat exposition degree and the fuel consumption. For comparison of our algorithms, λ will be set to 0.5. Finally, the optimal path planning of a UCAV can be obtained by minimizing Eq. 5 so that arbitrary sub-path of the trajectory must not: (a) touch any threatening environment; (b) intersect any threatening environment; (c) be entirely located inside of any threatened environments. Due to these constraints, it is clear that the entire trajectory will be outside of the dangerous areas, and thus ensuring a safe movement of a UCAV.

3   Adjusted GFWA for the UCAV Path Planning

Guided fireworks algorithm (GFWA) based on the FWA with dynamic resource allocation (dynFWA) presents a promising swarm intelligence algorithm [12]. GFWA presents a recent improvement of the original fireworks algorithm (FWA) proposed in 2010 by Tan and Zhu [14]. The latest version of guided fireworks algorithm introduces guided sparks for each of fireworks which produce a faster convergence rate. In the GFWA, fireworks were initially randomly deployed in search space. The explosion of a firework generates the specific number of sparks. For each spark that is located in search space objective function was calculated and these solutions were further used for position selection in the next generation. Exploitation and exploration were implemented by introducing two types of fireworks, well and poorly manufactured. Differences between two of them are in the number of produced sparks and their distribution around the explosion place. For each firework, the number of sparks and explosion amplitude is determined based on the objective function. Further, we provide in detail the implementation of the adjusted fireworks algorithm for the UCAV path planning problem:

Step 1. *Generation of the initial population:* Map old system XOY to new one X'O'Y'. Deploy a randomly distributed population $\mathbf{x}_i$ of n solutions (fireworks) to X'O'Y' system. Evaluate the fitness values of $\mathbf{x}_i$ due to Eq. 5, and set the variable *cycle* to one. Based on the feasibility rules, find the best firework (core firework, CF) $\mathbf{x}_{best}$ and its the best fitness value y_{min} before an iterative process.

Step 2. *Calculation of the explosion spark's number and explosion amplitudes:* Find out from the population $\mathbf{x}_i$ ($i = 1, 2, \cdots, n$) the worst fitness value y_{max}. Then, for each firework $\mathbf{x}_i$ determine the number of sparks s_i by the equation:

$$s_i = m \frac{y_{max} - f(\mathbf{x}_i) + \mu}{\sum_{i=1}^{n}(y_{max} - f(\mathbf{x}_i)) + \mu} \tag{6}$$

where parameter m controls the total number of sparks, f is the cost function, and μ is a small constant which is used to prevent the denominator from becoming zero. The amplitude of explosion which is used to calculate the position of the sparks was defined by the formula:

$$A_i = \hat{A} \frac{f(\mathbf{x}_i) + y_{min} + \mu}{\sum_{i=1}^{n}(f(\mathbf{x}_i) + y_{min}) + \mu} \tag{7}$$

where $\hat{A}$ is the highest value of the explosion amplitude. To produce better exploitation, the explosion amplitude for core firework was calculated as in [12]:

$$A_{CF}(t) = \begin{cases} A_{CF}(1), & \text{if } t = 1 \\ C_r A_{CF}(t-1), & \text{if } f(\mathbf{x}_{CF}(t)) = f(\mathbf{x}_{CF}(t-1)) \\ C_a A_{CF}(t-1), & \text{if } f(\mathbf{x}_{CF}(t)) < f(\mathbf{x}_{CF}(t-1)) \end{cases} \tag{8}$$

where t is current generation, while $C_a > 1$, $C_r < 1$ and $A_{CF}(1)$ are constants.

Step 3. *Calculation of new sparks:* Determine the initial spark's location $\mathbf{u}_j = \mathbf{x}_i$. For each location $\mathbf{u}_j$ generate k random dimensions by the formula $k = round(\sigma \cdot d)$, where $\sigma \in [0,1]$ is a uniform distributed number. Then, calculate the displacement $h_i = A_i \cdot \omega$, where ω is a random number from [0,1]. Move the spark $\mathbf{u}_j$ to a new place according to the formula $u_j^k = u_j^k + h$, where $j = 1, 2, \cdots, s_i$. In this computation step, the algorithm checks the bounds of the calculated solution $\mathbf{u}_j$. If the value u_j^k of the spark $\mathbf{u}_j$ overflows the allowed search space limits, then it is updated by $b_l^k + |u_j^k|\%(b_u^k - b_l^k)$, where b_l and b_u denote lower and upper boundaries of the spark, respectively.

Step 4. *Creation of a guiding spark for the firework* $\mathbf{x}_i$: For each firework $\mathbf{x}_i$ generate a new guiding spark $\mathbf{v}_l$ as follows:

$$v_l^k = \begin{cases} x_i^k, & \textbf{others} \\ a^k, & \text{if } r < C_f \end{cases} \tag{9}$$

where $r \in [0,1]$ is a random number drawn from uniform distribution, k is a dimension of vector, C_f is a constant, while the vector $\mathbf{a}$ presents the mean value of sparks generated by the firework $\mathbf{x}_i$. The vector $\mathbf{v}_l$ should be mapped to the potential space as done in Step 3. In the experiment we set $C_f = 0.7$.

Step 5. *Memorize the best current solution:* Evaluate the fitness values of the offsprings $\mathbf{x}_i$, $\mathbf{u}_j$, $\mathbf{v}_l$, and then record the best offspring as $\mathbf{x}_{best}$ with the best fitness value y_{min} related to feasibility rules. The best firework $\mathbf{x}_{best}$ is transferred to the next generation as the first (core) firework $\mathbf{x}_0$.

Step 6. *Update the positions of fireworks:* In this step the new vector $\mathbf{w}$ is created as a composition of fireworks $\mathbf{x}$, explosion sparks $\mathbf{u}$, and guiding sparks $\mathbf{v}$. The positions of fireworks $\mathbf{x}_i$ $(i = 1, \cdots, n)$ except the core firework (CF) $\mathbf{x}_0$ will be updated by random choosing $n - 1$ individuals from the vector $\mathbf{w}$.

Step 7. *Stopping criteria:* If the termination criterion is met or the variable *cycle* is equal to the maximum number of iterations, then the algorithm is finished. Otherwise, increase the variable *cycle* by one and go to Step 2.

Since the GFWA was not designed to deal with constraints as well the UCAV path planning is a multi-constrained problem, we propose the constraint-handling method based on feasibility rules. Before we underpin the mentioned rules, let us define some terms related to them. Each agent (spark, firework) which represents the path of UCAV will be punished by negative scores if and only if arbitrary sub-path of the trajectory violates any of three restriction conditions defined in Sect. 2.1. Therefore, a solution is feasible (a UCAV can safely move along the path) if the number of negative scores equals zero. Otherwise, it is infeasible. Similarly, as in Deb's mechanism [5], we introduce three feasibility rules. The first rule (R1) says that any feasible solution is preferred to any infeasible. Based on the second rule (R2), among two feasible solutions, the one having better fitness value is preferable. The last rule (R3) states that if both solutions are infeasible, the one with the lower sum of negative scores is preferred.

4   Experimental Analysis

In this part, we compare the performance of our GFWA method with other methods proposed in [7]. Our method has been implemented in C# programming language, and all experiments were conducted on a PC with Intel(R) Core(TM) i7-3770 K 3.5 GHz processor with 16 GB of RAM running under the Windows 10 x64 operating system. The experiments were organized in the same way as it was done in [7] where one flight environment was considered. The coordinates of both the starting point and target point are set as (10, 10) and (55, 100), respectively. In this paper, we included the simulation results of the methods presented in the papers [7,8,16] which encompassed performance analysis when the different dimensionality d of the problem was taken into consideration. The reported results were normalized in the same way as in [7]. The parameters for our GFWA method were set as follows: $n = 3$, $m = 20$, $\hat{A} = 100$, $C_r = 0.9$, $C_a = 1.2$ and the maximum number of sparks has bounded by 8. These parameters were set according to the recommendations in [12]. From the values of these parameters, the maximum population size of our GFWA method was set to 30 agents. The maximal number of fitness function evaluations (FFE) was set to 6000 to make it comparable with other mentioned algorithms. Algorithms were tested for the dimensions 5, 10, 15, 20, 25, 30, 35 and 40.

From the statistical data shown in Table 2, we can observe that our GFWA approach wins all algorithms for all dimensions and all statistical parameters except for dimensions $d = 25$, $d = 40$ and statistical parameter BEST, where BSO and MFA produced better performance. Based on the best-obtained results reached by our GFWA for almost all dimensions and all statistical parameters, as well as the smallest difference between the best and the worst solutions, it can be concluded that compared to the other algorithms, our proposed

Table 2. The best, worst and mean results generated over 100 runs with different problem dimensions and 6000 fitness function evaluations

D	Stat.	Algorithms								
		SGA	DE	PSO	ACO	MFA	BA	BAM	BSO	**GFWA**
5	BEST	9.9596	4.3568	5.6082	10.1164	4.3573	10.6909	4.3575	0.3843	**0.3685**
	WORST	22.6326	9.7959	9.7959	12.6928	12.4186	295.2557	10.2403	0.3873	**0.3850**
	MEAN	10.8836	8.0557	10.0765	11.4856	9.1673	56.4830	9.0542	0.3856	**0.37759**
10	BEST	1.5498	1.3952	2.1101	7.4746	1.3966	2.3600	1.3953	0.3690	**0.3689**
	WORST	5.7899	12.4821	23.2604	18.2565	3.7858	58.7386	10.7242	0.4372	**0.3940**
	MEAN	2.2813	3.1206	7.2212	12.5333	1.5740	19.4251	2.7075	0.4304	**0.3811**
15	BEST	0.9700	0.6204	3.2257	9.8297	0.6115	3.0757	0.6094	0.3774	**0.3756**
	WORST	9.9385	12.5250	28.0228	10.9917	3.8319	35.7454	10.1928	0.6771	**0.5228**
	MEAN	1.8973	2.3737	7.7362	10.2484	0.8967	13.6018	1.2318	0.4305	**0.4015**
20	BEST	0.8426	0.4913	2.3738	10.0836	0.4552	2.3950	0.4679	0.4541	**0.3917**
	WORST	11.6024	18.8897	34.7133	17.0266	2.0279	33.7068	3.7420	2.0108	**0.8572**
	MEAN	2.8621	3.0044	9.9091	16.3303	0.7004	13.6305	0.7609	0.7003	**0.5020**
25	BEST	1.3743	0.6265	2.3740	11.5490	0.4571	5.0173	0.4484	**0.3929**	0.4083
	WORST	16.0736	17.1415	31.6741	12.2373	3.7043	24.9265	3.5192	4.0914	**0.9603**
	MEAN	3.7238	4.6029	10.3315	11.5842	0.9987	14.9017	0.7093	0.6851	**0.5153**
30	BEST	1.5147	1.1301	3.6751	13.8615	0.5160	7.2470	0.4671	0.4411	**0.4265**
	WORST	14.0512	29.6529	35.6656	14.4647	8.3364	30.0844	10.2851	8.0677	**4.6736**
	MEAN	4.3798	11.4103	12.7964	13.9422	1.3568	16.6162	1.1067	1.0530	**1.0283**
35	BEST	1.5319	1.2849	5.4765	16.9476	0.4709	7.4484	0.4795	0.5979	**0.4652**
	WORST	15.6693	39.4435	38.0578	18.7271	5.8830	32.7374	8.8193	8.2009	**2.6918**
	MEAN	5.4943	19.1074	13.8799	18.3452	1.6009	17.7033	1.4617	1.4535	**1.3511**
40	BEST	1.9406	3.9617	5.5384	17.6142	**0.4506**	8.6500	0.6028	0.6003	0.5089
	WORST	22.5022	45.4130	35.5090	27.0641	7.7236	33.2634	8.4273	8.2159	**4.0936**
	MEAN	7.4237	28.7062	15.1555	24.7642	2.1978	19.9737	1.8769	2.0039	**1.8393**

GFWA algorithm was the most stable one and robust method. Also, to show that the differences between the obtained results are statistically significant, we have applied ANOVA statistics to three groups of data BEST, WORST, MEAN formed related to the number of control points d. We have tested two hypothesis H_0 and H_1. The null hypothesis states: "Differences between algorithms from Table 2 are not statistically significant", while the second hypothesis H_1 claims that "There are differences between them which are statistically significant". We have used statistical package SPSS to generate the results. One-way ANOVA in SPSS produced the following results: $F = 25.52$ (BEST group), $F = 10.74$, (MEAN group), $F = 3.41$ (WORST group), where $F_\alpha = 2.09$, and $\alpha = 0.05$. Based on these results we see that the test statistic F falls in the rejection region $F > F_\alpha$, which means that the sample evidence supports the alternative hypothesis H_1. Otherwords, we can reject the null hypothesis H_0 and conclude that there is sufficient evidence to indicate a difference between algorithms. Presented results prove the quality of the proposed adjusted guided fireworks algorithm for the UCAV path planning problem for all dimensions.

5   Conclusion

In this paper, we adopted a recently introduced a guided fireworks algorithm (GFWA) adopted for optimization of the UCAV path planning problem. The proposed method was tested on test environments from the literature with circular danger zones and different threat degrees, and it was compared to eight other methods from the literature. It has shown that our method was superior to all other used algorithms. Since GFWA can find a safe flight path with the smallest threat and minimum fuel cost, it establishes the GFWA as a promising approach to UCAV path planning problem with possible further improvements. Future work can include a third dimension into UCAV path planning.

References

1. Duan, H.B., Zhang, X.Y., Wu, J., Ma, G.J.: Max-min adaptive ant colony optimization approach to multi-UAVs coordinated trajectory replanning in dynamic and uncertain environments. J. Bionic Eng. **6**(2), 161–173 (2009)
2. Unmanned Aerial Vehicle (UAV) Market by Application, Class, System, UAV Type, Mode of Operation, Range, Point of Sale, MTOW, and Region - Global Forecast to 2025 (2018). https://www.researchandmarkets.com/research/thch75/global_unmanned?w=4
3. Alihodzic, A.: Fireworks algorithm with new feasibility-rules in solving UAV path planning. In: The 2016 International Conference on Soft Computing and Machine Intelligence (ISCMI 2016), pp. 53–57, November 2016
4. Alihodzic, A., Tuba, E., Capor-Hrosik, R., Dolicanin, E., Tuba, M.: Unmanned aerial vehicle path planning problem by adjusted elephant herding optimization. In: Proceedings of the 25th Conference on Telecommunications Forum Telfor (TELFOR 2017), pp. 804–807 (2017)
5. Brajevic, I., Ignjatovic, J.: An upgraded firefly algorithm with feasibility-based rules for constrained engineering optimization problems. J. Intell. Manuf., 1–30 (2018)
6. Brintaki, A.N., Nikolos, I.K.: Coordinated UAV path planning using differential evolution. Oper. Res. **5**(3), 487–502 (2005)
7. Dolicanin, E., Fetahovic, I., Tuba, E., Capor-Hrosik, R., Tuba, M.: Unmanned combat aerial vehicle path planning by brain storm optimization algorithm. Stud. Inform. Control **27**(1), 15–24 (2018)
8. Wang, G., Guo, L., Duan, H., Liu, L., Wang, H.: A modified firefly algorithm for UCAV path planning. Int. J. Hybrid Inf. Technol. **5**, 123 (2012)
9. Kabamba, P.T., Meerkov, S.M., Zeitz, F.H.: Optimal UCAV path planning under missile threats. In: IFAC Proceedings Volumes, vol. 38, no. 1, pp. 289–294 (2005)
10. Khatib, W., Fleming, P.J.: The stud GA: a mini revolution? In: Eiben, A.E., Bäck, T., Schoenauer, M., Schwefel, H.-P. (eds.) PPSN 1998. LNCS, vol. 1498, pp. 683–691. Springer, Heidelberg (1998). https://doi.org/10.1007/BFb0056910
11. Kurnaz, S., Cetin, O., Kaynak, O.: Adaptive neurofuzzy inference system based autonomous flight control of unmanned air vehicles. Expert Syst. Appl. **37**(2), 1229–1234 (2010)
12. Li, J., Zheng, S., Tan, Y.: The effect of information utilization: introducing a novel guiding spark in the fireworks algorithm. IEEE Trans. Evol. Comput. **21**(1), 153–166 (2017)

13. Li, S., Sun, X., Xu, Y.: Particle swarm optimization for route planning of unmanned aerial vehicles. In: 2006 IEEE International Conference on Information Acquisition, pp. 1213–1218 (2006)
14. Tan, Y., Zhu, Y.: Fireworks algorithm for optimization. In: Tan, Y., Shi, Y., Tan, K.C. (eds.) ICSI 2010. LNCS, vol. 6145, pp. 355–364. Springer, Heidelberg (2010). https://doi.org/10.1007/978-3-642-13495-1_44
15. Valavanis, K.P., Vachtsevanos, G.J.: Handbook of Unmanned Aerial Vehicles. Springer, Dordrecht (2015)
16. Wang, G., Guo, L., Duan, H., Liu, L., Wang, H.: A bat algorithm with mutation for UCAV path planning. Sci. World J. **2012**, 15 (2012)

Cuckoo Search Algorithm for Parameter Identification of Fermentation Process Model

Maria Angelova$^{(\boxtimes)}$, Olympia Roeva, and Tania Pencheva

Institute of Biophysics and Biomedical Engineering,
Bulgarian Academy of Sciences, Sofia, Bulgaria
{maria.angelova,olympia,tania.pencheva}@biomed.bas.bg

Abstract. Parameter identification of non-linear dynamic processes, among them fermentation ones, is rather difficult and non-trivial task to be solved. Failure of conventional optimization methods to provide a satisfactory solution provokes the idea some stochastic algorithms to be tested. As such, the promising metaheuristic algorithm Cuckoo search (CS) has been adapted and applied for a first time to a parameter identification of *S. cerevisiae* fed-batch fermentation process model. Aiming to improve the model accuracy and the algorithm convergence time, several pre-tests adjustments of CS have been done according to the specific optimization problem. Obtained results confirm the effectiveness and efficacy of the applied CS algorithm. In addition, a comparison between CS and simple genetic algorithm, proved as successful in parameter identification of fermentation process model, has been done. Algorithms advantages and disadvantages have been outlined and the more reliable one have been distinguished.

Keywords: Cuckoo search · Genetic algorithm
Parameter identification · Fermentation process

1 Introduction

Yeasts, and particularly *S. cerevisiae*, play a key role in contemporary production of medicines, foods and beverages. Also, *S. cerevisiae* are the microorganisms behind the most common type of fermentation. As a whole, fermentation processes (FP) have a numerous specific peculiarities that turned their modelling to a rather difficult to be solved task. As such, FP models have a complex structure described by nonlinear, dynamic systems with interdependent and time-varying process variables. That is why the choice of an appropriate method for model parameter identification is of a key importance [1, 2, 11, 13, 14].

The optimization techniques could be classified in two groups – exact and heuristic. The exact strategies work well for many problems and guarantee the optimal solution finding. However, for a large amount of real-world complex tasks, including parameter identification of FP models, the exact methods

© Springer Nature Switzerland AG 2019
G. Nikolov et al. (Eds.): NMA 2018, LNCS 11189, pp. 39–47, 2019.
https://doi.org/10.1007/978-3-030-10692-8_4

became ineffective, since they are time consuming and require higher computational resources. These problems could be solved using another group optimization techniques, namely efficient metaheuristic algorithms.

In the group of heuristic optimization techniques, there are a plenty of so called nature-inspired metaheuristic algorithms, such as genetic algorithms (GA) [5], ant colony optimization (ACO) [3], artificial bee colony (ABC) optimization [6], bat algorithm (BA) [18], particle swarm optimization (PSO) [7], etc. They have been developed and used for solving of wide range optimization problems [15], among them several successful applications in FP modelling and control [2,10,12]. However, the researchers still work for finding the more powerful, efficient and adequate modelling concepts for the considered problems.

Cuckoo search (CS) is one of the most promising optimization algorithm inspired by animal behaviour phenomena. The CS algorithm, proposed by Yang and Deb in 2009 [16], takes as a metaphor of cuckoo species reproduction strategy in the nature, so called brood parasitism. CS has been applied to solving many optimization problems [4,8,9] and has been proved as very efficient method that, in some cases, is even outperformed other metaheuristics, such as genetic algorithms [1,14,17]. Here, CS algorithm has been adapted and applied for a first time to a parameter identification of S. cerevisiae fed-batch FP model. In addition, a comparison between CS and simple GA has been done with the purpose of outlining the algorithms advantages and disadvantages.

2   Problem Formulation

For the purposes of a parameter identification of fed-batch process of S. cerevisiae, a cultivation has been conducted in the Institute of Technical Chemistry University of Hannover, Germany. Experimental data consist of on-line measurements of substrate (glucose) and dissolved oxygen, as well as off-line measurements of biomass and ethanol. The full description of process conditions and experimental data are given in [10].

According to the mass balance and considering mixed oxidative functional state [10], non-linear mathematical model of S. cerevisiae fed-batch FP is commonly described as follows:

$$\frac{dX}{dt} = \mu X - \frac{F}{V} X \tag{1}$$

$$\frac{dS}{dt} = -q_S X + \frac{F}{V} \left(S_{in} - S\right) \tag{2}$$

$$\frac{dE}{dt} = q_E X - \frac{F}{V} E \tag{3}$$

$$\frac{dO_2}{dt} = -q_{O_2} X + k_L^{O_2} a \left(O_2^* - O_2\right) \tag{4}$$

$$\frac{dV}{dt} = F \tag{5}$$

where X, S, E, and O_2, are concentrations, respectively of biomass [g/l], substrate [g/l], ethanol [g/l] and dissolved oxygen [%]; O_2^* – dissolved oxygen saturation concentration, [%]; F – feeding rate, [l/h]; V – volume of bioreactor, [l]; $k_L^{O_2}a$ – volumetric oxygen transfer coefficient, [1/h]; S_{in} – glucose concentration in the feeding solution, [g/l]; μ, q_S, q_E, q_{O_2} – specific growth/utilization rates of biomass, substrate, ethanol and dissolved oxygen, [1/h].

According to functional state modelling approach [10], specific rates in Eqs. (1)–(5) are as follows:

$$\mu = \mu_{2S}\frac{S}{S+k_S} + \mu_{2E}\frac{E}{E+k_E}, \quad q_S = \frac{\mu_{2S}}{Y_{SX}}\frac{S}{S+k_S}$$

$$q_E = -\frac{\mu_{2E}}{Y_{EX}}\frac{E}{E+k_E}, \quad q_{O_2} = q_E Y_{OE} + q_S Y_{OS}$$

(6)

where μ_{2S}, μ_{2E} are the maximum growth rates of substrate and ethanol, [1/h]; k_S, k_E – saturation constants of substrate and ethanol, [g/l]; Y_{ij} – yield coefficients, [g/g].

All functions in the model (Eqs. (1)–(6)) are continuous and differentiable, as well as all model parameters fulfil the non-zero division requirement.

For the considered here model (Eqs. (1)–(6)), the following vector including nine model parameters is going to be identified: $p = [\mu_{2S}, \mu_{2E}, k_S, k_E, Y_{SX}, Y_{EX}, k_L^{O_2}a, Y_{OS}, Y_{OE}]$.

As an optimization criterion, mean square deviation between the model output and the experimental data for biomass, substrate, ethanol and dissolved oxygen, obtained during the cultivation has been used:

$$J = \sum(Y - Y^*)^2 \rightarrow min,$$

(7)

where Y is the experimental data, Y^* – model predicted data, $Y = [X, S, E, O_2]$.

3 Cuckoo Search Algorithm

Cuckoos are amazing birds, known with their aggressive reproduction strategy [16]. Cuckoo species not only lay their eggs in alien nests, but they may also remove other birds eggs with the purpose to increase the hatching probability of their own generation. The specific cuckoo's brood parasitism reduces the possibility their eggs to be abandoned and thus increases the cuckoo's reproductivity. The interesting breeding behaviour is implemented in cuckoo search metaheuristic algorithm for optimization problems solving. In the CS algorithm, the eggs in the nest are assumed as a set of candidate solutions of an optimization problem, while the cuckoo egg is interpreted as a new coming solution. The ultimate goal of the method is to use iteratively these new and potentially better solutions for very good solution finding of the problem.

In CS algorithm switching parameter p_a control a balanced combination between local random walk and global explorative random walk. Equation (8) presents the local random walk, while Eq. (9) gives the global random walk, carried out using Lévy flights [14,16]:

$$x_i^{t+1} = x_i^t + \alpha s \otimes H(p_a - \varepsilon) \otimes (x_j^t - x_k^t)$$

(8)

$$x_i^{t+1} = x_i^t + \alpha L(s, \lambda) \qquad (9)$$

In Eq. (8), x_j^t and x_k^t are two different solutions, selected by random permutation, $H(u)$ is a Heaviside function, ε is a random number drawn from a uniform distribution, s is the step size, $\otimes$ means the entry-wise product of two vectors.

In Eq. (9), $\alpha > 0$ is the step size scaling factor, and

$$L(s, \lambda) = \frac{\lambda \Gamma(\lambda) \sin(\pi \lambda / 2)}{\pi} \frac{1}{s^{1+\lambda}}, \quad (s \gg s_0 > 0) \qquad (10)$$

The initial solution generation is given by

$$x = Lb + (Ub - Lb) * rand(size(Lb)), \qquad (11)$$

where *rand* is a random number generator uniformly distributed in the space $[0, 1]$ and Ub and Lb are respectively the upper and the lower boundary of the j-th nest.

The pseudo-code of CS algorithm, according to [16], is presented in Fig. 1.

```
begin
    Objective function F(x), x = (x_1, ..., x_d)^T
    Generate initial population of n host nests x_i, (i = 1, 2, ..., n)
        while (t < MaxGeneration) or (stop criterion) do
            Get a cuckoo randomly by Lévy flights
            Evaluate its quality or fitness value F_i
            Choose a nest among n (say, j) randomly
                if (F_i < F_j) then
                    Replace j by the new solution i
                end if
            A fraction (p_a) of worse nests are abandoned
            New solutions (nests) are built
            Keep the best solutions, i.e. nests with quality solutions
            Rank the solutions and find the current best
        end while
    Postprocess results and visualization
end
```

Fig. 1. Pseudo code of CS algorithm

4   Simple Genetic Algorithm

Genetic algorithms, as one of the most popular biologically inspired metaheuristic methods, have pointed more and more attention mainly due to the fact that they reach a good solution in the field of complex dynamic systems optimization, and particularly parameter identification of FP models [2,10]. Simple genetic algorithms (SGA), originally presented in Goldberg [5], searches a global optimal solution using three main genetic operators in a sequence selection, crossover and mutation over the individuals in one population. After the reproduction, SGA calculates the objective function for the offspring, and the best fitted individuals are selected to replace the parents, according to their objective function values. GA finish calculations when the stop criteria is fulfilled, in this case, when the total number of generations is reached.

5 Results and Discussion

5.1 Numerical Computations

CS and GA implementations for the purposes of parameter identification of the *S. cerevisiae* fed-batch fermentation model (Eqs. (1)–(6)) have been performed in *Matlab* environment.

All computational results have been done using Intel® Core™i3 CPU M 380 @ 2.53 GHz, 4 GB Memory (RAM), Windows 8 (64 bit) operating system.

Metaheuristic methods typically require a large number of parameters to be tuned, depending on the problem solving, while CS requires only two parameters, namely, the population size (number of nests) n and the probability rate of replacement p_a. This makes the CS parameters tuning much easier than other metaheuristic approaches. In current investigation, the CS algorithm parameters have been tuned taking into account the results in [1,4,14,16]. The population size n has been investigated for the following values: $n = \{10; 15; 20; 30; 60; 120\}$, while switching parameter p_a have been tried for different values as follows $p_a = \{0.05; 0.15; 0.25; 0.4; 0.5\}$. The results show that the algorithm performance is similar for considered here different p_a and n values. This fact confirms one reported in [4,16], that no fine adjustment is needed for the method to perform well. However, some model parameters values, as well as the objective function, became more adequate in case of increase the number of iterations. According to the several pre-test adjustments, in this paper CS parameters n and p_a have been set to 15 and 0.25, respectively. The following lower bounds (Lb) and upper bounds (Ub) of the identified parameters vector $p = [\mu_{2S}, \mu_{2E}, k_S, k_E, Y_{SX}, Y_{EX}, k_L^{O_2}a, Y_{OS}, Y_{OE}]$ of *S. cerevisiae* fed-batch fermentation model (Eqs. (1)–(6)) are going to be used [2], as in the case of GA implementation:
$Lb = [0.9\ 0.05\ 0.08\ 0.5\ 0.1\ 0.1\ 0.001\ 0.001\ 0.001]$;
$Ub = [1\ 0.15\ 0.15\ 0.8\ 1\ 3\ 300\ 1000\ 1000]$.

Going ahead, the results obtained with CS algorithm have been compared to those obtained when GA has been applied for parameter identification of the considered here *S. cerevisiae* fed-batch fermentation model (Eqs. (1)–(6)). The main GA operators and parameters, according to [2] (with only one difference - number of individuals set to 15 instead of 20, in order to be equal to the number of nests in CS), are summarized in Table 1.

Table 2 demonstrates the averaged (after 30 runs) results obtained when applying CS and GA for parameter identification of *S. cerevisiae* fed-batch fermentation model. Applied ANOVA test to measure the relative difference

Table 1. Main GA operators and parameters

Operator	Type
Fitness function	Linear ranking
Selection function	Roulette wheel selection
Crossover function	Double point
Mutation function	Bit inversion
Reinsertion	Fitness-based
Parameter	Value
Generation gap	0.8
Crossover probability	0.95
Mutation probability	0.05
Number of generations	100
Number of variables	9
Number of individuals	15

between two algorithms shows that the GA and CS achieve statistically different results. The p-value is 2.8000e–005. The ANOVA results are presented also in Fig. 2.

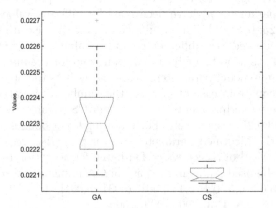

Fig. 2. Results from one-way analysis of variance

The results presented in Table 2 are obtained under identical conditions concerning the size of population: the number of nests in CS algorithm is 15, while in GA the number of individuals is 15. As shown in Table 2, the values of the optimization criterion obtained with two investigated here algorithms are very similar, varying between 0.0221 and 0.0223. But presented herewith CS algorithm yields slightly better results ($J = 0.0221$), than GA ($J = 0.0223$).

Table 2. Results from model parameters identification applying CS and GA

	Cuckoo search	Genetic algorithm
J	0.0221	0.0223
μ_{2S}	0.9180	0.9742
μ_{2E}	0.1470	0.1208
k_S	0.1230	0.1321
k_E	0.8000	0.7984
Y_{SX}	0.3968	0.4277
Y_{EX}	1.9703	1.6428
$k_L^{O_2}a$	95.6787	82.2360
Y_{OS}	762.0870	647.7941
Y_{OE}	743.4037	51.7464

The logical explanation of the fact that CS outperforms GA might be found in fewer parameters to be fine-tuned in CS, than in GA. Only two parameters, namely population size n and switching parameter p_a determine the CS algorithm's efficiency. Thus, a good balance of intensive local search and an efficient exploration of the search space is more easily achieved in CS. Also, such a balance leads to a more efficient algorithm performance [16]. Compared to CS, GA requires fine tuning of much more parameters and operators (Table 1) for a specific problem. This is the main disadvantage of GA compared to CS approach. However, it is worth noting that for different set of GA's parameters (for example, presented in [2, 14]) the algorithm might achieve better results.

Presented results show the effectiveness of CS algorithm for solving such a complex nonlinear problems, as *S. cerevisiae* parameter identification of FP model.

6   Conclusions

Here, for the first time, the promising Cuckoo search metaheuristic algorithm has been adapted and applied, after several pre-test for parameters adjustments, to the parameter identification of a *S. cerevisiae* fed-batch FP model. As a result, it has been confirmed that CS algorithm could be used as powerful and efficient tool for parameters identification of the non-linear dynamic model of a *S. cerevisiae* fed-batch FP. Further, CS and GA performance have been compared and CS algorithm has been distinguished as more reliable one, concerning the objective function value, for the considered here optimization problem. Also, an advantage in favour of CS approach is tuning of only two algorithm parameters, namely the population size n and the rate of replacement p_a. As future work, aiming at improvement of algorithms effectiveness, some CS hybridizations with modified GAs or other metaheuristic algorithms will be developed and tested.

Acknowledgements. The work is partially supported by the National Science Fund of Bulgaria under grants DM 07/1 "Development of New Modified and Hybrid Meta-heuristic Algorithms" and DN02/10 "New Instruments for Knowledge Discovery from Data, and Their Modelling".

References

1. Abdel-Baset, M., Hezam, I.: Cuckoo search and genetic algorithm hybrid schemes for optimization problems. Appl. Math. Inf. Sci. **10**(3), 1185–1192 (2016)
2. Angelova, M., Pencheva, T.: Tuning genetic algorithm parameters to improve convergence time. Int. J. Chem. Eng. **2011**, 7 (2011). https://doi.org/10.1155/2011/646917. Article ID 646917
3. Dorigo, M., Stutzle, T.: Ant Colony Optimization. MIT Press, Cambridge (2004)
4. Gálvez, A., Iglesias, A., Cabellos, L.: Cuckoo search with Lévy flights for weighted Bayesian energy functional optimization in global-support curve data fitting. Sci. W. J. **2014**, 11 (2014). https://doi.org/10.1155/2014/138760
5. Goldberg, D.E.: Genetic Algorithms in Search, Optimization and Machine Learning. Addison Wesley Longman, London (2006)
6. Karaboga, D.: An idea based on honey bee swarm for numerical optimization. Technical report-TR06, Erciyes University, Engineering Faculty, Computer Engineering Department (2005)
7. Kennedy, J., Eberhart, R.: Particle swarm optimization. In: Proceedings of IEEE International Conference on Neural Networks, vol. IV, pp. 1942–1948 (1995). https://doi.org/10.1109/ICNN.1995.488968
8. Nguyen, T.T., Vo, D.N.: Modified cuckoo search algorithm for short-term hydrothermal scheduling. Electr. Power Energy Syst. **65**, 271–281 (2015)
9. Ong, P.: Adaptive cuckoo search algorithm for unconstrained optimization. Sci. W. J. **2014**. https://doi.org/10.1155/2014/943403
10. Pencheva, T., Roeva, O., Hristozov, I.: Functional State Approach to Fermentation Processes Modelling. Prof. M. Drinov Academic Publishing House, Sofia (2006)
11. Petrov, M., Ilkova, T., Vanags, J.: Modelling of a batch whey cultivation of *Kluyveromyces marxianus var. lactis* MC 5 with investigation of mass transfer processes in the bioreactor. Int. J. Bioautomation **19**(1), S81–S92 (2015)
12. Roeva, O., Fidanova, S., Paprzycki, M.: Population size influence on the genetic and ant algorithms performance in case of cultivation process modeling. In: Fidanova, S. (ed.) Recent Advances in Computational Optimization. SCI, vol. 580, pp. 107–120. Springer, Cham (2015). https://doi.org/10.1007/978-3-319-12631-9_7
13. Roeva, O., Pencheva, T., Tzonkov, St., Hitzmann, B.: Functional state modelling of cultivation processes: dissolved oxygen limitation state. Int. J. Bioautomation **19**(1), S93–S112 (2015)
14. Roeva, O., Atanassova, V.: Cuckoo search algorithm for model parameter identification. Int. J. Bioautomation **20**(4), 483–492 (2016)
15. Yang, X.-S.: Nature-Inspired Optimization Algorithms. Elsevier Inc., London (2014)
16. Yang, X.-S., Deb, S.: Cuckoo search via Lévy flights. In: Proceedings of World Congress on Nature & Biologically Inspired Computing (NaBIC), USA, pp. 210–214. IEEE Publications (2009)

17. Yang, X.-S., Deb, S.: Multiobjective cuckoo search for design optimization. Comp. Oper. Res. **40**, 1616–1624 (2013)
18. Yang, X.S.: A new metaheuristic bat-inspired algorithm. In: González, J.R., Pelta, D.A., Cruz, C., Terrazas, G., Krasnogor, N. (eds.) NICSO 2010. SCI, vol. 284, pp. 65–74. Springer, Heidelberg (2010). https://doi.org/10.1007/978-3-642-12538-6_6

Optimization of String Rewriting Operations for 3D Fractal Generation with Genetic Algorithms

Todor Balabanov, Janeta Sevova, and Kolyu Kolev[✉]

Institute of Information and Communication Technologies, Bulgarian Academy of Sciences, acad. Georgi Bonchev Str., block 2, office 514, 1113 Sofia, Bulgaria
kkolev@iict.bas.bg
http://www.iict.bas.bg/

Abstract. String rewriting is a modification of the idea for the context-free grammar. Modification consists in the fact that there is not separation on terminal and nonterminal symbols. Each symbol in string rewriting, is considered as nonterminal and it can produce longer sequence. By this way infinite structures are created as fractals. In a video presentation by Jack Hodkinson the use of 2D images instead of text symbol pixels is suggested. Each pixel (geometric square) is divided in nine sub-squares. The color of the pixel determine the pattern in which the nine squares are arranged. By this way each pixel gives a rule for subdivision of the containing area. Hodkinson gives a method, to setup the fractal, to reproduce a particular 2D image (for example, the glyph of the Greek letter π). This problem is well known in the fractals theory as Fractal Inverse Problem (FIP). It is an optimization problem, so a good option is the use of genetic algorithm (GA), to assemble set of rules, for string rewriting with 2D pixels. Series of experiments with the size of substitution matrix (not only 3×3, but also 2×2 and 4×4) are done by Hodkinson. Also are made series of experiments with different number of colors, starting with black/white and reaching the final successful experiment with thirty shades of blue.

Keywords: Fractal Inverse Problem · Genetic Algorithms
String rewriting

1 Introduction

This study addresses FIP which is well known in the fractals theory [1,2]. There is a branch in the theory for the formal languages which is related to a context-free grammar (CFG). In such grammar there are set of production rules. Application of these rules, lead to generation of all possible strings in a specified formal language [3]. In CFG, a production rule is an operation of simple replacement. Rules are applied regardless of the context. The left hand side of a CFG rule is always nonterminal symbol. It means that nonterminal symbols are not part

© Springer Nature Switzerland AG 2019
G. Nikolov et al. (Eds.): NMA 2018, LNCS 11189, pp. 48–54, 2019.
https://doi.org/10.1007/978-3-030-10692-8_5

of the resulting string. The right hand side of a CFG rule, consists of terminal symbol, nonterminal symbols or combination of both. The main idea in CFGs is that nonterminal symbols are substituted with the given rules until there is no nonterminal symbols in the generated string.

String rewriting system (SRS) is a substitution system in which the rules does not contain nonterminal symbols. Each terminal symbol can appear on the left hand side (LHS) of the rule and on the right hand side (RHS) of the rule. With such definition SRS are useful for generation of infinite structures and they are mainly used for the generation of some fractals (for example: Box Fractal, Cantor Dust, Cantor Square Fractal and Sierpinski Carpet). Most of the fractals are generated with binary rules (black and white colors), but in the Hodkinson's research it is clearly shown that limit of the colors can be higher.

Fig. 1. Fractal generated by Jack Hodkinson.

The problem in 1D is well presented by [4], but in Hodkinson's research, the genetic algorithms (GAs) are used for 2D shape of Greek letter π, to be reconstructed with substitution rules of 30 shades of blue pixels (Fig. 1). The usage of differential evolution is also applicable [5], but not in Hodkinson's research, because the optimization is in a discrete space. The key problem in this research is which rules to be selected to achieved the final goal. GA is a promising approach for such kind of combinatorial optimization. A single rule has LHS pixel with particular color and nine pixels on the RHS as 3×3 substitution matrix. Image starts with a single pixel and it grows on size by 3 on each iteration.

This study extends the idea from 2D space into 3D space, which is the main contribution of the authors. Voxels are used instead of pixels. Square matrix of 3×3 is replaced with cube $3 \times 3 \times 3$. The paper is organized as follows: Sect. 1 introduces the problem; Sect. 2 presents a model and optimization approach; Sect. 3 gives experiment details and the results are shown; Sect. 4 concludes and some further ideas for research are pointed.

2   Model and Optimization

A finite 3D space is represented as three-dimensional array of voxels in the
model. Each voxel is encoded with a single long integer value, which consists of
the RGB (red, green, blue) information for the color of the voxel. In this model
32 different transition $3 \times 3 \times 3$ matrices are selected, each corresponding to the
one of the 32 colors. The steps for fractal generation are given in Algorithm 1.

Algorithm 1. Fractal generation algorithm.

1: **procedure** GENERATE($Level, Voxel$)
2:     **if** $Level \leqslant 0$ **then**
3:         return
4:     $v \leftarrow Split(Voxel)$
5:     **for** each v **do**
6:         Paint(v)
7:         Generate($Leve$-1, v)

The splitting of the voxel is done on 27 smaller voxels with 1/3 side size of
the original Step 4. Each sub-voxel is painted according to the color of transition
matrix pattern Step 6. A recursive call of generator procedure is done for each
sub-voxel Step 7. Recursive generation of the fractal stops when the recursion
level reached zero Step 3.

The individuals in the genetic algorithm are presented as 32 transition matrices each with size $3 \times 3 \times 3$ (Listing 1.1 in [8]). Long integer numbers are used
for RGB color representation for all 27 cells of the transition matrix.

Listing 1.1. Chromosomes encoding.

```
List<Chromosome> list = new LinkedList<Chromosome>();

for (int i = 0; i < 37; i++) {
        Long cells[] = new Long[COLORS.length * 3*3*3];

        for (int j = 0; j < cells.length; j++) {
                cells[j] =
                COLORS[PRNG.nextInt(COLORS.length)];
        }

        list.add(new TransitionsChromosome(cells,
        RECURSION_DEPTH, COLORS, target, start));
}
```

The order of the matrices shows which transition will be applied according to
used colors order. The steps of the optimization are illustrated in Algorithm 2.

Parents in the population are appointed by tournament selection with arity
of two (Listing 1.2 in [8]).

Algorithm 2. Genetic algorithm.

```
1: procedure OPTIMIZATION
2:     p ← Initialize()                              ▷ Initialize population.
3:     while TimeLimit > 0 do
4:         Selection(p)                              ▷ Select parents.
5:         Crossover(p)                              ▷ Crossover parents.
6:         Mutation(p)                               ▷ Mutate children.
7:         Evaluation(p)                             ▷ Evaluate offspring.
```

Listing 1.2. Genetic algorithm parameters.

```
GeneticAlgorithm optimizer = new GeneticAlgorithm(
new UniformCrossover<TransitionsChromosome >(0.5),  0.9,
new TransitionsMutation(COLORS),  0.01,
new TournamentSelection (2));
```

Uniform crossover is taken as the most symmetric crossover (Listing 1.2 in [8]). From each parent randomly are taken 50% of the elements. Random change of a color is used for the mutation operator (Listing 1.3 in [8]).

Listing 1.3. Color mutation of a single voxel.

```
List<Long> values = new
ArrayList<Long >(((TransitionsChromosome) original).
getRepresentation ());

values.set (PRNG. nextInt (values.size()),
colors [PRNG. nextInt (colors.length )]);
```

The evaluation is done by application of each 32 transition matrices with the certain level of recursion depth. As suggested in [1], an Euclidean distance between generated fractal and target shape is used as fitness function (Listing 1.4 in [8]).

Listing 1.4. Euclidean distance between voxels.

```
double result = 0;
for (int x = 0; x < a.length &&
                x < b.length; x++)
for (int y = 0; y < a[x].length &&
                y < b[x].length; y++)
for (int z = 0; z < a[x][y].length &&
                z < b[x][y].length; z++)
result += (a[x][y][z] − b[x][y][z]) *
          (a[x][y][z] − b[x][y][z]);

result = Math.sqrt (result);
```

The initial fractal is a single cube painted with the color with the first index in the order of the colors. Genetic algorithms are global heuristics optimization and this means that it does not matter what kind of color the initial voxel will have.

3   Experiments and Results

All experiments are done in Java programming language with Apache Genetic Algorithms Framework [6] as an optimization approach and Printing in 3D - JavaSCAD [7] as 3D visualization tool. The source code used for all experiments is an open source and can be found at [8] for the interested readers who would like to reproduce the experiments and results. Increasing the space dimensions from 2D to 3D greatly increases the complexity of the optimization process. Because of the increased complexity less ambitious shape than the Greek letter π is taken, in our case it is a voxelized sphere (Fig. 2). The sphere is one of the simplest 3D objects and that is why it was selected as optimization target. For visualization purposes the first color in the set of colors (1 of 32 in this case) is accepted as transparent. The goal of the experiment is form a single side cube to find such $3 \times 3 \times 3$ set of transitions which will lead to fractal generation of a target 3D object (in this case sphere) (Table 1).

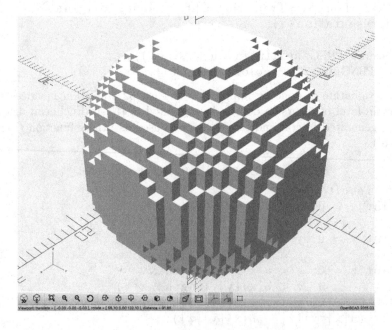

Fig. 2. Voxelized sphere as fractal optimization target object.

The genetic algorithm is used with the following parameters:

Table 1. Genetic algorithm parameters.

Parameter	Value
Elitism rate	0.10
Crossover rate	0.90
Mutation rate	0.01
Number of individuals	37
Number of variables	864
Optimization minutes	30

Fig. 3. Intermediate snapshots during the optimization process.

The optimization process converges much more slower in the 3D space than what was achieved by Hodkinson in the 2D space. As it is shown in Fig. 3, starting form a single cube it goes to the desired shape too slowly. The target shape is sphere, but intermediate shapes looks more as cubes than the desired target object.

4   Conclusions

In this study the idea for string rewriting is extended in the 3D space and instead of pixels voxels are used with $3 \times 3 \times 3$ substitution matrix. The study is only on virtual models but with the usage of industrial tomography, 3D objects can be scanned and with 3D color printer generated fractals can be printed. The extension of the idea for fractal generation from 2D to 3D space shows that in 3D space the process is much more time consuming than in the 2D space. Some speed-ups are possible by parallel implementation of the genetic algorithm [9]. Voxel based representation of 3D objects consumes a lot of computational resources and generated objects are with limited usefulness. A possible application is compression of 3D images, which is similar to the idea of 2D images compression [10,11]. Achieved results are promising for further research of the fractal nature of the world in which all we are living.

Acknowledgements. This work was supported by private funding of Velbazhd Software LLC.

References

1. Guerin, E., Tosan, E.: Fractal inverse problem: approximation formulation and differential methods. In: Levy-Vehel, J., Lutton, E. (eds.) Fractals in Engineering, pp. 271–285. Springer, London (2005). https://doi.org/10.1007/1-84628-048-6_17. ISBN 978-1-84628-047-4
2. Nettleton, D.J., Garigliano, R.: Evolutionary algorithms and a fractal inverse problem. Biosystems **33**(3), 221–231 (1994). ISSN 0303-2647
3. Ochoa, G.: On genetic algorithms and lindenmayer systems. In: Eiben, A.E., Bäck, T., Schoenauer, M., Schwefel, H.-P. (eds.) PPSN 1998. LNCS, vol. 1498, pp. 335–344. Springer, Heidelberg (1998). https://doi.org/10.1007/BFb0056876
4. Shonkwiler, R., Mendivil, F., Deliu, A.: Genetic algorithms for the 1-D fractal inverse problem. In: Proceedings of the Fourth International Conference on Genetic Algorithms, pp. 495–501. Morgan Kaufmann (1991)
5. Zelinka, I.: Inverse fractal problem. In: Price, K.V., Storn, R.M., Lampinen, J.A. (eds.) Differential Evolution. NCS, pp. 479–498. Springer, Heidelberg (2005). https://doi.org/10.1007/3-540-31306-0_17. ISBN 978-3-540-20950-8
6. The Apache Software Foundation: Genetic Algorithms - Apache Commons (2018). http://commons.apache.org/proper/commons-math/userguide/genetics.html
7. PrintingIn3D: Printing in 3D - JavaSCAD (2018). http://www.printingin3d.eu/javascad
8. Balabanov, T.: The idea of Jack Hodkinson for fractal generation implemented in 3D version (2018). http://github.com/TodorBalabanov/Jack-Hodkinson-3D-Fractal
9. Shonkwiler, R.: Parallel genetic algorithms. In: Proceedings of the Fifth International Conference on Genetic Algorithms, pp. 199–205. Morgan Kaufmann, CA (1993)
10. Vences, L., Rudomin, I., Carretera, K., Guadalupe, L.: Genetic algorithms for fractal image and image sequence compression. In: Proceedings of Comptacion Visual, pp. 35–44 (1997)
11. Al-Bundi, S.S., Al-Saidi, N.M., Al-Jawari, N.J.: Crowding optimization method to improve fractal image compressions based iterated function systems. Int. J. Adv. Comput. Sci. Appl. **7**(7), 392–401 (2016)

Quantifying the Effectiveness of First-Hop Redundancy Protocols in IP Networks

Paul Bourne[1]([⊠])[iD], Neville Palmer[1], and Jan Skrabala[2]

[1] Solent University, East Park Terrace,
Southampton, Hampshire SO14 0YN, UK
{paul.bourne,neville.palmer}@solent.ac.uk
[2] Vostron Ltd, 32B Castle Way, Southampton, Hampshire SO14 2AW, UK
jan@vostron.com

Abstract. First hop redundancy protocols are an essential tool for improving the availability of IP networks. Several protocols exist to rapidly configure redundant paths in the event of a link failure but a review of literature provides little guidance on their relative performance. This paper proposes a method for appraising the effectiveness of three common first hop redundancy protocols within IP networks and critically analyses the approach. Results from a typical network topology are presented as well as a discussion around the validity of the results and potential sources of error. Consideration is given to the potential of advanced numerical methods to minimise error, particularly in large scale networks. Recommendations are made regarding the optimal configurations to be used when comparing first hop redundancy protocols, as well as further research required to improve the accuracy of the results obtained.

Keywords: FHRP · HSRP · GLBP · VRRP · Network redundancy
Gateway protocol · Failover · Network performance measurement

1 Introduction

First Hop Redundancy Protocols (FHRP) provide an essential tool for increasing availability in critical switched IP networks. They provide a mechanism for fast failover to the next hop from a primary path to a secondary path within a group of backup routers. The process is faster than waiting for spanning tree or dynamic routing protocols to converge on a new path due to the limited scope and pre-configuration of the FHRP. Essentially two or more routers are able to share the default gateway at OSI layer 3, which provides an alternative route or may even be used for rudimentary load balancing. This does necessitate the use of multilayer switches but can be implemented at the Access Layer or Distribution Layer [1, 2]. YanHua and WeiZhe [3] have shown that such protocols are suitable for use within cable television IP networks when combined with device redundancy, although they caution against diminishing returns as the network complexity increases.

Cisco has developed two major proprietary protocols Hot Standby Routing Protocol (HSRP) and Gateway Load Balancing Protocol (GLBP). Another common protocol is

© Springer Nature Switzerland AG 2019
G. Nikolov et al. (Eds.): NMA 2018, LNCS 11189, pp. 55–63, 2019.
https://doi.org/10.1007/978-3-030-10692-8_6

the Virtual Router Redundancy Protocol (VRRP), which is available as an open IEFT standard RFC5798. It is similar to HSRP in operation but not compatible [4, 5].

HSRP is configured for an interface using the standby command and allows the user to configure a virtual gateway for the connected hosts to use. Priorities are specified such that an active router is allocated with one or more standby routers sharing the virtual address with the active router as shown by Fig. 1. Packets are forwarded based on an IP/MAC address pair and standby routers monitor the status of the active router to promote a backup router in the case of a link failure on the active path. Tracking objects can be used to monitor interfaces or Service Level Agreement (SLA) tracking can monitor connectivity beyond the first hop. Either method can update the router priorities to determine the active path. Different priorities can also be assigned to different Virtual Local Area Networks (VLAN) to implement basic load balancing, although this may become unwieldy on large networks [2, 6].

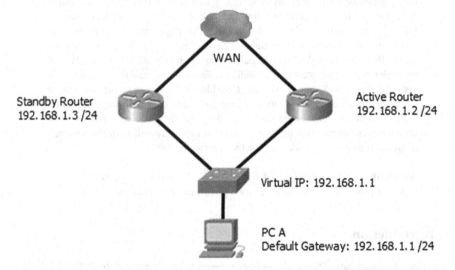

Fig. 1. A typical HSRP configuration with the host using a common virtual gateway IP address.

GLBP uses multiple gateways simultaneously, which enables more effective load balancing and therefore uses all the bandwidth within the topology. Routers within a GLBP group may be the Active Virtual Gateway (AVG), an Active Virtual Forwarder (AVF) or the Standby Virtual Gateway (SVG). The AVG assigns virtual MAC address to the other group members. Up to four AVFs, including the AVG, are able to forward packets and the SVG is ready to take over from the AVG based on a similar priority system to HSRP with decrements based on tracking objects. GLBP is implemented on an interface using the glbp command and load balancing can be achieved within the group by assigning packets to the MAC addresses of the AVFs via an equal round-robin, by weighting certain paths or based on the host [2, 7].

VRRP is very similar to HSRP in that is uses a single virtual gateway that is shared between a master and one or more backup router. It is implemented on the interface using the `vrrp` command and supports object tracking to determine failures. Fewer IP addresses may be used by VRRP than HSRP as the physical IP address for the master router may also be used as the virtual IP address [8].

Several articles have been published that outline the configuration options for HSRP, GLBP and VRRP [5, 9–11]. These discuss how the options affect the underlying algorithm but there appears to be little guidance on how to optimise the parameters or how the common protocols compare under similar conditions. Ibrahimi et al. [11] and Rahman et al. [12] demonstrate a basic method, using continuous Internet Control Message Protocol (ICMP) echo requests to show the duration of a link failure. Pavlik et al. [2] show a more accurate method to determine the interval between missing and restored replies using timestamps from a packet sniffer. None of the studies investigate the time taken to restore a link after the primary link recovers.

2 Method for Protocol Comparison

There are several network simulation packages that may be used to rapidly analyse the behavior of different configurations. Common simulators include Cisco's Packet Tracer, open source application GNS3 and Riverbed Modeler. However for these to produce accurate results, they would require detailed implementations of the software and protocols running on the network devices as well as accurate models of the hardware performance. Their accuracy cannot be verified so it is preferred to initially characterise the protocol on real hardware.

A testbed was created to emulate a typical business Internet connection as shown by Fig. 2. The customer had an edge router, which connected to their Local Area Network (LAN). This had redundant links to the external network via primary and secondary routers, which were then connected to the main router at the Internet Service Provider (ISP). The switch simplified the configuration of the customer router by removing the need for two interfaces on the same subnet. The primary link utilised a Gigabit Ethernet connection whereas the secondary was only Fast Ethernet. Secondary links are often metered in practice so load balancing was not implemented. To simulate a link failure, the interface G0/0 was shut down on the primary router with a tracking object to promote the secondary link based on the line-protocol state. The FHRPs were configured to decrement the priority of the primary router below that of the secondary router if the line-protocol went down; this causes the secondary router to preempt the primary router and traffic would be rerouted. An alternative method would be to check the reachability of the loopback interface on the destination network using ICMP requests to decrement the priority upon failure. This would be a more meaningful detection method within a real network but the non-deterministic nature of packet generation and propagation may distort the results, which should be focused on the responsiveness of the redundancy protocol itself.

Three methods were considered for measuring the interval between the interface changing state and the FHRP responding.

1. Continuous ICMP echo requests could be sent with a 100 ms interval to the loopback interface on the main router from a PC connected to the customer LAN.

This could be achieved using the Packet INternet Groper (PING) utility packaged with the Ubuntu operating system. The number of dropped requests would infer the time taken for the protocol to failover to the nearest 200 ms.

2. The same PING messages could be captured using a packet sniffer such as Wireshark running on the PC. The timestamps could be used to calculate the time between the first PING request not to be followed by a reply and the first request to be followed by a reply. The precision is based on the PING interval so would again be to the nearest 200 ms.

3. Debug messages on the primary and customer routers could be used to determine the interval between the primary link failure and establishment of the secondary path. The devices could be time-synchronised using the Network Time Protocol (NTP) and debugging logging could be configured to millisecond precision.

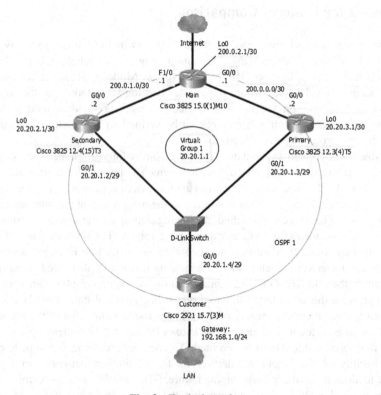

Fig. 2. Testbed topology.

Method three was used as it provided better precision than ICMP messaging. The interval between PINGs could be reduced to improve the precision of the other methods but the accuracy of packet generation could not be guaranteed due to the non-deterministic nature of PC applications and domestic-grade network interface cards. Method one was also used to provide a comparison to previous studies. The test was performed five times for each FHRP protocol – HSRP, GLBP and VRRP – with a default (D) and rapid (R) configuration. The minimum, maximum and mean average failover and restoration times were recorded. The restoration time was considered important as real networks often use metered secondary connections, which incur a charge based on the amount of data carried.

By default HSRP sets the time between hello messages to 3 s and the hold time to be 10 s. The hold time is the interval after which the active router is declared to be down once hello messages are not acknowledged. The hold timer must be greater than the hello timer to avoid the active router being declared down between hello messages – a factor of at least three is usually employed [13, 14]. GLBP uses the same system of hello and hold timers with identical default and minimum values [15]. HSRP and GLBP timers can be set with millisecond granularity but a shorter timer increases the amount of overhead traffic and can cause the system to behave erratically as it becomes too sensitive [16]. For the rapid test the hello timer will be set to 1 s and the hold timer to 3 s; this will allow the effect of smaller timers to be observed without creating instability in the system.

VRRP advertisements are sent every second by default with a failover delay calculated by three times the advertisement interval plus the router's skew time. The skew time is based on the inverse of the priority. The standard doesn't include shorter advertisement intervals but Cisco has implemented this in their Internetwork Operating Systems down to 20 ms [9, 17]. For the rapid test the priorities were configured as high as possible to reduce the skew time.

A summary of configurations is included in Table 1. It should be noted that pre-emption delay should normally be configured to allow for the boot time of the equipment [16]. HSRP and VRRP do not load balance by default and GLBP had load-balancing disabled for the test.

Table 1. Summary HSRP configurations.

	Hello interval	Hold timer	Preemption delay	Priority
HSRP D	3	10	0	-
HSRP R	1	3	0	-
GLBP D	3	10	30	-
GLBP R	1	3	3	-
VRRP D	1	-	0	100/110
VRRP R	1	-	0	253/254

3 Analysis

The results from the investigation are summarised in Tables 2 and 3 and presented in Figs. 3 and 4.

Table 2. Failover times for FHRP in ss.ms

	Min	Max	Mean	Dropped
HSRP D	03.390	05.643	04.780	55
HSRP R	03.818	05.561	05.067	55
GLBP D	32.456	34.392	33.339	55
GLBP R	03.604	04.020	03.858	55
VRRP D	02.757	03.389	03.112	55
VRRP R	02.316	02.969	02.665	55

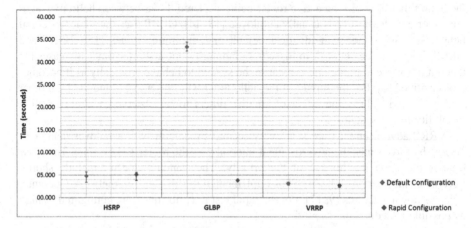

Fig. 3. Mean failover times with minimum and maximum variance.

Table 3. Restoration times for FHRP in ss.ms

	Min	Max	Mean	Dropped
HSRP D	25.133	34.931	29.635	0
HSRP R	25.312	29.521	27.876	0
GLBP D	32.507	36.524	34.253	0
GLBP R	03.284	04.295	04.040	0
VRRP D	02.372	03.363	02.989	0
VRRP R	02.236	02.970	02.583	0

It can be seen from the results that there is a reasonably significant difference in performance between the protocols. With the default configurations, VRRP is the fastest to both failover and restore the primary link followed by HSRP and GLBP. VRRP was also the fastest to failover and restore the primary link with the rapid configuration, although this was followed by GLBP then HSRP. The rapid configuration gave the most dramatic increase in sensitivity to GLBP as the protocol has an

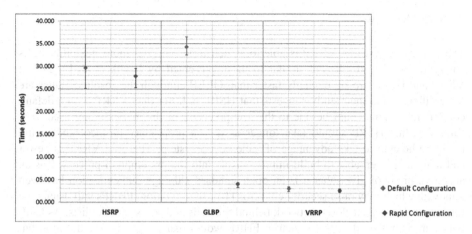

Fig. 4. Mean restoration times with minimum and maximum variance.

additional pre-emption delay of 30 s by default. HSRP appears to be very slow to restore the primary link even with shorter timers configured.

It's interesting that HSRP took longer to failover when given the rapid configuration. This could be linked to flooding the interfaces with hello messages but would require further investigation.

Another point of interest is that every test resulted in 11 sets of 5 PING responses being dropped before the replies continued. This equates to a consistent outage of approximately 5.5 s that is at odds with the variable delays calculated from the log messages. This could be because the OSPF routing protocol is actually quite effective within such a simple network and was able to reconverge faster than the redundancy protocols under some circumstances. However, this doesn't explain why there weren't fewer losses when the FHRP was faster.

Initially the time delay was calculated between the line-protocol and FHRP state change messages within the logs on the secondary router. When VRRP was configured this lead to a negative delay and closer inspection of the logs revealed that the tracking objects logged the line-protocol state change about 3 s before the line-protocol itself. As such the tracking protocol logs were used as the reference. This could reveal limitations within the logging process – whilst it is possible to configure millisecond resolution, the behaviour within the software stack is likely to be non-deterministic. In this case it is possible that the tracking logs are prioritised over those of the line-protocol within the software hence the discrepancy.

One final observation was made when comparing the logs of the primary and secondary routers. The state changes of the secondary router were used as the timing reference for the investigation, but with HSRP the primary router appeared to be promoted to active before the secondary router returned to standby. Further investigation of the packet flow would be required to determine how the routers behave in the interim and therefore which timestamps most accurately indicate their sensitivity.

4   Conclusions

The results suggest that VRRP is able to switch faster and has more flexibility in its configuration. The results are similar to those presented by Pavlik et al. [2] in that VRRP and HSRP are significantly faster than GLBP although the absolute values are quite different and their study suggests that HSRP is faster than VRRP with a default configuration. This could be due to the low precision of the methodology used. It is surprising that HSRP should take longer as it is a proprietary protocol and would therefore be able to take advantage of a closed ecosystem. That said VRRP is newer and may have been able to build improvements into the pre-emption process. It is an attractive protocol given that it is an open standard although authentication has been removed, which may raise security concerns [16]. In mitigation, FHRPs will usually be implemented within a closed network behind firewalls and access control lists. As load-balancing is not used in this scenario, FHRP would make a good second option but networks with more complex load-balancing requirements may find GLBP more effective but the timers should be adjusted to improve sensitivity.

The investigation has produced a good indication of the sensitivity of the three major protocols but has also raised some questions that require further study. The majority of these questions revolve around the impact of other configuration parameters. The comparison between protocols appears to be valid, with increased precision compared to existing studies, but the absolute figures may contain some inaccuracy. As such the results may be used to inform network design and characterisation but further consideration should be given to the routing protocol configuration to understand its interaction with FHRPs. The timers could be reduced further to find the minimum failover time although it should be noted that in reality such short timers are unlikely to be used as they may cause instability in flapping connections as well as increasing the protocol overheads from excessive hello packets. The logging process should be investigated to determine the accuracy of timestamps and packet flows should be monitored to understand whether the timestamps accurately reflect the flow. This analysis may also help with understanding the ICMP behaviour. Overall the method appears to have been successful and could easily be scaled to systems with multiple backup routers by using a syslog server to collate logging messages from multiple devices. Once the accuracy of the test method has been verified, the protocols should be characterised and compared to results obtained from a simulation of the testbed. Null hypothesis testing may be used to verify the accuracy of the simulations so that complex configurations involving load balancing and multiple standby routers may be more rapidly optimised than is feasible using real hardware.

References

.1. CCNA Routing and Switching Practice and Study Guide: LAN Redundancy. http://www.ciscopress.com/articles/article.asp?p=2204384&seqNum=4. Accessed 19 Apr 2018
2. Pavlik, J., et al: Gateway redundancy protocols. In: 15th IEEE International Symposium on Computational Intelligence and Informatics, pp. 459–464. IEEE, Budapest (2014)

3. YanHua, Z., WeiZhe, M.: The design of cable television IP access network based on hot standby router protocol. In: International Conference on Image Analysis and Signal Processing, pp. 1–4. IASP, Hangzhou (2012)
4. Virtual Router Redundancy Protocol (VRRP) Version 3 for IPv4 and IPv6. https://tools.ietf.org/html/rfc5798. Accessed 19 Apr 2018
5. Oppenheimer, P.: Top-Down Network Design, 3rd edn. Cisco Press, Indianapolis (2011)
6. Hot Standby Router Protocol Features and Functionality. https://www.cisco.com/c/en/us/support/docs/ip/hot-standby-router-protocol-hsrp/9281-3.html. Accessed 19 Apr 2018
7. Cisco: First Hop Redundancy Protocols Configuration Guide, Cisco IOS Release 12.2SX. Cisco, San Jose (2011)
8. VRRP Overview. https://www.juniper.net/documentation/en_US/junose15.1/topics/concept/vrrp-overview.html. Accessed 03 June 2018
9. First Hop Redundancy Protocols Configuration Guide, Cisco IOS Release 15SY. https://www.cisco.com/c/en/us/td/docs/ios-xml/ios/ipapp_fhrp/configuration/15-sy/fhp-15-sy-book/fhp-vrrp.html. Accessed 19 Apr 2018
10. Configuring VRRP. https://www.juniper.net/documentation/en_US/junos/topics/example/vrrp-configuring-example.html. Accessed 01 June 2018
11. Ibrahimi, M., et al.: Deploy redundancy of internet using first hop redundancy protocol and monitoring it using IP service level agreements. Int. J. Eng. Sci. Comput. 7(10), 15320–15322 (2017)
12. Rahman, Z., et al.: Performance evaluation of first HOP redundancy protocols (HSRP, VRRP & GLBP). J. Appl. Environ. Biol. Sci. 7(3), 268–278 (2017)
13. Hot Standby Router Protocol (HSRP): Frequently Asked Questions. https://www.cisco.com/c/en/us/support/docs/ip/hot-standby-router-protocol-hsrp/9281-3.html. Accessed 19 Apr 2018
14. Standby-hold-timer. https://www.ibm.com/support/knowledgecenter/SS9H2Y_7.5.0/com.ibm.dp.doc/standby-hold-timer_interface.html. Accessed 19 Apr 2018
15. Timers (GLBP). https://www.cisco.com/c/m/en_us/techdoc/dc/reference/cli/n5k/commands/timers-glbp.html. Accessed 19 Apr 2018
16. Froom, R., Sivasubramanian, B., Frahim, E.: Implementing Cisco Switched Networks. Indianapolis, Cisco (2010)
17. VRRP failover-delay Overview. https://www.juniper.net/documentation/en_US/junos/topics/concept/vrrp-failover-delay-overview.html. Accessed 19 Apr 2018

Derivation of a Coordinate System of Three Laser Triangulation Sensors in a Plane

Ján Buša[(⊠)] [iD], Miroslav Dovica, and Lukáš Kačmár

Technical University in Košice, Letná 9, 04001 Košice, Slovakia
{jan.busa,miroslav.dovica,lukas.kacmar}@tuke.sk

Abstract. The paper discusses the derivation of an accurate coordinate measuring system based on the records of three fixed laser triangulation sensors done for a workpiece in movement.

The influence of the measurement and rounding inaccuracy on the identification accuracy using numerical simulation methods are assessed.

Keywords: Metrology · Laser triangulation sensor · Gauge block Sensors coordinate system · Minimization · GNU Octave

1 Introduction

The standard approach to the length measurements in coordinate metrology assumes a moving measuring system and a fixed workpiece [1]. In the present paper a quantitative characterization of the movement of a workpiece of interest is given based on records provided by a system of three fixed triangulation sensors [2,3] (see Fig. 1) using the laser triangulation principle described in [4,5]. To this aim, we assume that the output of a sensor interface is provided by the distance to the point target along the sensor laser beam.

The usage of stationary sensor system records asks for the accurate knowledge of the positions of the sensors and the corresponding directions of their laser beams. Here we describe a simple method for the identification of the fixed coordinate system bound to the sensors based on a set of measurements done for different positions of a gauge block [6] – a metal or ceramic block involving two parallel opposing faces the distance w between which is known with very high accuracy (Fig. 2).

The discussion is divided into four parts: the problem formulation (Sect. 2), the problem solution (Sect. 3), numerical case studies (Sect. 4), and conclusions (Sect. 5).

2 Formulation of the Problem

Given the set of three fixed sensors S_1, S_2, and S_3, with predefined but unknown angular directions of their laser beams, and a certain position of the gauge block

© Springer Nature Switzerland AG 2019
G. Nikolov et al. (Eds.): NMA 2018, LNCS 11189, pp. 64–71, 2019.
https://doi.org/10.1007/978-3-030-10692-8_7

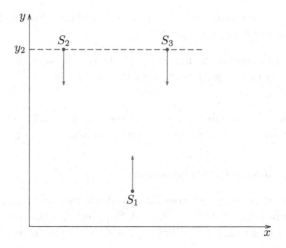

Fig. 1. Sensor positions and orientations in a world coordinate system.

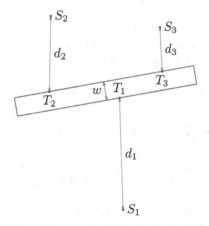

Fig. 2. Distances to the point targets.

(Fig. 2), a triplet of distances $(d_1; d_2; d_3)$ measured along the laser beam from each sensor to the block, is recorded and stored. The N times iteration of the measuring process for different positions of the rectangular gauge block of width w provides the input of the problem, $(d_{1_i}, d_{2_i}, d_{3_i})$, $i = 1, 2, \ldots, N$.

2.1   Uniqueness of the Sensor Triplet Coordinate System

A same triplet of distances $(d_1; d_2; d_3)$ can be obtained for different configurations of the arrangement shown in Fig. 2, e.g., under the rigid rotation of the sensors together with the gauge block.

To enforce solution uniqueness, the definition of the sensor coordinate system is done under the following constraints:

1. the position of the sensor S_1 in the (x, y) plane is fixed at $(0, 0)$;
2. a same value y_2 of the y-coordinate of the sensors S_2 and S_3 is chosen (as illustrated in Fig. 1).

We have then to determine six parameters (three position coordinates and three angles) for $S_1 = (0; 0; \alpha_1)$, $S_2 = (x_2; y_2; \alpha_2)$, and $S_3 = (x_3; y_2; \alpha_3)$.

2.2   Precision of Point Determination

In practice, instead of the exact coordinates of a sensor, $S = (\bar{x}, \bar{y}, \bar{\alpha})$, we may know only the approximate set $\tilde{S} = (\tilde{x}, \tilde{y}, \tilde{\alpha})$. Suppose that the true distance from the sensor S to a point target $T = (x_T, y_T)$ is $\bar{d}$, while the measured distance is $\tilde{d}$, such that

$$|\tilde{x} - \bar{x}| < \varepsilon_x, \qquad |\tilde{y} - \bar{y}| < \varepsilon_x, \qquad |\tilde{\alpha} - \bar{\alpha}| < \varepsilon_\alpha, \qquad |\tilde{d} - \bar{d}| < \varepsilon_d. \qquad (1)$$

Using the measured values, we infer an approximate point target $\tilde{T} = (x_{\tilde{T}}, y_{\tilde{T}})$ the deviation of which from the true point target T is characterized by the following upper bound to the inaccuracy[1]

$$\|\tilde{T} - T\|_\infty = \max \left(|x_{\tilde{T}} - x_T|, |y_{\tilde{T}} - y_T| \right) < \varepsilon_x + \varepsilon_d + \tilde{d} \cdot \varepsilon_\alpha. \qquad (2)$$

3   Solution of the Problem

Under the above constraints on the definition of the coordinate system, the only invariant feature associated to any position of the gauge block (see Fig. 2 for notations) follows from the fact that the targets T_2 and T_3 stay on a same face of the gauge block, which is opposite to that of the target T_1. Hence the distance from T_1 to the straight line $\overline{T_2 T_3}$ should equate w for all block placements.

Given the sensor parameters S_1, S_2, and S_3 defined in Sect. 2.1, the values $\tilde{w}_i = \tilde{w}_i(S_1, S_2, S_3)$ which denote the calculated distances from T_1 to the line $\overline{T_2 T_3}$ for the input triplets $(d_{1_i}, d_{2_i}, d_{3_i})$, $i = 1, 2, \ldots, N$, will be different from the true distance w and will depend on the accuracy of measurement of these triplets. If the number N of the measurements is large enough, the best approximating values of the six parameters $x_2; x_3; y_2; \alpha_1; \alpha_2; \alpha_3$ are obtained as the solution of the nonlinear least squares problem

$$(S_1^*, S_2^*, S_3^*) = \operatorname*{argmin}_{(S_1, S_2, S_3)} \min \sum_{i=1}^{N} \left[\tilde{w}_i(S_1, S_2, S_3) - w \right]^2. \qquad (3)$$

---

[1] An accuracy better than 0.01 mm could be considered as sufficient in practice.

To solve the problem (3) numerically it is not necessary to write an explicit formula for the function to be minimized. For given "exact" triplets values[2] we solved the problem using GNU Octave [7] – an open source alternative to the proprietary programming language MATLAB [8]. Our scripts and functions could be, of course, used in MATLAB as well.

The relevant part of the script `sensors_identification_cs.m` is:

```
dl=4; % lines distance in mm
S1=[0;0;pi/2]; S2=[-10;40;3*pi/2]; S3=[15;40;3*pi/2]; % initial S
x=[S1(3);S2;S3([1 3])]; % initial vector for minimization
opt=optimset('TolX',1.e-16,'TolFun',1.e-16,'MaxIter',50000,...
             'MaxFunEvals',50000,'Display','final');
[x,w]=fminsearch('dist_diff_cs',x,opt); % Octave
```

Here the function to be minimized is

```
function [dd]=dist_diff_cs(x)
% differences between actual distances and required by dl
global dl distances nf;
nf=nf+1; % function evaluations counter
S1=[0;0;x(1)]; S2=[x(2);x(3);x(4)]; S3=[x(5);x(3);x(6)];
cs1=[cos(S1(3));sin(S1(3))]; cs2=[cos(S2(3));sin(S2(3))];
cs3=[cos(S3(3));sin(S3(3))];
[m,n]=size(distances);
for k=1:m,
    T1=S1(1:2)+distances(k,1)*cs1;
    T2=S2(1:2)+distances(k,2)*cs2;
    T3=S3(1:2)+distances(k,3)*cs3;
    d_dif(k)=AtoBC(T1,T2,T3)-dl;
end
dd=norm(d_dif);
```

where `AtoBC(A,B,C)` is a function calculating the distance from the point `A` to the line defined by the points `B` and `C`.

4 Numerical Case Studies

4.1 Exact Distances Data

Let us suppose that the sensors are placed at the points $(0,0)$, $(-10,40)$, and $(15,40)$ (units are mm), and oriented similar to those shown in Fig. 1 with the corresponding beam angles 90, 270, and 270°. However, due to some placement inexactness the true parameters are instead:

$$S_1 = \left[0, 0, \frac{0.99 \cdot \pi}{2}\right]; \ S_2 = \left[-9.9, 40.1, \frac{3.02 \cdot \pi}{2}\right]; \ S_3 = \left[15.1, 39.9, \frac{2.99 \cdot \pi}{2}\right],$$

---

[2] Here the double precision data rounded to 10^{-7} mm (1 Å) are considered to be "exact".

such that the sensor positions and the beam orientations are slightly modified, the beam angles being 89.1, 271.8, 269.1°, respectively. The true sensor coordinates do not satisfy the constraint $y_2 = y_3$. To get a sensor coordinate system obeying to this constraint, we have to rotate the given system by an angle $\varphi = \arctan\big((y_2 - y_3)/(x_3 - x_2)\big)$.

In this case it is natural to start the minimization with the parameters supposed to be true. Below the results of using `sensors_identification_cs` command are given[3]:

```
Original sensors
   x          y             angle       dangle
   0          0           1.5550884       89.1
  -9.9       40.1         4.7438049      271.8
  15.1       39.9         4.6966810      269.1


Original triangle sides are 25.00080  42.66169  41.30399 mm
                and angles are 34.59698  75.67362  69.72940 degrees.


Original sensors rotated by angle 0.45835646 degrees
   x          y             angle       dangle
   0          0           1.5630882     89.55836
 -10.220473  40.019519    4.7518047    272.25836
  14.780327  40.019519    4.7046808    269.55836


Initial sensors
   x          y             angle       dangle
   0          0           1.5707963        90
 -10         40           4.7123890       270
  15         40           4.7123890       270


Initial triangle sides are 25.00000  42.72002  41.23106 mm
                and angles are 34.59229  75.96376  69.44395 degrees.


Before optimization
Test 1: straight lines distance is 3.9725118 mm.
...
Test 6: straight lines distance is 3.9848938 mm.


After optimization
Test 1: straight lines distance is 4 mm.
...
Test 6: straight lines distance is 4 mm.


Function evaluation number was 2385.


Final sensors
   x          y             angle       dangle
   0          0           1.5630884     89.55837
```

---

[3] The output format is slightly modified.

```
-10.220492   40.019519    4.7518052   272.25838
 14.780334   40.019519    4.7046808   269.55835
```

```
Final triangle sides are 25.00083  42.66169  41.30400 mm
        and angles are 34.59701  75.67359  69.72939 degrees.
```

```
Angles differences are -0.45837  -0.45838  -0.45835 degrees.
```

The final sensor parameters with a same y-coordinate match very well the original sensor triangle rotated by $0.45835646°$. The maximum difference in x coordinate is less than $0.02\,\mu$m. The maximum angular misfit is about $5 \cdot 10^{-7}$ rad or $2.62 \cdot 10^{-5}$ deg. Using (2) we get for, e.g., $\tilde{d} = 50$ mm:

$$\|\tilde{T} - T\|_\infty = \max\left(|x_{\tilde{T}} - x_T|, |y_{\tilde{T}} - y_T|\right) < \varepsilon_x + \varepsilon_d + \tilde{d} \cdot \varepsilon_\alpha < 1.5 \cdot 10^{-4}\,\text{mm}.$$

For practical purposes this precision is sufficient.

In this example, six distance triplets were sufficient for the identification of the sensor coordinate system. A combination of three different angles and two different positions of the gauge block was used in the simulated measurement:

```
for k=1:6, s=[3;mod(k,3)-1]; s=s(:); s=s/norm(s); n=[s(2);-s(1)];
    if n(2)<0, n=-n; end; p1=[-15;20+3*mod(k,2);s]; ... ;
end
```

4.2 Rounded Distances Data

If the distance triplets are not exact, then, of course, the distances from the approximate targets T_1 to the approximate lines $\overline{T_2 T_3}$ will, in general, be different from the width w (see the example below).

Precision $0.1\,\mu$m. Let us consider a solution to the system identification problem for input data – the distance triplets – rounded to $0.1\,\mu$m.

```
Original lines distances
Test 1: straight lines distance is 4.0000149 mm.
...
Test 5: straight lines distance is 3.9999307 mm.
Test 6: straight lines distance is 4.0000099 mm.
```

```
After optimization
Test 1: straight lines distance is 4 mm.
...
Test 6: straight lines distance is 4 mm.
```

```
Final sensors
     x           y          angle       dangle
     0           0        1.5626635    89.53402
-10.239850   40.019234   4.7522437   272.28351
 14.843013   40.019234   4.7033105   269.47984
```

```
Final triangle sides are 25.08286   42.68318   41.30852 mm
           and angles are 34.70218   75.64748   69.65033 degrees.
```

```
Angles differences are -0.43402   -0.48351   -0.37984 degrees.
```

The error in the x-coordinate, of 0.063 mm, it is unacceptably large. The error in the y-coordinate, of about 0.3 μm, is excellent. The maximum angle difference of about 0.0014 rad or 0.08° is poor. Using (2) we get for $\tilde{d} = 50$ mm:

$$\|\tilde{T} - T\|_\infty = \max\left(|x_{\tilde{T}} - x_T|, |y_{\tilde{T}} - y_T|\right) < 0.14 \text{ mm},$$

what is obviously poor.

In this case when 6 parameters have to be determined from 6 measurements, the "optimal" block width of 4 mm have been reached for all 6 measurements, however resulting parameters have a poor precision. Indeed, one can see that the "original" points corresponding to the rounded distances from sensors will not give the wall distances of exactly 4 mm. Using more distance triplets will result in a better precision. E.g., for 15 points we have got an upper bound of $\|\tilde{T} - T\|_\infty$ of about 0.012 mm, and using 21 points we have got:

```
After optimization
Test 1: straight lines distance is 3.9999932 mm.
...Test 21: straight lines distance is 4.0000058 mm.
```

```
Function evaluation number was 2117.
```

```
Final sensors
   x          y            angle      dangle
   0          0          1.5631659    89.56281
-10.22436   40.019506   4.7519152   272.26468
 14.778141  40.019506   4.7047695   269.56344
```

```
Final triangle sides are 25.00250   42.66092   41.30494 mm
and angles are 34.59946   75.66839   69.73215 degrees.
```

```
Angles differences are -0.46281   -0.46468   -0.46344 degrees.
```

```
S2 precision is 0.0038872816   1.2985923e-05 mm.
S3 precision is 0.0021865184   1.2985923e-05 mm.
Angles precision is -7.7699306e-05   -0.00011042672   -8.865779e-05 rad.
Angles precision is -0.0044518423   -0.0063269849   -0.0050797172 degree.
```

Using (2) we get for $\tilde{d} = 50$ mm:

$$\|\tilde{T} - T\|_\infty = \max\left(|x_{\tilde{T}} - x_T|, |y_{\tilde{T}} - y_T|\right) < 0.00951 \text{ mm}.$$

Such an accuracy is sufficient in practice.

5   Conclusions and Future Work

In this paper the 2D solution of the problem has been derived. We have found that using a sufficiently large number of measurement points results in a sensor coordinate system identification with good enough accuracy.

In practice, however, we could not expect that all laser beams have a common (horizontal) plane of propagation, and also we could not expect that all sensors are placed at the same vertical position. This will need the solution of the general 3D problem with the sensor specifications asking for 11 parameters:

$$S_1 = (0; 0; 0; \alpha_1; \beta_1), \quad S_2 = (x_2; y_2; z_2; \alpha_2; \beta_2), \quad S_3 = (x_3; y_2; z_3; \alpha_3; \beta_3).$$

A 3D representation of a gauge block will be used under the assumption that the block is moving on a plane (with a common vertical orientation given by the vector normal to this plane). The solution of this general problem is in progress.

Acknowledgements. Work supported within the project VEGA 1/0224/18 "Research and development of testing and measuring methods in coordinate metrology". The authors are grateful to Gheorghe Adam for fruitful discussion and advice.

References

1. Weckenmann A., Kraemer P., Hoffmann J.: Manufacturing metrology – state of the art and prospects. In: 9th International Symposium on Measurement and Quality Control (9th ISMQC), 21–24 November 2007, 7 p. IIT Madras (2007)
2. MTI Instruments. https://www.mtiinstruments.com/technology-principles/laser-triangulation-sensors/. Accessed 29 April 2018
3. Micro-epsilon sensors. https://www.micro-epsilon.com/. Accessed 29 April 2018
4. Berkovic, G., Shafir, E.: Optical methods for distance and displacement measurements. Adv. Opt. Photonics **4**, 441–471 (2012). https://doi.org/10.1364/AOP.4.000441
5. AZO Sensors. An Introduction to Laser Triangulation Sensors. https://www.azosensors.com/article.aspx?ArticleID=523. Accessed 29 April 2018
6. Gauge block. https://en.wikipedia.org/wiki/Gauge_block. Accessed 29 April 2018
7. GNU Octave. https://www.gnu.org/software/octave/. Accessed 1 May 2018
8. Matlab. https://www.mathworks.com/products/matlab.html. Accessed 1 May 2018

Optimisation Techniques in Wildfire Simulations. Test Case Kresna Fire August 2017

Nina Dobrinkova[1(✉)], Momchil Panayotov[2], and Peter Boyvalenkov[3]

[1] Institute of Information and Communication Technologies,
Bulgarian Academy of Sciences, Sofia, Bulgaria
ninabox2002@gmail.com
[2] Dendrology Department, University of Forestry, Sofia, Sofia, Bulgaria
panayotov.m@ltu.bg
[3] Institute of Mathematics and Informatics,
Bulgarian Academy of Sciences, Sofia, Bulgaria
peter@math.bas.bg

Abstract. After the year of 2000 wild land fires are with constant occurrence on Bulgarian territory and the fire season is having two picks one in March - April and another July - October periods. In 2017 wild land fires occurred all over south Bulgaria even in places where fires are not that common. Such place was Kresna Gorge where is located Struma Motorway part of the EU corridor IV connecting Vidin (North-west Bulgaria) with Thessaloniki (North-west Greece) areas. In this paper we evaluate the available parameters for the Kresna Gorge fire which happened in the period 24–29 August 2017. We are summarizing the novel approaches used in Wildfire Analyst (WFA) tool version 2.8 in comparison with FARSITE tool. WFA is not applied so far for Bulgarian fire events, but have very good potential for real time decision making and this is one of its first assessments for its usage.

1 Introduction

Wildfires became more common on the territory of EU in the last decades. Official reports published on the main web-site of the European Forest Fire Information System (EFFIS) [1] shows that wildfires are more frequent and more severe, especially in South EU countries. Particularly, officially published statistics from the ministry of Agriculture and Forests in Bulgaria reports figures about burned area in all affected zones in Bulgaria, which are in line with the EU trend. Fire regime is changing and fire behavior and spread are getting more extreme since 1970s until current days. After the year of 2000 wildfires on the Bulgarian territory are having two fire season windows: (1) early spring: March and April; (2) summer: from July to October. In 2017 wildfires affected Southern parts of Bulgaria, even in areas where fire occurrence is not usual. One of these areas was Kresna Gorge in the period 24–29 August 2017. The area where

© Springer Nature Switzerland AG 2019
G. Nikolov et al. (Eds.): NMA 2018, LNCS 11189, pp. 72–79, 2019.
https://doi.org/10.1007/978-3-030-10692-8_8

the fire has occurred was located east from Kresna Gorge, affecting vast area of forest lands and endangering the local population properties and lives. The fire started in grasslands located south from river Mechkulska and propagated further south approaching the villages of Old Kresna, Oshtava and Vlahi. The fire spread affected mainly grasslands, shrubs and conifer (pinus nigra type) vegetation. The reason of the fire occurrence and its one week propagation was caused because of the long period of dry, hot weather in July and August in this zone. An analysis of satellite images based on Landsat 8 has been done by GeoPolymorphic Cloud Company. It showed temperature anomalies for the area before the fire occurrence in the range $35 - 40\,^{\circ}$C at noon time [2]. The high temperature combined with low humidity extracted by the satellite data was presenting great increasement of the wildfire risk occurrence. The fire signal was filed on 24.08.2017 and the satellite data warning was confirmed. The paper is organized as follows. In Sect. 2 we present our initial investigations done for preparation of the fire simulations. In Sect. 3 we describe the area of the Kresna fire with special attention to its flora. Section 4 is devoted to description of the data preparation for simulations by Wildfire Analyst tool. We show previous experience and possible improvement in wild fire simulations in Bulgaria in Sect. 5 and give conclusions in Sect. 6.

2   Initial Analysis and Investigation

So far, statistics and investigation of the Kresna fire have not been presented officially by the state of Bulgaria through its ministry of Agriculture, Forests and Food. This is why we are using the satellite based analyses of private company with geoinformatics orientation - Geopolymorphic Cloud. This is done in order to describe as accurately as possible the state of the art of the available public data and all available (satellite) sources of information. The analysis of Geopolymorfic provides information that the VIIRS satellite detected first the fire when passing over the territory of Bulgaria at 13:48 (VIIRS Active Fire Product from 24.08.2017). The affected area where it gave its first ignition point is a pasture with south-southeast exposure. The analysis gave information about the number of satellites capturing each part of the earth's surface several times a day. The ones which gave information for surface temperature and the anomaly causing this fire were MODIS and VIIRS (Active Fire Products) for the period 24–29.08.2017 [2]. The total area described in the analysis of the affected area after fire suppression calculated on the basis of a satellite image of Sentinel 2 was 22 915 decares (approx. 2000 ha). The analysis gives as final result that nearly 65% of the burned area is in a forest fund. A total of 14 737 decares (1473,7 ha) were forests damaged by the fire in August 2017 in the Kresna Gorge and the rest were areas of flat lands, grasslands and urban areas. More details about this analysis as map distribution can be seen on Fig. 1. The analyzed data had been presented as yellow polygon of burned area (burned scar of the wildfire) and the rest of the colors inside the yellow polygon represent the types of forest species which had been affected and completely burned. More descriptions in maps will be presented in full extent combined with the initial outcomes and simulations of the burned area analyses under the project DISARM (see also [2]).

Fig. 1. Map of Kresna Gorge burned forestry area in the period 24–29 August 2017 [2]

3   Kresna Fire Description of the Location

The Kresna fire burnt from 24 to 29 August 2017. The terrain of the fire is hilly and low-mountain, situated at elevations from about 250 m a.s.l. to 750 m a.s.l. in Southwestern Bulgaria. The fire affected the eastern slopes of Kresna gorge above Struma river and roughly between the villages Mechkul (to the north) and Vlahi (to the south). The fire started in grassland-shrub community and transferred to plantations from Pinus nigra, where it quickly spread assisted by the dry conditions. The total affected area was about 22600 dka, of which 65% were forest territories. The fire burnt mostly plantations from Austrian pine (8700 dka), Scots pine (Pinus sylvestris), Robinia pseudoacacia (211 dka) (1260 dka), natural forests dominated by Quercus pubescens (2000 dka) and Quercus petraea (660 dka) and smaller territories of other species. The severity of burning was highest in the pine plantations, where the fire caused almost complete mortality. On steeper slopes above Struma River the severity of burning was lower, where some deciduous trees and single specimen of Juniperus excelsa survived. The territory partly lies in protected zones BG002003 from the Bulgarian Natura2000 network declared under the Birds directive and BG000366 declared under the Habitats Directive. The region is characterized with specific for Bulgaria flora, which is rich on species typically found in the Mediterranean biome and hosts numerous bird and reptile species. It is considered as one of the most important hotspots of biodiversity in Bulgaria. The typical forests for the region are dominated by oak species (mostly Quercus pubescens, Quercus frainetto and Quercus cerris), Carpinus orientalis, Crataegus pentagyna and Crataegus monogyna, Fraxinus ornus, Pistacia terebinthus, numerous bushes. In the vicinity is the biggest population of Juniperus excelsa in Bulgaria and Europe, which is protected in Tissata Strict Nature Reserve. However much of the oak forests were degraded due to long use by local people for firewood and pasturing of domestic animals such as goats and sheep. In the second half of the XX century large areas were re-planted with Pinus nigra and Pinus sylvestris in order to limit erosion processes. The climate in the area is transitional-Mediterranean characterized by dry and hot summers, relatively mild winters with short duration of the snow cover. The average annual temperature is $14\,°C$, the annual precipitation $514\,mm$ with precipitation maximum in the autumn-winter period. The average temperature of the coldest month (January) is $-1.3\,°C$. The absolute recorded temperature maximum is $+42.5\,°C$, while the absolute minimum is $-29.6\,°C$. It is interesting to note that the area, where the fire started and initially spread was affected also by at least two previous fires in the last 30 years, which caused the need for several subsequent re-plantings.

4   Data Preparation for Kresna Gorge Fire

In order to start specifically dedicated analyses on the Kresna Gorge parameters needed for wildfire simulations we had to start collecting information from different Bulgarian agencies. This is necessary because there is not one repository available for all simulation inputs combined in one single source. Thus the team which is working on data collection and preparation for its usage under the Wildfire Analyst (WFA; [3]) simulation is combining efforts from Bulgarian Academy of Sciences and Forestry University in Sofia. As a first step we had to estimate the area of the fire, which has as starting point Kresna Gorge, located north of Stara Kresna village. The fire spread was in direction south to the village of Vlahi. In the Western direction, the fire had descended on a steep slope almost to the Struma River, but there was a lower intensity of the fire spread. In this area north winds are very probable to happen, so local weather conditions created by the fire could be the case for the time frame when it was active. The general coordinates of the fire can be described as follows:

Starting Region as ignition point and its North Border (Approx.): 41.818961°; 23.179666° (for the first date of its occurrence 24th Aug. 2017) Western border: 41.814869°; 23.160966° Eastern border for the first two days of the fire spread: 41.801595°; 23.220162° Eastern border for the last days: 41.777043°; 23.224270° (this are the coordinates of the fire in its end phase on 29th Aug. 2017) Southern limit: 41.746235°; 23.234691° As next steps we are collecting and processing data for the local meteo conditions, relief and fuel load as Wildfire Analyst requirements are described. This as a final result will probably represent the most complete data set for a fire simulation in Bulgaria. All meteorological data from the local hydro meteorological stations will be elaborated and provided for the period of the fire occurrence (24th–29th August 2017). WFA requires for simulations predefined format with one hour time step. In fact, we will get the best possible in Bulgaria data by the National Institute of Meteorology and Hydrology of the Bulgarian Academy of Sciences. The local relief will be used from raster 30 × 30 DEM resolution available for the area of interest. This is good DEM for a start of the simulations, however lower resolution of 25 m or even 10 m can give even better simulation results. The most difficult part of the work is the creation of the fuel models of the burned biological species. In order to do this we will use Fire Behavior Fuel Models (FBFMs) by applying the Anderson (1982) [4] and Scott-Burgan (2005) [5] in order to obtain the burning materials. This data will be extracted from forestry maps provided by the local unit of the ministry of agriculture, forest and food of Bulgaria. Direct extraction from forest maps of burning materials in the FBFMs classifications known until now is very difficult, because in Bulgaria the only classifications done are for biological species, which requires further division for simulation purposes. Comparison with the actual final burned area (see, for example, Fig. 2, posted from the local press in Kresna Gorge as a drone driven picture) will be used for calibration and analysis.

Fig. 2. Kresna Gorge burned scar area one day before fire end (28.08.2017) [6]

5   Past and New Approaches for Wildfire Simulations in Bulgaria

In the framework of bilateral cooperation program between Greece and Bulgaria 2007–2013 a team from Institute of Mathematics and Informatics of the Bulgarian Academy of Sciences had the opportunity to work in the Zlatograd forestry department under the project OUTLAND. Great amount of data was collected and processed and several simulations of past fire events have been implemented. The used approach had been done by implementation of Rothermel fire spread model implemented in FARSITE [7]. The main scenarios used under the simulations had been oriented towards grasslands with scenario descriptions covering: Fuel moisture values: 6% (1-h), 7% (10-h), 9% (100-h), 45% ($liveherba - ceous$), and 75% ($livewoody$); Daily maximum temperatures: $17-21$ °C; Daily minimum relative humidity: 24–50%; Winds: generally from the west-southwest at 1-2 k h-1 Outcomes of these simulations have been oriented towards creation of custom Fire Behaviour Fuel Models (FBFMs) and more detailed application of FARSITE for Bulgarian test cases. The other main group of simulations which had been done for the zone of the Forestry department in Zlatograd was covering scenario similar to this one: Fuel moisture values: 3% (1-h), 4% (10-h), 5% (100-h), 40% ($liveherbaceous$) and 70% ($livewoody$); Daily maximum temperatures: $7 - 10$ °C; Daily minimum relative humidity: 36–40%; Winds: generally from the north-northeast at 10-2 k h-1 In this case the outcomes have been oriented towards generalization of the available FBFMs towards tree species that are available for Bulgaria and applicable in cases of simulations with FARSITE. More information on how the data has been collected elaborated and analyzed can be found in the papers [8], where Dobrinkova et all have worked out in details the methodology about how data elaboration and wild fire simulations for Bulgarian cases can be done. As mentioned above, we are going to use WildFire Analyst (WFA) for simulations of the Kresna Fire. WFA is software that provides real-time analysis of wildfire behavior and is capable to simulate the spread of wildfires in real time. Simulations could be completed quickly, in seconds, to support real time decision making. WFA is specifically designed to address this issue, providing analysis capabilities for a range of situations and users. WFA provides integration with ESRI's ArcGIS with no conversion or pre-processing required. This greatly increases usability, allowing users to concentrate on interpreting simulation outputs, and making important decisions about how and where to deploy firefighting resources. This software was designed to be used with a laptop or tablet at the incident command center, in the operations center, or directly on scene, providing outputs in less than a minute. The software has the option for usage of predefined weather scenarios, or current and forecasted weather obtained via web services. This is done to model fire behavior and provide outputs within seconds. For wildfires, time is of great importance in order Incident commanders to have good enough estimations well in advance and deploy their forces. The fire simulator have options to provide a range of analytical outputs, available as GIS maps and charts, for better decision making. The software have desktop platform, or web and mobile, which enabled applications, capabilities and results for wide range of users. Contemporary fire behavior software tools required a high degree of specialization, training and effort in the preparation and conversion of GIS data to use the software. Historically, this has been a limitation in using these tools for initial attack and real time applications. WFA have embeded different fire spread models like Rothermel (1972) [7] or Kitral (1998) [9] that is generally used in South America. Additionally, it has other models such as Nelson equations to estimate fuel moistures from weather data (2000) [10] or WindNinja (Forthofer et al. 2009) [11], a software that computes spatially varying wind fields for wildland fire applications in complex terrain. Input data required by the software is similar to other fire simulators such as FARSITE or BehavePlus due to the use of same fire spread models and equations. WFA introduces new simulation modes with innovative enhancements including real time processing performance, automatic rate of spread (ROS) adjustments based on observations used to create fire behavior databases, calculation of evacuation time zones (or 'firesheds'), and integration of simulation results for asset and economic impact analysis [12, 13]. Mathematically the inverse travel time method modifies the fire spread of the front called ROS in a certain direction θ given by the fire spread engine in the opposite direction represented by formula 1:

$$ROS_{Evac}(\theta, model\ inputs) = ROS_{Stand}(\theta + \pi, model\ inputs)\ for\ all\ angle\ \theta$$
(1)

WFA has been used operationally for diverse agencies worldwide, for instance by Military Emergency Unit and several regions in Spain, CONAF and private companies in Chile, Italy and US wildfire services.

6   Conclusion

Wildfires are problem on the Bulgarian territory which every year is happening more often. All experimental simulations with models that can provide the first responders with tools and information for fire behavior prediction are useful and needed. Every outcome is increasing the understanding of how this work from nowadays can be used in real time and as a prevention measure in future. Wildfire Analyst tool can be calibrated and proposed for implementation in test areas, where Bulgarian authorities can test it and use it for future. The first testing simulation results give more than 90% match with the real fire burned scar captured by the data from the available satellite images. More elaborated data and simulations are still in progress in order to prove the initial promising results.

Acknowledgements. This work has been partially supported by the project DN 12/5 called: Efficient Stochastic Methods and Algorithms for Large-Scale Problems and the Balkan-Med project "Drought and fire observatory and early warning system -DISARM".

References

1. Official web-site reports of EFFIS. http://effis.jrc.ec.europa.eu/reports-and-publications/annual-fire-reports//news.ibox.bg/news/id_1888536796
2. http://facebook.com/GeoPolymorphicCloud/. (In Bulgarian)
3. Dobrinkova, N., Hollingsworth, L., Heinsch, F.A., Dillon, G., Dobrinkov, G.: Bulgarian fuel models developed for implementation in FARSITE simulations for test cases in Zlatograd area. In: Wade, D.D., Fox, R.L. (eds.) Electronic Proceeding. Robinson ML (Comp). Proceedings of 4th Fire Behavior and Fuels Conference, 18–22 February 2013, Raleigh, NC and 1–4 July 2013, St. Petersburg, Russia. International Association of Wildland Fire, Missoula, pp. 513–521 (2014)
4. Anderson, H.E.: Aids to determining fuel models for estimating fire behavior. USDA Forest Service, Intermountain Forest and Range Experiment Station, Research Report INT-122 (1982)
5. Scott, J.H., Burgan, R.E.: Standard fire behavior fuel models: a comprehensive set for use with Rothermel's surface fire spread model. General Technical report RMRS-GTR-153. Fort Collins, CO: U.S. Department of Agriculture, Forest Service, Rocky Mountain Research Station, 72 p. (2005)
6. In Bulgarian local news about Fire spread in Kresna Gorge. http://trud.bg/municipality-Kresna-donors-campaign/
7. Rothermel, R.C.: A mathematical model for predicting fire spread in wildland fuels. Research Paper INT-115, pp. 1–40. US Department of Agriculture, Forest Service, Intermountain Forest and Range Experiment Station, Ogden, UT (1972)
8. Dobrinkov, G., Dobrinkova, N.: Input data preparation for fire behavior fuel modeling of Bulgarian test cases. In: Lirkov, I., Margenov, S.D., Waśniewski, J. (eds.) LSSC 2015. LNCS, vol. 9374, pp. 335–342. Springer, Cham (2015). https://doi.org/10.1007/978-3-319-26520-9_37
9. Pedernera, P., Julio, G.: Improving the Economic Efficiency of Combatting Forest Fires in Chile: The KITRAL System. USDA Forest Service General Technical report PSWGTR-173, pp. 149–155 (1999)
10. Nelson Jr., R.M.: Prediction of diurnal change in 10-h fuel stick moisture content. Can. J. For. Res. **30**, 1071–1087 (2000). https://doi.org/10.1139/CJFR-30-7-1071
11. Forthofer, J., Shannon, K., Butler, B.: Simulating diurnally driven slope winds with WindNinja (link is external). In: Proceedings of 8th Symposium on Fire and Forest Meteorological Society 13–15 October 2009, Kalispell, MT (2,037 KB), 13 p. (2009)
12. Ramirez, J., Monedero, S., Buckley, D.: New approaches in fire simulations analysis with Wildfire Analyst. In: The 5th International Wildland Fire Conference, Sun City, South Africa (2011)
13. Monedero, S., Ramirez, J., Molina-Terrén, D., Cardil, A.: Simulating wildfires backwards in time from the final fire perimeter in point-functional fire models. Environmental Model. Softw. **92**, 163–168 (2017)

Subtraction of Two 2D Polygons with Some Matching Vertices

Georgi Evtimov and Stefka Fidanova[✉]

Institute of Information and Communication Technologies,
Bulgarian Academy of Sciences, Acad. G. Bonchev Str., bl. 25A, 1113 Sofia, Bulgaria
gevtimov@abv.bg, stefka@parallel.bas.bg

Abstract. Cutting Stock Problem (CSP) is an important industrial problem. In this paper we focus on the variant arising in building construction, where a set of plates needs to be cut from rectangular stock, minimizing the waste. The CSP is known to be NP-hard combinatorial problem. The goal of this work is to propose an efficient way for subtracting of two polygons with some matching vertices.

1 Introduction

The cutting stock problem (CSP) arises in many industrial applications [8]. Most of the authors solve the simplified model when the items are rectangular. CSP with rectangular items appear in paper and glass industries [7], container loading, Very-large-scale integration (VLSI) design, and various scheduling tasks [9]. A more complicate version of the problem is when the items are not rectangular. This problem arises in building constructions in fasteners production, clothes production, shoes production and so on. In [7] the main topic is a two-dimensional orthogonal packing problem, where a fixed group of small rectangles must be fitted into a large rectangle and the unused area of the large rectangle is minimized. The algorithm combines a replacement method with a genetic algorithm. In [1] a greedy randomized adaptive search procedure is developed. In this study, there is a large primary stock that has to be cut into smaller pieces, so as to maximize the value of the pieces. Cintra et al. [3] propose an exact algorithm based on dynamic programming, which is appropriate for small problems, because the problem is NP-hard. Dusberger and Raidl [4,5] propose two metaheuristic algorithms based on variable neighborhood search. The above mentioned works solve the simplified problem with rectangular items.

In building constructions industry the cutting shapes are polygons, which may have irregular shape and can be convex or concave. Such diversity of shapes increases substantially the difficulty of the problem. The aim is to arrange all given polygons from the project in given minimum area. The polygons are often generated by CAD environment as a list of points which will be represent as tables with relevant attributes. Each polygon is a boundary of a bounded domain in the plane. The goal of this work is to design an algorithm that produces polygon (or polygons) that is intersection, union or subtraction of two domains determined by two polygons.

© Springer Nature Switzerland AG 2019
G. Nikolov et al. (Eds.): NMA 2018, LNCS 11189, pp. 80–87, 2019.
https://doi.org/10.1007/978-3-030-10692-8_9

2 Problem Formulation

The CSP where the items are arbitrary polygons is very difficult. Some authors try to solve the problem with arbitrary polygons by completing the items to rectangle, but in most of the cases it is not effective [2]. In our variant of CSP rectangular sheet with fixed width and unlimited length is done. The set $E = \{i_1, i_2, \ldots, i_n\}$ consists of ordered items - polygons, which can be convex or concave. Every item is specified with the coordinates of his nodes and number of orders d_i, for $i = 1, \ldots, n$, when positioning the items can be rotated.

The objective is to find a cutting pattern P with a minimal waste. The solution is an arrangement of the items from E on the stock sheet. The solution can be represented by cutting sequence and coordinates of the nodes of the cutting items.

3 Method for Intersecting Two Polygons

In scientific literature there are methods for intersecting of rectangle polygons [10]. But for any concave or convex polygons there is no relevant information about the above mentioned operations. Our algorithm uses the following operations:

1. Find the intersection point(s) of two line segments;
2. Remove the wasted points from a given polygon, i.e. the points that are not vertices of the polygon;
3. Identify the orientation of the polygon, namely, check if the tracing of the points of the polygons is in clockwise or anti-clockwise direction;
4. Check whether a given point is inside a given polygon.

Figure 1 illustrates various situations of polygons with some common vertexes.

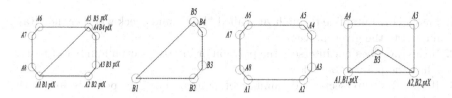

Fig. 1. Polygons A and B with some matching vertices.

Intersection of Polygons. In this article we shall consider various cases for cross points of two polygons, illustrated on Fig. 2.

For our purposes it is important to define precisely when two segments intersect. In our consideration the line segments include the end points, i.e. they are closed sets. We say that two line segments intersect if they have at least one common point. In the case when the line segments coincide, the intersection points are the end points. And finally, if one segment is contained in another, then the intersection points are the end points of the smaller one. Often CAD systems produce line segments with zero lengths. We need to handle such a case as well. In such situation we may face two possibilities: (1) One segment has positive length and the other has zero length; in this situation the algorithm will produce five names of one end point and one name of the other end as shown on Fig. 2(f); (2) both segments have zero lengths; in this situation the algorithm will produce six names of the point, shown on Fig. 2(d).

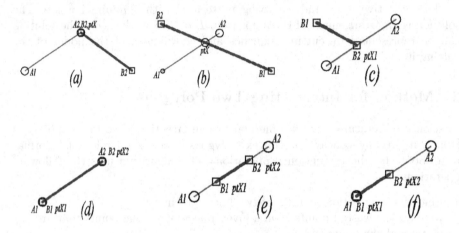

Fig. 2. Cases of crossing of two lines

How to Determine Whether a Given Point A0 is Inside of a Given Polygon. In our previous paper, [6], we described two methods: balanced sum of angles (BSA) and Ray crossing method (RCM). The BSA is a reliable and very informative method, but is slower than RC. Here is a short description of the RCM algorithm:

1. Pass a horizontal line through $A0$, called "Ray", and check whether the "Ray" intersects the given polygon.
2. If the Ray does not intersect the polygon or intersects even number of points. Then the point is outside.
3. If the Ray intersects odd number of points, then the point is inside the polygon.

If we apply the RCM directly to a given polygon, this may result computationally intensive search, since there may have thousands of line segments. To reduce the computational cost we use two steps. In the first step we embed the polygon into rectangular box and check whether the point is outside the box. This step is computationally cheap. If the point is inside the box then we will apply the algorithm to the original polygon (Fig. 3).

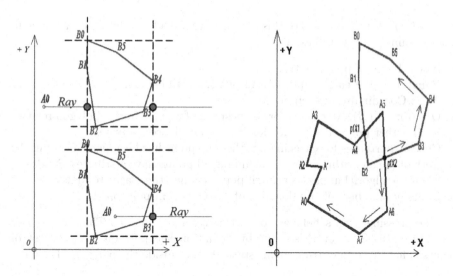

Fig. 3. Left: Checking whether a point is inside the polygon; Right: Interaction between two polygons

Orientation of a Given Polygon. Our algorithm checks if the orientations of the polygons A and B are opposite. That means that if we go through the table from the top to the bottom the vertices of A and the vertices of B are traced in opposite directions. We will consider polygons without self-cross points. In this case, the polygon bounds a connected domain.

Two polygons may have the following positions relative to each other: (1) the polygons do not intersect, (2) the polygons have nontrivial intersection; (3) one of the polygons is contained in the domain determined by the other.

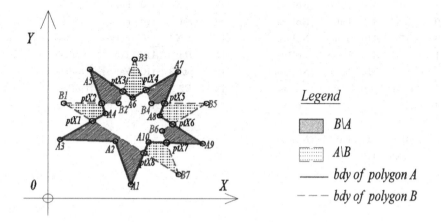

Fig. 4. Subtraction domains for the Polygons A and B

Subtracting Two Polygons. For this purposes we provide a table for each polygon. The columns of these tables are:

1. Index: Index of points $0, 1, 2, \ldots, n$;
2. Name: Label of each points $A1, A2, ptX1, \ldots$ and so on;
3. X, Y: Coordinates of each point;
4. Out from A (B): Attributes for the position of a point from a polygon relative to the other polygon – "True" for inside and "False" outside;
5. Check: Attributes for checking whether the point has been passed during the work of the algorithm – At the beginning all points have attributes "NoPass"; when the algorithm passes though point, change the status to "Pass".
6. Domain: Numbers of sub domains of cut Out from polygon A.

The crossing points belongs to both polygons A and B. That means two possible status True or False, but by definition the polygon is closed set of points. That gives to crossing point status False - inside the polygon (Table 1).

Table 1. Boolean table for polygons A and B shown in Fig. 4

Polygon A						Polygon B					
Index	Name	X, Y	From B	Check	Domain	Index	Name	X, Y	From A	Check	Domain
0	$A1$	X, Y	True	NoPass		0	$B1$	$A10$	True	NoPass	
1	$A2$	$A10$	True	NoPass		1	ptX1	$A10$	False	NoPass	
2	$A3$	$A10$	True	NoPass		2	ptX8	$A10$	False	NoPass	
3	ptX1	$A10$	False	NoPass		3	$B7$	$A10$	True	NoPass	
4	$A4$	$A10$	False	NoPass		4	ptX7	$A10$	False	NoPass	
5	ptX2	$A10$	False	NoPass		5	$B6$	$A10$	False	NoPass	
6	$A5$	$A10$	True	NoPass		6	ptX6	$A10$	False	NoPass	
7	ptX3	$A10$	False	NoPass		7	$B5$	$A10$	True	NoPass	
8	$A6$	$A10$	False	NoPass		8	ptX5	$A10$	False	NoPass	
9	ptX4	$A10$	False	NoPass		9	$B4$	$A10$	False	NoPass	
10	$A7$	$A10$	True	NoPass		10	ptX4	$A10$	False	NoPass	
11	ptX5	$A10$	False	NoPass		11	$B3$	$A10$	True	NoPass	
12	$A8$	$A10$	False	NoPass		12	ptX3	$A10$	False	NoPass	
13	ptX6	$A10$	False	NoPass		13	$B2$	$A10$	False	NoPass	
14	$A9$	$A10$	True	NoPass		14	ptX2	$A10$	False	NoPass	
15	ptX7	$A10$	False	NoPass							
16	$A10$	$A10$	False	NoPass							
17	ptX8	$A10$	False	NoPass							

Now we consider the problem for subtraction of two polygons - second case. So, we will analyze the above table as follows:

1. Rotate table (A) (cycling rotating) - find one point which is noPass and Status Out from B = True;
2. Get the first point - $A1$ - since it is not a cross point and Out from B= "True";
3. Get the next point - $A2$ - since it is not a cross point and Out from B= "True";
4. Get the next point - $A3$ - since it is not a cross point and Out from B= "True";
5. Do not get next point - ptX1 - switch to table B;
6. Find ptX1 - from Table B and now the process does in table B;
7. Do not get next point - ptX8 - switch to table A and find ptX8;
8. Get ptX8 - through the table A. Now the active table is A;
9. Get next point - $A1$ - the table is cycling list of points! And here the method stop because the first and the last point coincide.

Table 2. The results after the first pass of the algorithm

About Polygon A						About Polygon B					
Index	Name	$A10$	From B	Check	Domain	Index	Name	$A10$	From A	Check	Domain
0	$A1$	$A10$	True	Pass	1	0	$B1$	$A10$	True	NoPass	
1	$A2$	$A10$	True	Pass	1	1	ptX1	$A10$	False	Pass	1
2	$A3$	$A10$	True	Pass	1	2	ptX8	$A10$	False	Pass	1
3	ptX1	$A10$	False	Pass	1	3	$B7$	$A10$	True	NoPass	
4	$A4$	$A10$	False	NoPass		4	ptX7	$A10$	False	NoPass	
5	ptX2	$A10$	False	NoPass		5	$B6$	$A10$	False	NoPass	
6	$A5$	$A10$	True	NoPass		6	ptX6	$A10$	False	NoPass	
7	ptX3	$A10$	False	NoPass		7	$B5$	$A10$	True	NoPass	
8	$A6$	$A10$	False	NoPass		8	ptX5	$A10$	False	NoPass	
9	ptX4	$A10$	False	NoPass		9	$B4$	$A10$	False	NoPass	
10	$A7$	$A10$	True	NoPass		10	ptX4	$A10$	False	NoPass	
11	ptX5	$A10$	False	NoPass		11	$B3$	$A10$	True	NoPass	
12	$A8$	$A10$	False	NoPass		12	ptX3	$A10$	False	NoPass	
13	ptX6	$A10$	False	NoPass		13	$B2$	$A10$	False	NoPass	
14	$A9$	$A10$	True	NoPass		14	ptX2	$A10$	False	NoPass	
15	ptX7	$A10$	False	NoPass							
16	$A10$	$A10$	False	NoPass							
17	ptX8	$A10$	False	Pass	1						

This cycle contains the first domain of points of section - $A1, A3, ptX1,$ $ptX8, A1$. At end of the cycle we can give name of first domain - "Domain 1". The method is repeated while there is no points from table A which is Out from B = "True" and status = "NoPass" (Table 2).

The first point of the polygon is given in bold.

The second step of the algorithm ignores the points that have check "Pass". This cycle contain second sub domain of points of section - $A5,$ $ptX3, B2, ptX2, A5$.

The algorithm terminates when all points with attribute Out from B = "True" are given attribute Check = "Pass".

In Table 3 we get the information about all sub-domains. Here is the correct reading of the results:

sub-domain 1 is described by the points: $A1, A2, A3, ptX1, ptX8, A1$;
sub-domain 2 is described by the points: $A5, ptX3, B2, ptX2, A5$;
sub-domain 3 is described by the points: $A7, ptX5, B4, ptX4, A7$;
sub-domain 4 is described by the points: $A9, ptX7, B6, ptX6, A9$.

The algorithm has been tested on a variety of examples that are representative for the cutting stock problem applied to plates in construction projects. The proposed algorithm provide to be computationally less expensive, than the existing commercial products.

Table 3. The results after the algorithm terminates.

About Polygon A						About Polygon B					
Index	Name	X, Y	From B	Check	Domain	Index	Name	X, Y	From A	Check	Domain
0	$A1$	X, Y	True	Pass	**1**	0	$B1$	X, Y	True	NoPass	
1	$A2$	X, Y	True	Pass	1	1	$ptX1$	X, Y	False	Pass	1
2	$A3$	X, Y	True	Pass	1	2	$ptX8$	X, Y	False	Pass	1
3	$ptX1$	X, Y	False	Pass	1	3	$B7$	X, Y	True	NoPass	
4	$A4$	X, Y	False	NoPass		4	$ptX7$	X, Y	False	Pass	4
5	$ptX2$	X, Y	False	Pass	2	5	$B6$	X, Y	False	Pass	4
6	$A5$	X, Y	True	Pass	**2**	6	$ptX6$	X, Y	False	Pass	4
7	$ptX3$	X, Y	False	Pass	2	7	$B5$	X, Y	True	NoPass	
8	$A6$	X, Y	False	NoPass		8	$ptX5$	X, Y	False	Pass	3
9	$ptX4$	X, Y	False	Pass	3	9	$B4$	X, Y	False	Pass	3
10	$A7$	X, Y	True	Pass	**3**	10	$ptX4$	X, Y	False	Pass	3
11	$ptX5$	X, Y	False	Pass	3	11	$B3$	X, Y	True	NoPass	
12	$A8$	X, Y	False	NoPass		12	$ptX3$	X, Y	False	NoPass	2
13	$ptX6$	X, Y	False	Pass	4	13	$B2$	X, Y	False	Pass	2
14	$A9$	X, Y	True	Pass	**4**	14	$ptX2$	X, Y	False	Pass	2
15	$ptX7$	X, Y	False	Pass	4						
16	$A10$	X, Y	False	NoPass							
17	$ptX8$	X, Y	False	Pass	1						

4   Conclusion

The cutting stock problem is an important industrial problem. In this paper we focus on the subtraction of two polygons. We propose an algorithm which can subtract two polygons with arbitrary shape. The proposed methodology is evaluated on experimental tasks. The obtained results are presented and analyzed.

Acknowledgment. The work presented here is partially supported by the Bulgarian National Research Fund under Grants DFNI DN 12/5 and DFNI DN 02/10.

References

1. Alvarez-Valdes, R., Parajon, A., Tamarit, J.M.: A computational study of heuristic algorithms for two-dimensional cutting stock problems. In: 4th Metaheuristics International Conference (MIC 2001), pp. 16–20 (2001)
2. Alvarez-Valdes, R., Parreno, F., Tamarit, J.M.: A Tabu Search algorithm for two dimensional non-guillotine cutting problems. Eur. J. Oper. Res. **183**(3), 1167–1182 (2007)
3. Cintra, G., Miyazawa, F., Wakabayashi, Y., Xavier, E.: Algorithms for two-dimensional cutting stock and strip packing problems using dynamic programming and column generation. Eur. J. Oper. Res. **191**, 61–85 (2008)
4. Dusberger, F., Raidl, G.R.: A variable neighborhood search using very large neighborhood structures for the 3-staged 2-dimensional cutting stock problem. In: Blesa, M.J., Blum, C., Voß, S. (eds.) HM 2014. LNCS, vol. 8457, pp. 85–99. Springer, Cham (2014). https://doi.org/10.1007/978-3-319-07644-7_7
5. Dusberger, F., Raidl, G.R.: Solving the 3-staged 2-dimensional cutting stock problem by dynamic programming and variable neighborhood search. Electron. Notes Discrete Math. **47**, 133–140 (2015)
6. Evtimov, G., Fidanova, S.: Heuristic algorithm for 2D cutting stock problem. In: Lirkov, I., Margenov, S. (eds.) LSSC 2017. LNCS, vol. 10665, pp. 350–357. Springer, Cham (2018). https://doi.org/10.1007/978-3-319-73441-5_37
7. Gonçalves, J.F.: A hybrid genetic algorithm-heuristic for a two-dimensional orthogonal packing problem. Eur. J. Oper. Res. **183**(3), 1212–1229 (2007)
8. Parmar, K., Prajapati, H., Dabhi, V.: Cutting stock problem: a survey of evolutionary computing based solution. In: Proceedings of Green Computing Communication and Electrical Engineering (2014). https://doi.org/10.1109/ICGCCEE.2014.6921411
9. Lodi, A., Martello, S., Vigo, D.: Recent advances on two-dimensional bin packing problems. Discrete Appl. Math. **123**, 379–396 (2002)
10. Lin, Z., Li, Y.: An efficient algorithm for intersection, union and difference between two polygons. In: Proceedings of Computer Network and Multimedia Technology, Wuhan, China, pp. 1–4 (2009)

InterCriteria Analysis of Different Variants of ACO Algorithm for Wireless Sensor Network Positioning

Stefka Fidanova[1(✉)] and Olympia Roeva[2]

[1] Institute of Information and Communication Technologies – BAS,
Acad. G. Bonchev Str., bl. 25A, 1113 Sofia, Bulgaria
stefka@parallel.bas.bg
[2] Institute of Biophysics and Biomedical Engineering – BAS,
Acad. G. Bonchev Str., bl. 105, 1113 Sofia, Bulgaria
olympia@biomed.bas.bg

Abstract. Wireless sensor networks are formed by spatially distributed sensors, which communicate in a wireless way. This network can monitor various kinds of environment and physical conditions like movement, noise, light, humidity, images, chemical substances etc. A given area needs to be fully covered with minimal number of sensors and the energy consumption of the network needs to be minimal too. We propose several algorithms, based on Ant Colony Optimization, to solve the problem. We study the algorithms behaviour when the number of ants varies from 1 to 10. We apply InterCriteria analysis to study relations between proposed algorithms and number of ants and analyse correlation between them.

Keywords: Ant Colony Optimization · InterCriteria Analysis
Wireless Sensor Network

1 Introduction

Wireless Sensor Networks (WSN) allow the monitoring of large areas without the intervention of a human operator. The WSN can be used in areas where traditional networks fail or are inadequate. They find applications in a variety of areas such as climate monitoring, military use, industry and sensing information from inhospitable locations. Unlike other networks, sensor networks depend on deployment of sensors over a physical location to fulfil a desired task.

A WSN node contains several components including the radio, battery, microcontroller, analog circuit, and sensor interface. In battery-powered systems, higher data rates and more frequent radio use consume more power. There are several open issues for sensor networks such as signal processing [17], deployment [21], operating cost, localization and location estimation. The wireless sensors, have two fundamental functions: sensing and communicating. However, the sensors which are fare from the high energy communication node (HECN) can not

© Springer Nature Switzerland AG 2019
G. Nikolov et al. (Eds.): NMA 2018, LNCS 11189, pp. 88–96, 2019.
https://doi.org/10.1007/978-3-030-10692-8_10

communicate with him directly. The sensors transmit their data to this node, either directly or via hops, using nearby sensors as communication relays.

Jourdan [14] solved an instance of WSN layout using a multi-objective genetic algorithm – a fixed number of sensors had to be placed in order to maximize the coverage. In some applications most important is the network energy. In [12] is proposed Ant Colony Optimization (ACO) algorithm and in [20] is proposed evolutionary algorithm for this variant of the problem. In [7] is proposed ACO algorithm taking in to account only the number of the sensors. In [8] a multi-objective ACO algorithm, which solves the WSN layout problem is proposed. The problem is multi-objective with two objective functions – (i) minimizing the energy consumption of the nodes in the network, and (ii) minimizing the number of the nodes. The full coverage of the network and connectivity are considered as constraints. A mono-objective ant algorithm which solves the WSN layout problem is proposed in [9]. In [16] are proposed several evolutionary algorithms to solve the problem. In [15] is proposed genetic algorithm which achieves similar solutions as the algorithms in [16], but it is tested on small test problems.

The current research is an attempt to investigate the influence of the number of ants on the ACO algorithm performance, which solves the WSN layout problem, and quality of the achieved solutions and to find the minimal number of ants which are enough to achieve good solutions. For this purpose the InterCriteria Analysis (ICrA) approach is applied.

ICrA, proposed by [5], is a recently developed approach for evaluation of multiple objects against multiple criteria and thus discovering existing correlations between the criteria themselves. It is based on the apparatus of the index matrices (IMs) [1], and the intuitionistic fuzzy sets [2,3] and can be applied to decision making in different areas of knowledge. Various applications of the ICrA approach have been found in science and practice – e-learning [18], algorithms performance [11], medicine [19], etc. Data from series of ACO optimization procedures, published in [8–10], are used to construct IMs. ICrA is applied over the so defined IMs and the results are discussed.

The paper is organized as follows: in Sect. 2 is given the WSN layout problem formulation, in Sect. 3 is presented the background of the InterCriteria Analysis, in Sect. 4 the numerical results are presented and a discussions is provided. The concluding remarks are given in Sect. 5.

2 Problem Formulation

Each sensor node of WSN sense an area around itself called its sensing area, which is determined by the sensing radius. The communication radius determines how far the node can send his data. A special node in the WSN called High Energy Communication Node (HECN) is responsible for external access to the network. Every sensor node in the network must have communication with the HECN, connectivity of the network. The communication radius is often much smaller than the network size, they transmit their date by other nodes which are closer to the HECN.

In our formulation, sensor nodes has to be placed in a terrain providing full sensing coverage with a minimal number of sensors and minimizing the energy spent in communications by any single node. Minimal number of sensors means cheapest network for constructing. Minimal energy means cheapest network for exploitation. The energy of the network defines the lifetime of the network, how frequently the batteries need to be replaced. These are opposed objectives and we look for a good balance between number of sensors and energy consumption.

The WSN operates by rounds: In a round, every node collects the data and sends it to the HECN. Every node transmits the information packets to the neighbour that is closest to the HECN. If a node has n neighbours, each one receives 1/n of its traffic load. Every node has a traffic load equal to 1 (corresponding to its own sent data) plus the sum of all traffic loads received from neighbours.

3   InterCriteria Analysis

Following [5] we will obtain an Intuitionistic Fuzzy Pair (IFP) [6] as the degrees of "agreement" and "disagreement" between two criteria applied on different objects. We remind briefly that an IFP is an ordered pair of real non-negative numbers $\langle a, b \rangle$ such that $a + b \leq 1$.

For clarity, let us be given an Index Matrix (IM) (see [1]) whose index sets consist of the names of the criteria (for rows) and objects (for columns). The elements of this IM are further supposed to be real numbers (in the general case, this is not required). We will obtain an IM with index sets consisting of the names of the criteria (for rows and for columns) with elements IFPs corresponding to the "agreement" and "disagreement" of the respective criteria.

Further by O we denote the set of all objects $O_1, O_2, \ldots, O_n$ being evaluated, and by $C(O)$ the set of values assigned by a given criteria C to the objects, i.e.

$$O \overset{\text{def}}{=} \{O_1, O_2, \ldots, O_n\}, C(O) \overset{\text{def}}{=} \{C(O_1), C(O_2), \ldots, C(O_n)\}.$$

Let $x_i = C(O_i)$. Then the following set can be defined:

$$C^*(O) \overset{\text{def}}{=} \{\langle x_i, x_j \rangle | i \neq j \,\&\, \langle x_i, x_j \rangle \in C(O) \times C(O)\}.$$

Further, if $x = C(O_i)$ and $y = C(O_j)$, $x \prec y$ will be written iff $i < j$.

In order to compare two criteria we must construct the vector of all internal comparisons of each criteria, which fulfill exactly one of three relations R, $\overline{R}$ and $\tilde{R}$. In other words, we require that for a fixed criterion C and any ordered pair $\langle x, y \rangle \in C^*(O)$ it is true:

$$\langle x, y \rangle \in R \Leftrightarrow \langle y, x \rangle \in \overline{R}, \tag{1}$$

$$\langle x, y \rangle \in \tilde{R} \Leftrightarrow \langle x, y \rangle \notin (R \cup \overline{R}), \tag{2}$$

$$R \cup \overline{R} \cup \tilde{R} = C^*(O). \tag{3}$$

From the above it is seen that we need only consider a subset of $C(O) \times C(O)$ for the effective calculation of the vector of internal comparisons since from (1)–(3) it follows that if we know what is the relation between x and y we also know what is the relation between y and x. Thus we will only consider lexicographically ordered pairs $\langle x, y \rangle$. Let, for brevity: $C_{i,j} = \langle C(O_i), C(O_j) \rangle$.

Then for a fixed criterion C we construct the vector:

$$V(C) = \{C_{1,2}, C_{1,3}, \ldots, C_{1,n}, C_{2,3}, C_{2,4}, \ldots, C_{2,n}, C_{3,4}, \ldots, C_{3,n}, \ldots, C_{n-1,n}\}.$$

Further, to simplify our considerations, we replace the vector $V(C)$ with $\hat{V}(C)$, where for each $1 \leq k \leq \frac{n(n-1)}{2}$ for the k-th component it is true:

$$\hat{V}_k(C) = \begin{cases} 1 & \text{iff } V_k(C) \in R, \\ -1 & \text{iff } V_k(C) \in \overline{R}, \\ 0 & \text{otherwise.} \end{cases}$$

Then when comparing two criteria we determine the "degree of agreement" between the two as the number of matching components (divided by the length of the vector for normalization purposes).

4   Numerical Results and Discussion

In our previous works [7–10] we propose different variants of ACO algorithm to solve WSN problem. In [10] is applied multi-objective ACO algorithm. In [7] the problem is converted to mono-objective by multiplying the two objective functions, and in [9] the problem is converted to mono-objective by summing the two objective functions. We apply our algorithms on rectangular area consisting 500×500 points and the communication and coverage radius of every sensor cover 30 points. The number of used ants is from 1 to 10. For our current research we use results published in these papers.

The input IM and the full set of obtained numerical results could be found in http://intercriteria.net/studies/aco/.

ICA is applied over the data presented in the input IM, where different ACO algorithms ($ACOu$-mono-objective with multiplication, $ACOs$-mono-objective with sum and $ACOm$- multi-objective) are presented as criteria and number of sensors [223, 224, 225, 226, 227, 228, 229, 230, 231, 232, 233, 234, 235, 236, 237, 238, 239, 240, 241, 242, 243, 244] – as objects. The cross-platform software for ICrA approach, ICrAData, is used [13].

The results are analysed based on the scale proposed in [4], which defines the consonance and dissonance between the criteria pairs.

The obtained most significant and interesting ICrA results are presented in tables below.

Table 1 displays the criteria pairs that are in SPC. Only the relations between algorithms with different objective functions are shown. There are 24 more criteria pairs that are also in SPC. These criteria pairs correspond to the algorithms with the same objective function and close number of ants, for example

$ACOu8 - ACOu9$ or $ACOs2 - ACOs3$. As some exceptions, for the criteria pairs $ACOu1 - ACOu2$ and $ACOs4 - ACOs5$ a WPC is registered(monitored) and for the $ACOu5 - ACOu6$ – WD. It should be noted that these pairs correspond to the case of mono-criteria variant of the problem. These pairs are not shown in the table. There is only one criteria pair corresponding to the multi-criteria case – $ACOm6 - ACOm10$. The presented criteria pairs show that the SPC is observed for the four cases for $ACOs - ACOm$, and for one – $ACOu - ACOm$. There is not observed strong relation between results using $ACOu$ and $ACOs$. It could be summarized that SPC is observed between same mono-objective algorithms where the number of ant is closed, which is logical.

Table 1. ACO algorithms in SPC

Criteria pairs of ACO algorithms of ACO algorithms	Value of $\mu_{C,C'}$
$ACOs8 - ACOm8$	0.98
$ACOs5 - ACOm8$	0.96
$ACOs6 - ACOm8$	0.96
$ACOs7 - ACOm8$	0.96
$ACOu3 - ACOm3$	0.95
$ACOs5 - ACOm8$	0.96
$ACOs6 - ACOm8$	0.96
$ACOs7 - ACOm8$	0.96
$ACOs8 - ACOm8$	0.98
$ACOu3 - ACOm3$	0.95

In Table 2 again only the results from different objective functions, that are in PC, are displayed. In this case the pairs of multi-criteria algorithms with a close number of ants are appeared, as well as the mono-criteria with a bit larger difference between number of ants (e.g. $ACOs2 - ACOs10$, $ACOs3 - ACOs10$, $ACOu2 - ACOu10$, $ACOus - ACOu10$, etc.). The all obtained results are presented in http://intercriteria.net/studies/aco/.

From total 84 cases of criteria pairs between different algorithms 19 pairs are between $ACOs$ and $ACOu$ (mono-criteria) and the rest are between mono and multi-criteria ACO algorithms. That means that the relation between mono and multi-criteria algorithms are weaker than the relations between different mono-criteria algorithms. The last ones are mainly in the SPC.

The presented in Table 2 results show that the algorithms $ACOs - ACOu$ are more related than $ACOu - ACOm$.

163 criteria pairs are in WPC. 157 from them are between mono and multi-criteria ACO algorithms. Only 6 pairs between $ACOs - ACOs$ and 4 between $ACOu - ACOu$. It shows that the relation between mono and multi-criteria ACO algorithms are weaker. 137 of the criteria pair are between $ACOu - ACOs$

and $ACOu - ACOm$, i.e. $ACOu$ performance is more different in comparison with the performance of the $ACOs$ and $ACOm$.

64 criteria pairs are in WD. There are not criteria pairs $ACOs - ACOs$.

In Table 3 only the criteria pairs between the same objective function are presented. The results show the influence of the different number of ants on the ACO performance using the same objective function. The result show that $ACOs$ is less sensitive to the number of ants, than the $ACOu$ and $ACOm$. The larger difference of the performance are observed for $ACOu$ and $ACOm$ – 53% of all criteria pairs. In 14% of the cases the WD is observed for the algorithms

Table 2. ACO algorithms in PC

Criteria pairs of ACO algorithms	Value of $\mu_{C,C'}$	Criteria pairs of ACO algorithms	Value of $\mu_{C,C'}$
$ACOu6 - ACOm3$	0.93	$ACOs8 - ACOm9$	0.88
$ACOs9 - ACOm8$	0.92	$ACOu2 - ACOm2$	0.88
$ACOu4 - ACOm3$	0.92	$ACOu4 - ACOs1$	0.88
$ACOs8 - ACOm5$	0.92	$ACOu6 - ACOs9$	0.88
$ACOu1 - ACOm1$	0.92	$ACOs9 - ACOm3$	0.87
$ACOu3 - ACOs4$	0.92	$ACOu2 - ACOm4$	0.87
$ACOu4 - ACOm4$	0.92	$ACOu5 - ACOs1$	0.87
$ACOu5 - ACOm3$	0.92	$ACOs9 - ACOm7$	0.87
$ACOu5 - ACOm4$	0.91	$ACOu6 - ACOs10$	0.87
$ACOu3 - ACOm4$	0.91	$ACOs7 - ACOm6$	0.86
$ACOs7 - ACOm5$	0.90	$ACOs8 - ACOm10$	0.86
$ACOs9 - ACOm9$	0.90	$ACOu7 - ACOs10$	0.86
$ACOu10 - ACOm3$	0.90	$ACOs7 - ACOm10$	0.86
$ACOu7 - ACOm3$	0.90	$ACOu10 - ACOs9$	0.86
$ACOu8 - ACOm3$	0.90	$ACOu5 - ACOs10$	0.86
$ACOu9 - ACOm3$	0.90	$ACOu7 - ACOs9$	0.86
$ACOs7 - ACOm7$	0.90	$ACOu8 - ACOs9$	0.86
$ACOu1 - ACOm4$	0.90	$ACOu9 - ACOs9$	0.86
$ACOs7 - ACOm9$	0.89	$ACOu1 - ACOs4$	0.85
$ACOs8 - ACOm7$	0.89	$ACOu2 - ACOm5$	0.85
$ACOu4 - ACOs4$	0.89	$ACOu2 - ACOs10$	0.85
$ACOu2 - ACOm3$	0.89	$ACOu4 - ACOs10$	0.85
$ACOu3 - ACOs1$	0.89	$ACOs9 - ACOm6$	0.85
$ACOu5 - ACOs4$	0.89	$ACOu2 - ACOs4$	0.85
$ACOs8 - ACOm6$	0.88	$ACOu3 - ACOs9$	0.85
$ACOs9 - ACOm5$	0.88	$ACOu6 - ACOs1$	0.85

Table 3. ACO algorithms in WPC

Criteria pairs of ACO algorithms of ACO algorithms	Value of $\mu_{C,C'}$
$ACOs4 - ACOs10$	0.77
$ACOs4 - ACOs5$	0.82
$ACOs4 - ACOs6$	0.82
$ACOs4 - ACOs7$	0.82
$ACOs4 - ACOs8$	0.81
$ACOs4 - ACOs9$	0.84
$ACOu1 - ACOu2$	0.78
$ACOu1 - ACOu3$	0.81
$ACOu1 - ACOu4$	0.82
$ACOu1 - ACOu5$	0.82

$ACOu - ACOu$ and $ACOm - ACOm$ with bigger difference between number of ants (more than three), which that $ACOu$ and $ACOm$ highly influenced of the number of ants.

5   Conclusion

The InterCriteria analysis is a powerful tool for studying relations between different objects. We study three variants of ACO algorithm applied on WSN problem. Every variant is tested with various number of ants, between 1 and 10. We search the correlation between variants of ACO and number of ants. WSN problem is a multi-objective problem. When it is converted to mono-objective by summing the two objective functions, the algorithm is less sensitive to the number of used ants. When the problem is solved like multi-objective, we observe bigger difference of algorithm performance according to the number of ants. There is greater similarity between performance of the two mono-objective variants, than between some of mono-objective and multi-objective variants. In a future we will continue with study of the influence of the ACO parameters values on algorithm performance.

Acknowledgments. Work presented here is partially supported by the Bulgarian National Scientific Fund under Grants DFNI DN 12/5 "Efficient Stochastic Methods and Algorithms for Large-Scale Problems" and DN 02/10 "New Instruments for Knowledge Discovery from Data, and their Modelling".

References

1. Atanassov, K.: Index Matrices: Towards an Augmented Matrix Calculus. Studies in Computational Intelligence, vol. 573. Springer, Basel (2014). https://doi.org/10.1007/978-3-319-10945-9

2. Atanassov, K.: Intuitionistic Fuzzy Sets, VII ITKR Session, Sofia, 20–23 June 1983 (1983). Reprinted: Int J Bioautomation, 20(S1), 2016, S1–S6

3. Atanassov, K.: Review and New Results on Intuitionistic Fuzzy Sets, Mathematical Foundations of Artificial Intelligence Seminar, Sofia (1988). Preprint IM-MFAIS-1-88, Reprinted: Int J Bioautomation, 20(S1), 2016, S7–S16

4. Atanassov, K., Atanassova, V., Gluhchev, G.: InterCriteria analysis: ideas and problems. Notes Intuitionistic Fuzzy Sets **21**(2), 81–88 (2015)

5. Atanassov, K., Mavrov, D., Atanassova, V.: Intercriteria decision making: a new approach for multicriteria decision making based on index matrices and intuitionistic fuzzy sets. Issues IFSs GNs **11**, 1–8 (2014)

6. Atanassov, K., Szmidt, E., Kacprzyk, J.: On intuitionistic fuzzy pairs. Notes Intuitionistic Fuzzy Sets **19**(3), 1–13 (2013)

7. Fidanova, S., Marinov, P., Alba, E.: Ant algorithm for optimal sensor deployment. In: Madani, K., Correia, A.D., Rosa, A., Filipe, J. (eds.) Computational Intelligence, Studies of Computational Intelligence, vol. 399, pp. 21–29. Springer, Heidelberg (2012). https://doi.org/10.1007/978-3-642-27534-0_2

8. Fidanova, S., Marinov, P., Paprzycki, M.: Influence of the number of ants on multiobjective ant colony optimization algorithm for wireless sensor network layout. In: Lirkov, I., Margenov, S., Waśniewski, J. (eds.) LSSC 2013. LNCS, vol. 8353, pp. 232–239. Springer, Heidelberg (2014). https://doi.org/10.1007/978-3-662-43880-0_25

9. Fidanova, S., Marinov, P., Paprzycki, M.: Influence of the number of ants on multiobjective ant colony optimization algorithm for wireless sensor network layout. In: Lirkov, I., Margenov, S., Waśniewski, J. (eds.) LSSC 2013. LNCS, vol. 8353, pp. 232–239. Springer, Heidelberg (2014). https://doi.org/10.1007/978-3-662-43880-0_25

10. Fidanova, S., Shindarov, M., Marinov, P.: Wireless sensor positioning using ACO algorithm. In: Sgurev, V., Yager, R.R., Kacprzyk, J., Atanassov, K.T. (eds.) Recent Contributions in Intelligent Systems. SCI, vol. 657, pp. 33–44. Springer, Cham (2017). https://doi.org/10.1007/978-3-319-41438-6_3

11. Fidanova, S., Roeva, O., Paprzycki, M., Gepner, P.: InterCriteria analysis of ACO start startegies. In: Proceedings of the 2016 Federated Conference on Computer Science and Information Systems, pp. 547–550 (2016)

12. Hernandez, H., Blum, C.: Minimum energy broadcasting in wireless sensor networks: an ant colony optimization approach for a realistic antenna model. J. Appl. Soft Comput. **11**(8), 5684–5694 (2011)

13. Ikonomov, N., Vassilev, P., Roeva, O.: ICrAData - software for InterCriteria analysis. Int. J. Bioautomation **22**(1), 1–10 (2018)

14. Jourdan, D.B.: Wireless sensor network planning with application to UWB localization in GPS-denied environments, Massachusets Institute of Technology, Ph.D. thesis (2000)

15. Konstantinidis, A., Yang, K., Zhang, Q., Zainalipour-Yazti, D.: A multiobjective evolutionary algorithm for the deployment and power assignment problem in wireless sensor networks. J. Comput. Netw. **54**(6), 960–976 (2010)

16. Molina, G., Alba, E., Talbi, E.G.: Optimal sensor network layout using multi-objective Metaheuristics. Univers. Comput. Sci. **14**(15), 2549–2565 (2008)
17. Pottie, G.J., Kaiser, W.J.: Embedding the internet: wireless integrated network sensors. Commun. ACM **43**(5), 51–58 (2000)
18. Krawczak, M., Bureva, V., Sotirova, E., Szmidt, E.: Application of the InterCriteria decision making method to universities ranking. In: Atanassov, K.T., et al. (eds.) Novel Developments in Uncertainty Representation and Processing. AISC, vol. 401, pp. 365–372. Springer, Cham (2016). https://doi.org/10.1007/978-3-319-26211-6_31
19. Todinova, S., Mavrov, D., Krumova, S., Marinov, P., Atanassova, V., Atanassov, K., Taneva, S.G.: Blood plasma thermograms dataset analysis by means of InterCriteria and correlation analyses for the case of colorectal cancer. Int. J. Bioautomation **20**(1), 115–124 (2016)
20. Wolf, S., Merz, P.: Evolutionary local search for the minimum energy broadcast problem. In: van Hemert, J., Cotta, C. (eds.) EvoCOP 2008. LNCS, vol. 4972, pp. 61–72. Springer, Heidelberg (2008). https://doi.org/10.1007/978-3-540-78604-7_6
21. Xu, Y., Heidemann, J., Estrin, D.: Geography informed energy conservation for ad hoc routing. In: Proceedings of the 7th ACM/IEEE Annual International Conference on Mobile Computing and Networking, Italy, pp. 70–84 (2001)

Description of Dynamics of Ellipsoidal Estimates of Reachable Sets of Nonlinear Control Systems with Bilinear Uncertainty

Tatiana F. Filippova[✉][iD]

Department of Optimal Control,
Krasovskii Institute of Mathematics and Mechanics, Russian Academy of Sciences,
16 S. Kovalevskaya str., Ekaterinburg 620990, Russian Federation
ftf@imm.uran.ru

Abstract. The state estimation problems for control systems with unknown but bounded uncertainties with a set-membership description of uncertain parameters and functions are studied. The modified state estimation approaches based on special structure of nonlinearity and uncertainty that are simultaneously present in the control system are developed. The studies are motivated by numerous modeling problems for dynamical systems with uncertainty and nonlinearity arising in different fields such as physical engineering problems, economical modeling, ecological problems. This investigation continues previous researches and a more complicated case is considered here, when the dynamical equations of control system contain two types of nonlinearities, one of which is of quadratic type and another one contains uncertain matrix parameters. Such models may arise in applications related, in particular, to satellite control problems with nonlinearity and disturbances in the model description. The main new results consist in deriving the dynamical equations for the ellipsoidal estimates of reachable sets of the control system under study. Related numerical algorithms and simulation results are also given.

Keywords: Nonlinear control system · Reachable set
Estimation under uncertainty

1 Introduction

The research continues the study of estimation problems for uncertain dynamical systems in the case when a probabilistic description of noise and errors is not available, but only bounds on them are known [1–6]. Mathematical models of this kind appear in studies of dynamical systems under uncertainty in many applications such as physics, cybernetics, biology, economics and other

Supported by the Russian Science Foundation (RSF Project No. 16-11-10146).

G. Nikolov et al. (Eds.): NMA 2018, LNCS 11189, pp. 97–105, 2019.
https://doi.org/10.1007/978-3-030-10692-8_11

areas [7–10]. Therefore, the development of analytical methods and numerical schemes for the analysis of control systems with nonlinearity and uncertainty is an important goal for both theory and applications.

In this paper the modified state estimation techniques which use the special type of nonlinearity of a control system and also take into account state constraints are proposed. We assume that the system nonlinearity is generated by the combination of two types of functions in related differential equations, one of which is bilinear and the other one is quadratic. The additional state constraints (of ellipsoidal type) are also imposed.

The paper further develops results of [11–20], namely we give here the continuous time version of the upper estimates of reachable sets of the control system with uncertainty, nonlinearity and state constraints.

2   Problem Formulation

2.1   Basic Notations

Let $\mathrm{I\!R}^n$ denote the n–dimensional Euclidean space and $x'y$ is the usual inner product of $x, y \in \mathrm{I\!R}^n$ with the prime as a transpose and with $\|x\| = (x'x)^{1/2}$. We use the symbol comp $\mathrm{I\!R}^n$ for the variety of all compact subsets $A \subset \mathrm{I\!R}^n$ and the symbol conv $\mathrm{I\!R}^n$ for the variety of all compact convex subsets $A \subset \mathrm{I\!R}^n$.

Let us denote the set of all closed convex subsets $A \subseteq \mathrm{I\!R}^n$ by the symbol clconv $\mathrm{I\!R}^n$. Let $\mathrm{I\!R}^{n \times m}$ stands for the set of all real $n \times m$-matrices, diag $\{v\}$ denotes a diagonal matrix with the elements of vector v on the main diagonal. Denote by $I \in \mathrm{I\!R}^{n \times n}$ the identity matrix and by Tr (A) the trace of $n \times n$-matrix A (the sum of its diagonal elements).

We denote also by $B(a, r) = \{x \in \mathrm{I\!R}^n : \|x - a\| \leq r\}$ the ball in $\mathrm{I\!R}^n$ with a center $a \in \mathrm{I\!R}^n$ and a radius $r > 0$ and denote by

$$E(a, Q) = \{x \in \mathrm{I\!R}^n : (Q^{-1}(x - a), (x - a)) \leq 1\}$$

the *ellipsoid* in $\mathrm{I\!R}^n$ with a center $a \in \mathrm{I\!R}^n$ and with a symmetric positive definite $n \times n$-matrix Q.

2.2   Problem Description

Consider the following nonlinear control system

$$\dot{x} = A(t)x + f(x)d + u(t), \quad x_0 \in \mathcal{X}_0, \quad t \in [t_0, T], \tag{1}$$

where $x, d \in \mathrm{I\!R}^n$, $\|x\| \leq K$ $(K > 0)$, $f(x)$ is the nonlinear function, which is quadratic in x, that is $f(x) = x'Bx$, with a given symmetric and positive definite $n \times n$-matrix B.

Control function $u(t)$ is assumed to be Lebesgue measurable on $[t_0, T]$ and it satisfies the constraint $u(t) \in \mathcal{U}$ for a.e. $t \in [t_0, T]$ where $\mathcal{U}$ is a given set belonging to comp $\mathrm{I\!R}^n$. We will assume further that $\mathcal{U} = E(\hat{a}, \hat{Q})$.

We assume that the $n \times n$-matrix function $A(t)$ in (1) has the form

$$A(t) = A^0 + A^1(t). \tag{2}$$

Here A^0 is a given constant $n \times n$-matrix, the measurable and $n \times n$-matrix $A^1(t)$ is unknown but bounded, $A^1(t) \in \mathcal{A}^1$ ($t \in [t_0, T]$).

Therefore we have the constraint

$$A(t) \in \mathcal{A} = A^0 + \mathcal{A}^1, \tag{3}$$

where we assume that

$$\mathcal{A}^1 = \{ \ A = \{a_{ij}\} \in R^{n \times n} : a_{ij} = 0 \ \text{ for } \ i \neq j, \ \text{ and } \\ a_{ii} = a_i, \ i = 1, \ldots, n, \ a = (a_1, \ldots, a_n), \ a'Da \leq 1 \ \}, \tag{4}$$

with $D \in \mathbb{R}^{n \times n}$ being a symmetric and positive definite matrix.

We assume that $\mathcal{X}_0$ in (1) is an ellipsoid, $\mathcal{X}_0 = E(a_0, Q_0)$, with a symmetric and positive definite matrix $Q_0 \in \mathbb{R}^{n \times n}$ and with a center $a_0 \in \mathbb{R}^n$.

One of the important issues in the theory of control under uncertainty is how to specify the set of all solutions $x(t)$ to (1) that satisfy the additional state constraints (the "viability" constraint [1,2])

$$x(t) \in Y(t), \quad t_0 \leq t \leq T \tag{5}$$

where $Y(t) \in \text{conv } \mathbb{R}^n$ ($t \in [t_0, T]$).

In particular, the viability constraint (5) may be induced by the so-called measurement equation

$$y(t) = G(t)x + w(t),$$

where $y(t)$ is a p-vector function corresponding to measurement results which are obtained with unknown but bounded "noises" $w(t)$

$$w \in Q^*(t), \quad Q^*(t) \in \text{comp } \mathbb{R}^p,$$

here $Q^*(t)$ is a given set-valued function, $G(t)$ is a given $p \times n$–matrix function, (some earlier problem settings and previous results in this field may be found in [2]).

Here we consider the case when the state constraint of type (5) is defined by the ellipsoid,

$$x(t) \in Y = E(\tilde{a}, \tilde{Q}), \quad t_0 \leq t \leq T, \tag{6}$$

with the center $\tilde{a} \in \mathbb{R}^n$ and the positive definite $n \times n$–matrix $\tilde{Q}$. We will assume further that there exists at least one solution $x^*(t)$ of (1) that satisfies the condition (6).

Let the absolutely continuous function $x(t) = x(t; u(\cdot), A(\cdot), x_0)$ be a solution to dynamical system (1)–(6) with initial state $x_0 \in \mathcal{X}_0$, with admissible control

$u(\cdot)$ and with a matrix $A(\cdot)$ satisfying (2)–(4). The reachable set $\mathcal{X}(t)$ at time t $(t_0 < t \leq T)$ of system (1)–(5) is defined as the set

$$
\begin{aligned}
\mathcal{X}(t) = \big\{ x \in \mathbb{R}^n : \ &\exists\, x_0 \in \mathcal{X}_0, \ \exists\, u(\cdot) \in \mathcal{U}, \ \exists\, A(\cdot) \in \mathcal{A}, \\
&x = x(t) = x\big(t; u(\cdot), A(\cdot), x_0\big), \ x(s) \in Y, \ t_0 \leq s \leq t \big\}
\end{aligned}
\tag{7}
$$

It should be noted that the problem of exact construction of reachable sets of control systems is very difficult even for linear systems [3]. So instead of exact problem solution the different estimation approaches were developed in many researches [1–5]. The main problem studied here is to find external ellipsoidal estimates (with respect to inclusion of sets) for reachable sets $\mathcal{X}(t)$ $(t_0 < t \leq T)$ and to study the dynamics of such upper estimates in time t. We investigate here a more complicated case than in [20], since we assume now that we have an additional state constraint on the trajectories of the control system, which significantly complicates the problem analysis.

3   Main Results

Earlier some approaches were proposed to obtain differential equations describing dynamics of external (and in some cases internal) ellipsoidal estimates for reachable sets of control system under uncertainty, e.g., in [21] the authors studied estimation problems for systems with uncertain matrices in dynamical equations, but additional nonlinear terms in dynamics were not considered there.

Differential equations of ellipsoidal estimates for reachable sets of a nonlinear dynamical control system were derived in [13] for the case when system state velocities contain quadratic forms but in that case the uncertainty in matrix coefficients was not assumed.

Later, in [15], differential equations for external ellipsoidal estimates of reachable sets of a control system with nonlinearity and with uncertain matrix were derived under assumption that all elements $\{a_{ij}\}$ of the matrix A were bounded in modulus.

Here we investigate the case different from the above mentioned results when we assume the quadratic-type constraints on unknown matrix $A(t)$ included in the system dynamics and assume also the presence of additional constraints on the system states. In this case the related analysis of the dynamical properties of proposed ellipsoidal estimates is more complicated and was not carried out before.

3.1   Auxiliary Estimate

We use here the idea of [2] for the so-called elimination of state constraints in the construction of reachable sets of the system (1)–(6) (see also related results in [22]). Consider the following auxiliary differential inclusion with $n \times n$–matrix parameter L,

$$
\begin{aligned}
\dot{z} \in (A^0 - L + \mathcal{A}^1)z \ &+ \ f(z) \cdot d \ + E(\hat{a}, \hat{Q}) + L \cdot E(\tilde{a}, \tilde{Q}), \\
&t_0 \leq t \leq T, \ z_0 \in X_0 = E(a_0, Q_0).
\end{aligned}
\tag{8}
$$

Denote by $Z(t; t_0, X_0, L)$ ($t \in [t_0, t_1]$) the trajectory tube to (8) for a fixed matrix parameter L. Let $\mathcal{L}$ denotes the class of all $n \times n$-matrix L.

We will need the following result which may be used as a basis for the constraints elimination procedure.

Lemma 1. ([2]) *The following estimate is true*

$$X(t) \subseteq \bigcap_{L \in \mathcal{L}} Z(t; t_0, X_0, L), \quad t_0 \le t \le T. \tag{9}$$

Remark 1. It was proven in [2] that more precise upper estimates similar to (9) may be obtained if we will use a wider class of matrices L (for example, if we consider all matrices depending on time, $L = L(t)$), here for simplicity we consider only constant matrices L.

Using this approach and results of [13, 15, 16, 19, 20] we can find the upper ellipsoidal estimates for reachable sets $Z(t) = Z(t; t_0, X_0, L)$ of the nonlinear system (8).

3.2   Main Theorem

The following result describes the dynamics of the external ellipsoidal estimates of the reachable set $X(t) = X(t; t_0, X_0)$ ($t_0 \le t \le T$) of the system (1)–(6).

Denote the maximal eigenvalue of the matrix $B^{1/2} Q_0 B^{1/2}$ as k^2, therefore k^2 is the smallest positive number for which the inclusion

$$X_0 = E(a_0, Q_0) \subseteq E(a_0, k^2 B^{-1})$$

is satisfied (see also related constructions in [19]).

Theorem 1. *The inclusion*

$$X(t; t_0, X_0) \subseteq E(a_L^+(t), r_L^+(t) B^{-1}) \tag{10}$$

is true for any $t \in [t_0, T]$ and for any $L \in \mathcal{L}$ where functions $a_L^+(t)$, $r_L^+(t)$ are the solutions of the following system of ordinary differential equations

$$\dot{a}_L^+(t) = (A^0 - L)a_L^+(t) + ((a_L^+(t))' B a_L^+(t) \tag{11}$$
$$+ r_L^+(t))d + \hat{a} + L\tilde{a}, \quad t_0 \le t \le T,$$
$$\dot{r}_L^+(t) = \max_{\|l\|=1} \left\{ l'(2 r_L^+(t) B^{1/2}(A^0 - L) \right.$$
$$+ 2d(a_L^+(t))' B) B^{-1/2}$$
$$\left. + q^{-1}(r_L^+(t)) B^{1/2} \hat{Q}_L^* B^{1/2})) l \right\} + q(r_L^+(t)) r_L^+(t),$$
$$q(r) = ((nr)^{-1} \text{Tr}(B\hat{Q}^*))^{1/2},$$

where the positive definite matrix $\hat{Q}^*$ *is such that*

$$\mathcal{A}^1 a_0 + E(0, \hat{Q}) + k_0 D^{1/2} B^{1/2} B(0, 1) + L E(0, \tilde{Q}) \subseteq E(0, \hat{Q}^*) \qquad (12)$$

and the initial states for $a_L^+(t)$ *and* $r_L^+(t)$ *are*

$$a_L^+(t_0) = a_0, \quad r_L^+(t_0) = k^2.$$

Proof. The above estimates are derived following the scheme of the proof of Theorem 2 in [16] with the necessary corrections done according to new class of constraints on unknown parameters and functions included in the system description.

Corollary 1. *The following estimate is true for any* $t \in [t_0, T]$

$$X(t; t_0, X_0) \subseteq \bigcap_{L \in \mathcal{L}} E(a_L^+(t), r_L^+(t) B^{-1}).$$

Proof. The inclusion follows directly from Theorem 1.

Remark 2. The numerical scheme and the related algorithm for constructing upper estimates of reachable sets of the system under consideration may be also formulated similar to algorithms described in [13–15].

4   Numerical Simulation

The following example illustrates the results.

Example 1. Consider the following nonlinear control system

$$\begin{cases} \dot{x}_1 = a_1 x_1 + x_1^2 + x_2^2 + u_1, \\ \dot{x}_2 = a_2 x_2 + u_2, \\ x_0 \in X_0, \qquad t_0 \le t \le T. \end{cases} \qquad (13)$$

Here $t_0 = 0$, $T = 0.4$. The uncertain initial state x_0 belongs to the ball $X_0 = B(0, 1)$, uncertain parameters $\{a_1, a_2\}$ satisfy the constraint $a_1^2 + a_2^2 \le 1$, admissible control functions belong to the set $U = B(0, 0.1)$.

The state constraint is defined by the ellipsoid $E(\tilde{a}, \tilde{Q})$ where $\tilde{a} = 0$ and $\tilde{Q} = \text{diag}\{0.64, 4\}$.

The reachable set $X(t)$ with the estimating ellipsoids $E(a_{L_i}(t), Q_{L_i}(t))$ found by Theorem 1 are shown in Fig. 1 for different choices of matrices L_i ($i = 1, 2, 3$, $t = 0.1$). Two estimating ellipsoidal tubes described in Theorem 1 are shown in Fig. 2.

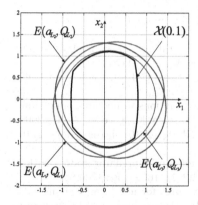

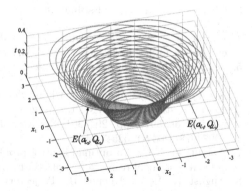

Fig. 1. Ellipsoidal estimates for reachable set X(t) (2d-picture in the plane of state variables).

Fig. 2. Ellipsoidal tubes containing reachable set X(t) (3d-picture in the plane of state variables and time).

5   Conclusions

The paper deals with the problems of state estimation for uncertain dynamical control systems under the assumption that the initial system state and some parameters in dynamical equations are unknown but bounded with given constraints.

The focus of this research was done on deriving differential equations which describe the dynamics of the ellipsoidal estimates of reachable sets of the systems under study.

Basing on the results of ellipsoidal calculus developed earlier we present the new state estimation results which use the special bilinear–quadratic structure of nonlinearity and uncertainty of the control system and allow to construct the external ellipsoidal estimates of reachable sets. Examples and numerical results related to procedures of set-valued approximations of trajectory tubes and reachable sets are also given.

The results obtained here may be used in further theoretical and applied researches in optimal control and state estimation for dynamical systems with more complicated classes of uncertainty of set-membership and other types.

References

1. Kurzhanski, A.B., Varaiya, P.: Dynamics and Control of Trajectory Tubes: Theory and Computation. Systems & Control, Foundations & Applications, vol. 85. Basel, Birkhäuser (2014)
2. Kurzhanski, A.B., Filippova, T.F.: On the theory of trajectory tubes - A mathematical formalism for uncertain dynamics, viability and control. In: Kurzhanski, A.B. (ed.) Advances in Nonlinear Dynamics and Control: a Report from Russia. Progress in Systems and Control Theory, vol. 17, pp. 122–188. Birkhäuser, Boston (1993)

3. Chernousko, F.L.: State Estimation for Dynamic Systems. CRC Press, Boca Raton (1994)
4. Polyak, B.T., Nazin, S.A., Durieu, C., Walter, E.: Ellipsoidal parameter or state estimation under model uncertainty. Automatica **40**, 1171–1179 (2004)
5. Schweppe, F.C.: Uncertain Dynamical Systems. Prentice-Hall, Englewood Cliffs (1973)
6. Krastanov, M.I., Veliov, V.M.: High-order approximations to nonholonomic affine control systems. In: Lirkov, I., Margenov, S., Waśniewski, J. (eds.) LSSC 2009. LNCS, vol. 5910, pp. 294–301. Springer, Heidelberg (2010). https://doi.org/10.1007/978-3-642-12535-5_34
7. Apreutesei, N.C.: An optimal control problem for a prey-predator system with a general functional response. Appl. Math. Lett. **22**(7), 1062–1065 (2009)
8. August, E., Lu, J., Koeppl, H.: Trajectory enclosures for nonlinear systems with uncertain initial conditions and parameters. In: Proceedings of 2012 American Control Conference, Fairmont Queen Elizabeth, Montréal, Canada, June, pp. 1488–1493 (2012)
9. Boscain, U., Chambrion, T., Sigalotti, M.: On some open questions in bilinear quantum control. In: European Control Conference (ECC), Zurich, Switzerland, July 2013, pp. 2080–2085 (2013)
10. Ceccarelli, N., Di Marco, M., Garulli, A., Giannitrapani, A.: A set theoretic approach to path planning for mobile robots. In: Proceedings of 43rd IEEE Conference on Decision and Control, Atlantis, Bahamas, December, pp. 147–152 (2004)
11. Filippova, T.F.: Set-valued solutions to impulsive differential inclusions. Math. Comput. Modell. Dyn. Syst. **11**(2), 149–158 (2005)
12. Filippova, T.F.: Construction of set-valued estimates of reachable sets for some nonlinear dynamical systems with impulsive control. Proc. Steklov Inst. Math. **269**(S.2), 95–102 (2010)
13. Filippova, T.F.: Differential equations of ellipsoidal state estimates in nonlinear control problems under uncertainty. Discrete and Continuous Dynamical Systems, Supplement 2011, Dynamical Systems, Differential Equations and Applications, vol. 1, pp. 410–419. American Institute of Mathematical Sciences, Springfield (2011)
14. Filippova, T.F.: State estimation for uncertain systems with arbitrary quadratic nonlinearity. In: Proceedings of PHYSCON 2015, Istanbul, Turkey, pp. 1–6 (2015)
15. Filippova, T.F.: Estimates of reachable sets of impulsive control problems with special nonlinearity. In: AIP Conference Proceedings, vol. 1773, 100004, pp. 1–8 (2016). https://doi.org/10.1063/1.4964998
16. Filippova, T.F.: Differential equations of ellipsoidal state estimates for bilinear-quadratic control systems under uncertainty. J. Chaotic Model. Simul. (CMSIM) **1**, 85–93 (2017)
17. Filippova, T.F., Berezina, E.V.: On state estimation approaches for uncertain dynamical systems with quadratic nonlinearity: theory and computer simulations. In: Lirkov, I., Margenov, S., Waśniewski, J. (eds.) LSSC 2007. LNCS, vol. 4818, pp. 326–333. Springer, Heidelberg (2008). https://doi.org/10.1007/978-3-540-78827-0_36
18. Filippova, T.F., Matviychuk, O.G.: Reachable sets of impulsive control system with cone constraint on the control and their estimates. In: Lirkov, I., Margenov, S., Waśniewski, J. (eds.) LSSC 2011. LNCS, vol. 7116, pp. 123–130. Springer, Heidelberg (2012). https://doi.org/10.1007/978-3-642-29843-1_13
19. Filippova, T.F., Matviychuk, O.G.: Estimates of reachable sets of control systems with bilinear-quadratic nonlinearities. Ural Math. J. **1**(1), 45–54 (2015)

20. Filippova, T.F.: Estimation of star-shaped reachable sets of nonlinear control systems. In: Lirkov, I., Margenov, S. (eds.) LSSC 2017. LNCS, vol. 10665, pp. 210–218. Springer, Cham (2018). https://doi.org/10.1007/978-3-319-73441-5_22
21. Chernousko, F.L., Rokityanskii, D.Y.: Ellipsoidal bounds on reachable sets of dynamical systems with matrices subjected to uncertain perturbations. J. Optimiz. Theory Appl. **104**(1), 1–19 (2000)
22. Gusev, M.I.: Application of penalty function method to computation of reachable sets for control systems with state constraints. In: AIP Conference Proceedings, vol. 1773, 050003, pp. 1–8 (2016). https://doi.org/10.1063/1.4964973

Evaluation of Serial and Parallel Shared-Memory Distance-1 Graph Coloring Algorithms

Lukas Gnam[1(✉)], Siegfried Selberherr[2], and Josef Weinbub[1]

[1] Christian Doppler Laboratory for High Performance TCAD,
Institute for Microelectronics, TU Wien, Vienna, Austria
{gnam,weinbub}@iue.tuwien.ac.at
[2] Institute for Microelectronics, TU Wien, Vienna, Austria
selberherr@iue.tuwien.ac.at

Abstract. Within the scope of computational science and engineering, the standard graph coloring problem, the distance-1 coloring, is typically used to select independent sets on which subsequent parallel computations can be guaranteed. As graph coloring is an active field of research, various algorithms are available, each offering advantages and disadvantages. We compare several serial as well as parallel shared-memory graph coloring algorithms for the standard graph coloring problem based on reference graphs. Our investigation covers well established as well as recent algorithms and their support for balanced and unbalanced approaches. An overview on speedup, used number of colors, and their respective population for different test graphs is provided. It is shown that the parallel approaches produce similar results as the serial methods, but for specific cases the serial algorithms still remain a good option, when certain properties (e.g., balancing) are of major importance.

Keywords: Graph coloring · Shared-memory · Distance-1 coloring
Parallel algorithm

1 Introduction

The decomposition of computational tasks into independent sets, which pave the way for a subsequent parallelization step, is a widely used approach to exploit parallel computing resources. Examples of such use cases are community detection [9], mesh adaptation [7], and linear algebra [10] algorithms.

In this work, we consider the standard graph coloring problem, the distance-1 coloring. For a general graph $G(V, E)$ a distance-1 coloring is a coloring, where any two adjacent vertices receive different colors. Hence, each color represents an independent set for possible subsequent parallel processing. Usually, the goal of a distance-1 graph coloring problem is to use as few colors as possible: One fact which is often neglected is the population of the resulting sets. Considering the standard formulations of such algorithms, they mostly result in highly

© Springer Nature Switzerland AG 2019
G. Nikolov et al. (Eds.): NMA 2018, LNCS 11189, pp. 106–114, 2019.
https://doi.org/10.1007/978-3-030-10692-8_12

unbalanced populations. A heavily unbalanced coloring can lead to *colors* which contain insufficient workload to achieve acceptable parallel efficiency, thus leading to an undesired bottleneck in a parallel workflow. Therefore, graph coloring algorithms typically aim to use as few colors as possible to enable the subsequent workflow to achieve proper scalability.

Although there have been several efforts to compare different graph coloring algorithms in the past [1,6,9], we pick up on recent developments in this field and provide an overview on some of the newest distance-1 graph coloring algorithms. We show their difference in the number of colors used for different graphs, as well as a comparison of the resulting color populations. Additionally, we investigate the overhead in execution time experienced for different balancing approaches. For the parallel algorithms we also provide speedup and strong scalability data.

In Sect. 2, we briefly discuss the related work and present the coloring algorithms we used in our evaluation, followed by the actual evaluation of the algorithms in Sect. 3.

2   Coloring Algorithms

The most widely used approach to achieve a coloring of a graph is the *Greedy* coloring algorithm [5]. It iterates the graph and assigns the smallest color permissible to the active graph vertex, by checking the color assigned to its neighbors (usually colors are denoted using integers). Thus, for any graph with maximum degree d^1 this algorithm uses at most $d + 1$ colors. One major drawback of this algorithm is, that the highest colors are the ones assigned the least, leading to a skewness in the population of the colors. The *Greedy* algorithm is part of our study.

To alleviate this skewness, the *Greedy* algorithm can be adapted such that it assigns the least used color permissible, leading to a more balanced coloring of the graph [9,11]. This algorithm is also part of our study and henceforth denoted as *Greedy-LU*.

Based on an approach from Gebremedhin and Manne [4], a parallel algorithm for speculative graph coloring was introduced by Çatalyürek et al. [1], which follows a two-step strategy. The first step is to color the graph vertices in parallel without checking for any possible conflicts. In a second step the previously colored graph vertices are checked and, if conflicts occur, the corresponding graph vertices are marked for recoloring in the next iteration. Hence, the number of occurring conflicts defines the number of total iterations, which could lead to performance drawbacks. Catalyürek's algorithm, from now on referred to as *Parallel*, acts as our baseline for the shared-memory parallel coloring approach.

Recently, Lu et al. presented several serial shared-memory parallel algorithms for distance-1 graph coloring [9]. Following their results, we selected the *Scheduled Reverse* algorithm to be included in our study. This particular approach uses the *Greedy* algorithm to obtain an initial coloring and improves the balancing by moving vertices from colors with high population to colors with low

---

[1] The degree of a vertex of a graph is the number of incident edges [3].

population, without introducing new colors. We chose to limit the algorithm to three iterations, following the results and suggestions presented in [9], which is a reasonable compromise between color balancing and computational overhead. Within the remainder of this work, we will refer to this algorithm as *Parallel Recolor*.

3 Evaluation

3.1 Benchmark Platform

We used a single compute node of the Vienna Scientific Cluster 3 (VSC-3). A node offers two Intel Xeon E5-2650v2 Ivy Bridge EP processors running with 2.6 GHz and a total of 64 GB of main memory. Hence, 16 physical and 32 logical cores are available. The benchmarks were compiled using Intel's C++ compiler, version 17.0.4, with -O3 optimization. Additionally, we made use of Intel's thread-core affinity capabilities using the *KMP_AFFINITY* environment variable to ensure proper thread pinning to avoid thread migration. We used a static OpenMP loop scheduling, as the problems do not pose a load-balancing issue.

3.2 Test Graphs

For our investigations we used four different graphs, where three of them were created with the parallel graph generation software PaRMAT [8] following the approach from Catalyürek [1]. We varied the graph parameters *a*, *b*, and *c* (see Table 1) while keeping a constant total number of vertices of 16 777 216. The first graph, RMAT-ER, is a graph belonging to the so-called Erdös-Rényi class with a normal degree distribution, whereas the other two, RMAT-G and RMAT-B, have multiple local maxima of the degree distribution (for more details see [1]). The highest number of neighbors for a vertex (maximum vertex degree) is observed in the RMAT-B graph with 49 212. The fourth test graph is taken from the University of Florida Sparse Matrix Collection [2] and is a mesh modeling a BMW 3 series car. An overview on the graph properties is shown in Table 1.

Table 1. Properties of the four graphs used in this evaluation study as well as the graph parameters for the RMAT graphs used within PaRMAT.

Graph	Vertices	Avg. degree	Max. degree	a	b	c
RMAT-ER	16 777 216	16	42	0.25	0.25	0.25
RMAT-G	16 777 216	16	41 938	0.45	0.15	0.25
RMAT-B	16 777 216	16	49 212	0.55	0.15	0.15
BMW	227 362	48.65	335	-	-	-

3.3 Coloring Quality

In Fig. 1 we show the maximum number of colors used by each of the individual algorithms for the four input graphs. For the RMAT-ER graph the *Greedy-LU* algorithm uses the most colors, i.e., 23, compared to 12 for all others, including the *Parallel* algorithm with 32 threads. Due to its recoloring approach the *Parallel Recolor* algorithm uses always the same number of colors as the *Greedy* algorithm, because the latter acts as the initial input coloring and no colors are added while balancing. This is independent of the number of threads being used. Additionally, it can be observed that the actual number of colors is always way below the maximum vertex degree, and often also in the range of the nearly optimal *Greedy* coloring algorithm. In the case of the RMAT-G graph, *Greedy-LU* produces an output using 70 colors, whereas the other algorithms use only 26. The resulting colorings require about 1600 times less colors than the maximum vertex degree occurring in this graph, except of the *Greedy-LU* algorithm which uses nearly 600 times less colors. The difference results from the *Greedy* coloring approach applied in the two parallel algorithms, which is unbalanced in contrast to the *Greedy-LU* algorithm. Considering the RMAT-B graph, the *Greedy-LU* algorithm requires the most colors, i.e., 395, which is about four times more than the demand from the *Parallel* algorithm using 32 threads. The results for the three-dimensional BMW mesh show that *Parallel* with 32 threads uses the least number of colors. This is a result of the parallel execution of the *Greedy* coloring approach in its tentative coloring phase, yielding a better outcome in the number of used colors and their population for this specific graph.

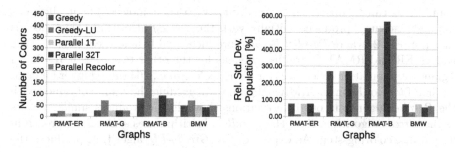

Fig. 1. Maximum number of colors (left) and relative standard deviation of the color population (right) resulting from the investigated algorithms for each of the input graphs. For the *Parallel* algorithm the results obtained with 1 (1T) and 32 threads (32T) are depicted. Note that the single-threaded version produces the same results as the *Greedy* algorithm. Since the maximum number of colors used by the *Parallel Recolor* algorithm does not depend on the number of threads only one result bar is shown in each figure. (Color figure online)

To compare the population of the colors produced by the implemented algorithms we use the relative standard deviation of the coloring results (see Fig. 1). The different distributions of vertex degrees in each graph (see Table 1) strongly influence the resulting color populations: Vertices with a high degree compared to the graph's average degree lead to a higher number of colors with very low population. This is especially observable for the RMAT graphs, where for RMAT-ER the deviations range between 11–75%, for RMAT-G between 0–270%, and for RMAT-B between 0–565%. For the BMW graph the deviations are between 26–74%. As can be seen, the best results (i.e., the least deviation) are obtained using the *Greedy-LU* algorithm, since it initially tries to balance the colors. In case of the three RMAT graphs the *Parallel Recolor* algorithm returns the second best deviation results, followed by the *Greedy* and the *Parallel* algorithm. The *Parallel* algorithm using 32 threads produces similar results as *Greedy*, except for the RMAT-B graph, where its deviation is larger than the deviation resulting from the *Greedy* algorithm. For the BMW graph the *Parallel* algorithm using 32 threads produces the best deviation results after the *Greedy-LU* algorithm, followed by the *Parallel Recolor* algorithm. Because it assigns the smallest color permissible, the *Greedy* algorithm produces the highest skewness. Regarding the color population for the BMW graph, *Parallel* also performs better than *Parallel Recolor*. Nevertheless, the relative standard deviation must not be viewed as a single quality metric for the population deviation of the used colors, because there can still be very large differences between specific colors. Figure 2 shows these differences in the color populations occurring for the different graphs. Since the *Greedy* algorithm uses the smallest color permissible for coloring a vertex, the color population decreases for increasing color indices when applying the *Greedy* as well as the *Parallel* algorithm. Therefore, we observe a high skewness in the results produced by the *Greedy* and the *Parallel* algorithm. Since the *Parallel Recolor* algorithm does not add new colors, strong *jumps* of color populations can be observed, especially for high color indices, but it alleviates the high skewness produced initially with the *Greedy* algorithm. This effect is shown in Fig. 2, where *Parallel Recolor* produces well balanced colorings for the RMAT-ER and RMAT-G graphs, whereas for the other two cases it results in significant population differences for higher colors (e.g., 264 times higher for color 54 than for color 53 in the RMAT-B graph). Regarding the results obtained with the *Parallel* algorithm, our investigations show that it produces similar results as the *Greedy* algorithm, since the *Parallel* algorithm uses a *Greedy* approach in its parallel coloring step. As expected, the *Greedy-LU* algorithm produces the best balanced colorings for all test graphs but at the price of using 1.5 to almost 5 times more colors than the unbalanced *Greedy* algorithm.

3.4   Strong Scaling Analysis

Since not only the resulting number of colors and their respective population can have a strong impact on the overall application performance, but also the execution time of the coloring algorithm itself, we additionally investigated the strong scaling capabilities of the *Parallel* and the *Parallel Recolor* algorithm based on the reference graphs. For the latter there is, in addition to the initial use of the *Greedy* coloring algorithm, also some serial part in the algorithm which prepares the data for parallel execution. Therefore, it can be expected

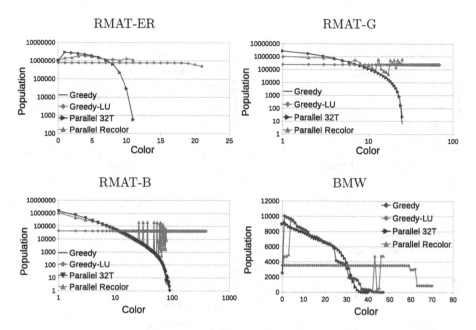

Fig. 2. Resulting color populations for the four test graphs evaluated in our study. For the *Parallel* algorithm we show the results obtained using 32 threads (32T), since the single threaded version produces the same output as the *Greedy* algorithm. The coloring results for the *Parallel Recolor* algorithm are independent from the number of threads. (Color figure online)

that the recoloring approach of the *Parallel Recolor* algorithm is most likely to experience parallel performance limitations. Figure 3 and Table 2 show this expected behavior. All timings are averaged based on three iterations.

The execution times for the single-threaded versions show major differences for the three RMAT test graphs. The *Parallel Recolor* algorithm is nearly two times faster than the *Parallel* algorithm for the RMAT-ER graph, and 1.3 times faster for the other two RMAT graphs, because the *Parallel* algorithm iterates at least twice over the graph (coloring and checking), whereas *Parallel Recolor* retains the nearly optimal initial *Greedy* coloring. As shown in Table 2, the single-threaded execution times for the BMW graph are similar, due to the smaller number of vertices: 0.132 s for the *Parallel* algorithm and 0.137 s for the *Parallel Recolor* algorithm. As expected, when increasing the number of threads *Parallel* outperforms *Parallel Recolor* for more than 2 threads for the RMAT-ER and the RMAT-G graph and for more than 4 threads for the RMAT-B graph. However, for the BMW graph the *Parallel Recolor* algorithm scales best for up to 16 threads, albeit breaking down for 32 threads, because of the increased number of recoloring conflicts occurring with a higher number of threads for this graph.

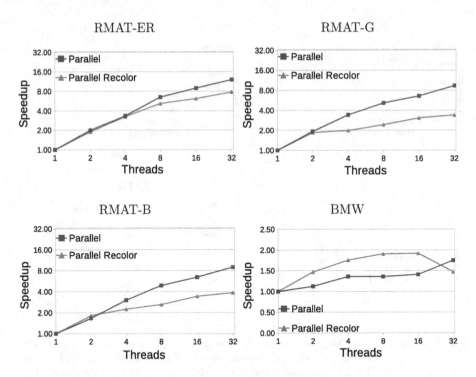

Fig. 3. Speedups of the parallel algorithms for the different test graphs.

Table 2. Execution times in seconds of the algorithms for the test graphs, with the fastest time for each graph in bold. For the *Parallel* and *Parallel Recolor (Par.Rec.)* algorithms the results obtained with 1 (1T) and 32 threads (32T) are shown.

Graph	Greedy	Greedy-LU	Parallel 1T	Parallel 32T	Par.Rec.1T	Par.Rec.32T
RMAT-ER	7.213	12.32	26.71	**2.21**	13.97	10.03
RMAT-G	5.765	140.21	19.55	**2.10**	15.11	9.53
RMAT-B	6.372	29.02	19.50	**2.17**	15.43	10.21
BMW	**0.036**	0.378	0.132	0.075	0.137	0.242

4   Conclusions

As shown in this work, the resulting colorings depend heavily on the type and properties of the respective input graph (e.g., see Figs. 1 and 2). In order to make an adequate choice it is therefore necessary to determine the requirements of the specific task for which the coloring should be used. If a balanced color distribution is the primary metric of interest, then the *Greedy-LU* algorithm is the best choice. However, if the application in mind has to repeatedly execute the coloring algorithm, it is likely that the execution time spent in coloring becomes more and more dominant: As the timings in Table 2 indicate, there are cases where the *Greedy-LU* algorithm takes about 24 times longer than the unbalanced serial *Greedy* algorithm. This results from the fact, that the *Greedy-LU* algorithm has to maintain a list of how often each color has already been assigned, such that the least used color permissible is picked.

This introduces the execution performance - and by extension parallel scalability - as a potential core metric. As shown in Figs. 2 and 3, as well as in Table 2, in most cases the use of a parallel coloring strategy, like the *Parallel* algorithm using 32 threads, can save up to 70% of time spent in coloring compared to the *Greedy* algorithm while producing similar results. In our test cases the speed of the *Parallel Recolor* algorithm suffers from its serial preparation step, yielding inferior execution performance than the *Parallel* algorithm. Nevertheless, if balancing of the colors *and* execution time are of importance, the recoloring approach remains still a reasonable choice, because in some cases it is more than 14 times faster than the *Greedy-LU* algorithm and guarantees that the number of colors used is the same as with the *Greedy* algorithm.

In all cases, the *Greedy-LU* algorithm proves to give the best balanced colorings, while in most cases the *Parallel* algorithm guarantees a proper tradeoff between execution time and the number of colors. For the BMW graph, the serial *Greedy* algorithm provides a well-balanced tradeoff between all three performance parameters.

Acknowledgements. The financial support by the *Austrian Federal Ministry for Digital and Economic Affairs* and the *National Foundation for Research, Technology and Development* is gratefully acknowledged. The computational results presented have been achieved using the Vienna Scientific Cluster (VSC).

References

1. Çatalyürek, Ü.V., Feo, J., Gebremedhin, A.H., Halappanavar, M., Pothen, A.: Graph coloring algorithms for multi-core and massively multithreaded architectures. Parallel Comput. **38**(10), 576–594 (2012)
2. Davis, T.A., Hu, Y.: The University of Florida sparse matrix collection. ACM Trans. Math. Softw. **38**(1), 1:1–1:25 (2011)
3. Diestel, R.: Graph Theory, 5th edn. Springer, Heidelberg (2017)
4. Gebremedhin, A.H., Manne, F.: Scalable parallel graph coloring algorithms. Concurr. Pract. Exp. **12**(12), 1131–1146 (2000)
5. Gyárfás, A., Lehel, J.: On-line and first fit colorings of graphs. J. Graph Theory **12**(2), 217–227 (1988)
6. Hawick, K., Leist, A., Playne, D.: Parallel graph component labelling with GPUs and CUDA. Parallel Comput. **36**(12), 655–678 (2010)
7. Ibanez, D., Shephard, M.: Mesh adaptation for moving objects on shared memory hardware. In: Proceedings of the International Meshing Roundtable (2016)
8. Khorasani, F., Gupta, R., Bhuyan, L.N.: Scalable SIMD-efficient graph processing on GPUs. In: Proceedings of the International Conference on Parallel Computing Technologies, pp. 39–50 (2015)
9. Lu, H., Halappanavar, M., Chavarría-Miranda, D., Gebremedhin, A.H., Panyala, A., Kalyanaraman, A.: Algorithms for balanced graph colorings with applications in parallel computing. IEEE Trans. Parallel Distrib. Syst. **28**(5), 1240–1256 (2017)

10. Manne, F.: A parallel algorithm for computing the extremal eigenvalues of very large sparse matrices. In: Kågström, B., Dongarra, J., Elmroth, E., Waśniewski, J. (eds.) PARA 1998. LNCS, vol. 1541, pp. 332–336. Springer, Heidelberg (1998). https://doi.org/10.1007/BFb0095354
11. Manne, F., Boman, E.: Balanced Greedy colorings of sparse random graphs. In: Proceedings of the Norwegian Informatics Conference, pp. 113–124 (2005)

Solving Function Approximation Problems Using the L^2-Norm of the Log Ratio as a Metric

Ivan D. Gospodinov$^{(\boxtimes)}$, Stefan M. Filipov, and Atanas V. Atanassov

Department of Computer Science, University of Chemical Technology
and Metallurgy, blvd. Kl. Ohridski 8, 1756 Sofia, Bulgaria
gospodinov@uctm.edu

Abstract. This article considers the following function approximation problem: Given a non-negative function and a set of equality constraints, find the closest to it non-negative function which satisfies the constraints. As a measure of distance we propose the L^2-norm of the logarithm of the ratio of the two functions. As shown, this metric guarantees that (i) the sought function is non-negative and (ii) to the extent to which the constraints allow, the magnitude of the difference between the sought and the given function is proportional to the magnitude of the given function. To solve the problem we convert it to a finite dimensional constrained optimization problem and apply the method of Lagrange multipliers. The resulting nonlinear system, together with the system for the constraints, are solved self-consistently by applying an appropriate iterative procedure.

Keywords: Non-negativity · Relative difference
Constrained optimization · Lagrange multipliers

1 Introduction

Consider the following function approximation problem [1–6]: The function $u : [a, b] \rightarrow \mathbb{R}$ is given along with a set of equality constraints that need to be met. We seek a function $u^* : [a, b] \rightarrow \mathbb{R}$ that meets the constraints and at the same time is as close to u as possible. The functional dependence u may come from a theoretical model, experimental results, numerical solution of a differential equation, etc. For example, $u(t)$, $t \in [a, b]$ could be the concentration of an enzyme or the kinetic energy of a moving body as a function of time (or position). The constraints may come, for example, from experimental observations, overall mass or energy balance considerations, theoretical requirements for the functional dependence, etc.

Usually, the function u^* is defined as the function that minimizes, subject to the given constraints, the L^2-norm of the absolute difference between u^* and u. This approach is a particular case of the well-known Least Squares method [7–9]. It distributes the magnitude of the absolute difference $u^* - u$ as uniformly

© Springer Nature Switzerland AG 2019
G. Nikolov et al. (Eds.): NMA 2018, LNCS 11189, pp. 115–124, 2019.
https://doi.org/10.1007/978-3-030-10692-8_13

across the entire interval $[a, b]$ as possible. Thus, small functional values of u are usually displaced by roughly the same distance as large functional values. In certain problems this could be undesirable. For example, a small concentration u of a substance at a certain point in time t, after imposing the constraints and performing the optimization, may turn into a negative concentration u^* at that time t. To avoid negative results, usually, the non-negativity constraint $u^* \geq 0$ is imposed. To solve the corresponding constrained optimization problem by the method of Lagrange multipliers [10], the Karush–Kuhn–Tucker (KKT) approach [10–12], which allows inequality constraints, can be used. To incorporate the inequality constraint the barrier method or the penalty method [13,14] can also be employed. In the barrier method the objective function (i.e. the function that has to be minimized) is modified so that it grows rapidly as u^* approaches the forbidden region and becomes infinity at the boundary. In the penalty method a penalty function is added to the objective function in order to "penalize" when u^* is outside the allowed region. After the solution is found, the barrier/penalty is made steeper/higher and the problem is solved again. Thus, iteratively, a solution satisfying the non-negativity requirement is reached. Other approaches, including some recent ones, to constrained optimization problems with equality and inequality constraints are [15–17].

To avoid negative results we propose a different approach to the considered function approximation problem. Instead of using the L^2-norm of $u^* - u$ as a metric we propose the L^2-norm of $\ln(u^*/u)$ [18,19]. In the sequel, we will refer to $\ln(u^*/u)$ as the *log ratio* (*of u^* to u* or *between u^* and u*). This approach, as shown in the following sections, ensures non-negativity of the result u^* for any non-negative original function u. In addition, when the deviation $u^* - u$ is small, the approach also ensures that the magnitude of the relative difference $(u^* - u)/u$ is as uniformly distributed across the entire interval $[a, b]$ as possible. The proposed approach is appropriate when the function u represents intrinsically non-negative physical quantity such as mass, density, concentration, kinetic energy, etc. It is not appropriate when u represents a quantity that can take negative as well as positive values, e.g. position along an axis, velocity component, force component, potential energy, etc.

This paper is structured as follows: In Sect. 2 the absolute and the relative difference, and their connection to the log ratio, are investigated. In Sect. 3, using the proposed log ratio based metric, the function approximation problem is formulated. Then, in Sect. 4, the problem is discretized and the resulting constrained optimization problem is solved, in general form, using the method of Lagrange multipliers. Similar solutions for minimizing the H^1 semi-norm of the absolute difference or the log ratio and applications can be found in [20–24]. In Sect. 5, an iterative procedure for self-consistent solution of the obtained nonlinear system together with the constraints, when the constraints are linear, is proposed. Finally, in Sect. 6, examples are presented and discussed.

2   Absolute Difference, Relative Difference, and Log Ratio

The *absolute difference* δ between the functions u^* and u at point t is

$$\delta(t) = u^*(t) - u(t), \quad t \in [a, b]. \tag{1}$$

If $\delta(t) = const$ throughout the interval $[a, b]$, then the functions u^* and u are equally separated at each point t. An example is shown in Fig. 1, where $\delta(t) = -0.1$ in the whole interval $[0, 2]$.

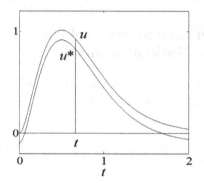

Fig. 1. The functions u^* and u are equally separated from each other at each point t in an absolute sense

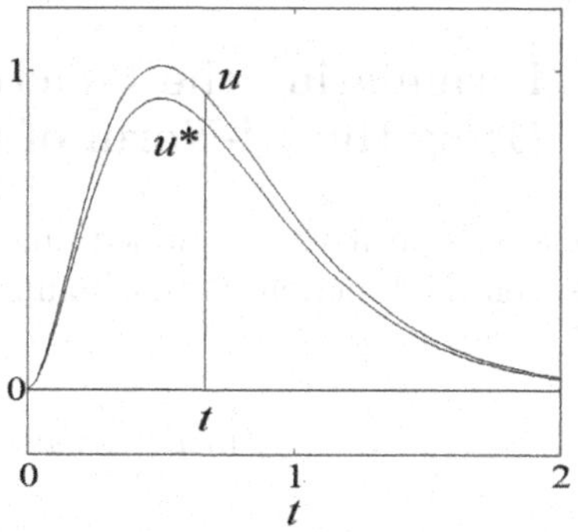

Fig. 2. The functions u^* and u are equally separated from each other at each point t in a relative sense

If $\delta(t) = 0$ for all $t \in [a, b]$, then $u^*(t) \equiv u(t)$, i.e. the two functions are identical. Therefore, a natural way to require closeness between the functions u^* and u is to require $\delta(t)$ to be as close to zero at each point t as possible and at the same time $|\delta(t)|$ to be as uniformly distributed across the interval as possible. Thus, the distance between u^* and u is defined as the L^2-norm of the difference between the functions, i.e. $||u^* - u||$. The function u^* that minimizes this norm, subject to the imposed constraints, is the closest function to u in an absolute sense. The square of the difference $u^* - u$ ensures that $|\delta(t)|$ is as uniformly distributed throughout $[a, b]$ as possible.

Sometimes, as discussed in the introduction, it is preferable to work with the relative distance between the two functions and avoid negative results. Consider the ratio $\mu(t) = u^*(t)/u(t)$, $t \in [a, b]$. If $\mu(t) = const$ throughout the interval $[a, b]$, then the functions u^* and u are equally separated at each point t in a relative sense. An example is shown in Fig. 2, where $\mu(t) = 0.9$ in the interval $t \in [0, 2]$.

If $\mu(t) = 1$ for all $t \in [a, b]$, then $u^*(t) \equiv u(t)$. Introducing the notation $\delta_L(t) = \ln \mu(t)$, and taking the logarithm of $\mu(t)$ we obtain the log ratio between the two functions at point t:

$$\delta_L(t) = \ln \frac{u^*(t)}{u(t)} = \ln u^*(t) - \ln u(t), \quad t \in [a, b]. \tag{2}$$

Expanding $\ln u^*$ in (2) in Taylor series around u gives $\delta_L = \delta/u + O(\delta^2)$, where δ is the absolute difference (1). Since $\delta(t)/u(t)$ is the *relative difference* at t, it follows that for small values of δ the function $\delta_L(t)$ is a measure of the relative difference between the two functions at point t. Why the relative difference can be estimated by $\ln(u^*/u)$ is also explained in [19]. If $\delta_L(t) = 0$ for all $t \in [a, b]$, then $u^*(t) \equiv u(t)$, i.e. the two functions are identical. Hence, one possible way to require relative closeness between u^* and u is to require that $\delta_L(t)$ be as close to zero as possible at each point t and $|\delta_L(t)|$ be as uniformly distributed across the interval as possible.

3   Formulating the Function Approximation Problem Using the L^2-Norm of the Log Ratio as a Metric

Let u be a positive real-valued function of a real independent variable $t \in [a, b]$. We seek the function u^* that minimizes the L^2-norm of $\ln u^* - \ln u$:

$$\|\ln u^* - \ln u\| = \sqrt{\int_a^b (\ln u^* - \ln u)^2 \, dt}, \qquad (3)$$

and at the same time satisfies some given constrains. In this work we consider only linear constraints for the function u^*, for example: linear combinations of functional values at certain points, e.g. linear boundary conditions, and linear integral constraints like $\int_a^b f(t) u^*(t) \, dt = 1$. Note, that the function u^* minimizes the distance (3) if and only if it minimizes

$$const\|\ln u^* - \ln u\|^2, \quad const > 0. \qquad (4)$$

Since $|\ln u^*|$ goes to infinity as u^* goes to zero, the requirement for a minimal distance (3) will keep u^* "away" from zero.

4   Solving the Discretized Problem Using the Method of Lagrange Multipliers

The formulated function approximation problem is in fact an infinite dimensional constrained optimization problem. To solve the problem we first discretize it as in [20–22] to obtain a corresponding finite dimensional optimization problem. Partitioning the interval $[a, b]$ by N mesh-points t_i into $N - 1$ intervals of equal size defines a uniform mesh on the interval:

$$\{t_i = a + (i - 1)h, \quad h = (b - a)/(N - 1), \quad i = 1, 2, \ldots, N\}. \qquad (5)$$

Any function defined on the mesh (5) will be called a *mesh function*. Let the function u be given as a mesh function $\{u_i = u(t_i), \quad i = 1, 2, \ldots, N\}$. We want the function u^*, also a mesh function, to satisfy the following $M < N$ linear constraints:

$$\mathbf{A}.\mathbf{u}^* = \mathbf{c}, \tag{6}$$

where $\mathbf{u}^* = [u_1^*, u_2^*, \ldots, u_N^*]^T$ is the $N \times 1$ column-vector of the unknown values of u^*, $\mathbf{A}$ is an $M \times N$ matrix with components A_{ji}, $j = 1, 2, \ldots, M$, $i = 1, 2, \ldots, N$, and $\mathbf{c}$ is an $M \times 1$ column-vector $\mathbf{c} = [c_1, c_2, \ldots, c_M]^T$. Note, that any linear integral constraint could be incorporated in (6) after replacing the integral by a sum. We want the function u^* to be as close to u as possible in the sense of the defined metric (3). In order to define distance between the mesh functions u and u^* expression (4) is discretized. Replacing the integral by a sum and omitting the constant factor, which does not affect the minimization, the following objective function is obtained:

$$I = \sum_{i=1}^{N} (\ln u_i^* - \ln u_i)^2. \tag{7}$$

If u_i^*, $i = 1, 2, \ldots, N$ minimize I and at the same time satisfy (6), then the function u^* will be a solution to the discretized function approximation problem.

Finding the minimum of I (7), subject to constraints (6), is a constrained optimization problem [10]. To solve the problem the method of Lagrange multipliers [10] is employed. First, the matrix equation (6) is rearranged by transferring the left-hand side term to the right-hand side. Then, the j-th equation from the system of equations (6) is multiplied by the Lagrange multiplier λ_j, summed over j, and the result is added to the objective function I to obtain:

$$J = I + \sum_{j=1}^{M} \lambda_j \left(c_j - \sum_{i=1}^{N} A_{ji} u_i^* \right). \tag{8}$$

Then, the derivatives of J with respect to the unknowns $u_1^*, u_2^*, \ldots, u_N^*$ are equated to zero. From the equations $\partial J / \partial u_k^* = 0$, $k = 1, 2, \ldots, N$ the following system is obtained:

$$2(\ln u_k^* - \ln u_k) \frac{1}{u_k^*} - \sum_{j=1}^{M} \lambda_j A_{jk} = 0, \quad k = 1, 2, \ldots, N. \tag{9}$$

To write (9) in a matrix form, the following definition is introduced:

Definition 1. *Let* $\mathbf{x}$ *and* $\mathbf{y}$ *be two* $N \times 1$ *column-vectors:* $\mathbf{x} = [x_1, x_2, \ldots, x_N]^T$, $\mathbf{y} = [y_1, y_2, \ldots, y_N]^T$. *The no-sign product between the two column-vectors is defined as* $\mathbf{x}\mathbf{y} = [x_1 y_1, x_2 y_2, \ldots, x_N y_N]^T$. *The exponent of a column-vector is defined as* $\exp(\mathbf{x}) = [e^{x_1}, e^{x_2}, \ldots, e^{x_N}]^T$.

Note that if $\mathbf{H}$ is an $N \times N$ matrix, then $\mathbf{H}.(\mathbf{x}\mathbf{y}) = (\mathbf{H}.\mathbf{x})\mathbf{y}$. Now the system (9) can be written as

$$\mathbf{u}^* = \mathbf{u} \exp \left(\frac{1}{2} \mathbf{u}^* (\mathbf{A}^T . \boldsymbol{\lambda}) \right), \tag{10}$$

where $\mathbf{u}^*$ and $\mathbf{u} = [u_1, u_2, \ldots, u_N]^T$ are $N \times 1$ column-vectors, $\boldsymbol{\lambda} = [\lambda_1, \lambda_2, \ldots, \lambda_M]^T$ is an $M \times 1$ column-vector, while $\mathbf{A}^T$ is the transposed matrix $\mathbf{A}$. To obtain $\mathbf{u}^*$ Eq. (10) and the equation for the constraints (6) should be solved together. The two equations constitute a system of $N + M$ equations for the $N + M$ unknowns u_i^*, $i = 1, 2, \ldots, N$ and λ_j, $j = 1, 2, \ldots, M$.

That $\mathbf{u}^*$ satisfies (6) means that $\mathbf{u}^*$ satisfies the given constraints. That $\mathbf{u}^*$ satisfies (10) means that $\mathbf{u}^*$ minimizes (7), hence u^* is the closest function to u in the relative sense described above. Note, that Eq. (10) guarantees non-negativity of the function u^* for any non-negative original function u.

5   Self-consistent Iterative Procedure

The column-vector $\mathbf{u}^*$ from (10) is substituted into the equation for the constraints (6) to get:

$$\mathbf{A}.\mathbf{u}\exp\left(\frac{1}{2}\mathbf{u}^*(\mathbf{A}^T.\boldsymbol{\lambda})\right) = \mathbf{c}. \tag{11}$$

To obtain the unknown $\mathbf{u}^*$ and $\boldsymbol{\lambda}$ the two equations (10) and (11) should be solved simultaneously. We propose the following method for self-consistent solution of (10) and (11). First, choose a starting guess for $\boldsymbol{\lambda}$ and solve (10) iteratively using the fixed-point (simple iteration) method to obtain the first approximation for $\mathbf{u}^*$. It is convenient to use fixed-point iteration since $\mathbf{u}^*$ in (10) is expressed as a function of $\mathbf{u}^*$. Substitute the obtained $\mathbf{u}^*$ into Eq. (11) and solve the equation iteratively, using the Newton's method, to obtain the first approximation for $\boldsymbol{\lambda}$. Substitute this $\boldsymbol{\lambda}$ into (10) and solve (10) again to obtain the second $\mathbf{u}^*$ approximation. The process is repeated until convergence. If $\boldsymbol{\lambda} = [0, 0, \ldots, 0]^T$ is chosen as a starting guess, then the first approximation for $\mathbf{u}^*$ will be just $\mathbf{u}$.

In order to solve Eq. (11) for $\boldsymbol{\lambda}$ by the Newton's method we transfer the left-hand side term to the right and introduce the $M \times 1$ column-vector $\mathbf{e}(\boldsymbol{\lambda})$ with components:

$$e_j = c_j - \sum_{i=1}^{N} A_{ji} u_i \exp\left(\frac{1}{2}u_i^*(\mathbf{A}^T.\boldsymbol{\lambda})_i\right), \quad j = 1, 2, \ldots, M. \tag{12}$$

Solving Eq. (11) is equivalent to solving $\mathbf{e}(\boldsymbol{\lambda}) = 0$. First, an expression for the $M \times N$ Jacobian matrix $\partial\mathbf{e}(\boldsymbol{\lambda})/\partial\boldsymbol{\lambda}$ is obtained:

$$\frac{\partial e_j}{\partial \lambda_m} = -\frac{1}{2}\sum_{i=1}^{N} A_{ji} u_i \exp\left(\frac{1}{2}u_i^*(\mathbf{A}^T.\boldsymbol{\lambda})_i\right) u_i^* A_{mi}, \quad j, m = 1, 2, \ldots, M. \tag{13}$$

Then, for any given approximation $\mathbf{u}^*$, the equation $\mathbf{e}(\boldsymbol{\lambda}) = 0$ is solved iteratively, obtaining each new value for $\boldsymbol{\lambda}$ from the previous value using the Newton's formula:

$$\boldsymbol{\lambda}_{new} = \boldsymbol{\lambda} - [\partial\mathbf{e}(\boldsymbol{\lambda})/\partial\boldsymbol{\lambda}]^{-1}.\mathbf{e}(\boldsymbol{\lambda}). \tag{14}$$

In (14) $[\partial\mathbf{e}(\boldsymbol{\lambda})/\partial\boldsymbol{\lambda}]^{-1}$ is the inverse of the Jacobian matrix.

6   Numerical Examples and Discussion

In this section two particular problems are solved by using both metrics: the L^2-norm of the absolute difference and the L^2-norm of the log ratio. The presented examples demonstrate the usefulness of the proposed metric.

Example 1. Consider the mesh function $\{u_i = 30t_i^2 \exp(-4t_i), \quad i = 1, 2, \ldots, N\}$ defined on the uniform mesh (5) with $a = 0$, $b = 3$, and $N = 41$. This function could, for example, represent the change of the concentration of an enzyme over time.

Find the non-negative mesh function u^* that is closest to the function u in the relative sense described above and satisfies the following two linear constraints:

$$\sum_{i=1}^{N} u_i^* = 10.62, \quad \sum_{i=1}^{N} \exp(t_i) u_i^* = 25.63. \tag{15}$$

Compare the results with the mesh function ua that is closest to u in an absolute sense and satisfies the same constraints.

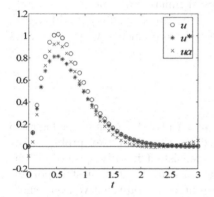

Fig. 3. Solution to Example 1

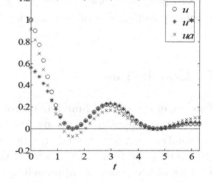

Fig. 4. Solution to Example 2

We solve (10) together with constraints (15) by the method described in Sect. 5. Starting from $\boldsymbol{\lambda} = [0, 0]^T$ and the corresponding column-vector $\mathbf{u}$, the solution $\boldsymbol{\lambda} = [-0.3851, -0.0899]^T$ and the corresponding vector $\mathbf{u}^*$ are reached in 6 steps within precision $|\delta\boldsymbol{\lambda}| < 10^{-6}$ Fig. 3 shows the original function u, the relatively closest to it function u^*, and the absolutely closest to it function ua. The function ua was found by minimizing the L^2-norm of the absolute difference. The figure shows that ua starts from a negative value and is also negative in the interval $2.025 \le t \le 2.4$. The function increases toward the end of the interval, in contrast to the behavior of the original function u. The results show that the values of ua from about $t = 1.8$ on are inconsistent with what is expected from the non-negative, exponentially decreasing (in this part of the domain) function u. On the contrary, the function u^* is non-negative everywhere and the overall behavior of u^* agrees very well with that of the original function u.

Example 2. Consider the mesh function $\{u_i = \cos^2(t_i)\exp(-t_i/2),\quad i = 1, 2, \ldots, N\}$ defined on the uniform mesh [5] with $a = 0$, $b = 2\pi$, and $N = 41$. This could, for example, be the kinetic energy of an oscillating body in a viscous media as a function of time.

Find the non-negative mesh function u^* that is closest to the original function u in the relative sense described above, and satisfies the following two linear constraints:

$$\sum_{i=1}^{N} u_i^* = 0.8 \sum_{i=1}^{N} u_i, \quad \sum_{i=1}^{N}(t_i^2 - 1)u_i^* = 1.2 \sum_{i=1}^{N}(t_i^2 - 1)u_i. \tag{16}$$

Compare the results with the mesh function ua that is closest to u in an absolute sense and satisfies the same constraints.

Equation (10) along with constraints (16) is solved by the method described in Sect. 5. Starting from $\boldsymbol{\lambda} = [0, 0]^T$ and the corresponding column-vector $\mathbf{u}$, the solution $\boldsymbol{\lambda} = [-1.7530, 0.2899]^T$ and the corresponding vector $\mathbf{u}^*$ are reached in 14 steps within precision $|\delta\boldsymbol{\lambda}| < 10^{-6}$.

The original function u, the relatively closest function u^*, and the absolutely closest function ua, found by minimizing the L^2-norm of the absolute difference, are shown in Fig. 4. Again, u^* meets all imposed requirements, whereas ua does not.

7 Conclusion

This paper considered function approximation with the L^2-norm of the log ratio as a metric. General equations were derived for the discretized problem. For the solution of the obtained equations a self-consistent iterative procedure was proposed. Examples were presented that demonstrate the two most important features of the proposed approach: (i) the result u^* is non-negative everywhere and (ii) to the extent to which the constraints allow, the magnitude of the absolute difference between u^* and u is proportional to the magnitude of u.

References

1. Achiezer, N.I.: Theory of Approximation (translated by C. J. Hyman). Ungar, New York (1956)
2. Timan, A.F.: Theory of Approximation of Functions of a Real Variable. Pergamon (Translated from Russian) (1963)
3. Korneichuk, N.P., Ligun, A.A., Doronin, V.G.: Approximation with Constraints, Kiev (1982). (in Russian)
4. Milovanović, G.V., Wrigge, S.: Least squares approximation with constraints. Math. Comput. **46**(174), 551–565 (1986). AMS
5. Lorentz, G. G.: Approximation of Functions. AMS (2005)
6. Pedregal, P.: Optimization and Approximation. UNITEXT, vol. 108. Springer, Cham (2017). https://doi.org/10.1007/978-3-319-64843-9

7. Gauss, C.F.: Theory of the Combination of Observations Least Subject to Errors. Society for Industrial and Applied Mathematics (Translated from original 1820 manuscript by G.W. Stewart) (1995)
8. Lawson, C.L., Hanson, R.J.: Solving Least Squares Problems. SIAM, Philadelphia (1995)
9. Bjorck, A.: Numerical Methods for Least Squares Problems. Society for Industrial and Applied Mathematics, Philadelphia (1996)
10. Bertsekas, D.P.: Constrained Optimization and Lagrange Multiplier Methods. Massachusetts Institute of Technology, Athena Scientific, Belmont (1996)
11. Kuhn, H. W., Tucker, A. W.: Nonlinear programming. In: Proceedings of 2nd Berkeley Symposium, Berkeley, pp. 481–492, MR 0047303. University of California Press (1951)
12. Karush, W.: Minima of Functions of Several Variables with Inequalities as Side Constraints. M.Sc. Dissertation, Department of Mathematics, University of Chicago, Chicago, Illinois (1939)
13. Luenberger, D.G., Ye, Y.: Penalty and Barrier Methods. In: Linear and Nonlinear Programming. International Series in Operations Research & Management Science, vol. 228. Springer, Cham (2016). https://doi.org/10.1007/978-3-319-18842-3_13
14. Freund, R.M.: Penalty and barrier methods for constrained optimization. Lecture Notes, Massachusetts Institute of Technology (2004)
15. Fukushima, M.: Solving inequality constrained optimization problems by differential homotopy continuation methods. J. Math. Anal. Appl. **133**(1), 109–121 (1988)
16. Li, J., Yang, Z.: A QP-free algorithm without a penalty function or a filter for nonlinear general-constrained optimization. Appl. Math. Comp. **316**, 52–72 (2018)
17. Strekalovsky, A. S., Minarchenko, I. M.: A local search method for optimization problem with d.c. inequality constraints. Appl. Math. Model (2017). https://doi.org/10.1016/j.apm.2017.07.031
18. Graff, C.: Expressing relative differences (in percent) by the difference of natural logarithms. J. Math. Psych. **60**, 82–85 (2014)
19. Graff, C.: Why estimating relative differences by Ln(A/B) in percentage and why naming it geometric difference. In: Conference: ICPS, Amsterdam, Netherlands (2015). https://hal.archives-ouvertes.fr/hal-01480972/document
20. Filipov, S.M., Atanassov, A., Gospodinov, I.D.: Constrained functional similarity by minimizing the H^1 seminorm and applications to engineering problems. J. Sci. Eng. Educ. **1**(1), 61–67 (2016). University of Chemical Technology and Metallurgy. http://dl.uctm.edu/see/node/jsee2016-1/12-Filipov_61-67.pdf
21. Filipov, S.M., Atanasov, A.V., Gospodinov, I.D.: Constrained relative functional similarity by minimizing the H1 semi-norm of the logarithmic difference. In: International Conference Automatics 2016, Proceedings of Technical University of Sofia, vol. 66, no. 2, pp. 349–358. Technical University of Sofia (2016). http://proceedings.tu-sofia.bg/volumes/Proceedings_volume_66_book_2_2016.pdf
22. Gospodinov, I.G., Krumov, K., Filipov, S.M.: Laplacian preserving transformation of surfaces and application to boundary value problems for Laplace's and Poisson's equations. Math. Model. **1**(1), 14–17 (2017). STUME. https://stumejournals.com/journals/mm/2017/1/14

23. Filipov, S.M., Gospodinov, I.D., Faragó, I.: Shooting-projection method for two-point boundary value problems. Appl. Math. Lett. **72**, 10–15 (2017). https://doi.org/10.1016/j.aml.2017.04.002

24. Filipov, S.M., Gospodinov, I.D., Angelova, J.: Solving two-point boundary value problems for integro-differential equations using the simple shooting-projection method. In: Dimov, I., Fidanova, S., Lirkov, I. (eds.) NMA 2014. LNCS, vol. 8962, pp. 169–177. Springer, Cham (2015). https://doi.org/10.1007/978-3-319-15585-2_19

Application of Parallel and Hybrid Metaheuristics for Graph Partitioning Problem

Zbigniew Kokosiński$^{(\boxtimes)}$ and Marcin Pijanowski

Faculty of Electrical and Computer Engineering,
Cracow University of Technology, ul. Warszawska 24, 31-155 Kraków, Poland
zk@pk.edu.pl

Abstract. In this paper parallel and hybrid metaheuristics for graph partitioning are compared taking into account their efficiency in terms of a cost function and computation time. Seventeen methods developed on the basis of evolutionary algorithm, simulated annealing and tabu search are implemented and tested against graph instances computed on the basis of queen graphs from DIMACS repository and a class of random R–MAT graphs. These graphs are supposed to model a class of digital circuits being subject of decomposition into a given number of modules. In partitioning process several additional constraints have to be satisfied in order to enable composition of original circuits from subcircuits by means of VLSI/FPGA modules.

Keywords: Graph partitioning · Circuit partitioning
Parallel metaheuristics · Hybrid metaheuristics
Approximate algorithms · DIMACS graphs · R–MAT graphs

1 Introduction

The graph partitioning problem for an undirected graph $G = (V, E)$ is a division of V into k pairwise disjoint subsets (partitions) such that all partitions are of approximately equal size and the edge–cut, i.e., the total number of edges having their incident nodes in different subdomains, is minimized.

The partitioning of a digital circuit arises when the technology used for its implementation imposes constraints related to the circuit size, available chip resources (macro cells, interconnections), I/O pins, clock distribution, energy dissipation etc. In real design problems a straightforward approach is to decompose the original circuit into a number of sub–circuits (partition blocks) satisfying certain design requirements [4,9,12,15]. Usually the expected number of blocks is known and the number of interconnections between blocks have to be minimized. In some cases the implementation of a circuit in many heterogenious FPGA can cause problems not only with splitting the original circuit but also with partitioning and modifying the existing test benchmarks [2,7].

© Springer Nature Switzerland AG 2019
G. Nikolov et al. (Eds.): NMA 2018, LNCS 11189, pp. 125–132, 2019.
https://doi.org/10.1007/978-3-030-10692-8_14

Digital circuits are usually modeled by graphs and many design problems can be performed by methods developed in graph theory. However, the task of circuit partitioning falls into category of graph clustering/partitioning problems which are known to be intractable, i.e. they belong to the class of NP-hard problems. Therefore, the search for an optimal solution is not efficient and heuristics/metaheuristics providing an approximate solution are to be applied [10,11].

Graph partitioning and clustering was the topic of the 10th DIMACS Implementation Challenge [8] and included both theoretical and real world problems. The Chalange goals were: identifying a standard set of benchmark instances and generators, establishing the most appropriate problem formulations and objective functions for a variety of applications, comparison of present methods in hopes of identifying the most effective algorithmic innovations that have been proposed.

In the article seventeen approximate methods developed on the basis of evolutionary algorithm (EA), simulated annealing (SA) and tabu search (TS) are investigated. Independently, in [4], a mixed SA and TS algorithm was successfully applied for topological partitioning in a parallel test-pattern generator.

The resulting computer application was used for testing these metaheuristics against a set of constructed problem instances and compared taking into account their efficiency in terms of cost function and computation time. Realistic modeling of benchmarks was the key issue in the conducted research. For the first time a family of modified DIMACS *queen* graphs [6] as well as recursively defined random R–MAT graphs [5] with various parameters were applied for testing. These graphs are supposed to adequately model a class of digital circuits being subject of decomposition into a given number of modules. Approximate solving of graph partitioning problem with a realistic cost function being minimized enables efficient decomposition of an original digital circuit by means of VLSI/FPGA modules.

2   Graph Partitioning

Graph Partitioning Problem (GPP) relies on clustering of graph $G(V, E)$ vertices into partition blocks (clusters). Let us intruduce the notation used throughout the paper.

A partition $C = \{C_1, C_2, \ldots, C_k\}$ of V is called clustering of G and C_i, $1 \leq i \leq k$, are called clusters. C is called trivial if either $k = 1$, or all clusters C_i contain only one element. We will indentify cluster C_i with the induced subgraph G_i of G, i.e. $G_i = (C_i, E(C_i))$, where $E(C_i) = \{\{u, v\} \in E : u, v \in C_i\}$. Hence, $E(C) = \sum_{i=1}^{k} E(C_i)$ is the set of intra–cluster edges, and $E \backslash E(C)$—the set of inter–cluster edges. The graph density is denoted by $d(G) = 2|E|/|V|(|V| - 1)$.

The purpose of the optimization problem is to find a clustering of G into k clusters providing that the edge–cut $ext = |E \backslash E(C)|$ is minimal. Sometime there are additional constraints on the maximum number of inter–cluster edges E_i coming from a single cluster C_i. In some cases clustering is expected to be equitable in the sense of either the number of $|V_i|$ or $|E_i|$.

Constructing a k–clustering with a fixed number of k, $k \geq 3$, is NP–hard [3].

2.1   Definition of the Cost Function

According to the design requirements solution quality is defined by the following cost function (subject of minimization):

$$f = a \cdot ext + b \cdot bn + c \cdot |bnp|^3 + d \cdot bsp^2 + e \cdot pp^2, \qquad (1)$$

where:

a, b, c, d – are integer coefficients;
ext – is the sum if inter–cluster edges (the edge–cut);
bn – is the assumed number of clusters called *block number*;
bnp – is the difference between bn and the actual number of blocks called *block number penalty*;
bsp – is the difference between the assumed block size bs and the actual block size called *block size penalty*;
pp – is the difference between the assumed number of block' I/O pins pn and the actual number of block' I/O pins called *pin penalty*.

Exponents of bp and pp were determined experimentally and reflect the relative importance of the corresponding terms of cost function f. In any solution satisfying the design assumptions bp, bsp and pp should be zeroed.

Terms of the cost function f provide the designer additional informations about solution quality. When design constraints are not fully satisfied this information helps the designer to choose a right way to complete the design process.

2.2   Test Instances

Graph coloring instances [6] were originally designed and collected in DIMACS repository for the purpose of testing and comparing graph coloring algorithms. In graph coloring problem (GCP) partition blocks are assumed to be independent sets (ISs). For most DIMACS graphs their chromatic numbers are already known. In the complementary problem, i.e. partitioning into cliques (PIC), partition blocks are cliques. In both problems the number of partition blocks k is minimized. For the given graph $G(V, E)$ and its complementary graph $G'(V, E) = G(V, E')$ solutions for GCP and PIC, respectively, are equivalent with minimum $\chi(G) = k$.

PIC problem resambles GPP (circuit partitioning) in which intra–connection density in any partition block is maximal. The PIC problem instance G' is obtained from the instance G of GCP. Under assuption $k \in \{\chi(G) - 1, \chi(G) - 2\}$, the partition into k cliques is not existing, making the corresponding GPP even harder to solve.

The queen graph G_q of size $n \times n$ has the squares of two-dimentional chessboard for its vertices and two such vertices are adjacent if, and only if, queens placed on the two squares attack each other. For interesting introduction to queen graph coloring one may refere to Chvátal's article [3]. This class of graphs was chosen for generation of our test instances.

Basic DIMACS graphs selected for construction of test instances were queen graphs [6]:

1. *queen6.6*, $|V| = 36$, $|E| = 290$, $\chi(G) = 7$, $d(G) = 0,460$;
2. *queen7.7*, $|V| = 49$, $|E| = 475$, $\chi(G) = 7$, $d(G) = 0,404$;
3. *queen8.8*, $|V| = 64$, $|E| = 728$, $\chi(G) = 9$, $d(G) = 0,361$;
4. *queen9.9*, $|V| = 81$, $|E| = 1056$, $\chi(G) = 9$, $d(G) = 0,326$;
5. *queen10.10*, $|V| = 100$, $|E| = 1470$, $\chi(G) = 11$, $d(G) = 0,297$.

The actual test instances are the corresponding complementary graphs G', that have the following characteristics:

Q1: *queen6.6′*, $|V| = 36$, $|E'| = 340$, $d(G) = 0,640$, $k = 5$, $bs = 8$, $pn = 110$;
Q2: *queen7.7′*, $|V| = 49$, $|E'| = 701$, $d(G) = 0,596$, $k = 6$, $bs = 9$; $pn = 200$;
Q3: *queen8.8′*, $|V| = 64$, $|E'| = 1288$, $d(G) = 0,639$, $k = 7$, $bs = 10$; $pn = 350$;
Q4: *queen9.9′*, $|V| = 81$, $|E'| = 2184$, $d(G) = 0,674$, $k = 8$, $bs = 11$; $pn = 520$;
Q5: *queen10.10′*, $|V| = 100$, $|E'| = 3480$, $d(G) = 0,703$, $k = 9$, $bs = 12$; $pn = 800$.

In addition a number of random R–MAT graphs was used in the experimental part of the paper [5]. R–MAT graphs with $|V| = 2^b$ are generated recursively with the required density in ($b - 1$ steps. Initially, adjacency matrix is zeroed. Then the generation algorithm determines the position of a consecutive "1" by random choosing the input matrix (submatrix) quater (NW, NE, SW, SE) with given quater probabilities a, b, c and d, all greater then 0, which sum $a + b + c + d = 1$. Depending of the distribution of probalibilities a graph G with the corresponging distribution of vertices with a given degree is generated [5].

The set of R–MAT graphs G has the following characteristics:

R1: *rmat50_35*, $|V| = 50$, $|E| = 429$, $d(G) = 0,350$, $k = 5$, $bs = 10$, $pn = 150$;
R2: *rmat50_40*, $|V| = 50$, $|E| = 490$, $d(G) = 0,400$, $k = 5$, $bs = 10$, $pn = 250$;
R3: *rmat50_65*, $|V| = 50$, $|E| = 796$, $d(G) = 0,650$, $k = 5$, $bs = 10$, $pn = 270$;
R4: *rmat50_70*, $|V| = 50$, $|E| = 858$, $d(G) = 0,700$, $k = 5$, $bs = 10$, $pn = 300$;
R5: *rmat75_84*, $|V| = 75$, $|E| = 2331$, $d(G) = 0,840$, $k = 5$, $bs = 15$, $pn = 800$;
R6: *rmat100_90*, $|V| = 100$, $|E| = 4455$, $d(G) = 0,900$, $k = 5$, $bs = 20$, $pn = 1500$.

2.3 Metaheuristics

The basis heuristics, their combinations and parallel/hybrid versions established a testbed for experimental part of our research [1,13]. The tested algorithms are:

sSA—sequential Simulated Annealing (SA),
mirSA—parallel SA with Multiple Independent Runs (MIR),
aSA—asynchronous SA,
sTS—sequential Tabu Search (TS),
mirTS—parallel TS with MIR,

aTS—asynchronous TS,
sEA—sequential Evolutionary Algorithm (EA),
mirEA—parallel EA with MIR,
iEA—island EA without migration,
ibmEA—island EA with migration of best individuals,
irmEA—island EA with migration of random individuals,
sSA-TS—sequential SA-TS,
mirSA-TS—parallel SA-TS with MIR,
aSA-TS—asynchronous SA-TS,
aEA-SA—asynchronous EA-SA,
aEA-TS—asynchronous EA-TS,
aEA-SA-TS—asynchronous EA-SA-TS.

Pseudocodes of the above algorithms are available from WWW site [16].

Initial values of basic parameters used in the above algorithms are the following: Number of iterations in a single step = 15, Stop criterion = 15, Initial temperature (SA) = 10, Size of the tabu list (TS) = 7, Number of candidates (TS) = 8, Population size (EA) = 30, Offspring number (EA) = 5, Crossover probability (EA) = 0,8, Mutation probability (EA) = 0,1. The assumed cost function f coefficients are: a = 1, b = 1, c = 5, d = 5, e = 5.

Parameters of parallel algorithms are: Number of processors (mir) = 6, Communication rate = 10, Number of islands (iEA, ibmEA, irmEA) = 6, Migration size (ibmEA, irmEA) = 18, Migration rate (ibmEA, irmEA) = 10.

3   Computational Experiments

The main purpose of the experimental part is graph $G(V, E)$ partitioning satysfying design assumptions related to the number of blocks (clusters) and simultaneously minimizing the number of interconnections between partition blocks.

The primary objective is to minimize the cost function f. The secondary objective is to minimize the computation time. In parallel algorithms computation times of parallel processors are added up (parallel execution of the algorithm is simulated).

In the first experiment we are searching for minimal values of f and min–cut ext. The esssential results for the four queen graphs are reported in Table 1. The best results of f and ext, the methods winning for at least one graph and the corresponding computation times are shown in the bold font. The iteration details and terms of f are not reported due to lack of space.

For Q1 graph many methods produce equivalent results, but the best computation time is obtained by sSA-TS. Relatively low running times are required for mirSA-TS and mirTS. For Q3 graph iEA provides the best f while mirEA finds a solution the best ext. However, mirEA uses only 42,5 % of iEA computation time. Many other parallel and hybrid algorithms can find quite satisfying solutions in a shorter time. For Q4 graph mirSA-TS outperforms EA-based methods in terms of computation time, providing optimal f and suboptimal ext. Similarly, the best solution quality for Q5 graph provide iEA and ibmEA but their

Table 1. Computational test results for graph instances Q1, Q3, Q4 and Q5.

Graph	Q1			Q3			Q4			Q5		
Algorithm	f	ext	time [s]	f	ext	time [s]	f	ext	time [s]	f	ext	time [s]
sSA	286	254	8,12	1168	1093	4,81	1993	1898	9,57	3200	3079	5,07
mirSA	290	255	2,09	1160	1089	12,07	1932	1867	67,04	3140	4412	92,56
aSA	288	255	4,07	1172	1094	8,15	2000	1902	9,96	3211	3083	15,40
sTS	262	242	10,43	1096	1058	14,75	1902	1852	43,00	3074	3015	43,71
mirTS	**261**	**242**	**24,32**	1094	1056	73,25	1888	1845	204,4	3058	3009	327,3
aTS	263	243	13,10	1091	1054	58,67	1906	1851	193,5	3082	2987	161,4
sEA	**261**	**242**	**244,4**	1095	1050	595,5	1887	**1836**	**629,5**	3085	3011	1414
mirEA	**261**	**242**	**93,23**	1080	**1048**	**263,9**	1887	**1836**	**626,8**	3063	3000	1364
iEA	**261**	**242**	**95,79**	**1079**	1049	**620,8**	1881	1840	1229	**3056**	**2974**	**2811**
ibmEA	**261**	**242**	**102,9**	1092	1054	525,3	1889	1837	861,3	**3056**	2982	**2043**
irmEA	264	243	98,44	1082	1049	597,1	1895	1840	731,9	3060	2976	1355
sSA-TS	**261**	**242**	**2,57**	1093	1056	15,59	1886	1844	15,81	3082	2995	40,57
mirSA-TS	**261**	**242**	**15,45**	1088	1051	46,54	**1880**	1838	**94,40**	3070	3007	242,3
aSA-TS	**261**	**242**	**20,09**	1088	1051	53,43	1885	1842	142,6	3078	3016	244,4
aEA-SA	**261**	**242**	**31,45**	1086	1051	126,8	1903	1844	113,3	3073	3005	284,3
aEA-TS	263	243	34,70	1083	1050	233,9	1903	1844	274,3	3061	2992	592,2
aEA-SA-TS	**261**	**242**	**82,78**	1097	1058	136,9	1887	1843	465,1	3087	3020	122,0

Table 2. Computational test results for graph instances R3, R4, R5 and R6.

Graph	R3			R4			R5			R6		
Algorithm	f	ext	time [s]	f	ext	time [s]	f	ext	time [s]	f	ext	time [s]
sSA	715	640	2,15	740	685	1,12	1924	1862	1,34	3681	3681	3,40
mirSA	703	643	19,5	740	685	4,42	1932	1866	13,40	3679	3612	24,23
aSA	711	638	7,50	748	686	4,04	1938	1872	12,50	3681	3616	20,76
sTS	679	622	6,93	708	669	11,04	1842	1824	13,12	3597	3574	33,75
mirTS	**677**	**621**	**85,57**	**704**	**667**	**67,78**	1840	1823	107,1	3587	3569	149,2
aTS	683	624	48,12	708	669	35,79	1844	1825	92,31	3583	3567	184,5
sEA	681	623	54,50	714	672	73,26	1854	1830	139,8	3595	3573	263,6
mirEA	681	623	346,5	710	670	406,7	1840	1823	1338	3597	3574	2755
iEA	679	622	380,3	708	669	276,8	1846	1826	1307	3745	2585	2274
ibmEA	683	624	320,8	712	671	202,5	1842	1824	1005	3579	3579	3132
irmEA	**677**	**621**	**463,4**	710	670	206,7	1854	1830	889,3	3589	3570	2028
sSA-TS	681	623	6,28	706	668	7,47	**1836**	**1821**	**38,73**	3591	3571	29,75
mirSA-TS	679	622	41,20	**704**	**667**	**38,76**	1840	1823	83,48	**3573**	**3562**	**257,6**
aSA-TS	**677**	**621**	**51,79**	706	668	47,48	1838	1822	97,09	3579	3565	259,1
aEA-SA	681	623	103,3	714	672	60,81	1840	1823	282,3	3601	3576	458,9
aEA-TS	681	623	151,8	708	669	115,8	1842	1824	324,8	3591	3571	265,9
aEA-SA-TS	683	624	91,68	706	668	186,3	1846	1826	255,3	3593	3572	626,4

computation times are hardly acceptable. Good alternatives with a significantly shorter computation times are mirTS and aEA-TS. Other results are more distant from the best solution found.

In the second experiment, devoted to R–MAT graphs, we are also searching for minimal values of f and min–cut ext. The esssential results for four input graphs R3, R4, R5 and R6 are reported in Table 2. The best results of f and ext, the methods winning for at least one graph and the corresponding computation times are shown in the bold font.

For relatively easy R3 graph instance three methods find best values of f and ext: aSA-TS, mirTS and irmEA. They are listed in the increasing order of computation times. For R4 graph two methods: mirTS and mirSA-TS produce equivalent results in a reasonable time, but mirSA-TS is almost two times faster. The worst solution is found by aSA. For R5 graph the single winner is sSA-TS with acceptable computation time, a good alternative is sTS which is three times faster then sSA-TS still ensuring competitive resuls. Other efficient methods: aSA-TS, mirSA-TS, aEA-SA and aEA-TS require a longer time. The best solution quality for R6 graph ensures mirSA-TS within acceptable computing time. Other methods are worse in terms of f and ext.

Computer application *Electronic Circuit Decomposition* used for the presented research was written in C++/CLI within Visual C++ 2008, Express Edition environment. GUI was made in Windows Forms Application. For program execution .NET Framework 3.5 package is needed, supplied by ZedGraph library.

4   Conclusions

From the reported research one can conclude that for graph partitioning problem with modified queen and R–MAT graph instances the most valuable components for hybrid algorithms are TS and EA, which can be combined within one algorithm. This confirms earlier results on different set of benchmarks reported in [7]. EA usually involves longer computation time. The essential part of parallel algorithms is Multiple Independent Runs (MIR) model except mirSA algorithm.

In general, the differences in computation time by various methods can be extremal, while the quality of all graph partitioning methods was good, usually not exceeding several percent of the values f and ext of the best solution. The research may be continued with focus on the outstanding mixed methods as well as larger and harder graph partitioning instances.

It would be also desirable to apply the best methods to circuit benchmarks resulting from engineering practice and verify their efficiency on real design problems [14].

Acknowledgements. This work was supported by the research grant No. E–3/611/2017/DS from Cracow University Technology.

References

1. Alba, E. (ed.): Parallel Metaheuristics: A new Class of Algorithms. Wiley-Interscience, New Jersey (2005)
2. Chadha, A.: Benchmark Creation for Circuit Partitioning Algorithms. Gauging the Performance of Circuit Partitioning Algorithms. Lambert Academic Publishing (2015)
3. Chvátal, V.: Colouring the queen graphs. http://users.encs.concordia.ca/chvatal/queengraphs.html
4. Bhuvaneswari, M.C., Jagadeeswari, M.: Circuit partitioning for VLSI layout. In: Bhuvaneswari, M.C. (ed.) Application of Evolutionary Algorithms for Multi-objective Optimization in VLSI and Embedded Systems, pp. 37–46. Springer, New Delhi (2015). https://doi.org/10.1007/978-81-322-1958-3_3
5. Chakrabarti, D., Zhan, Y., Faloutsos, C.: R-MAT: a recursive model for graph mining. In: Proceedings of 2004 SIAM International Conference Data Mining, 2004 (2004). https://doi.org/10.1137/1.9781611972740.43
6. COLOR web site. http://mat.gsia.cmu.edu/COLOR/instances.html
7. Gil, C., Ortega, J., Montoya, M.G., Banos, R.: A Mixed Heuristic for Circuit Partitioning. Comput. Optim. Appl. **23**, 321–340 (2002). https://doi.org/10.1023/A:1020551011615
8. Bader, A., et al. (ed.): Graph Partitioning and Graph Clustering, 10th DIMACS Implementation Challenge, Atlanta, 12–13 February 2012, vol. 588. AMS, Contemporary Mathematics (2013)
9. Khang, A., Liening, J., Markov, I., Hu, J.: VLSI Physical Design: From Graph Partitioning to Timing Closure, pp. 33–54. Springer, Dordrecht (2011). https://doi.org/10.1007/978-90-481-9591-6
10. Kernighan, B.W., Lin, S.: An efficient heuristics procedure for partitioning graphs. Bell Syst. Tech. J. **49**, 291–307 (1970)
11. Kokosiński, Z., Bała, M.: Solving graph partitioning problems with parallel meta-heuristics. In: Fidanova, S. (ed.) Recent Advances in Computational Optimization. SCI, vol. 717, pp. 89–105. Springer, Cham (2018). https://doi.org/10.1007/978-3-319-59861-1_6
12. Kozieł, S., Szczęśniak, W.: Evolutionary algorithm for electronic system partitioning and its applications in VLSI design. In: Proceedings of 6th IEEE International Conference on Electronics, Circuits and Systems, ICECS 1999, Pafos, Cyprus, 5–8 September 1999, vol. 3, pp. 1412–1414 (1999)
13. Sadiq, S., Habib, Y.: Iterative Computer Algorithms with Applications in Engineering. Solving Combinatorial Optimization Problems. Wiley - IEEE Computer Society Press (2000)
14. Swethaa, R.R., Devi, K.A.S., Yousef, S.: Hybrid partitioning algorithm for area minimization in circuits. Procedia Comput. Sci. **48**, 692–698 (2015). https://doi.org/10.1016/j.procs.2015.04.203
15. Szczęśniak, W.: Application of adaptive circuit partitioning algorithms to reductions of interconnections length between elements of VLSI circuit. In: Proceedings of 9th IEEE International Conference on Electronics, Circuits and Systems, ICECS 2002, Dubrownik, Croatia, 15–18 September 2002. https://doi.org/10.1109/ICECS.2002.1046259
16. Pseudocodes of algorithms from section 2.3. http://www.pk.edu.pl/~zk/pubs/NMA18.zip

Monte Carlo Approach for Modeling and Optimization of One-Dimensional Bimetallic Nanostructures

Vladimir Myasnichenko[1]($\boxtimes$), Nickolay Sdobnyakov[1],
Leoneed Kirilov[2]($\boxtimes$), Rossen Mikhov[2], and Stefka Fidanova[2]

[1] Tver State University, Tver, Russia
viplabs@yandex.ru
[2] Institute of Information and Communication Technologies,
Bulgarian Academy of Sciences, Sofia, Bulgaria
l_kirilov_8@abv.bg

Abstract. In this paper we present a method for optimizing of metal nanoparticle structures. The core of the method is a lattice Monte-Carlo method with different lattices combined with an approach from molecular dynamics. Interaction between atoms is calculated using multi-particle tight-binding potential of Gupta – Cleri&Rosato. The method allows solving of problems with periodic boundary conditions. It can be used for modeling of one-dimensional (nanowire, tube) and two-dimensional (nano-film) structures. If periodic boundary conditions are not given, we assume finite dimensions of the model lattice. In addition, automatic relaxation of the crystal lattice can be performed in order to minimize further the potential energy of the system. Both stretching and compressing of the lattice is permitted. A computer implementation of the method is developed. It allows easy and efficient operation. It uses the commonly accepted XYZ format for describing metal nanoparticles. The parameters of the method, such as number and type of metal atoms, temperature of the system, etc. are entered on a separate command line. The method is tested extensively on a large set of examples.

1 Introduction

Metal, including gold, nanowires (filamentary nanocrystals, nanofibers) is a rapidly expanding field of research. Gold nanowires can be used in transparent electrodes for flexible displays [1, 2]. A particularly important point related to electrodes is the stability of nanowires at a thermal load. The minimization of surface energy caused by thermally activated diffusion leads to the rupture of nanowires. This was observed for copper [3], for silver [4], for gold [5], and also for platinum [6].

The behavior of nanostructures at elevated temperatures can be very different from the macroscopic material. It is well known that small nanoparticles/nanowires will melt at a much lower temperature, which depends on their size [7].

A combination of simulation tools for thermodynamic properties and stability of nanosystems is proposed in [8]; mainly parallel Monte Carlo algorithms for icosahedral, multilayer Pd–Pt clusters. The model is on a 3D cubic lattice. In [9], the chemical

G. Nikolov et al. (Eds.): NMA 2018, LNCS 11189, pp. 133–141, 2019.
https://doi.org/10.1007/978-3-030-10692-8_15

ordering in "magic-number" Pd-Ir nanoalloys is studied. The density functional theory is compared with the results of the free energy concentration expansion method. In [10], the problem for stable structures of alloy nanoparticles is investigated. A two-step search strategy is proposed. The first strategy is based on extensive global optimization search and is combined with an empirical potential with density-functional local relaxation. The structure and thermodynamics of Cu-Ni nanoalloys is studied in [11]. An atomic model is described in the framework of a potential based on the second-moment approximation of the tight binding potential. In [12], a novel structure for free Co-Pt nanoalloys is developed. Three computational methodologies have been combined. The energetic stability of the novel structure has been checked.

The structure and energetics of Pd–Pt nanoalloys are studied in [13]. The model is based on the second-moment approximation to tight binding theory. To solve the problem the authors apply a genetic algorithm. Schebarchov and Wales [14] study the problem of structure predicting of multi-component systems. The system is represented as a generalized graph. They apply a Kernighan and Lin heuristic procedure to find locally optimal partitions of an arbitrary graph. The same problem is studied in [15]. A local optimization technique combined with multiple local-neighborhood search is applied. In [16] is proposed a parallel modification of the Birmingham cluster genetic algorithm for global optimization of nanoalloy clusters using a pool strategy. The method is illustrated for global optimization of the $Au_{10}Pd_{10}$ cluster using the Gupta potential. The structure of different AuCu clusters is studied in [17] by means of a Parallel Excitable Walkers algorithm and molecular dynamics.

Computer simulation by the molecular dynamics method has been widely used to study the structural defects and the melting temperature of nanowires and gold nanostructures [18, 19] as well as their elasticity and plasticity [20, 21]. The same problems have been studied by means of molecular statics method [22]. In these studies, the significant role of surface tension was determined. The kinetic Monte Carlo was used in [23] for modeling structural transitions and atomic diffusion in gold nanoparticles. In [24] is presented a reliable way to construct a rigid lattice barrier parameterization of face-centered and body-centered cubic metal lattices for the Kinetic Monte Carlo model. Three different barrier sets for Cu and one for Fe are produced that can be used for Kinetic Monte Carlo simulations.

2 The Monte Carlo Approach

The searching for stable configurations of metal and alloy nanostructures is not a trivial problem from computational point of view. It is equivalent to minimization of the Potential Energy Surface. This function has an extremely large number of local minima. Thus the question for developing efficient and effective methods arises.

The proposed method has four distinctive features. First, a lattice Monte Carlo method with different lattices is used. Second, automatic stretching/compressing of the lattice is done in order to find better optimal solution. Third, Periodic Boundary Conditions (PBCs) are implemented so that the method can be used for modeling of one-dimensional (nanowire, tube) and two-dimensional (nano-film) structures. If periodic boundary conditions are not given, we assume finite dimensions of the model lattice. Fourth, the resulting nanoparticle structures are relaxed at low temperature within molecular dynamics, choosing one of them as an approximation of the global minimum.

Interaction between atoms is calculated using the multi-particle tight-binding potential of Gupta – Cleri & Rosato [25, 26]. The total potential energy of the system is defined as follows:

$$E = \sum_i \left(\sum_{j \neq i} E_{ij}(a, b) - \sqrt{\sum_{j \neq i} B_{ij}(a, b)} \right) \tag{1}$$

$$E_{ij}(a, b) = A_{ab} \exp \left(-p_{ab} \left(\frac{r_{ij}}{r_{0,ab}} - 1 \right) \right) \tag{2}$$

$$B_{ij}(a, b) = \xi_{ab}^2 \exp \left(-2q_{ab} \left(\frac{r_{ij}}{r_{0,ab}} - 1 \right) \right), \tag{3}$$

where i ranges over all atoms; j ranges over all atoms other than i but within distance R_{cut} from i; a and b represent the species of the atoms i and j; $E_{ij}(a, b)$ is the repulsive component of the potential due to the atoms i and j; $B_{ij}(a, b)$ is the binding component of the potential due to the atoms i and j; r_{ij} is the distance between the atoms; $r_{0,ab}$, A_{ab}, p_{ab}, ξ_{ab}, q_{ab} are parameters that depend only on the species of the atoms. R_{cut} is the maximum distance beyond which the interaction is assumed to be zero.

There is no temperature effect increasing the potential energy because comparison of the resulting energy occurs after cooling by the Molecular Dynamics method down to 0,01 K.

According to our method, Periodic Boundary Conditions are defined as follows:

$$r_{ij} = \sqrt{|\Delta x_{ij}|^2 + |\Delta y_{ij}|^2 + |\Delta z_{ij}|^2}, \qquad |\Delta x_{ij}| = \min \{ |x_i - x_j|, |L_x - |x_i - x_j|| \}, \tag{4}$$

where x_i, x_j are the x-coordinates of the atoms i and j inside the periodic cell, and L_x is the size of the periodic cell along the x axis. If the y or z axis is also periodic, then $|\Delta y_{ij}|$, $|\Delta z_{ij}|$ are also computed in a similar way. The non-periodic case corresponds to $L_x = \infty$.

The complete algorithm is as follows (Fig. 1):

Step 1. Read the input data: initial positions of the atoms, the dimensions of the window for Periodic Boundary Conditions, other control parameters, etc. Atoms that do not have their initial positions given are placed at random.

Step 2. For all nodes, pre-compute the lists of neighbors, vicinities and other information that is known ahead of time.

Step 3. Check the exit criteria. The cycle stops when either the requested number of iterations is exceeded, or the system has reached equilibrium. If Yes, go to Step 11.

Step 4.   Adjust the temperature according to the following formula:

$$T = \max\ \{1, T_0 + s\Delta T\}, \tag{5}$$

where T_0 and ΔT are constants, and s is the iteration number. This check is performed once every several thousand iterations.

Step 5.   Choose an atom at random.

Step 6.   Choose a neighboring empty node at random. If there are no empty neighbors, return to Step 3.

Step 7.   Calculate the potential energy difference for the atom moving into the selected empty node, taking into account Periodic Boundary Conditions.

Step 8.   If the energy would not increase, perform the jump and return to Step 3.

Step 9.   Otherwise, calculate the jump probability $P = \exp(-\Delta E/kT)$ and generate a random number p $(0 \leq p < 1)$.

Step 10.   If the number is smaller than the probability, perform the jump, otherwise do nothing. Either way, return to Step 3.

Step 11.   Perform iteratively stretching and compressing of the lattice along each axis with step 0.01 (or using another value from an input parameter) to minimize the potential energy further.

3 Computer Implementation

We solve several important issues during its development resulting from the characteristics of the proposed method and from the nature of the problem solved.

We minimize the required computation during the main loop, since it has to be executed for millions of iterations. To this end, we have made extensive use of pre-computation and memorization, inspired by the approach discussed in [27].

A. For each node i, compute the list of nodes j within distance R_{cut} from it. As a sub-list, remember the list of immediate neighbors of i.

B. For each node i, each node j within distance R_{cut} from i, and each combination of a and b, use (2) and (3) to compute the values of $E_{ij}(a, b)$ and $B_{ij}(a, b)$. From this point on, we can forget about Cartesian coordinates and work exclusively with indexes and these pre-computed values using only (1).

C. For each atom i, compute $\sum_{j \neq i (r_{ij} \leq R_{cut})} B_{ij}(a, b)$. This value is remembered and kept updated throughout the algorithm.

These pre-computations allow each step of the main loop to run in constant time. For the energy difference calculation (Step 7), doing it in the naïve way according to (1) would require iterating over not only the R_{cut}-vicinities of the chosen atom and of the chosen empty node, but also over the R_{cut}-vicinities of their R_{cut}-vicinities (since the values under the square roots need to be modified). Pre-computation C allows us to calculate the correct energy by iterating over only the first vicinities.

The data structures are organized as follows. The nodes are kept in two arrays N and A, where N gives the index of a node into A and A gives the index of a node into N. N is sorted in input order, while A is sorted to begin with the atoms and end with the empty nodes. This allows us to query information about nodes and atoms, select atoms at random, add, remove and move atoms around, all in constant time.

Input is given as a list of nodes, specified by their Cartesian coordinates. Each node can either be empty or contain an atom. The distance between adjacent nodes in the input data may slightly vary (within 15%). The data format is given in the commonly accepted XYZ format for describing metal nanoparticles.

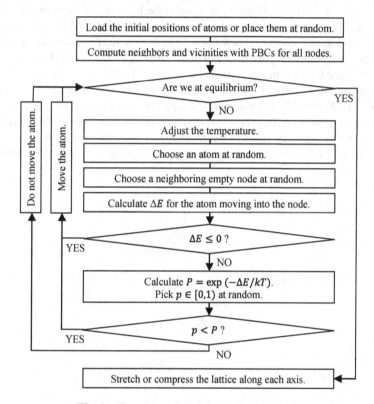

Fig. 1. Flowchart of the Monte Carlo method

Other aspects of the problem, such as the number and type of metal atoms, the periodic or non-periodic boundary conditions, the temperature of the system, etc., are specified as additional parameters. These can be given either on the command line, or on the second line ("comment" line) of the input XYZ file. The output file format is identical to the input file format, with the values for the energy of the system and all other parameters shown on the "comment" line. This allows multiple successive runs of the program on the same data. Output of intermediate results can also be requested.

The program is written in the C programming language and is tested on Windows and Linux platforms. Care has been taken to organize its core functions in a simple way in order to permit compiler optimizations.

Since the proposed method is non-deterministic, during testing and using the algorithm, several runs are performed and its behavior is stable.

4 Numerical Experiments

The method is demonstrated in the following example: fcc lattice with 6400 nodes of total size 22.55 × 2.10 × 2.31 nm. The X axis (the longest one, corresponding to the direction of [110]) has periodic boundary conditions. We model 1D structures of 1100, 3000 and 5000 atoms, cooling down from 2200 K to 1 K. Two variants of chemical composition are presented: pure gold and gold-silver in a ratio of 1:1. For comparison, 8 different Monte Carlo simulation modes were used: 1 million, 2 million, 4 million, 8 million, 16 million, 32 million, 64 million and 128 million iterations (Figs. 2, 3 and 4).

For size 1100 (the thinnest nanowire), a separation into individual nanoparticles of round shape is observed. This is a consequence of the Rayleigh-Plateau instability. Minimizing surface energy leads to the breakage of nanowires of both compositions. For other sizes, the layers/nodes along the Y and Z axes are not enough to obtain this effect.

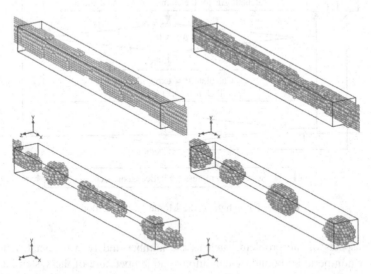

Fig. 2. Au3000 (top left), Au1500Ag1500 at T = 1 K (top right), Au550Ag550 at T = 1980 K (bottom left), Au550Ag550 at T = 1 K (bottom right).

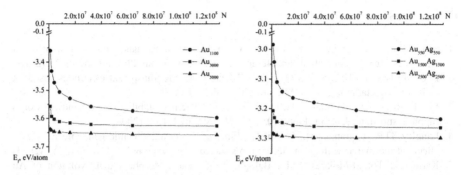

Fig. 3. Dependency of the final potential energy (per atom) on the number of Monte Carlo iterations in 1D gold nanowires (left) and in gold-silver bimetal nanowires (right).

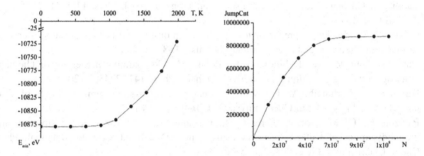

Fig. 4. An Au3000 cluster cooling in 128 million iterations: dependency of the minimum energy on the temperature (left) and dependency of the number of atom jumps on the iteration number (right).

5 Conclusion

A new method for modelling one-dimensional bimetallic nanostructures is proposed. It has four distinctive features: a lattice Monte Carlo method with different lattices is applied; automatic stretching/compressing of the lattice is done in order to find better optimal solution; Periodic Boundary Conditions (PBCs) are implemented so that the method can be used for modeling of 1D and 2D nanostructures; the resulting nanoparticle structures are relaxed at low temperature within molecular dynamics.

Acknowledgments. This research is supported by the Russian Foundation for Basic Research project No. 18-38-00571 mol_a and the Bulgarian NSF under the grant DFNI-DN 12/5.

References

1. Sannicolo, T., Lagrange, M., Cabos, A., et al.: Metallic nanowire-based transparent electrodes for next generation flexible devices: a review. Small **12**(44), 6052–6075 (2016)
2. Luo, M., Liu, Y., Huang, W., Qiao, W.: Towards flexible transparent electrodes based on carbon and metallic materials. Micromachines **8**(1), 12 (2017)
3. Li, H., Biser, J.M., Perkins, J.T., Dutta, S., et al.: Thermal stability of Cu nanowires on a sapphire substrate. J. Appl. Phys. **103**(2), 024315-1–024315-9 (2008)
4. Langley, D.P., Lagrange, M., Giusti, G., Jiménez, C., et al.: Metallic nanowire networks: effects of thermal annealing on electrical resistance. Nanoscale **6**(22), 13535–13543 (2014)
5. Karim, S., Toimil-Molares, M.E., Balogh, A.G., et al.: Morphological evolution of Au nanowires controlled by Rayleigh instability. Nanotechnology **17**(24), 5954–5959 (2006)
6. Rauber, M., Muench, F., Toimil-Molares, M.E., Ensinger, W.: Thermal stability of electrodeposited platinum nanowires and morphological transformations at elevated temperatures. Nanotechnology **23**(47), 475710 (2012)
7. Granberg, F., Parviainen, S., Djurabekova, F., Nordlund, K.: Investigation of the thermal stability of Cu nanowires using atomistic simulations. J. Appl. Phys. **115**(21), 213518-1–213518-5 (2014)
8. Calvo, F.: Solid-solution precursor to melting in onion-ring Pd-Pt nanoclusters: a case of second-order-like phase change? Faraday Discuss. **138**, 75–88 (2008)
9. Davis, J., Johnston, R., Rubinovich, L., Polak, M.: Comparative modelling of chemical ordering in palladium-iridium nanoalloys. J. Chem. Phys. **141**, 224307 (2014)
10. Ferrando, R., Fortunelli, A., Johnston, R.: Searching for the optimum structures of alloy nanoclusters. Phys. Chem. Chem. Phys. **10**, 640–649 (2008)
11. Panizon, E., Olmos-Asar, J., Peressi, M., Ferrando, R.: The study of the structure and thermodynamics of CuNi nanoalloys using a new DFT-fitted atomistic potential. Phys. Chem. Chem. Phys. **17**, 28068–28075 (2015)
12. Parsina, I., DiPaola, C., Baletto, F.: A novel structural motif for free CoPt nanoalloys. Nanoscale **4**, 1160–1166 (2012)
13. Paz-Borbón, L., Mortimer-Jones, Th, Johnston, R., Posada-Amarillas, A., et al.: Structures and energetics of 98 atom Pd–Pt nanoalloys: potential stability of the Leary tetrahedron for bimetallic nanoparticles. Phys. Chem. Chem. Phys. **9**, 5202–5208 (2007)
14. Schebarchov, D., Wales, D.: A new paradigm for structure prediction in multicomponent systems. J. Chem. Phys. **139**(22), 221101 (2013). https://doi.org/10.1063/1.4843956
15. Schebarchov, D., Wales, D.: Quasi-combinatorial energy landscapes for nanoalloy structure optimization. Phys. Chem. Chem. Phys. **17**, 28331–28338 (2015)
16. Shayeghi, A., Götz, D., Davis, J.B.A., Schäfer, R., Johnston, R.L.: Pool-BCGA: a parallelised generation-free genetic algorithm for the ab initio global optimisation of nanoalloy clusters. Phys. Chem. Chem. Phys. **17**, 2104 (2015)
17. Toai, T.J., Rossi, G., Ferrando, R.: Global optimisation and growth simulation of AuCu clusters. Faraday Discuss. **138**, 49–58 (2008). https://doi.org/10.1039/b707813g
18. Bilalbegović, G.: Structures and melting in infinite gold nanowires. Solid State Commun. **115**, 73–76 (2000)
19. Liu, W., Chen, P., Qiu, R., Khan, M., et al.: A molecular dynamics simulation study of irradiation induced defects in gold nanowire. Nucl. Instrum. Methods Phys. Res. Sect. B Beam Interact. Mater. Atoms. **405**, 22–30 (2017)
20. Diao, J., Gall, K., Dunn, M.L., Zimmerman, J.A.: Atomistic simulations of the yielding of gold nanowires. Acta Mater. **54**(3), 643–653 (2006)

21. Zepeda-Ruiz, L.A., Sadigh, B., Biener, J., Hodge, A.M., et al.: Mechanical response of freestanding Au nanopillars under compression. Appl. Phys. Lett. **91**(10), 101907-1–101907-3 (2007)
23. He, X., Cheng, F., Chen, Z.-X.: The lattice kinetic Monte Carlo simulation of atomic diffusion and structural transformation for gold. Sci. Rep. **6**(1), 33128 (2016)
24. Baibuz, E., Vigonski, S., Lahtinena, J., Zhao, J., Jansson, V., Zadin, V., Djurabekova, F.: Migration barriers for surface diffusion on a rigid lattice: challenges and solutions. Comput. Mater. Sci. **146**, 287–302 (2018). https://doi.org/10.1016/j.commatsci.2017.12.054
25. Cleri, F., Rosato, V.: Tight-binding potentials for transition metals and alloys. Phys. Rev. B **48**, 22–33 (1993)
26. Sutton, A., Chen, J.: Long-range Finnis-Sinclair potentials. Philos. Mag. Lett. **61**, 139–146 (1990)
27. Myshlavtsev, A.V., Stishenko, P.V.: Modification of the metropolis algorithm for modeling metallic nanoparticles. Omsk Sci. Newsp. **1**(107), 21–25 (2012). (in Russian)

Factors for Search Methods Scalability

Kalin Penev[(✉)]

School Media Arts and Technology, Solent University, East Park Terrace,
Southampton SO14 0YN, UK
Kalin.Penev@solent.ac.uk

Abstract. Scalability of systems performance becomes a challenge for modern digital systems. Achievement of system ability to complete wide range of tasks in terms of computational performance and effective use of resources require substantial research. Significant efforts are directed towards design of large scale hardware systems. However resolving scalable tasks require also scalable software capable of completion both many small simple tasks and large complex tasks using effectively available hardware, energy and time. This article discusses factors for successful search and optimisation in particular search methods scalability. Discussion is illustrated with an overview of publications.

Keywords: Optimisation · Algorithms scalability

1 Introduction

According to recent, published analyses modern communications and remote information services combined with high quality visual user interface lead to an extreme data traffic and growing demand of computational resources [6, 7, 14]. Computer Systems, as a core component of communication systems and services, face a dilemma – maximal speed and minimal delays versus minimal energy consumption and maximal efficiency. In practice this dilemma could be managed with scalable computational systems capable of minimal energy consumption and maximal efficiency in particular for low workload and in the same time able to guarantee minimal delays for high workload. Published statistics suggest growing computer serves demand [14], which requires substantial engineering and research efforts and initiatives [8], directed towards large scale computational systems. Design and implementation of large scale hardware, however, does not seem to be sufficient to guarantee minimal delays without efficient scalable software. Consideration of computational workload could identify and distinguish two types – first one is based on variable number of relatively small and simple for resolving tasks and second variable number of complex resource consuming tasks which in the same time require minimal delays and efficient use of available hardware and other resources. This article discusses factors essential for achieving software scalability on complex, resource consuming tasks and in particular on methods for search and optimisation.

Evaluation of different aspect of various methods on complex tasks is already subject of many publications. One aspect is the performance of the multi-agent

G. Nikolov et al. (Eds.): NMA 2018, LNCS 11189, pp. 142–149, 2019.
https://doi.org/10.1007/978-3-030-10692-8_16

algorithms. Evaluation identifies that there is clearly a need to develop effective test functions to evaluate various global optimization algorithms [12].

Proposed and explored are new variants of evolutionary algorithms. Experimental studies for high-dimensional test functions identify difficulties of fitness evaluation and a need for future improvements on fitness evaluation [22].

In a recent study [13] are explored implications, often masked at low dimensions, which increase significantly, when extended the domain of application to higher dimensions. It concludes that further detailed studies are needed to examine the relative behaviour of search methodologies in high dimensional spaces that could lead to a striking improvement in performance.

For resolving high dimensional multi-objective tasks, distributed parallel computing technologies with novel variable decomposition and optimization strategies ere explored [4]. This study confirms that when the objectives number is large and the number of variables is huge, the optimization process will be extremely time-consuming. Based on the distributed parallel strategies, the optimization tasks can be allocated to large amount of computation units, which can substantially improve the running efficiency [4]. Such efficiency should be subject of rigorous evaluation in terms of communications' delays between distributed systems and use of energy.

A recent study [3] on correlation between pairs of variables in dependence on the problem dimensionality identifies that computational feasibility of the experiments imposes an extremely slow search in case of high dimensions. Stated beliefs of this publication [3] are that: - an exponential increase of budget and population with the dimensionality is still practically impossible; and - an algorithm can quickly improve upon initial guesses if it integrates the knowledge that an apparent weak correlation between pairs of variables occurs, regardless the nature of the problem. This position could face arguments of other publications [1, 19]. An exploration of heuristic rules for nonlinear function local minimisation applied on a number of benchmark functions achieves very positive results [11]. It recommends application of explored techniques to constrained optimisation problems and its use as local optimiser in global optimisation schemes using multi-start version, replacement of the analytical calculation of partial derivatives with an automatic differentiation scheme and design of ease to use tool for practical applications, which is in line with the intention of this article.

A very detailed and comprehensive study of swarm algorithms enhanced with adaptive competition strategy and swarm heterogeneity recommends improvement of exploration abilities [5]. This could contribute towards scalable to large scale search and optimisation methods. An evaluation of combined several algorithms shows very good results on some high dimension tasks [15]. It concludes that automated tuning can be very effective on high-dimensional functions and the intention is to apply extensively automatic algorithm configuration techniques, integrated into algorithm engineering process, to develop new state-of-the-art algorithms for continuous optimization. This should be encouraged. A challenging paper [1] proposes a hypercube optimization algorithm for global optimization with a faster convergence, applied to high dimension benchmarking functions. Achieved results deserve attention. It would be of interest to see this algorithm applied to other heterogeneous benchmarks including tasks with unknown optimal solution.

An investigation on extremely high-dimensional optimization [2], applies MapReduce parallel programming model to popular tests for 1000, 500 000, 1 000 000, 3 000 000 and 10 000 000 dimensions. It uses GPU (Graphics Processing Unit) accelerated version of Memetic algorithm based on the CUDA programming model for NVIDIA GPUs, which allows scaling to problems with millions of variables. MapReduce demonstrates to be an effective approach to scale optimization algorithms on extremely high-dimensional problems [2]. This is supported with large number of results. A comparison to other published results [1, 19] shows abilities of this method.

Modified line search method which makes use of partial derivatives and re-starts the search process after a given number of iterations by modifying the boundaries based on the best solution obtained at the previous iteration (or set of iterations) is explored against several high dimensional benchmark functions [9]. Empirical results illustrate scalability of this approach tested to high dimensional functions. It would be nice to see how this concept manages with hard rigorous tests.

According to an advanced article [16], which reviews metaheuristics accuracy and efficiency, exploration of optimisation problems should consider also level of precision of the achieved results, in particular when dealing with high dimension. An increase of the results' precision, leads to an exponential growth of the number of possible solutions, which should be analysed, and this reflects on time for calculation and required computation resources. Similar issues are already explored and published [18, 19, 21, 22].

Reviewed studies explore various search methods and their essential aspects, which could contribute towards development of high scalable optimisation methods. The following section discusses factors which are reflected to scalability of Free Search method published in the literature [15, 16, 21].

2 Factors for Search Methods Scalability

Capability of a search method to resolve successfully optimisation tasks with various parameters number will be called in this article - search method scalability. Methods which could manage within the range of 100 dimensions could be classified as scalable methods. Methods which could find a solution with an arbitrary precision, for acceptable period of time, using limited computational resources for tasks with 1000 and more parameters could be classified as highly scalable methods. Achieving search methods scalability to various parameters number could be classified as ability for adaptation to multi-dimensional spaces. Very often search and optimisation methods utilise concepts which are observed in nature [4, 5, 12, 15, 18, 19, 21–23]. In the nature, due to many reasons, it could be perceived and apprehended, only, natural behaviour within 2 dimensions for most of the creatures or within 3 dimensions for sea creatures and birds. This makes difficult to get ideas from nature how to explore multidimensional spaces. During the design, implementation and evaluation of the adaptive heuristic Free Search were identified several factors, which could support successful search within multi-dimensional spaces. Typical for Free Search ability for adaptation to tasks with heterogeneous landscape can be interpreted as an adaptation to large number of dimensions without changes or tuning of the search method. These factors are discussed within the following sections.

2.1 Knowledge

One of the most critical factor for successful adaptation is knowledge – according to the reviewed publications [1, 3–5, 9, 11–13, 15, 17, 21, 23] common understanding is - requirement for little, if any, prior knowledge relating to the search environment. Most of the techniques can therefore successfully negotiate search spaces described by a wide range of model types and structures e.g., discrete, continuous, mixed-integer, quantitative, qualitative, etc. In Free Search as a factor for adaptation is accepted that prior knowledge requirements should be limited to search space dimensions, search space borders and constraints, only. For adaptation to dimension changes requirement for knowledge about search space dimensions should be limited to technical requirements and should not restrict the search process. Algorithm should be able to: (1) abstract valuable for the search process knowledge; (2) evaluate and assess this knowledge; (3) store the knowledge in efficient manner, namely within minimal storage space and minimal delays for access and use; (4) use the knowledge for process guidance and improvement (Instead to relay on users' tuning.); (5) update and improve the knowledge during the process of search.

An evidence how this reflect on search scalability and adaptation to dimensions changes is illustrated with experimental results in particular on heterogeneous hard, tests including global and constrained with unknown optimal solutions [18, 19, 22].

2.2 Exploratory Capabilities

Other factor for search methods scalability is excellent exploratory capabilities especially where tasks is considered with hard global and most importantly unknown solution similarly to practical tasks [17]. Exploratory capabilities should harmonise convergence and divergence in a complete process guided by acquired during the process of search knowledge. An example of how this could be implemented in practice is adaptive heuristic method called Free Search [18, 19]. During the search process this algorithm explores locations from the search space. It evaluates their quality against the objective functions and stores the knowledge about this quality. An original peculiarity of Free Search, which has no analogue in other optimisation methods, is a variable called sense, which analyses stored knowledge and uses it for further guidance of the search process. The relation sense & locations quality plays the role of a tool for regulation of divergence and convergence within the search process.

A consideration of three idealised general states of sensibility distribution – uniform, enhanced and reduced related with locations quality distribution can clarify the decision-making policy and ability for adaptation of the algorithm. In case of uniformly distributed sensibility and locations quality, low level of sensibility can lead to selection of any location for start position. A high level sensibility can lead to selection for start position locations with high quality and will ignore locations with low quality. It is assumed that during a stochastic process within a stochastic environment any deviation could lead to non-uniform changes of the process. The achieved results play a role of deviator. The enhancement of the sensibility urges the search around the area of the best-found solution. This situation appears naturally when the locations qualities are very different and stochastic generation of the sensibility produces high values.

External adding of a constant or a variable to sensibility could make an enforced enhancement of the sensibility. An enhanced sensibility will lead to selection and more precise differentiation of locations with higher quality and will ignore these with lower quality. In this manner algorithm naturally converges to the best locations. By reducing the sensibility will be allowed exploration around locations with low quality. This situation naturally appears when the locations qualities are very similar and randomly generated sensibility is low. In this case low quality locations can be selected with high probability, which indirectly will decrease the probability for selection of high quality locations. In this manner algorithm naturally diverges across the search space. Subtracting of a constant or a variable from sensibility could make an enforced reduction of the sensibility.

By using sense algorithm builds knowledge about the quality of the search space and in the same time creates skills how to recognise further, higher or lower quality locations. The cognition and the skills are abstracted from the achieved results, only. This guarantees that the method is task independent and could adapt to various tasks and dimensions.

2.3   Avoid and Escape from Local Optima

Essential factor for search process success is ability to avoid local optima. The stochastic nature of the various algorithms combined with continuing random sampling of the search space can prevent convergence upon local sub-optima. Most of the methods attempt to avoid local optima and achieve it with varying success, which is illustrated by many publications [1, 3–5, 9, 11–13, 15, 18, 19, 21–23].

The ability to escape from trapping in local sub-optima within continuous search space leads to better performance, reliability and scalability [18, 19].

2.4   Identification and Provision of Multiple Good Solutions

From practical point of view good factor for search scalability is identification and provision of multiple good solutions [17]. This factor is partially explored using Step test [18]. Appropriate benchmarks with multiple equal optima are two dimensional Himmelblau [10] and Shubert [20] test functions. Implementation and evaluation of this capability require development of scalable benchmarks with multiple equal optima and needs more research.

2.5   Ability to Handle High Dimensionality

Although it may look obvious it should be stated that methods scalability explicitly requires ability to handle high dimensionality. According to the literature [17] performance of many search techniques can rapidly deteriorate as dimensionality increases whereas the characteristic sampling of the design space of the search algorithms can maintain relatively efficient search of high-dimensional domains. Ability to handle high dimensionality should be interpreted not only as an ability to converge to a good solution on some task. This should be an ability to achieve global solution with arbitrary precision within acceptable period of time using limited (or minimal) computational resources and energy. Published results highlight how different methods handle high dimensionality [1–5, 9, 11–13, 15, 18, 19, 21–23].

2.6   Time

Time is perhaps the most critical factor for search methods scalability. Consideration of time factor can distinguish two components – time used for objective function calculation and time for algorithm's activities.

2.6.1   Time for Objective Function Calculation
Time used for single calculation of the objective function is dependent on tasks complexity and dimensionality and cannot be avoided. For different tasks this dependence varies. It could be linear with low or high proportion, exponential etc. Initial evaluation on these variations is published [19]. A possible way to decrees overall search process time, is to decrease the number of objective function calculations required for identification of the optimal solution. This particularly applies for time consuming functions. This factor requires further research.

2.6.2   Time for Algorithm's Activities
Time for algorithm's activities is dependent on the algorithm itself. Different methods consume different time for processing the same number of objective function evaluation [18]. Time used for algorithm's activities could be reduced and minimised by improvement of the operations, reducing time consuming events and processes. For this factor extensive use of computational resources does not seem to be an acceptable solution for acceleration of the search process.

2.7   Energy

Use of energy is other most critical factor for scalability. Published research on search methods evaluations does not explore sufficiently use of energy. Very often, intention to accelerate optimisation process and to reduce delays lead to an extensive use of hardware such as parallel processing, distributed computing, GPU accelerated computing, etc. [2, 4]. All these approaches increase energy consumption and therefore cannot be classified as efficient. Published [19] measurement, evaluation and analysis of energy consumption and cost suggests that use of energy is dependent on tasks complexity and dimensions number. This dependence becomes much more tangible for tasks with high number of parameters. Further investigation should evaluate rigorously and assess variety of methods against energy consumption.

3   Conclusion

This article reviews optimisation of high dimensional numerical tests and discusses factors for search methods scalability. It could be summarised that scalability of available methods is partially explored and needs additional evaluation and analysis. Further investigation should focus on evaluation and measure of time and computational resources sufficient for completion of hard, global and constrained multidimensional tasks. Algorithms analysis and improvement, minimisation of energy consumption and time delays require also further research.

References

1. Abiyev, R.H., Tunay, M.: Optimization of high-dimensional functions through hypercube evaluation. Comput. Intell. Neurosci. **2015** (2015). http://dx.doi.org/10.1155/2015/967320. Article ID 967320, 11 pages
2. Cano, A., García-Martínez, C., Ventura, S.: Extremely high-dimensional optimization with MapReduce: scaling functions and algorithm. Inf. Sci. **415-416**, 110–127 (2017)
3. Caraffini, F., Neri, F., Iacca, G.: Large scale problems in practice: the effect of dimensionality on the interaction among variables. In: Squillero, G., Sim, K. (eds.) EvoApplications 2017. LNCS, vol. 10199, pp. 636–652. Springer, Cham (2017). https://doi.org/10.1007/978-3-319-55849-3_41
4. Cao, B., et al.: Distributed parallel particle swarm optimization for multi-objective and many-objective large-scale optimization. IEEE Access **5**, 8214–8221 (2017)
5. Chu, X., et al.: AHPS2: an optimizer using adaptive heterogeneous particle swarms. Inf. Sci. **280**, 26–52 (2014). https://doi.org/10.1016/j.ins.2014.04.043
6. CISCO, The Zettabyte Era - Trends and Analysis - Cisco, White Papers (2017a). http://www.cisco.com/c/en/us/solutions/collateral/service-provider/visual-networking-index-vni/vni-hyper connectivity-wp.html. Accessed 11 June 2018
7. CISCO, Cisco Visual Networking Index: Forecast and Methodology, 2016–2021- Cisco, White Papers (2017b). https://www.cisco.com/c/en/us/solutions/collateral/service-provider/visual-networking-index-vni/complete-white-paper-c11-481360.html. Accessed 11 June 2018
8. ETP4HPC, 2018 European Technology Platform for High Performance Computing. http://www.etp4hpc.eu/missionvision.html. Accessed 11 June 2018
9. Grosan, C., Abraham, A., Hassainen, A.: A line search approach for high dimensional function optimization. Telecommun. Syst. **46**(3), 217–243 (2011)
10. Himmelblau, D.: Applied Non-linear Programming. McGraw-Hill, New York (1972)
11. Kotsialos, A.: Nonlinear optimisation using directional step lengths based on RPROP. Optim. Lett. **8**(4), 1401–1415 (2014)
12. Liang, J., et al.: Performance evaluation of multiagent genetic algorithm. Nat. Comput. **5**(1), 83–96 (2006)
13. Macnish, C., Yao, X.: Direction matters in high-dimensional optimisation, pp. 2372–2379. IEEE (2008)
14. van der Meulen, R., Costello, K.: Gartner Says Worldwide Server Revenue Grew 33.4 Percent in the First Quarter of 2018, While Shipments Increased 17.3 Percent, STAMFORD, Conn., June 11, 2018 (2018). https://www.gartner.com/newsroom/id/3878666. Accessed 11 June 2018
15. Oca, M.D., Aydin, D., Stützle, T.: An incremental particle swarm for large-scale continuous optimization problems: an example of tuning-in-the-loop (re)design of optimization algorithms. Soft. Comput. **15**(11), 2233–2255 (2011)
16. Nesmachnow, S.: An overview of metaheuristics: accurate and efficient methods for optimisation. Int. J. Metaheuristics **3**(4), 320–347 (2014)
17. Parmee, I.C.: Evolutionary and Adaptive Computing in Engineering Design. Springer, London (2001). https://doi.org/10.1007/978-1-4471-0273-1

18. Penev, K.: Performance evaluation on optimisation of 200 dimensional numerical tests - results and issues. In: The 28-th International Scientific Conference of the Faculty of Industrial Technology of TU-Sofia, FIT 2015, pp. 461–468. TU Sofia, Bulgaria (2015). ISBN 978-619-167-178-6

19. Penev, K.: Free search in multidimensional space M. In: Lirkov, I., Margenov, S. (eds.) LSSC 2017. LNCS, vol. 10665, pp. 399–407. Springer, Cham (2018). https://doi.org/10. 1007/978-3-319-73441-5_43

20. Schwefel, H.-P.: Evolution strategy in numerical optimisation. Ph.D. Thesis, Technical University of Berlin (1975)

21. Yin, J., Wang, Y., Hu, J.: Free search with adaptive differential evolution exploitation and quantum-inspired exploration. J. Netw. Comput. Appl. **35**(3), 1035–1051 (2012)

22. Yang, Z., Tang, K., Yao, X.: Differential evolution for high-dimensional function optimization, pp. 3523–3530. IEEE (2007)

23. Zhongda, T., et al.: A prediction method based on wavelet transform and multiple models fusion for chaotic time series. Chaos Solitons Fractals Interdiscip. J. Nonlinear Sci. Nonequilibrium Complex Phenom. **98**, 158–172 (2017)

The Statistic Analysis of Conjunctive Adverbs Used in the First Bulgarian School Books in Mathematics (from the First Half of XIX c.)

Velislava Stoykova[✉]

Bulgarian Academy of Sciences, Institute for Bulgarian Language,
52, Shipchensky proh. str., bl. 17, 1113 Sofia, Bulgaria
vstoykova@yahoo.com

Abstract. The paper presents an approach to extract and analyze conjunctive adverbs in the first Bulgarian school books in mathematics published during the first half of XIX c. in Serbia. It applies Information Retrieval (IR) approach and the Sketch Engine software to search electronic readable format of the books (without normalizing their graphical representation). The methodology uses statistically-based search techniques to evaluate related queries (keywords) and further, the query search is optimized by limiting the scope of the search using options according to related search criteria. The search results are analyzed both with respect to the syntactic distribution (generally as logical connectives and transitions) and to the semantics they express.

Keywords: Data mining · Big data · Knowledge discovery

1 Introduction

The contemporary formal syntactic theories regard the conjunctive adverbs as a connection between two clauses that convert the clause they introduce into adverbial modifier. Often, the conjunctive adverbials are referred to as connectives, and they modify the verb, the adjective, or another adverb in the main clause, and in that way, they modify the previously expressed logical predication. The conjunctive adverb functions as an adverbial connective (also known as a logical transition) which is used within a second clause, so to show its logical relationship to the first. That relations can represent sequence, contrast, cause and effect, purpose or reason, etc.

Some authors tend to analyze conjunctive adverbs at the text level [9] introducing the term *transitions* and defining them as providing the text cohesion by making text more explicit or by signaling how ideas relate to one another. The transitions are divided into coordinating, subordinating, temporal, etc.

Nevertheless, for both analyses of conjunctive adverbs (as logical connectives or as transitions) a formal syntactic encoding is required, so to process (to parse)

© Springer Nature Switzerland AG 2019
G. Nikolov et al. (Eds.): NMA 2018, LNCS 11189, pp. 150–157, 2019.
https://doi.org/10.1007/978-3-030-10692-8_17

the electronic language resources. In our research, we use a statistically-based approach instead, so to extract the conjunctive adverbs which are specific for the first mathematical texts written in Bulgarian language.

The approach is based on the use of the Distributional Semantic Model frameworks which regard the use of statistical similarity (or distance) as relevant for semantic similarity (or distance) [2]. Thus, we evaluate statistically-significant words as semantically-relevant and we extract most common words which function as typical for that mathematical texts conjunctive adverbs showing additional information about their variability and functionality.

The techniques used adopt metrics for extraction of different types of word semantic relations by estimation of word similarity measure [1]. Further, we are going to present such techniques and to discuss the received results of using statistical functions of the Sketch Engine software [7] for keywords extraction, concordances and collocations generation in searching electronic educational resource which contains the first mathematical school books in Bulgarian language published during the first half of XIX c. in Serbia.

2   The Sketch Engine (SE)

The SE software allows approaches to extract semantic properties of words and most of them are with multilingual application. Generating keywords is a widely used technique to extract terms of particular studied domain. Also, semantic relations can be extracted by generation of related word contexts through word concordances which define context in quantitative terms and a further work is needed to be done to extract semantic relations by searching for co-occurrences and collocations of related keyword.

Co-occurrences and collocations are words which are most probably to be found with a related keyword. They assign the semantic relations between the keyword and its particular collocated word which might be of a similarity or of a distance. The statistical approaches used by SE to search for co-occurrence and collocated words are based on defining the probability of their co-occurrences and collocations. We use techniques of $T - score$, $MI - score$ and $MI^3 - score$ for corpora processing and searching. For all, the following terms are used: N – corpus size, f_A – number of occurrences of keyword in the whole corpus (the size of concordance), f_B – number of occurrences of collocated keyword in the whole corpus, f_{AB} – number of occurrences of collocate in the concordance (number of co-occurrences). The related formulas for defining $T - score$, $MI - score$ and $MI^3 - score$ are as follows:

$$T - Score = \frac{f_{AB} - \frac{f_A f_B}{N}}{\sqrt{f_{AB}}} \tag{1}$$

$$MI - Score = log_2 \frac{f_{AB} N}{f_A f_B} \tag{2}$$

$$MI^3 - Score = log_2 \frac{f_{AB}^3 N}{f_A f_B} \tag{3}$$

The $T - score$, $MI - score$ and $MI^3 - score$ are applicable for processing diachronic corpora as well [6].

Collocations have been regarded as statistically similar words which can be extracted by using techniques for estimation the strength of association between co-occurring words. Recent developments improved that techniques with respect to various application areas.

Further, we shall present and analyse the search results for extracting conjunctive adverbs using the SE software and shall compare related results with respect to their semantic types.

3   The Bulgarian Diachronic Mathematical Resource (BDMR)

The diachronic text corpora are designed to study predominantly how the grammar has changed over time. They typically use the annotation schemes which encode the grammar relations by regarding them as a relatively constant over certain period of time. The approach has been extensively used in diachronic corpora for languages like English, German, Spanish, etc. [4] allowing different types of semantic search. However, recent advances in IR offer the use of statistically-based approaches which allow comparison of different corpora with respect to several search criteria outlining both lexical and syntactic differences or changes without use of syntactic annotations.

The Bulgarian Diachronic Mathematical Resource (BDMR) is the first collection of mathematical texts in Bulgarian language from the first half of XIX c. It contains the first original mathematical school books written in Bulgarian language which were published in Serbia.

The texts included are limited according to the related historical period and are representative both for the language of that period and for the level of mathematical knowledge and terminology at that time. The BDMR contains two authoring school books: (i) 'Аритметика или наука числителна' (*Aritmetics or Science about Numbers*) by Chr. Pavlovich, published in 1833 in Belgrade [8] – consisting of almost 12 000 words, and (ii) 'Аритметическое руководство   за наставление на болгарските юноши' (*Arithmetic Guide for Bulgarian Adolescents*) by N. Bozveli and Em. Vaskidovich, published in 1835 in Kraguevac [3] – consisting of almost 7 000 words. The BDMR uses electronic readable format of the books which are uploaded in their original graphical representation and without normalizing the phonetic alternations according to contemporary spelling rules.

The resource was created to analyze the grammar and lexical features as well as to study the mathematical terminology of Bulgarian language for the related period. The BDMR was uploaded into the SE allowing the use of its incorporated options for storing, sampling, searching and filtering the texts according to different criteria. The resource is open for further development and enlargement.

4   Search Results

For our search experiments, we use the techniques for corpora comparison. The main idea is to use the BDMR as a basic resource which to compare to arbitrary reference corpus consisting of texts from standard contemporary Bulgarian language. In that way, we use the *keyness score* measurement [5], so to extract the non-content words (possibly conjunctive adverbials or conjunctions) which have to be specific both for the time period (the first half of XIX c.) and for the related domain (mathematics). The results from processing BDMR(i) and BDMR(ii) are presented at Fig. 1 and are similar.

Single-word		Score	F	RefF
перст	W	4,516.41	83	2
день	W	4,454.58	90	86
кольку	W	2,836.64	52	0
быва	W	2,673.05	49	0
толькү	W	2,563.98	47	0
сумма	W	2,500.56	46	3
остаток	W	2,343.08	43	1
третій	W	2,236.79	41	0
внетрешен	W	2,127.73	39	0
Примѣр	W	2,127.73	39	0
вторый	W	1,909.61	35	0
став	W	1,827.59	85	1,296
тамо	W	1,767.51	46	354
сирѣчь	W	1,746.01	32	0
знаменатель	W	1,746.01	32	0

Single-word		Score	F	RefF
тѣхъ	W	4,703.90	64	13
послѣ	W	4,104.87	55	0
отъ	W	3,437.25	98	951
тогда	W	2,591.89	36	31
тако	W	2,307.72	41	275
съ	W	2,037.35	106	2,431
чрезъ	W	1,986.59	28	44
подъ	W	1,896.44	27	53
ока	W	1,754.06	33	341
сирѣчь	W	1,717.16	23	0
какъ	W	1,595.26	22	25
гро	W	1,507.17	21	34
мѣсяцы	W	1,493.31	20	0

Fig. 1. The *keyness score* top-down search results from BDMR(i) and BDMR(ii) for non-content words.

They both include the word сирѣчь (*or*) which is considered as specific for both the time period and for the mathematical domain because it has 32 hits for the BDMR(i), 23 hits for the BDMR(ii), and 0 hits for the reference corpus used. The later results displayed for the BDMR(i) are given at Fig. 2.

They outline also another non-content word слѣдователно (*consequently, hence, therefore*) which has 21 hits for BDMR(i) and 0 hits for the reference corpus used. However, the *keyness score* measurement do not give any additional information about the extracted keywords' syntactic distribution, so to evaluate their syntactic functions.

For that, we use the approach already presented in [10] and used for cross-lingual mathematical terminology extraction. The main techniques applied are

☐ ставы	w 1,418.82	26	0
☐ найда	w 1,380.93	52	889
☐ тѣх	w 1,364.29	25	0
☐ человѣк	w 1,309.76	24	0
☐ раздробленіе	w 1,309.76	24	0
☐ онай	w 1,300.51	24	6
☐ четвертый	w 1,255.23	23	0
☐ круг	w 1,162.11	22	28
☑ слѣдователно	w 1,146.16	21	0

Fig. 2. The *keyness score* top-down search results from BDMR(i) for non-content words (continuation).

concordances generation and collocations generation. The concordances present all occurrences of a related keyword within all its related quantitative contexts. Thus, we generate concordances for the keyword сирѣчь for both BDMR(i) and BDMR(ii).

The results received are presented at Figs. 3 and 4, respectively, and display all related contexts that can be analyzed with respect to the syntactic variability and distribution of that keyword.

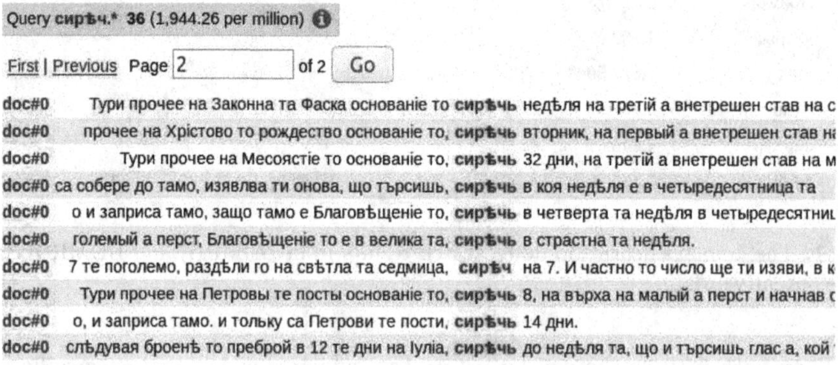

Query сирѣч.* **36** (1,944.26 per million) ⓘ

First | Previous Page 2 of 2 Go

doc#0   Тури прочее на Законна та Фаска основаніе то **сирѣчь** недѣля на третій а внетрешен став на с

doc#0   прочее на Хрістово то рождество основаніе то, **сирѣчь** вторник, на первый а внетрешен став н

doc#0   Тури прочее на Месоястіе то основаніе то, **сирѣчь** 32 дни, на третій а внетрешен став на м

doc#0   са собере до тамо, изявлва ти онова, що търсишь, **сирѣчь** в коя недѣля е в четыредесятница та

doc#0   о и записа тамо, защо тамо е Благовѣщеніе то, **сирѣчь** в четверта та недѣля в четыредесятни

doc#0   големый а перст, Благовѣщеніе то е в велика та, **сирѣчь** в страстна та недѣля.

doc#0   7 те поголемо, раздѣли го на свѣтла та седмица, **сирѣч** на 7. И частно то число ще ти изяви, в к

doc#0   Тури прочее на Петровы те посты основаніе то, **сирѣчь** 8, на върха на малый а перст и начнав с

doc#0   о, и записа тамо. и тольку са Петрови те пости, **сирѣчь** 14 дни.

doc#0   слѣдувая броенѣ то преброй в 12 те дни на Іуліа, **сирѣчь** до недѣля та, що и търсишь глас а, кой

Fig. 3. The concordances of the keyword сирѣчь from BDMR(i).

The context analysis shows that the word сирѣчь which is an adverb is used to assign a syntactic relations of coordination (often replaceable by the coordinating conjunction или (*or*) or by тоест (*that is, namely*). It marks a relation of logical equivalence between the coordinating phrases or clauses and can be regarded as an adverbial logical transition maintaining the text coherence. Its semantics is to show similarity, to signal restatement, etc.

The search results obtained for concordances generation of the keyword слѣдователно from BDMR(i) are presented at Fig. 5 and show the syntactic variability and distribution of that keyword.

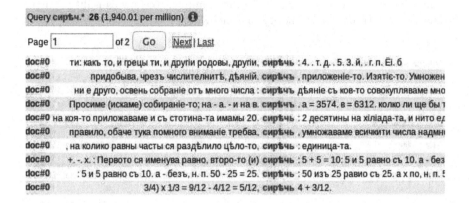

Fig. 4. The concordances of the keyword сирѣчь from BDMR(ii).

The results contain contexts which present interesting distribution and variability of the keyword. To analyze them in more details, we need to use a search optimization that is aimed at further query profiling by generation of keyword's collocations. The related results are presented at Fig. 6. They outline a variability of the keyword's use in combination with a conjunction и (*and*) and така (*thus*) which are the most frequent collocations.

Query **слѣдователно** 21 (1,134.15 per million) ⓘ

| Page |1 | of 2 | **Go** | Next | Last |

doc#0   , то есть прави я да знаменува хілляды. И така **слѣдователно** десятини на хіллядо то и стоти

doc#0   с тоя знак (.), сирѣчь с хіллядный а знак. И така **слѣдователно** прави промѣнявая знакове те:

doc#0   о, с четыри прачицы, така (""), и така **слѣдователно** . На реченно то

doc#0   с тѣх думая 2 и 8, 10: и 2, 12: и 3, 15. И така **слѣдователно** нека прави, и ще найде сумма

doc#0   то с части те на слѣдователный а род. и така **слѣдователно** . Те

doc#0   то составляват слѣдователный а погорен. И така **слѣдователно** . Те

doc#0   с едно собственно име раздробленія, кое то и ніе **слѣдователно** употребляваме.

doc#0   остаток раздѣлявася вторый о дѣлитель, и така **слѣдователно** докле буде остаток о нулла, и є

doc#0   пъть, и нахождася и друг четвертый член. И така **слѣдователно** ще са найде найпослѣдній о нє

doc#0   , на среднй а Маій, на четвертый а Іуній, и така **слѣдователно** , преходя от перст на перст, из

Fig. 5. The concordances of the keyword слѣдователно from BDMR(i).

However, a more detailed semantic analysis of the keyword's syntactic distribution shows that the syntactic function of the keyword is still far from its contemporary state. The majority of contexts show that the word слѣдователно (especially in combination with и and така) is used instead of послѣдователно which outlines its functioning as a logical connective. Thus, it can be regarded as subordinating transition which semantics is to introduce an item in a series (sequence), to introduce an example, to show causality (purpose or reason), to introduce a summary or conclusion.

Collocation candidates

	Cooccurrence count	Candidate count	T-score	MI	MI3
P|N така	13	104	3.561	6.369	13.770
P|N и	17	426	3.966	4.721	12.896
P|N разрѣшавашь	5	11	2.228	8.231	12.875
P|N ,	23	1,178	4.424	3.690	12.737
P|N Найди	5	13	2.227	7.990	12.634
P|N направи	5	13	2.227	7.990	12.634
P|N а	12	280	3.341	4.824	11.994
P|N .	18	1,065	3.863	3.482	11.822

Fig. 6. The collocation candidates of the keyword слѣдователно from BDMR(i).

5   Conclusion

We presented an IR approach which uses statistically-based techniques to search electronic collection of diachronic mathematical texts containing the first school books written in Bulgarian language during the first half of XIX c. The aim was to extract and study the syntactic phenomena, specific both to that period of time and to the domain of mathematics. We have extracted conjunctive adverbs by using keyness score measurement and have used concordances search and collocations generation, so to optimize the search. Later, we have analyzed the related results both with respect to their syntactic distribution and variability.

The obtained results outlined the words сирѣчь and слѣдователно which were analyzed from the point of view of text coherence, and were regarded as logical connectives (transitions). The former was evaluated as a coordinating transition whereas the later was evaluated as a subordinating transition, and the related semantics was defined for both, respectively.

Finally, from the point of view of diachronic syntax, the adverbial сирѣчь is not used any more in contemporary Bulgarian language and is regarded as stylistically archaic, whereas the adverbial слѣдователно is very actively used in contemporary academic writing, and not only in the domain of mathematics. Obviously, this is probably one of the first use of that adverbial in Bulgarian language (with the above described syntactic functions and semantics), since it defines the semantic relations between the connected phrases or clauses explicitly and maintains their logical interconnections.

References

1. Baroni, M., Evert, S.: Statistical methods for corpus exploitation. In: Lüdeling, K., Kytö, M. (eds.) Corpus Linguistics: An International Handbook, vol. 2, pp. 777–803. Mouton De Gruyter, Berlin/New York (2008)
2. Baroni, M., Lenci, A.: Distributional memory: a general framework for corpus-based semantics. Comput. Linguist. **36**(4), 673–721 (2010)
3. Бозвели, Н., Васкидович, Ем.: Аритметическое руководство за нас тавление на болгарските юноши. Княжеско-Сербска типография, Крагуевац. (1835) (in Bulgarian language)
4. Davies, M., Chapman, D.: The effect of representativeness and size in historical corpora: an empirical study of changes in lexical frequency. In: Chapman, D., Moore, C., Wilcox, M. (eds.) Studies in the History of the English Language VII: Generalizing vs. Particularizing Methodologies in Historical Linguistic Analysis, pp. 131–150. De Gruyter/Mouton, Berlin (2016)
5. Gabrielatos, C.: Keyness analysis: nature, metrics and techniques. In: Taylor, C., Marchi, A. (eds.) Corpus Approaches to Discourse: A critical review, pp. 225–258. Routledge, Oxford (2018)
6. Killgarriff, A., Husak, M., Woodrow, R.: The Sketch Engine as infrastructure for historical corpora. In: Jancsary, J. (ed.) Empirical Methods in Natural Language Processing, Proceedings of the Conference on Natural Language Processing 2012, pp. 351–356 (2012)
7. Killgarriff, A., et al.: The Sketch Engine: Ten Years On. Lexicography **1**, 17–36 (2014)
8. Павлович, Хр.:   Аритметика или наука числителна.   Княжеско Сербска типография, Белград. (1833) (in Bulgarian language)
9. Rappaport, B.: Using the elements of rhythm, flow, and tone to create a more effective and persuasive acoustic experience in legal writing. J. Legal Writ. Inst. **16**(1), 65–116 (2010)
10. Stoykova, V., Stankovic, R.: Using query expansion for cross-lingual mathematical terminology extraction. In: Silhavy, R. (ed.) CSOC2018 2018. AISC, vol. 764, pp. 154–164. Springer, Cham (2019). https://doi.org/10.1007/978-3-319-91189-2_16

Index Matrices as a Decision-Making Tool for Job Appointment

Velichka Traneva[1]([envelope]) [ORCID], Vassia Atanassova[2] [ORCID], and Stoyan Tranev[1] [ORCID]

[1] "Prof. Asen Zlatarov" University,
"Prof. Yakimov" Blvd, 8000 Bourgas, Bulgaria
veleka13@gmail.com,tranev@abv.bg
[2] Institute of Biophysics and Biomedical Engineering,
Bulgarian Academy of Sciences, Sofia, Bulgaria
vassia.atanassova@gmail.com,
http://www.btu.bg
http://biomed.bas.bg

Abstract. The paper explores the process of decision making, related to the appointment of the human factor in an incomplete information environment. We propose for the first time a new approach to optimization of the process of appointment and reappointment, based on partial knowledge about the values of evaluation criteria of the human resources over time, using the apparatuses of index matrices and of intuitionistic fuzzy sets.

In the paper, the 3-dimensional optimal appointment problem is formulated and an algorithm for its optimal solution is proposed, where the evaluations of candidates against criteria formulated by several experts in a certain time (or location), are intuitionistic fuzzy pairs. The proposed algorithm for the solution takes into account the ratings of the experts and the weighting coefficients of the assessment criterion according to its priority for the respective position.

Keywords: Appointment · Index matrix · Intuitionistic fuzzy pair

1 Introduction

The paper explores the process of decision-making related to the appointment of the human factor in an incomplete information environment, considered over time. We propose for the first time a new approach to optimization of the process for appointment (reappointment), based on partial knowledge about the values of evaluation criteria of the human resources in a time (or a location), using the apparatuses of index matrices (IMs), introduced in 1984 in [2], and of intuitionistic

Supported by the Bulgarian National Science Fund under Ref. No. DN-02-10 "New Instruments for Knowledge Discovery from Data, and their Modelling" and the project of Asen Zlatarov University under Ref. No. NIX-401/2017 "Modern methods of optimization and business management".

G. Nikolov et al. (Eds.): NMA 2018, LNCS 11189, pp. 158–166, 2019.
https://doi.org/10.1007/978-3-030-10692-8_18

fuzzy sets (IFSs), introduced in 1983 [1]. Problems of appointment to vacant positions in an organization by IFSs have been investigated in [11], and in [8] a two-dimensional model of the appointment process is presented in terms of IMs.

The rest of this paper is structured as follows. In Sect. 2, we present the preliminary concepts of IMs and of intuitionistic fuzzy pairs (IFPs). In Sect. 3, we define the intuitionistic fuzzy appointment problem, and suggest an algorithm for its optimal solution, in which evaluations of candidates against the criteria set by several experts in a certain moment time (or location) are IFPs.

2   Basic Definitions

This section provides some remarks on IFPs from [3] and on IMs from [4,12].

2.1   Short Notes on Intuitionistic Fuzzy Pairs

The **IFP** is an object with the form $\langle a, b \rangle$, where $a, b \in [0, 1]$ and $a + b \leq 1$, that is used as an evaluation of some object or process [6] and which components are interpreted as degrees of membership and non-membership.

Let us have two IFPs $x = \langle a, b \rangle$ and $y = \langle c, d \rangle$. In [6] are defined operations:

$$\neg x = \langle b, a \rangle; \qquad x \wedge_1 y = \langle \min(a, c), \max(b, d) \rangle;$$
$$x \vee_1 y = \langle \max(a, c) \rangle, \min(b, d) \rangle; \; x \wedge_2 y = x + y = \langle a + c - a.c, b.d \rangle;$$
$$x \vee_2 y = xy = \langle ac, b + d - b.d \rangle; \qquad x - y = \langle \min(a, d), \max(b, c) \rangle$$

and relations

$$x < y \text{ iff } a < c \text{ and } b > d; \; x \leq y \text{ iff } a \leq c \text{ and } b \geq d;$$
$$x = y \text{ iff } a = c \text{ and } b = d.$$

Let a set E be fixed. An **Intuitionistic Fuzzy Set** (IFS) A in E is an object of the following form (see [3]):

$$A = \{\langle x, \mu_A(x), \nu_A(x) \rangle | x \in E\},$$

where functions $\mu_A : E \to [0, 1]$ and $\nu_A : E \to [0, 1]$ define the degree of membership and the degree of non-membership of the element $x \in E$, respectively, and for every $x \in E$: $0 \leq \mu_A(x) + \nu_A(x) \leq 1$.

2.2   Definition of 3D-IFIM, Operations and Relations

Let $\mathcal{I}$ be a fixed set. By three-dimensional intuitionistic fuzzy index matrix (3D-IFIM) with index sets K, L and H ($K, L, H \subset \mathcal{I}$), we denote the object:

$$[K, L, H, \{\langle \mu_{k_i, l_j, h_g}, \nu_{k_i, l_j, h_g} \rangle\}]$$

$$\equiv \begin{array}{c|ccccc}
h_g \in H & l_1 & \cdots & l_j & \cdots & l_n \\
\hline
k_1 & \langle \mu_{k_1, l_1, h_g}, \nu_{k_1, l_1, h_g} \rangle & \cdots & \langle \mu_{k_1, l_j, h_g}, \nu_{k_1, l_j, h_g} \rangle & \cdots & \langle \mu_{k_1, l_n, h_g}, \nu_{k_1, l_n.h_g} \rangle \\
\vdots & \vdots & \cdots & \vdots & \cdots & \vdots \\
k_m & \langle \mu_{k_m, l_1, h_g}, \nu_{k_m, l_1, h_g} \rangle & \cdots & \langle \mu_{k_m, l_j, h_g}, \nu_{k_m, l_j, h_g} \rangle & \cdots & \langle \mu_{k_m, l_n, h_g}, \nu_{k_m, l_n, h_g} \rangle
\end{array},$$

where for every $1 \leq i \leq m, 1 \leq j \leq n,\ 1 \leq g \leq f$:

$$0 \leq \mu_{k_i,l_j,h_g}, \nu_{k_i,l_j,h_g}, \mu_{k_i,l_j,h_g} + \nu_{k_i,l_j,h_g} \leq 1.$$

Let $\mathcal{X}, Y, Z, U$ be fixed sets. Let operations "$*$" and "$\circ$" be defined so that: $* : \mathcal{X} \times \mathcal{Y} \to \mathcal{Z}$ and $\circ : \mathcal{Z} \times \mathcal{Z} \to \mathcal{U}$. Following [4,12], we recall some operations over IMs. Let us have two 3D-IFIMs $A = [K, L, H, \{\langle \mu_{k_i,l_j,h_g}, \nu_{k_i,l_j,h_g}\rangle\}]$ and $B = [P, Q, E, \{\langle \rho_{p_r,q_s,e_d}, \sigma_{p_r,q_s,e_d}\rangle\}]$.

Addition-(max, min)

$A \oplus_{(\max,\min)} B = [K \cup P, L \cup Q, H \cup E, \{\langle \phi_{t_u,v_w,x_y}, \psi_{t_u,v_w,x_y}\rangle\}]$, where

$$\langle \phi_{t_u,v_w,x_y}, \psi_{t_u,v_w,x_y}\rangle$$

$$= \begin{cases} \langle \mu_{k_i,l_j,h_g}, \nu_{k_i,l_j,h_g}\rangle, & \begin{array}{l} \text{if } t_u = k_i \in K, v_w = l_j, x_y = h_g \in H - E \\ \text{or } t_u = k_i \in K, v_w = l_j \in L - Q, x_y = h_g \in H \\ \text{or } t_u = k_i \in K - P, v_w = l_j \in L, x_y = h_g \in H; \end{array} \\[2em] \langle \rho_{p_r,q_s,e_d}, \sigma_{p_r,q_s,e_d}\rangle, & \begin{array}{l} \text{if } t_u = p_r \in P, v_w = q_s \in Q, x_y = e_d \in E - H \\ \text{or } t_u = p_r \in P, v_w = q_s \in Q - L, x_y = e_d \in E \\ \text{or } t_u = p_r \in P - K, v_w = q_s \in Q, x_y = e_d \in E; \end{array} \\[2em] \langle \max(\mu_{k_i,l_j,h_g}, \rho_{p_r,q_s,e_d}), & \text{if } t_u = k_i = p_r \in K \cap P, v_w = l_j = q_s \in L \cap Q \\ \min(\nu_{k_i,l_j,h_g}, \sigma_{p_r,q_s,e_d})\rangle, & \text{and } x_y = h_g = e_d \in H \cap E; \\[1em] \langle 0, 1\rangle, & \text{otherwise.} \end{cases}$$

Termwise multiplication-$(\min, \max)$

$$A \otimes_{(\min,\max)} B = [K \cap P, L \cap Q, H \cap R, \{\langle \phi_{t_u,v_w,x_y}, \psi_{t_u,v_w,x_y}\rangle\}],$$

where $\langle \phi_{t_u,v_w,x_y}, \psi_{t_u,v_w,x_y}\rangle = \langle \min(\mu_{k_i,l_j,h_g}, \rho_{p_r,q_s,e_d}), \max(\nu_{k_i,l_j,h_g}, \sigma_{p_r,q_s,e_d})\rangle$.

Multiplication with a constant α

$$\alpha A = [K, L, H\{\alpha\langle \mu_{k_i,l_j,h_g}, \nu_{k_i,l_j,h_g}\rangle\}],$$

If α is a real number then following [4,9]

$$\alpha\langle \mu_{k_i,l_j,h_g}, \nu_{k_i,l_j,h_g}\rangle = \langle 1 - (1 - \mu_{k_i,l_j,h_g})^\alpha, \nu_{k_i,l_j,h_g}^\alpha\rangle.$$

If $\alpha = \langle a, b\rangle$ is an IFP, then

$$\alpha\langle \mu_{k_i,l_j,h_g}, \nu_{k_i,l_j,h_g}\rangle = \langle a.\mu_{k_i,l_j,h_g}, b + \nu_{k_i,l_j,h_g} - b.\nu_{k_i,l_j,h_g}\rangle.$$

Multiplication

$$A \odot_{(\circ,*)} B = [K \cup (P - L), Q \cup (L - P), H \cup R, \{\langle \phi_{t_u,v_w,x_y}, \psi_{t_u,v_w,x_y}\rangle\}],$$

where

$$\langle \phi_{t_u,v_w,x_y}, \psi_{t_u,v_w,x_y} \rangle$$

$$= \begin{cases} \langle \mu_{k_i,l_j,h_g}, \nu_{k_i,l_j,h_g} \rangle, & \text{if } t_u = k_i \in K \\ & \& \ v_w = l_j \in L - P - Q \ \& \ x_y = h_g \in H \\ & \text{or } t_u = k_i \in K - P - Q \\ & \& \ v_w = l_j \in L \ \& \ x_y = h_g \in H; \\[2ex] \langle \rho_{p_r,q_s,r_d}, \sigma_{p_r,q_s,r_d} \rangle, & \text{if } t_u = p_r \in P \\ & \& \ v_w = q_s \in Q - K - L \ \& \ x_y = r_d \in R \\ & \text{or } t_u = p_r \in P - L - K \\ & \& \ v_w = q_s \in Q \ \& \ x_y = r_d \in R; \\[2ex] \langle \underset{l_j=p_r\in L\cap P}{\circ} (*(\mu_{k_i,l_j,h_g}, \rho_{p_r,q_s,r_d})), & \text{if } t_u = k_i \in K \ \& \ v_w = q_s \in Q \\ \quad \underset{l_j=p_r\in L\cap P}{*} (\circ(\nu_{k_i,l_j,h_g}, \sigma_{p_r,q_s,r_d}))\rangle, & \& \ x_y = h_g = r_d \in H \cap R; \\[2ex] \langle 0,1 \rangle, & \text{otherwise.} \end{cases}$$

where $\langle \circ, * \rangle \in \{\langle max, min \rangle, \langle min, max \rangle\}$.

Projection: Let $M \subseteq K$, $N \subseteq L$ and $U \subseteq H$. Then,

$$pr_{M,N,U}A = [M, N, U, \{b_{k_i,l_j,h_g}\}],$$

and for each $k_i \in M, l_j \in N$ and $h_g \in U, b_{k_i,l_j,h_g} = a_{k_i,l_j,h_g}$.

Aggregation operation by one dimension
Let $h_0 \notin H$. Let $\circ : \mathcal{X} \times \mathcal{X} \longrightarrow \mathcal{X}$, $* : \mathcal{X} \times \mathcal{X} \longrightarrow \mathcal{X}$ and
$\langle \circ, * \rangle \in \{\langle min, max \rangle, \langle max, min \rangle, \langle average, average \rangle\}$.

The definition of the aggregation operation by the dimension H [14] is:

$$\alpha_{(H,\circ,*)}(A, h_0) = \left\{ \begin{array}{c|c} l_j & h_0 \\ \hline \\ k_1 & \langle \underset{1\le g\le f}{\circ} \mu_{k_1,l_j,h_g}, \ \underset{1\le g\le f}{*} \nu_{k_1,l_j,h_g} \rangle \\[1ex] k_2 & \langle \underset{1\le g\le f}{\circ} \mu_{k_2,l_j,h_g}, \ \underset{1\le g\le f}{*} \nu_{k_2,l_j,h_g} \rangle \ | \ l_j \in L \\ \vdots & \vdots \\ k_m & \langle \underset{1\le g\le f}{\circ} \mu_{k_m,l_j,h_g}, \ \underset{1\le g\le f}{*} \nu_{k_m,l_j,h_g} \rangle \end{array} \right\}.$$

3   Index Matrices as a Decision-Making Tool for Job Appointment

Here, we will formulate a new type of three-dimensional intuitionistic fuzzy appointment problem, and will propose an algorithm for its optimal solution, based on the concepts of IMs and IFSs.

3.1   Problem Formulation

An organization has u vacant positions $\{v_1, \ldots, v_e, \ldots, v_u\}$. Employees who have never worked in the organization apply to these positions, but there are those who are currently working or have ever worked in it. A system for evaluating staff by criteria $\{c_1, \ldots, c_j, \ldots, c_n\}$ in a time-moment (location) h_g (for $1 \leq g \leq f$) by the experts $\{d_1, \ldots, d_s, \ldots, d_D\}$ operates in the organization. The candidates $\{k_1, \ldots, k_i, \ldots, k_m\}$ for the positions occupying or having similar positions in the organization have been periodically evaluated according to the criteria for the fulfillment of their official duties and their estimates es_{k_i, c_j, d_s} (for $1 \leq i \leq m, 1 \leq j \leq n, 1 \leq s \leq D$) are IFPs. At the moment of applying for vacancies, all candidates are evaluated by the experts according to the criteria for job hiring, and their evaluations es_{k_i, c_j, d_s} (for $1 \leq i \leq m, 1 \leq j \leq n, 1 \leq s \leq D$) are IFPs. The ratings of the experts $\{r_1, \ldots, r_w, \ldots, r_D\}$ be given, as well as the weighting coefficients of the assessment criteria c_j (for $1 \leq j \leq n$) according to their priority for the respective position v_e (for $1 \leq e \leq u$) $- pk_{c_j, v_e}$. The aim of the problem is to optimally allocate vacancies among the candidates for them.

3.2   An Algorithm for Finding an Optimal Solution to the Problem

Let us create the following 3D-IFIM, in accordance with the above problem:

$$ES[K, C, D, \{es_{k_i, c_j, d_s}\}]$$

$d_s \in D$	c_1	$\cdots$	c_j	$\cdots$	c_n
k_1	$\langle \mu_{k_1,c_1,d_s}, \nu_{k_1,c_1,d_s} \rangle$	$\cdots$	$\langle \mu_{k_1,c_j,d_s}, \nu_{k_1,c_j,d_s} \rangle$	$\cdots$	$\langle \mu_{k_1,c_n,d_s}, \nu_{k_1,c_n,d_s} \rangle$
$\vdots$	$\vdots$	$\ddots$	$\vdots$	$\ddots$	$\vdots$
k_m	$\langle \mu_{k_m,c_1,d_s}, \nu_{k_m,c_1,d_s} \rangle$	$\cdots$	$\langle \mu_{k_m,c_j,d_s}, \nu_{k_m,c_j,d_s} \rangle$	$\cdots$	$\langle \mu_{k_m,c_n,d_s}, \nu_{k_m,c_n,d_s} \rangle$

where $K = \{k_1, k_2, \ldots, k_m\}$, $C = \{c_1, c_2, \ldots, c_n\}$, $D = \{d_1, d_2, \ldots, d_D\}$ and the element $\{es_{k_i,c_j,d_s}\} = \langle \mu_{k_i,c_j,d_s}, \nu_{k_i,c_j,d_s} \rangle$ (for $1 \leq i \leq m, 1 \leq j \leq n, 1 \leq s \leq D$) is the estimate of the d_s-th expert for the k_i-th employee by the c_j-th criterion.

Note: The evaluation of the candidate $k_i \in K$ (for $1 \leq i \leq m$) on the criteria $c_j \in C$ (for $1 \leq j \leq n$) by the expert $d_s \in D$ may be changed through adding the hesitancy degrees to the membership degrees μ_{k_i,c_j,d_s}, if the forecast is optimistic. As a result, the evaluation of the candidate $k_i \in K$ against the criteria $c_j \in C$ by the expert $d_s \in D$ is expressed with a value from the interval $[\mu_{k_i,c_j,d_s}; \mu_{k_i,c_j,d_s} + \pi_{k_i,c_j,d_s}][10]$.

If the experts have a rating of the set $\{r_1, \ldots, r_w, \ldots, r_D\}$, it is necessary before the aggregate evaluation of the candidates to perform the following operations in order to obtain a rating matrix that takes the ratings:

$$ES^*[K, C, D, \{es^*_{k_i,c_j,d_s}\}]$$
$$= r_1 pr_{K,C,d_1} ES \oplus_{(max,min)} r_2 pr_{K,C,d_2} ES \ldots \oplus_{(max,min)} r_D pr_{K,C,d_D} ES,$$
$$ES := ES^*(es_{k_i,l_j} = es^*_{k_i,l_j}, \; \forall k_i \in K, \forall l_j \in L).$$

Let us apply the α_D-th aggregation operation to find the aggregate value of the k_i-th candidate against the c_j-th criterion in a time-moment h_g, (for $1 \le i \le m, 1 \le j \le n, 1 \le g \le f$) as follows:

$$\alpha_{(D,\circ,*)}(ES, h_g) = \left\{ \begin{array}{c|cc} c_j & \multicolumn{2}{c}{h_g} \\ \hline k_1 & \langle \; \underset{1 \le s \le D}{\circ} \; \mu_{k_1,c_j,d_s}, & * \; \underset{1 \le s \le D}{\nu_{k_1,c_j,d_s}} \\ \vdots & & \\ k_m & \langle \; \underset{1 \le s \le D}{\circ} \; \mu_{k_m,c_j,d_s}, & * \; \underset{1 \le s \le D}{\nu_{k_m,c_j,d_s}} \rangle \end{array} \;\middle|\; c_j \in C \right\},$$

where $h_g \notin D$ and $\langle \circ, * \rangle \in \{\langle min, max \rangle, \langle max, min \rangle, \langle average, average \rangle\}$.

Let us create the 3D-IFIM $A[K, C, H, \{a_{k_i,c_j,h_g}\}]$
$$= \alpha_{(D,\circ,*)}(ES, h_1) \oplus_{(max,min)} \alpha_{(D,\circ,*)}(ES, h_2) \ldots \oplus_{(max,min)} \alpha_{(D,\circ,*)}(ES, h_f)$$

$$= \begin{array}{c|ccccc} h_g \in D & c_1 & \cdots & c_j & \cdots & c_n \\ \hline k_1 & a_{k_1,c_1,h_g} & \cdots & a_{k_1,c_j,h_g} & \cdots & a_{k_1,c_n,h_g} \\ \vdots & \vdots & \ddots & \vdots & \ddots & \vdots \\ k_m & a_{k_m,c_1,h_g} & \cdots & a_{k_m,c_j,h_g} & \cdots & a_{k_m,c_n,h_g} \end{array},$$

where $K = \{k_1, k_2, \ldots, k_m\}$, $C = \{c_1, c_2, \ldots, c_n\}$, $H = \{h_1, h_2, \ldots h_f\}$ and for $1 \le i \le m, 1 \le j \le n, 1 \le g \le f : \{a_{k_i,c_j,h_g}\} = \langle \mu_{k_i,c_j,h_g}, \nu_{k_i,c_j,h_g} \rangle$ is the estimate of the k_i-th candidate for the c_j-th criterion in a time-moment h_g. If $\{a_{k_i,c_j,h_g}\}$ is an empty cell, then $\{a_{k_i,c_j,h_g}\} = \langle 0, 1 \rangle$. Then we apply the aggregation operation by the dimension H to find the aggregated evolutions of the k_i-th candidate for the c_j-th criterion for the whole period in which one worked or is still working in the organization (if one has not worked in the organization, one's aggregate score is the same as that applied for one's application for employment):

$$\alpha_{(H,\circ,*)}(A, h_0) = \left\{ \begin{array}{c|cc} c_j & \multicolumn{2}{c}{h_0} \\ \hline k_1 & \langle \; \underset{1 \le s \le D}{\circ} \; \mu_{k_1,c_j,d_s}, & * \; \underset{1 \le s \le D}{\nu_{k_1,c_j,d_s}} \rangle \\ \vdots & & \\ k_m & \langle \; \underset{1 \le s \le D}{\circ} \; \mu_{k_m,c_j,d_s}, & * \; \underset{1 \le s \le D}{\nu_{k_m,c_j,d_s}} \rangle \end{array} \;\middle|\; c_j \in C \right\} \qquad (1)$$

where $h_0 \notin H$ and

$$\langle \circ, * \rangle \in \{\langle min, max \rangle, \langle max, min \rangle, \langle average, average \rangle\}. \tag{2}$$

If the $\langle min, max \rangle$-pair is used in the aggregation operation (1), then we accept the pessimistic forecast for a candidate's evaluation by a given criterion. With $\langle max, min \rangle$, we get the optimistic forecast for a candidate by the criterion and with the pair $\langle average, average \rangle$, we get the average estimate of a candidate.

Let us define the 2D-IM PK of the weighting coefficients of the assessment criterion according to its priority to the corresponding position v_e $(1 \leq e \leq u)$

$$PK[C, V, \{pk_{c_j, v_e}\}] = \begin{array}{c|ccccc} & v_1 & \cdots & v_e & \cdots & v_u \\ \hline c_1 & pk_{c_1, v_1} & \cdots & pk_{c_1, v_e} & \cdots & pk_{c_1, v_u} \\ \vdots & \vdots & \ddots & \vdots & \ddots & \vdots \\ c_j & pk_{c_j, v_1} & \cdots & pk_{c_j, v_e} & \cdots & pk_{c_j, v_u} \\ \vdots & \vdots & \ddots & \vdots & \ddots & \vdots \\ c_n & pk_{c_n, v_1} & \cdots & pk_{c_n, v_e} & \cdots & pk_{c_n, v_u} \end{array},$$

where $C = \{c_1, c_2, \ldots, c_n\}$, $V = \{v_1, v_2, \ldots, v_u\}$ and for $1 \leq j \leq n, 1 \leq e \leq u$: pk_{c_j, v_e} are IFPs. Then we create 2D-IFIM $B[K, V, \{b_{k_i, v_e}\}]$:

$$B = \alpha_{(H, max, min)}(A, h_0) \odot_{(\circ, *)} PK,$$

which contains the cumulative estimates of the k_i-th candidate (for $1 \leq i \leq m$) for the v_e-th vacancy (for $1 \leq e \leq u$) and $\langle \circ, * \rangle$ are the same as (2). After this operation, we apply the aggregation operation $\alpha_{(K, max, min)}(B, k_0)$ by the dimension K to find the most suitable candidate for the vacant position v_e.

$$\alpha_{(K, max, min)}(B, k_0)$$

$$= \left\{ \begin{array}{c|ccc} & v_1 & \cdots & v_u \\ \hline k_0 & \langle \max\limits_{1 \leq i \leq m} \mu_{k_i, v_1}, \min\limits_{1 \leq i \leq m} \nu_{k_i, v_1} \rangle \cdots \langle \max\limits_{1 \leq i \leq m} \mu_{k_i, v_u}, \min\limits_{1 \leq i \leq m} \nu_{k_i, v_u} \rangle \end{array} \right\},$$

where $k_0 \notin K$. After applying the intuitionistic fuzzy hungarian algorithm [15] with the cost IM B, we will get the optimal allocation of the candidates to the jobs. In practice, evaluation criteria can be primary and secondary. The algorithm for selecting the most suitable candidate for some vacancy can be applied to the primary criteria. If several candidates have the same aggregate grade on the core criteria for a vacancy, then the same algorithm needs to be applied to the secondary factors. Prior to finding the optimal allocation of candidates to the vacancies, one of the level-operators to IMs, setting a threshold for job applicants [13] may be used.

4 Conclusion

We presented here a new approach to decision making in the process of appointment of the human factor by integrating incomplete information from independent experts, based on the concepts of IFSs and IMs. This method can be used in real life applications, where data quality is a matter of concern.

The outlined approach to supporting decision making, related to recruitment, has the following advantages: it can be applied to both the assignment problem with crisp parameters and with fuzzy ones; the algorithm can be extended in order to obtain the optimal solution for other types of multidimensional appointment problems.

Our future research will be related to the staff recruitment process, in which expert evaluations of candidates under the criteria are interval valued IFSs [5].

References

1. Atanassov K. T.: Intuitionistic Fuzzy Sets, VII ITKR Session, Sofia, 20–23 June 1983 (Deposed in Centr. Sci.-Techn. Library of the Bulg. Acad. of Sci., 1697/84) (in Bulgarian). Reprinted: Int. J. Bioautom. **20**(S1), S1–S6 (2016)
2. Atanassov, K.: Generalized index matrices. Comptes rendus de l'Academie Bulgare des Sciences **40**(11), 15–18 (1987)
3. Atanassov, K.: On Intuitionistic Fuzzy Sets Theory. STUDFUZZ, vol. 283. Springer, Heidelberg (2012). https://doi.org/10.1007/978-3-642-29127-2
4. Atanassov, K.: Index Matrices: Towards an Augmented Matrix Calculus. Studies in Computational Intelligence, vol. 573. Springer, Cham (2014). https://doi.org/10.1007/978-3-319-10945-9
5. Atanassov, K.: Interval Valued Intuitionistic Fuzzy Sets. World Scientific (2018, in press)
6. Atanassov, K., Szmidt, E., Kacprzyk, J.: On intuitionistic fuzzy pairs. Notes Intuitionistic Fuzzy Sets **19**(3), 1–13 (2013)
7. Bratton, J., Gold, J.: Human Resource Management: Theory and Practice, 5th edn. Macmillan Press, London (2012)
8. Bureva, V., Sotirova, E., Moskova, M., Szmidt, E.: Solving the problem of appointments using index matrices. In: Proceedings of the 13th International Workshop on Generalized Nets, vol. 29, pp. 43–48 (2012)
9. De, S.K., Biswas, R., Roy, A.R.: Some operations on intuitionistic fuzzy sets. Fuzzy sets Syst. **114**(4), 477–484 (2000)
10. Deng-Feng, L.: Decision and Game Theory in Management with Intuitionistic Fuzzy Sets. Studies in Fuzziness and Soft Computing, vol. 308. Springer, Heidelberg (2014). https://doi.org/10.1007/978-3-642-40712-3
11. Ejegwa, P., Kwarkar, L., Ihuoma, K.N.: Application of intuitionistic fuzzy multisets in appointment process. Int. J. Comput. Appl. (0975–8887) **135**(1), 1–4 (2016)
12. Traneva, V., Tranev, S.: Index Matrices as a Tool for Managerial Decision Making. Publ. House of the Union of Scientists, Bulgaria (2017). (in Bulgarian)
13. Traneva, V., Tranev, S., Atanassov, K.: Level Operators Over Intuitionistic Fuzzy Index Matrix (2018, in press)

14. Traneva, V., Sotirova, E., Bureva, V., Atanassov, K.: Aggregation operations over 3-dimensional extended index matrices. Adv. Stud. Contemp. Math. **25**(3), 407–416 (2015)
15. Traneva, V., Tranev, S., Atanassova, V.: An Intuitionistic Fuzzy Approach to the Hungarian Algorithm (2018, in press)

An Intuitionistic Fuzzy Approach
to the Hungarian Algorithm

Velichka Traneva$^{1(\boxtimes)}$ ⓘ, Stoyan Tranev^1ⓘ, and Vassia Atanassova^2ⓘ

^{1} "Prof. Asen Zlatarov" University, "Prof. Yakimov" Blvd., 8000 Bourgas, Bulgaria
veleka13@gmail.com, tranev@abv.bg
^{2} Institute of Biophysics and Biomedical Engineering,
Bulgarian Academy of Sciences, Sofia, Bulgaria
vassia.atanassova@gmail.com
http://www.btu.bg
http://www.biomed.bas.bg

Abstract. In the paper a new type of assignment problem is formulated, in which the costs of assigning tasks to candidates are intuitionistic fuzzy pairs. Additional constraints are formulated to the problem: an upper limit to the cost of assigning a particular resource to perform a particular activity and preferences defined in advance for assigning the resources by an index matrix. We propose for the first time the Hungarian algorithm for finding an optimal solution of this new type of assignment problem, based on the concept of index matrices.

Keywords: Assignment problem · Decision making
Hungarian algorithm · Index matrix · Intuitionistic fuzzy pair

1 Introduction

An assignment problem is a special type of linear programming problem which deals with assigning various activities to an equal number of resources on one-to-one basis in such a way so that the total time or total cost involved is minimized and total sale or total profit is maximized. The basic combinatorial (nonsimplex) method for the assignment problem is the Hungarian method, which is developed by Kuhn in 1955 [7] and solves the problem for $O(n^4)$ time. In today's dynamic market environment, parameters are increasingly fuzzy and unclear. Therefore the use of intuitionistic fuzzy set theory proposed by Atanassov [1] in 1983 is more appropriate to model the real problems involving imprecise parameters. The latest scientific results, associated with finding an optimal solution of assignment problem with triangular intuitionistic fuzzy costs (special case of intuitionistic fuzzy numbers), are in the paper [8]. In our paper a new type of assignment problem is formulated, in which the costs of assigning tasks to candidates are intuitionistic fuzzy pairs (IFPs) and also is defined the hungarian

Supported by the project of Asen Zlatarov University under Ref. No. NIX-401/2017 "Modern methods of optimization and business management".

algorithm for its optimal solution, based on the theory of index matrices (IMs, introduced in 1984 in [2]).

2   Basic Definitions

This section provides some remarks on intuitionistic fuzzy pairs from [3,6] and on index matrix tools from [4,10].

2.1   Short Notes on Intuitionistic Fuzzy Logic

The **IFP** is an object of the form $\langle a, b \rangle = \langle \mu(p), \nu(p) \rangle$, where $a, b \in [0, 1]$ and $a + b \leq 1$, that is used as an evaluation of a proposition p [5,6]. $\mu(p)$ and $\nu(p)$ respectively determine the "truth degree" and "falsity degree".

Let us have two IFPs $x = \langle a, b \rangle$ and $y = \langle c, d \rangle$. In [6] are defined operations and relations:

$$\neg x = \langle b, a \rangle; \qquad\qquad x \wedge_1 y = \langle \min(a, c), \max(b, d) \rangle;$$
$$x \vee_1 y = \langle \max(a, c), \min(b, d) \rangle; \qquad x \wedge_2 y = x + y = \langle a + c - a.c, b.d \rangle;$$
$$x \vee_2 y = x.y = \langle a.c, b + d - b.d \rangle; \qquad \alpha.x = \langle 1 - (1 - a)^\alpha, b^\alpha \rangle (\alpha \in R)$$
$$x - y \;\; = \langle \max(0, a - c), \min(1, b + d, 1 - a + c) \rangle$$

and relations

$$x \geq y \text{ iff } a \geq c \text{ and } b \leq d; x \leq y \text{ iff } a \leq c \text{ and } b \geq d;$$
$$x = y \text{ iff } a = c \text{ and } b = d.$$

The IFP x is an **"intuitionistic fuzzy false pair"** (**IFFP**) if and only if $a \leq b$, while x is a **"false pair"** (**FP**) iff $a = 0, b = 1$.

Let a set E be fixed. An **"intuitionistic fuzzy set"** (**IFS**) A in E is an object of the following form (see [3]): $A = \{\langle x, \mu_A(x), \nu_A(x) \rangle | x \in E\}$, where functions $\mu_A : E \to [0, 1]$ and $\nu_A : E \to [0, 1]$ define the degrees of membership and the non-membership of the $x \in E$, respectively, and for every $x \in E$: $0 \leq \mu_A(x) + \nu_A(x) \leq 1$.

2.2   Definition, Operations and Relations over 2D-IFIMs

Let $\mathcal{I}$ be a fixed set. By two dimensional intuitionistic fuzzy index matrix (2D-IFIM) with index sets K and L $(K, L \subset \mathcal{I})$, we denote the object:

$$[K, L, \{\langle \mu_{k_i, l_j}, \nu_{k_i, l_j} \rangle\}]$$

$$\equiv \begin{array}{c|ccccc} & l_1 & \cdots & l_j & \cdots & l_n \\ \hline k_1 & \langle \mu_{k_1, l_1}, \nu_{k_1, l_1} \rangle & \cdots & \langle \mu_{k_1, l_j}, \nu_{k_1, l_j} \rangle & \cdots & \langle \mu_{k_1, l_n}, \nu_{k_1, l_n} \rangle \\ \vdots & \vdots & \ddots & \vdots & \ddots & \vdots \\ k_m & \langle \mu_{k_m, l_1}, \nu_{k_m, l_1} \rangle & \cdots & \langle \mu_{k_m, l_j}, \nu_{k_m, l_j} \rangle & \cdots & \langle \mu_{k_m, l_n}, \nu_{k_m, l_n} \rangle \end{array},$$

where for every $1 \leq i \leq m, 1 \leq j \leq n$: $0 \leq \mu_{k_i,l_j}, \nu_{k_i,l_j}, \mu_{k_i,l_j} + \nu_{k_i,l_j} \leq 1$.

Following [4], we recall some operations over two IMs $A = [K, L, \{\langle \mu_{k_i,l_j}, \nu_{k_i,l_j} \rangle\}]$ and $B = [P, Q, \{\langle \rho_{p_r,q_s}, \sigma_{p_r,q_s} \rangle\}]$.

Addition-(max,min) $A \oplus_{(\max,\min)} B = [K \cup P, L \cup Q, \{\langle \phi_{t_u,v_w}, \psi_{t_u,v_w} \rangle\}]$, where

$$\langle \phi_{t_u,v_w}, \psi_{t_u,v_w} \rangle$$

$$= \begin{cases} \langle \mu_{k_i,l_j}, \nu_{k_i,l_j} \rangle, & \text{if } t_u = k_i \in K \text{ and } v_w = l_j \in L - Q \\ & \text{or } t_u = k_i \in K - P \text{ and } v_w = l_j \in L; \\ \langle \rho_{p_r,q_s}, \sigma_{p_r,q_s} \rangle, & \text{if } t_u = p_r \in P \text{ and } v_w = q_s \in Q - L \\ & \text{or } t_u = p_r \in P - K \\ & \text{and } v_w = q_s \in Q; \\ \langle \max(\mu_{k_i,l_j}, \rho_{p_r,q_s}), & \text{if } t_u = k_i = p_r \in K \cap P \\ \min(\nu_{k_i,l_j}, \sigma_{p_r,q_s}) \rangle, & \text{and } v_w = l_j = q_s \in L \cap Q; \\ \langle 0, 1 \rangle, & \text{otherwise.} \end{cases}$$

Termwise Multiplication-$(\min, \max)$

$$A \otimes_{(\min,\max)} B = [K \cap P, L \cap Q, \{\langle \phi_{t_u,v_w}, \psi_{t_u,v_w} \rangle\}],$$

where $\langle \phi_{t_u,v_w}, \psi_{t_u,v_w} \rangle = \langle \min(\mu_{k_i,l_j}, \rho_{p_r,q_s}), \max(\nu_{k_i,l_j}, \sigma_{p_r,q_s}) \rangle$

Reduction: We use symbol "$\perp$" for lack of some component in the separate definitions. In some cases, it is suitable to change this symbol with "0". The operations $(k, \perp)$-reduction of a given IM A is defined by:

$$A_{(k,\perp)} = [K - \{k\}, L, \{c_{t_u,v_w}\}],$$

where $c_{t_u,v_w} = a_{k_i,l_j}$ for $t_u = k_i \in K - \{k\}$ and $v_w = l_j \in L$.

Projection: Let $M \subseteq K$ and $N \subseteq L$. Then, $pr_{M,N} A = [M, N, \{b_{k_i,l_j}\}]$, where for each $k_i \in M$ and each $l_j \in N$, $b_{k_i,l_j} = a_{k_i,l_j}$.

Substitution: Let IM $A = [K, L, \{a_{k,l}\}]$ be given. Local substitution over the IM is defined for the couples of indices (p, k) and/or (q, l), respectively, by

$$\left[\frac{p}{k}; \perp\right] A = [(K - \{k\}) \cup \{p\}, L, \{a_{k,l}\}], \quad \left[\perp; \frac{q}{l}\right] A = [K, (L - \{l\}) \cup \{q\}, \{a_{k,l}\}].$$

Index type operations: $AGIndex_{(\min,\max),(\not{L})}(A) = \langle k_i, l_j \rangle$, which finds the index of the minimum element of A that has no empty value.

$$Index_{(\not{L})}(A) = \{\langle k_1, l_{v_1} \rangle, \ldots, \langle k_i, l_{v_i} \rangle, \ldots, \langle k_m, l_{v_m} \rangle\},$$

where $\langle k_i, l_{v_i} \rangle$ (for $1 \leq i \leq m$) is the index of the element of A, whose cell is full.

$$Index_{(\max \nu),k_i}(A) = \{\langle k_i, l_{v_1} \rangle, \ldots, \langle k_i, l_{v_x} \rangle, \ldots, \langle k_i, l_{v_V} \rangle\},$$

where $\langle k_i, l_{v_x} \rangle$ (for $1 \leq i \leq V, 1 \leq x \leq n$) is the indices of the IFFP of k_i-th row of A, for which $\nu_{k_i,l_{v_x}}$ is maximum.

$$Index_{(\max \nu),l_j}(A) = \{\langle k_{w_1}, l_j \rangle, \ldots, \langle k_{w_y}, l_j \rangle, \ldots, \langle k_{w_W}, l_j \rangle\},$$

where $\langle k_{w_y}, l_j \rangle$ (for $1 \leq y \leq W, 1 \leq j \leq n$) are the indices of the IFFP of l_j-th column of A, for which $\nu_{k_{w_y}, l_j}$ is maximum.

Aggregate global internal operation: $AGIO_{\oplus(\max,\min)}(A)$

This operation find the "$\oplus_{(\max,\min)}$"-operation of all the matrix elements.

Internal subtraction of IMs' components ([9,10]):

$$IO_{-(\max,\min)}(\langle k_i, l_j, A \rangle, \langle p_r, q_s, B \rangle) = [K, L, \{\langle \gamma_{t_u, v_w}, \delta_{t_u, v_w} \rangle\}],$$

where $k_i \in K$, $l_j \in L$; $p_r \in P$, $q_s \in Q$ and $\langle \gamma_{t_u, v_w}, \delta_{t_u, v_w} \rangle$

$$= \begin{cases} \langle \mu_{t_u, v_w}, \nu_{t_u, v_w} \rangle, & \text{if } t_u \neq k_i \in K, \\ & v_w \neq l_j \in L; \\ \langle \max(0, \mu_{k_i, l_j} - \rho_{p_r, q_s}), & \text{if } t_u = k_i \in K, \\ \min(1, \nu_{k_i, l_j} + \sigma_{p_r, q_s}, 1 - \mu_{k_i, l_j} + \rho_{p_r, q_s}) \rangle & v_w = l_j \in L. \end{cases}$$

The non-strict relation "inclusion about value" is defined by:

$$A \subseteq_v B \text{ iff } (K = P) \ \& \ (L = Q) \ \& \ (\forall k \in K)(\forall l \in L)(a_{k,l} \leq b_{k,l})$$

3   An Intuitionistic Fuzzy Approach to the Hungarian Algorithm of a New Type of Assignment Problem

Here, we solve the two-dimensional form of a new type of an intuitionistic fuzzy assignment problem (IFAP) by the Hungarian algorithm, based on the concept of IMs. Its formulation is: m candidates (workers, resources) $\{k_1, \ldots, k_i, \ldots, k_m\}$ need to be assigned for n activities or $\{l_1, \ldots, l_j, \ldots, l_n\}$. The cost for assigning $c_{k_i, l_j}(\forall k_i, \forall l_j)$ of the k_i-th candidate for the l_j-th job is presented as IFPs. The additional constraints to the problem are: an intuitionistic fuzzy upper limit to the cost of assigning a particular candidate to perform some activity and preferences, defined in advance for assigning resources. The purpose of the problem is to minimize the total cost of completing all jobs of the candidates.

The mathematical model of the above problem is as follows:

$$\text{minimize } \sum_{i=1}^{m} \sum_{j=1}^{n} c_{k_i, l_j} x_{k_i, l_j}$$

$$\text{Subject to: } \sum_{j=1}^{n} x_{k_i, l_j} = \langle 1, 0 \rangle, \qquad i = 1, 2, \ldots, m; \tag{1}$$

$$\sum_{i=1}^{m} x_{k_i, l_j} = \langle 1, 0 \rangle, \qquad j = 1, 2, \ldots, n;$$

$$x_{k_i, l_j} \in \{\langle 0, 1 \rangle \text{ or } \langle 1, 0 \rangle\}, \qquad \text{for } \ 1 \leq i \leq m, \qquad 1 \leq j \leq n.$$

The additional constraints to the problem (1) are:

- c_{pl, l_j}, for $1 \leq j \leq n$ – an intuitionistic fuzzy upper limit to the cost of assigning a particular candidate to perform l_j-th activity;
- preferences defined in advance for assigning resources.

Note: The operations "addition" and "multiplication", used in the problem (1) are those for IFPs, defined in Sect. 2.1.

Let us construct the following IM, in accordance with the problem (1):

$$
C[K,L] = \begin{array}{c|cccc}
 & l_1 & \cdots & l_n & R \\
\hline
k_1 & \langle \mu_{k_1,l_1}, \nu_{k_1,l_1} \rangle & \cdots & \langle \mu_{k_1,l_n}, \nu_{k_1,l_n} \rangle & \langle \mu_{k_1,R}, \nu_{k_1,R} \rangle \\
\vdots & \vdots & \ddots & \vdots & \vdots \\
k_m & \langle \mu_{k_m,l_1}, \nu_{k_m,l_1} \rangle & \cdots & \langle \mu_{k_m,l_n}, \nu_{k_m,l_n} \rangle & \langle \mu_{k_m,R}, \nu_{k_m,R} \rangle \\
Q & \langle \mu_{Q,l_1}, \nu_{Q,l_1} \rangle & \cdots & \langle \mu_{Q,l_n}, \nu_{Q,l_n} \rangle & \langle \mu_{Q,R}, \nu_{Q,R} \rangle \\
pl & \langle \mu_{pl,l_1}, \nu_{pl,l_1} \rangle & \cdots & \langle \mu_{pl,l_n}, \nu_{pl,l_n} \rangle & \langle \mu_{pl,R}, \nu_{pl,R} \rangle
\end{array},
$$

where $K = \{k_1, k_2, \ldots, k_m, Q, pl\}$, $L = \{l_1, l_2, \ldots, l_n, R\}$ and for $1 \leq i \leq m$, $1 \leq j \leq n$, $\{c_{k_i,l_j}, c_{k_i,R}, c_{Q,l_j}, c_{pl,l_j}\}$ are IFPs. Let we denote by $|K| = m + 2$ the number of elements of the set K; then $|L| = n + 1$. We also define

$$
X[K*, L*] = \begin{array}{c|ccccc}
 & l_1 & \cdots & l_j & \cdots & l_n \\
\hline
k_1 & x_{k_1,l_1} & \cdots & x_{k_1,l_j} & \cdots & x_{k_1,l_n} \\
\vdots & \vdots & \ddots & \vdots & \ddots & \vdots \\
k_m & c_{k_m,l_1} & \cdots & c_{k_m,l_j} & \cdots & c_{k_m,l_n}
\end{array},
$$

$K* = \{k_1, k_2, \ldots, k_m\}$, $L* = \{l_1, l_2, \ldots, l_n\}$, and for $1 \leq i \leq m$, $1 \leq j \leq n$:

$$
x_{k_i,l_j} = \langle \rho_{k_i,l_j}, \sigma_{k_i,l_j} \rangle = \begin{cases} \langle 1, 0 \rangle, & \text{if candidate } k_i \text{ is assigned the job } l_j \\ \langle 0, 1 \rangle, & \text{otherwise} \end{cases}
$$

and let in the beginning of the algorithm x_{k_i,l_j} is an empty cell $\forall k_i \in K*$, $\forall l_j \in L*$.

Additionally, it is necessary to create the IM of preferences $PREF[K*, L*]$ for assigning resources, with the same structure as X, where $K* = \{k_1, k_2, \ldots, k_m\}$, $L* = \{l_1, l_2, \ldots, l_n\}$ and

$$
pref_{k_i,l_j} = \begin{cases} \langle 1, 0 \rangle, & \text{if } k_i\text{-th candidate wishes to assign the } l_j\text{-th activity} \\ \langle 0, 1 \rangle, & \text{if } k_i\text{-th candidate does not wish to assign the } l_j\text{-th activity} \end{cases}.
$$

Let us we define the following auxiliary matrices: $S = [K, L, \{s_{k_i,l_j}\}]$, such that $S = C$ i.e. ($s_{k_i,l_j} = c_{k_i,l_j} \ \forall k_i \in K, \forall l_j \in L$);

$$
D[K*, L*] = \begin{array}{c|ccccc}
 & l_1 & \cdots & l_j & \cdots & l_n \\
\hline
k_1 & d_{k_1,l_1} & \cdots & d_{k_1,l_j} & \cdots & d_{k_1,l_n} \\
\vdots & \vdots & \ddots & \vdots & \ddots & \vdots \\
k_m & d_{k_m,l_1} & \cdots & d_{k_m,l_j} & \cdots & d_{k_m,l_n}
\end{array},
$$

where $K* = \{k_1, k_2, \ldots, k_m\}$, $L* = \{l_1, l_2, \ldots, l_n\}$, and for $1 \leq i \leq m$, $1 \leq j \leq n$: $d_{k_i,l_j} = \{1 \text{ or } 2\}$, if the elements s_{k_i,l_j} of S are crossed out with 1 or 2 lines.

$$RC[K*, e_0] = \begin{array}{c|c} & e_0 \\ \hline k_1 & rc_{k_1,e_0} \\ \vdots & \vdots \\ k_m & rc_{k_m,e_0} \end{array},$$

where $K* = \{k_1, k_2, \ldots, k_m\}$ and for $1 \leq i \leq m$: $rc_{k_i,l_j} = \{0 \text{ or } 1\}$ depending on whether the k_i-th row of the matrix S is crossed out.

$$CC[r_0, L*] = \begin{array}{c|ccccc} & l_1 & \cdots & l_j & \cdots & l_n \\ \hline r_0 & ccd_{r_0,l_1} & \cdots & cc_{r_0,l_j} & \cdots & cc_{r_0,l_n} \end{array},$$

where $L* = \{l_1, l_2, \ldots, l_n\}$, $1 \leq j \leq n$: $cc_{k_i,l_j} = \{0 \text{ or } 1\}$ depending on whether the l_j-th row of the matrix S is crossed out. At the beginning of the algorithm $rc_{k_i,e_0} = 0, cc_{r_0,l_j} = 0, x_{k_i,l_j} = \langle 0, 1 \rangle$ $(\forall k_i \in K*, \forall l_j \in L*)$.

We will propose a new approach for the algorithm for finding the optimal solution of the assignment problem (1) with intuitionistic fuzzy data using the Hungarian method [7], interpreted with the tools of IMs. A part of Microsoft Visual Studio.NET 2010 C project's program code [11] is used in the algorithm.

Step 1. We compare the number of rows with the number of columns in C.

Step 1.1. If the number of rows is greater than the number of columns, then a dummy column $l_{n+1} \in \mathcal{I}$ is entered in the matrix C, in which all prices $c_{k_i,l_{n+1}} (i = 1, \ldots, m)$ are equal to $\langle 1, 0 \rangle$; otherwise, go to the *Step 1.2.* For this purpose, the following operations are executed:

- we define the 2D-IM $C_1[K/\{Q, pl\}, \{l_{n+1}\}, \{c_{1_{k_i,l_{n+1}}}\}]$, whose elements are equal to $\langle 1, 0 \rangle$;
- the new cost matrix is C obtained by:
  $C := C \oplus_{(\max,\min)} C_1; s_{k_i,l_j} = c_{k_i,l_j}, \forall k_i \in K, \forall l_j \in L$, Go to *Step 2*.

Step 1.2. If the number of columns is greater than the number of rows, then a dummy row $k_{m+1} \in \mathcal{I}$ is entered in the cost matrix, in which all prices are equal to $\langle 1, 0 \rangle$. Similar operations to those in *Step 1.1.* are performed.

Step 2. Let us create IM DC, such $DC := C_{(\{Q,pl\},\{R\})} \otimes_{(\min,\max)} PREF$.
for (int $i = 0; i < m; i++$)
for (int $j = 0; j < n; j++$)
{if $dc_{k_i,l_j} = \langle 0, 1 \rangle$, then $\{dc_{k_i,l_j} := \perp; c_{k_i,l_j} := dc_{k_i,l_j}\}$
If $\left(\left[\frac{k_i}{pl}; \perp \right] pr_{pl,l_j} C \right) \sqsubset_v pr_{k_i,l_j} C$, then $c_{k_i,l_j} = \perp$. }
Let $EG = Index_{\not\perp}(C_{(\{Q,pl\},\{R\})}) = \{\langle k_{i_1}, l_{j_1} \rangle, \langle k_{i_2}, l_{j_2} \rangle, \ldots, \langle k_{i_\phi}, l_{j_\phi} \rangle\}$;
If $|EG| < m$, then the problem has not a solution and end of the algorithm.
Let us construct IM $S = [K, L, \{s_{k_i,l_j}\}]$ such that $S = C$, i.e. $s_{k_i,l_j} = c_{k_i,l_j}$ $(\forall k_i \in K, \forall l_j \in L)$.

Step 3. In each row of the S, the smallest element is found and it is subtracted from all elements in the appropriate row. Go to *Step 4*.

Step 3.1. For each row of the matrix S, the smallest element is found and is recorded to the right of the row, in column R:

for (int $i = 0; i < m; i + +$)

for (int $j = 0; j < n; j + +$)

$\{AGIndex_{(\min,\max),(\not\perp)}\left(pr_{k_i,L}S\right) = \langle k_i, l_{v_j}\rangle;$

If $pr_{k_i,l_{v_j}}S \subseteq_v \left(\left[\frac{k_i}{pl}; \perp\right] pr_{pl,l_{v_j}}S\right)$, then

$S_1[k_i, l_{v_j}] = pr_{k_i,l_{v_j}}S; S_2 = \left[\perp; \frac{R}{l_{v_j}}\right] S1; S := S \oplus_{(\max,\min)} S_2.\}$

Step 3.2. From each element of the matrix S, subtract the smallest element in the same row:

for (int $i = 0; i < m; i + +$)

for (int $j = 0; j < n; j + +$)

If $s_{k_i,l_j} \neq \perp$ then $\{IO_{-_{(\max,\min)}}\left(\langle k_i, l_j, S\rangle, \langle k_i, R, pr_{K,R}S\rangle\right)\}.$

Step 4. For each column of the matrix S, the smallest element is found and it is subtracted from all elements in the corresponding column and go to *Step 5*.

Step 4.1. For each column of the matrix S, the smallest element is found and is recorded down at the bottom of the column, in line Q:

for (int $j = 0; j < n; j + +$)

$\{AGIndex_{(\min,\max),(\not\perp)}\left(pr_{K,l_j}S\right) = \langle k_{w_i}, l_j\rangle,$

Let us create two 2D-EIMs S_3 $S_4 : S_3[k_{w_i}, l_j] = pr_{k_{w_i},l_j}S; S_4 = \left[\frac{Q}{k_{w_i}}; \perp\right] S_3;$

$S := S \oplus_{(\max,\min)} S_4.\}$

Step 4.2. for (int $j = 0; j < n; j + +$)

for (int $i = 0; i < m; i + +$)

If $s_{k_i,l_j} \neq \perp$ then $\{IO_{-_{(\max,\min)}}\left(\langle k_i, l_j, S\rangle, \langle Q, l_j, pr_{Q,L}S\rangle\right)\}.$

Step 5. Cross out all elements $\langle 0, 1\rangle$, in the S with the minimum possible number of lines (horizontal, vertical or both). If there is no element $\langle 0, 1\rangle$ in a given row or column, then the element with the maximal falsity degree is cross out from that row or column. If the number of these lines is m, go to *Step 7*. If the number of lines is less than m, go to *Step 6*. This step introduces IM $D[K*, L*]$, which has the same dimensions as the X matrix. We use it to mark whether an element in the S is covered with a horizontal or vertical line, or both. If $d[k_i, l_j] = 1, s[k_i, l_j]$ is covered with 1 line, if $d[k_i, l_j] = 2$, the $s[k_i, l_j]$ element is covered with 2 lines.

We create two matrices $CC[r_0, L*]$ and $RC[K*, e_0]$, in which it is recorded that the element is covered by a line in a row or column in the S matrix.

for (int $i = 0; i < m; i + +$)

for (int $j = 0; j < n; j + +$)

- If $s[k_i, l_j] = \langle 0, 1\rangle$ (or $\langle k_i, l_j\rangle \in Index_{(\max \nu),k_i}(S)$) and $d[k_i, l_j] = 0$, then $\{rc[k_i, e_0] = 1; d[k_i, l_j] = 1 \ \forall l_j; S_{(k_i,\perp)}\}$
- If $s[k_i, l_j] = \langle 0, 1\rangle$ (or $\langle k_i, l_j\rangle \in Index_{(\max \nu),k_i}(S)$) and $d[k_i, l_j] = 1$, then $\{d[k_i, l_j] = 2; cc[r_0, l_j] = 1; l_j \ d[k_i, l_j] = 1 \ \forall k_i; S_{(\perp,l_j)}\}.$

Then we count the covered rows and columns (number of units) in the IMs CC and RC through the operations:

$Index_{(1)}(RC) = \{\langle k_{u_1}, e_0\rangle, \ldots, \langle k_{u_i}, e_0\rangle, \ldots, \langle k_{u_x}, e_0\rangle\};$

$Index_{(1)}(CC) = \{\langle r_0, l_{v_1}\rangle, \ldots, \langle r_0, l_{v_j}\rangle, \ldots, \langle r_0, l_{v_y}\rangle\}.$

If $\text{count}(Index_{(1)}(RC)) + \text{count}(Index_{(1)}(CC)) = m = n$, then go to *Step 7*, otherwise to *Step 6*.

Step 6. We find the smallest element of the S that is not crossed by the lines in *Step 5*, and subtract it from from each of its uncovered elements, and we add it to each of its elements that is covered by two lines. We return to *Step 5*.

$AGIndex_{(\min,\max)}(S) = \langle k_x, l_y \rangle; \ IO_{-(\max,\min)}(\langle S \rangle, \langle k_x, l_y, S \rangle)$.

for (int $i = 0; i < m; i++$)

for (int $j = 0; j < n; j++$)

{if $d[k_i, l_j] = 2$ then

create $S_1 = pr_{k_x, l_y} C; S_2 = pr_{k_i, l_j} C \oplus_{(\max,\min)} \left[\frac{k_i}{k_x}; \frac{l_j}{l_y} \right] S_1; \ S := S \oplus_{(\max,\min)} S_2;$

if $d[k_i, l_j] = 1$ then $S := S \oplus_{(+)} pr_{k_i, l_j} C$}. Go to *Step 5*.

Step 7. An optimal solution has been found, which assignments of the candidates to the activities are located where the elements $\langle 1, 0 \rangle$ in the X, with a single such element in each row or column.

{for (int $i = 0; i < m; i++$)

for (int $j = 0; j < n; j++$)

if $(\langle k_i, l_j \rangle \in Index_{(\max \nu), k_i}(S))$ and $d[k_i, l_j] <> 4)$ then $x[k_i, l_j] = \langle 1, 0 \rangle;$

for (int $i = 0; i < m; i++$) $d[k_i, l_j] = 4;$

for (int $j = 0; j < n; j++$) $d[k_i, l_j] = 4;$}

The optimal solution is $X_{opt}[K*, L*, \{x_{k_i, l_j}\}]$ and the optimal assignment cost is:

$$AGIO_{\oplus_{(\max,\min)}} \left(C_{(\{Q, pu\}, \{R\})} \otimes_{(\min,\max)} X_{opt} \right).$$

4 Conclusion

In this paper we have defined the hungarian algorithm for a new type of intuitionistic fuzzy assignment problem with additional restrictions by using apparatus of IMs and IFPs. The time complexity of the proposed algorithm is comparable with that of standard hungarian algorithm. Its main advantages are that can be applied to problems with imprecise parameters and can be extended in order to obtain the optimal solution for other types of multidimensional problems.

References

1. Atanassov, K.T.: Intuitionistic Fuzzy Sets, VII ITKR Session, Sofia, 20–23 June 1983 (Deposed in Centr. Sci.-Techn. Library of the Bulg. Acad. of Sci. 1697/84) (2016). (in Bulgarian). Reprinted: Int. J. Bioautom. **20**(S1), S1–S6

2. Atanassov, K.: Generalized index matrices. Comptes rendus de l'Academie Bulgare des Sciences **40**(11), 15–18 (1987)

3. Atanassov, K.: On Intuitionistic Fuzzy Sets Theory. STUDFUZZ, vol. 283. Springer, Heidelberg (2012). https://doi.org/10.1007/978-3-642-29127-2

4. Atanassov, K.: Index Matrices: Towards an Augmented Matrix Calculus. Studies in Computational Intelligence, vol. 573. Springer, Cham (2014). https://doi.org/10.1007/978-3-319-10945-9

5. Atanassov, K.: Intuitionistic Fuzzy Logics. Studies in Fuzziness and Soft Computing, vol. 351. Springer, Cham (2017). https://doi.org/10.1007/978-3-319-48953-7
6. Atanassov, K., Szmidt, E., Kacprzyk, J.: On intuitionistic fuzzy pairs. Notes Intuitionistic Fuzzy Sets **19**(3), 1–13 (2013)
7. Kuhn, H.: On certain convex polyhedra. Bull. Am. Math. Soc. **61**, 557–558 (1955)
8. Rajaraman, K., Sophia Porchelvi, R., Irene Hepzibah, R.: Ones assignment method to solve intitionistic fuzzy assignment problems. J. Appl. Sci. Comput. **5**(8), 252–261 (2018)
9. Traneva, V.: Internal operations over 3-dimensional extended index matrices. Proc. Jangjeon Math. Soc. **18**(4), 547–569 (2015)
10. Traneva, V., Tranev, S.: Index Matrices as a Tool for Managerial Decision Making. Publ. House of the Union of Scientists, Bulgaria (2017). (in Bulgarian)
11. Munkres' Assignment Algorithm. http://csclab.murraystate.edu/~bob.pilgrim/445//munkres.html. Accessed 8 May 2018

Problem-Driven Numerical Method: Motivation and Application

Spectral Collocation Solutions to a Class of Pseudo-parabolic Equations

Călin-Ioan Gheorghiu$^{(\boxtimes)}$ (iD)

Romanian Academy, "Tiberiu Popoviciu" Institute of Numerical Analysis,
Str. Fantanele 57/67-68, 400320 Cluj-Napoca, Romania
ghcalin@ictp.acad.ro
http://www.ictp.acad.ro/gheorghiu

Abstract. In this paper we solve by method of lines (MoL) a class of pseudo-parabolic PDEs defined on the real line. The method is based on the sinc collocation (SiC) in order to discretize the spatial derivatives as well as to incorporate the asymptotic behavior of solution at infinity. This MoL casts an initial value problem attached to these equations into a stiff semi-discrete system of ODEs with mass matrix independent of time. A TR-BDF2 finite difference scheme is then used in order to march in time.

The method does not truncate arbitrarily the unbounded domain to a finite one and does not assume the periodicity. These are two omnipresent, but non-natural, ingredients used to handle such problems.

The linear stability of MoL is proved using the pseudospectrum of the discrete linearized operator. Some numerical experiments are carried out along with an estimation of the accuracy in conserving two invariants. They underline the efficiency and robustness of the method. The convergence order of MoL is also established.

Keywords: Pseudo-parabolic equation · Infinite domain
Camassa-Holm · Peakon · Sinc collocation · TR-BDF2
Linear stability · Pseudospectrum

1 Camassa-Holm Equation

Camassa and Holm [4] introduce the equation,

$$u_t + 2\kappa u_x - u_{xxt} + 3uu_x = 2u_x u_{xx} + uu_{xxx}, \ x \in \mathbb{R}, \ t > 0, \tag{1}$$

discuss its analytical properties and sketch its derivation. Here u is the fluid velocity in the x direction, κ is a constant related to the critical shallow-water wave speed, and subscripts denote partial derivatives.

This work was supported by a mobility grant of the Romanian Ministry of Research and Innovation, CNCS - UEFISCDI, project number PN-III-P1-1.1-MC-2018-1869, within PNCDI III.

G. Nikolov et al. (Eds.): NMA 2018, LNCS 11189, pp. 179–186, 2019.
https://doi.org/10.1007/978-3-030-10692-8_20

Actually, the equation is a new completely integrable dispersive shallow water equation that is bi-Hamiltonian. For $\kappa := 0$ the authors quoted above showed that this equation has solitary water waves of the form $c\exp\left(-|x - ct|\right)$ which they called "peakons". Dropping both high-order terms from the r. h. s. of (1) leads to the familiar Benjamin-Bona-Mahony (BBM) equation [2], or at the same order, to the Korteweg-de Vries (KdV) equation. Nevertheless, these extra terms profoundly alter the behavior of solitary waves. Thus, this extension of the BBM equation possesses *soliton solutions* whose limiting forms as $\kappa \to 0$ have peaks where first derivatives are discontinuous, i.e. *peakons*. The lack of smoothness at the peak of the peakon introduces high-frequency dispersive errors into the calculation. This is one of the main numerical challenges.

There is a fairly rich literature gathered around the Camassa-Holm equation. For instance, in [3] J. P. Boyd derives a perturbation series solution for general κ which converges even at peakon limit and also gives three analytical formulations for spatially representations of the peakons. He also observes that although the Camassa-Holm equation is integrable for general κ, it appears that imbricating solitary waves generates an exact spatially periodic solution only for the special cases $\kappa = 0$ and $\kappa/c = 1/2$. Camassa, Holm and Hyman [5] present *periodic* numerical solutions to Eq. (1) and a large discussion to this equation as a Hamiltonian system. They focus on the case $\kappa = 0$. Actually, we also have some performing methods, like Runge-Kutta discontinuous Galerkin, for periodic case or bounded domains obtained by empiric truncation.

The main motivation to design this MoL-SiC scheme is to avoid the arbitrary truncation of the domain or periodicity assumptions. Instead, we suggest an *asymptotic truncation* of the integration domain through the scaling factor of SiC. The present paper provides accurate numerical results for arbitrary κ case. We use a MoL based on SiC in order to discretize the spatial derivatives and TR-BDF2 in order to march in time.

The boundary conditions for solution $u(x, t)$ will be taken to be regular at infinity, i.e.

$$u_x^{(n)}(x, t) \to 0, \text{ as } |x| \to \infty, \tag{2}$$

for at least $n = 0, 1, 2$ where $u_x^{(n)}(x)$ denotes the n th derivative with respect to x. They are here *behavioral (natural)* rather than numerical and have been inspired by the analysis in [2].

The *pure initial-value problem* for the above equation is to ask for a solution u defined for $(x, t) \in \mathbb{R} \times \mathbb{R}^+$ having a specified initial configuration namely

$$u(x, 0) = u_0(x), \, x \in \mathbb{R}. \tag{3}$$

A typical situation arises wherein the initial disturbance has sensibly finite extent. Thus, we suppose a datum $u_0(x)$ being drawn from some classes of functions having limit at $\pm\infty$, i.e., satisfying at least the first condition in (2). In such circumstances the pure initial value problem for (1) is *well-posed* in Hadamard's classical sense (see [1]). That is, corresponding to a suitable initial datum $u_0(x)$, there is a unique function $u(x, t)$ defined on $\mathbb{R} \times \mathbb{R}^+$ satisfying the differential equation there, and for which (2) holds. Moreover, if $u_0(x)$ is slightly perturbed in its function class, then the solution u changes only slightly in response.

2   MoL Discretization to Camassa-Holm Equation

The MoL solution to the Eq. (1) reads

$$u_N(x,t) := \sum_{j=1}^{N} S(j,h)(x) u_j(t). \tag{4}$$

where $S(j,h)(x)$ are the *sinc discrete functions* introduced and analyzed by F. Stenger in [10] and $u_j(t)$ for $j := 1, \ldots, N$ are the nodal time unknowns. $N \in \mathbb{N}$ is the order of approximation.

The functions $S(j,h)(x)$ are defined by

$$S(j,h)(x) := \frac{\sin(\pi(x - x_j/h))}{\pi(x - x_j/h)},$$

where x_j are the equispaced nodes with spacing h, symmetric with respect to the origin and $\mathbf{x}_{Si} := (x_1, \ldots, x_N)^T$.

It is important to observe that SiC approximation $u_N(x,t)$ defined by (4) satisfies the regularity condition at infinity (2). The sinc differentiation collocation matrices on $\mathbf{x}_{Si}$ are denoted by $\mathbf{D}_{Si}^{(n)}$, $n = 1, 2, 3$. In order to implement the MoL we have used the form of these matrices provided by Weideman and Reddy in [12].

The Camassa-Holm PDE is one of the form

$$\mathcal{M}(u)u_t = \mathcal{L}(u) + \mathcal{N}(u), \tag{5}$$

where the operators $\mathcal{M}$ and $\mathcal{L}$ are linear and $\mathcal{N}$ is a nonlinear one. Once we discretize the spatial part of this PDE we get a system of ODEs,

$$\mathbf{M}(\mathbf{U})\mathbf{U}_t = \mathbf{L}(\mathbf{U}) + \mathbf{N}(\mathbf{U}), \tag{6}$$

where the vector $\mathbf{U}$ is defined by $\mathbf{U} := (u_1(t), \ldots, u_N(t))^T$.

The particular form of the discrete formulation (6) for Camassa-Holm equation reads

$$\left(\mathbf{D}_{Si}^{(2)}\mathbf{U} - \mathbf{U}\right)_t = -2\kappa\mathbf{D}_{Si}^{(1)}\mathbf{U} - 3\left(\mathbf{D}_{Si}^{(1)}\mathbf{U}\right).\mathbf{U} + 2\left(\mathbf{D}_{Si}^{(1)}\mathbf{U}\right).\left(\mathbf{D}_{Si}^{(2)}\mathbf{U}\right) + \mathbf{U}.\left(\mathbf{D}_{Si}^{(3)}\mathbf{U}\right), \tag{7}$$

where the *dot*, as in MATLAB, signifies the element wise product of two vectors.

For our dispersive problem it is known that the eigenvalues of the linear operator are close to the imaginary axis (see the analysis in [9]). They are depicted in Fig. 2 and are situated inside or very close to the region of stability of TR-BDF2 depicted in Fig. 1. This is the main reason in choosing this finite difference scheme to solve the ODEs system (7).

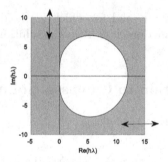

Fig. 1. The absolute stability region (in cyan) of TR-BDF2 finite difference scheme. The region is unlimited and span in the directions shown by the arrows. (Color figure online)

3 Linear Stability of MoL

We adopt the strategy from [11] in order to establish the linear (numerical) stability of MoL. We have successfully used this strategy in our contribution [7] in solving numerically the BBM equation which is another pseudo-parabolic one. The linearization of the discrete Eq. (7) is simply

$$\left(\mathbf{D}_{Si}^{(2)}-\mathbf{I}\right)\mathbf{U}_t = -2\kappa\mathbf{D}_{Si}^{(1)}\mathbf{U}, \tag{8}$$

where $\mathbf{I}$ is the unit matrix of order N. Its pseudospectrum is depicted in Fig. 2 and shows that indeed all eigenvalues are on the imaginary axis.

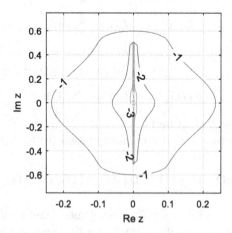

Fig. 2. The Λ_ε pseudospectrum for $\varepsilon = 10^{-1}, 10^{-2}, 10^{-3}$ for the discrete linearized operator (8), i.e., of the matrix $2\kappa\mathbf{D}_{Si}^{(1)}\left(\mathbf{I}-\mathbf{D}_{Si}^{(2)}\right)^{-1}$ with $\kappa := 0.5$ when SiC uses $N := 500$ and $h := 0.1$. The eigenvalues are situated on the imaginary axis.

If we overlap Fig. 2 and the region of stability of TR-BDF2 from Fig. 1 we observe the ε-pseudospectrum lies within a distance $O(\varepsilon) + O(\Delta t)$ of the stability region as $\varepsilon \to 0$ and $\Delta t \to 0$. Typically the time step has been of order 10^{-2} when (1) has been solved.

Thus we can conclude that the MoL-SiC scheme along with TR-BDF2 is numerically stable irrespective of values of parameter $\kappa \neq 0$. In fact, our numerical experiments have provided reasonable numerical results for the case $\kappa = 0$ even if the above analysis of stability does not literally apply to this case.

4   Numerical Experiments

Some numerical experiments will be reported in this section. We consider the *interaction of two peakons* which emerge from the initial data

$$u_0(x) := u_1^0(x) + u_2^0(x) = 2\exp(-|x+5|) + 0.5\exp(-|x+.5|). \qquad (9)$$

Actually we solve the Cauchy's problem (1)–(9). The interaction of two peakons as *singular soliton solutions* is depicted in Fig. 3. The evolution of initial data (9) to its final form of two separated and reversed in order peakons is provided in Fig. 4. These solitons travel with a speed proportional to their height and remain coherent after their collision (a short animation is instructive in this respect). In spite of the fact that peakons are *nonanalytic* solitons, a careful inspection of Figs. 3 and 4 show that no oscillations (like Gibbs' ones) appear when resolving the cusp of peakons.

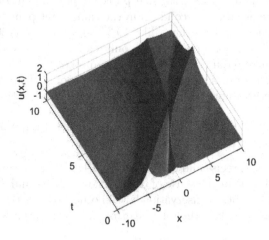

Fig. 3. The interaction of two peakons with $\kappa := 0.1164$ when SiC uses $N := 500$ and $h := 0.1$.

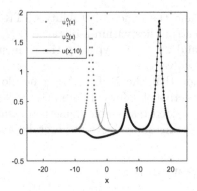

Fig. 4. The initial, i.e., $u_1^0(x)$ and $u_2^0(x)$ from (9), and final form of two peakons with $\kappa := 0.1164$ when SiC uses $N := 500$ and $h := 0.1$. The value $\kappa := 0.1164$ is the critical one from [3].

4.1   The Accuracy of MoL

We are now concerned with the accuracy of our outcomes (see also our contribution [8], Sect. 2.1). Thus in a log-linear plot we display in the b) panel of Fig. 5 the absolute values of the coefficients $u_k(t)$ of SiC solutions at several time moments.

We recall an important aspect of SiC, namely *the formulation of the collocation and the Galerkin methods for a boundary value problem using sinc functions is perfectly synonymous.* Consequently, two important conclusions can be inferred from this figure. First, as time proceed the computations are carried out with the same accuracy. Second, from the same right panel (b), we observe that the coefficients $|u_k(t)|$ behave like 10^{mk} with $m = -1/20$ which means the coefficients of SiC solution decrease only exponentially rather than super exponentially (which requires $m < -1$) as time proceed.

Constantin and Strauss [6] give a very simple proof of the orbital stability of the peakons in the H^1 norm using the following two conservation laws

$$I_1(t) = \int_{\mathbb{R}} \left(u^2 + u_x^2 \right) dx, \quad I_2(t) = \int_{\mathbb{R}} \left(u^3 + u\, u_x^2 \right) dx. \tag{10}$$

Now we can be more precise and call $u \in C\left([0,T]; H^1(\mathbb{R})\right)$ a solution to (1) if u is a solution to (1)–(3) in the sense of distributions and I_1 and I_2 are conserved for $t \geq 0$. We will use these conservation laws in order to asses the accuracy of our outcomes. We now introduce the relative errors for the invariants I_1 and I_2 with

$$Err^{rel}(I_m)(t_i) := \frac{|(I_m)(t_i) - (I_m)(0)|}{(I_m)(0)}, \; m = 1, 2, \; i = 1, 2, \ldots, 1248, \tag{11}$$

and display them in the left panel of Fig. 5. Corresponding to the initial data (9) we have $I_1(0) = 8.206559$ and $I_2(0) = 10.300063$. Moreover, the value 1248 in (11) stands for the number of integration steps used by TR-BDF2. We carry

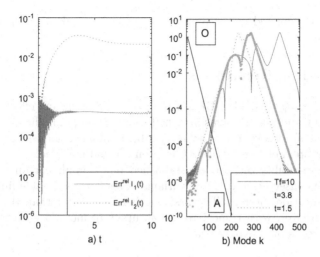

a) t   b) Mode k

Fig. 5. (a) The relative errors (11) in a log-linear plot. We assume the initial data (9) and integrate between $x_1 = -24.95$ and $x_N = 24.95$. (b) In a log-linear plot the absolute values of coefficients of solution for two peakons interaction with $\kappa := 0.1164$ when SiC uses $N := 500$ and $h := 0.1$. The solutions are evaluated at $Tf := 10$, $t := 3.8306$ and $t := 1.4926$. The straight line OA has the slope $m = -1/20$.

out the integration in (10) using a trapezoidal quadrature between the extreme nodes in SiC grid. In order to validate once more the above MoL we solve Eq. (1) with the initial data

$$u_0\,(x) := \begin{cases} c\,\exp\,(x+K)\,, & x \le -K, \\ c, & |x| < K, \\ c\,\exp\,(-x+K)\,, & x \ge K, \end{cases} \quad c := 0.6,\ K := 5. \qquad (12)$$

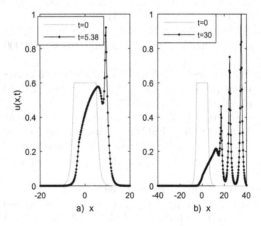

a) x   b) x

Fig. 6. The break up of the plateau traveling wave. (a) zoom in, (b) the dispersive behavior at the final momentum. We assume the initial data (12).

The solution is depicted in Fig. 6 and is in perfect accordance with well established results.

5   Concluding Remarks

The MoL-SiC along with TR-BDF2 proved to be a reliable and effective method to solve some Cauchy's problems attached to a pseudo-parabolic equation. The computational effort is kept in fairly reasonable limits, i.e., some seconds in order to solve the system of semi-discrete ODEs on a usual machine.

The linear stability of MoL is established but a nonlinear one is an open problem. The conservation of two invariants is kept in reasonable limits during the time integration interval. The first one behaves better than the second as expected. Moreover, with respect to the accuracy of method we establish that the convergence is only exponentially.

References

1. Amick, C.J., Bona, J.L., Schonbek, M.E.: Decay of solutions of some nonlinear wave equations. J. Differ. Equations **81**, 1–49 (1989). https://doi.org/10.1016/0022-0396(89)90176-9
2. Benjamin, T.B., Bona, J.L., Mahony, J.J.: Model equations for long waves in nonlinear dispersive systems. Philos. Trans. Roy. Soc. Lond. Ser.A. **272**, 47–78 (1972). https://doi.org/10.1098/rsta.1972.0032
3. Boyd, J.P.: Peakons and coshoidal waves: traveling wave solutions of the Camassa-Holm equation. Appl. Math. Comput. **81**, 173–187 (1997). https://doi.org/10.1016/0096-3003(95)00326-6
4. Camassa, R., Holm, D.D.: An integrable shallow water equation with peaked solitons. Phys. Rev. Lett. **71**, 1661–1664 (1993). https://doi.org/10.1103/PhysRevLett.71.1661
5. Camassa, R., Holm, D.D., Hyman, J.M.: A new integrable shallow water equation. In: Wu, T.-Y., Hutchinson, J.W. (eds.) Advances in Applied Mechanics, vol. 31, pp. 1–31. Academic Press, New York (1994)
6. Constantin, A., Strauss, A.W.: Stability of Peakons. Commun. Pure Appl. Math. **53**, 603–610 (2000). 10.1002/(SICI)1097-0312(200005)53:5<603::AID-CPA3>3.0.CO;2-L
7. Gheorghiu, C.I.: Stable spectral collocation solutions to a class of Benjamin Bona Mahony initial value problems. Appl. Math. Comput. **273**, 1090–1099 (2016). https://doi.org/10.1016/j.amc.2015.10.078
8. Gheorghiu, C.I.: Spectral Collocation Solutions to Problems on Unbounded Domains. Casa Cărţii de Ştiinţă Publishing House, Cluj-Napoca (2018)
9. Kassam, A.-K., Trefethen, L.N.: Fourth-order time-stepping for stiff PDEs. SIAM J. Sci. Comput. **26**, 1214–1233 (2005). https://doi.org/10.1137/S1064827502410633
10. Stenger, F.: Summary of sinc numerical methods. J. Comput. Appl. Math. **121**, 379–420 (2000). https://doi.org/10.1016/S0377-0427(00)00348-4
11. Trefethen, L.N.: Spectral Methods in MATLAB. SIAM Philadelphia (2000)
12. Weideman, J.A.C., Reddy, S.C.: A MATLAB differentiation matrix suite. ACM T. Math. Softw. **26**, 465–519 (2000). https://doi.org/10.1145/365723.365727

The Statistical Property in Finite Element Model of Elastic Contact Problems

Zhaocheng Xuan[✉]

Department of Computer Science, Tianjin University of Technology
and Education, Tianjin 300222, China
xuanzc@tute.edu.cn

Abstract. When the continuum elastic bodies in a contact problem are discretized by finite elements, we may look some of the nodes as an ensemble to use the concept or method of statistical physics to solve the mechanics problem. Each potential contact node of an elastic structure along with the normalized contact force on the node is considered as a system and all potential contact nodes together with their normalized contact forces are considered as a canonical ensemble, with the normalized contact force of each node representing the microstate of the node. The product of non-penetration conditions for potential contact nodes and the normalized nodal contact forces then act as an expectation that its value will be zero, and maximizing the entropy under the constraints of the expectation and the minimum potential energy principle results in an explicit probability distribution for the normalized contact forces that shows the relation between contact forces and displacements in a formulation similar to the formulation for particles occupying microstates in statistical physics. Moreover, an iterative procedure that solves a series of isolated systems to find the contact forces is presented. Finally, an example is examined to verify the correctness and efficiency of the procedure.

Keywords: Finite elements · Statistical property · Contact · Entropy

1 Introduction

The contact mechanics of continuum elasticity is foundational to the field of mechanical engineering; it provides necessary information for the safe and energy efficient design of technical systems and for the study of tribology and indentation hardness [1,2]. It is one type of highly nonlinear mechanics and studies the deformation of solids that touch each other at one point or more. The basic solutions for contact problem we want are the displacement and contact stress in

Supported by the National Natural Science Foundation of China under Grant No. 11772228.

the state of equilibrium of the structure. *Equilibrium* is a popular word in both the macrostate and microstate of physics. When the continuum elastic bodies in a contact problem are discretized by finite elements, can we look the finite element nodes or elements as the microscope particles? If we look the nodes on the potential contact part of the finite element model as an ensemble, can some core concepts related to the equilibrium of statistical physics, such as entropy etc., be used for solving the contact problems? Entropy has been used in various fields of science and engineering and received much attention in determining the *equilibrium, order (disorder), possibility, uncertainty* in a physical ensemble [3–6]. The motivation of this paper is to find if there are counterparts of the concepts, such as ensemble, microstate, possibility etc., in the finite element model of contact problems.

In this paper, we will discuss the statistical property in the contact problem, and verify it with an example of elastic contact problem by solving the contact forces of elastic bodies, focusing mainly on the normal direction, i.e., on frictionless contact mechanics. The elastic bodies discretized by finite elements are considered as an environment, each potential contact node (or node pair, *sic passim*) along with the normalized contact force on it is considered as a system, and all potential contact nodes along with their normalized contact forces are considered as an ensemble, with the normalized contact forces representing the microstates of the contact nodes. Using the maximum entropy principle, the relationship between the normalized contact forces and the displacements can be constructed. The displacements of the elastic body can be treated as the *temperature* source for the non-isolated environment. Given this formulation, we present an iterative procedure that can then solve a series of isolated ensembles for finding the contact forces, with the monotonicity of the potential energy used as a termination condition.

2   Entropy in Statistical Physics

Let us define an ensemble as consisting of m systems or particles, where the probability of system i occupying a microstate is p_i. A canonical ensemble is an ensemble defined by the probability $\mathbf{p} = \{p_1, p_2, \cdots, p_m\}^T$ that characterizes the microstates of a system in equilibrium with an environment at temperature T. Maximizing the Gibbs entropy $S = -\kappa \mathbf{p}^T \ln \mathbf{p}$ subject to the normalization condition $\mathbf{1}^T \mathbf{p} = 1$ and an expectation value of energy $\langle E \rangle = \mathbf{e}^T \mathbf{p}$ is equivalent to finding the stationery condition of the following Lagrangian function:

$$L(\mathbf{p}, \gamma, \delta) = -\kappa \mathbf{p}^T \ln \mathbf{p} + \gamma(\mathbf{e}^T \mathbf{p} - \langle E \rangle) + \delta(\mathbf{1}^T \mathbf{p} - 1), \qquad (1)$$

where κ is a coefficient related to the ensemble, γ and δ are Lagrange multipliers corresponding to the two constraints, namely, the normalization condition and the expectation, respectively, and $\mathbf{e} = \{e_1, e_2, \cdots, e_m\}^T$ is the energy of the particles. In this paper, we follow the convention of denoting matrices or vectors with bold letters and logarithmic functions or exponential functions in compact forms, e.g., $\mathbf{1} = \{1, \cdots, 1\}^T$, $\ln \mathbf{p} = \{\ln p_1, \cdots, \ln p_m\}^T$,

$\exp(\mathbf{p}) = \{\exp(p_1), \cdots, \exp(p_m)\}^T$. Letting the differential of equation (1) with respect to $\mathbf{p}$ be zero, that is,

$$- \kappa \ln \mathbf{p} - \kappa \mathbf{1} + \gamma \mathbf{e} + \delta \mathbf{1} = \mathbf{0}, \tag{2}$$

we can express $\mathbf{p}$ in terms of $\mathbf{e}$ as follows:

$$\mathbf{p} = \exp\left(\frac{\delta}{\kappa}\mathbf{1} - 1 + \frac{\gamma}{\kappa}\mathbf{e}\right) = \exp\left(\frac{\delta}{\kappa} - 1\right)\exp\left(\frac{\gamma}{\kappa}\mathbf{e}\right). \tag{3}$$

Using the normalization condition $\mathbf{1}^T\mathbf{p} = 1$, we then obtain

$$\mathbf{p} = \frac{\exp\left(\frac{\gamma}{\kappa}\mathbf{e}\right)}{Z}, \tag{4}$$

where Z is the system partition function, a sum over the microstates of the system,

$$Z = \mathbf{1}^T \exp(\frac{\gamma}{\kappa}\mathbf{e}). \tag{5}$$

The relationship between the entropy S, system energy E, and temperature T can be captured as $\frac{1}{T} = \frac{\partial S}{\partial E}$, and so for this case of a non-isolated system, we have

$$\frac{1}{T} = \frac{\partial S}{\partial \langle E \rangle}. \tag{6}$$

Now let us use it to eliminate the Lagrange multiplier γ. After some manipulations, we have the relationship between $\mathbf{p}$, E and T as follows:

$$\mathbf{p} = \frac{\exp\left(-\frac{1}{\kappa T}\mathbf{e}\right)}{\mathbf{1}^T \exp\left(-\frac{1}{\kappa T}\mathbf{e}\right)}. \tag{7}$$

The probability is obtained by maximizing the entropy for a system, and it has been proved that the entropy for a system is simply $\frac{1}{m}$ of the entropy of the ensemble, so maximizing the entropy of the ensemble will also produce the same results as Eq. (7). More details on entropy in statistical physics can be found in [6] and its references.

3 Entropy in Contact Mechanics

Let us consider an elastic structure of two bodies coming into contact; see Fig. 1. The boundary of the elastic bodies consists of the displacement boundary Γ_v, the external loads boundary Γ_t, and the potential contact boundary Γ_c. The elastic bodies are discretized by finite elements, and the finite element nodes are separated into two groups: one includes the potential contact nodes on the potential contact boundary, and the other includes the nodes on the rest of the body. We consider the contact part as a canonical ensemble, as in statistical physics. Table 1 presents a comparison of entropy in statistical physics and in contact mechanics.

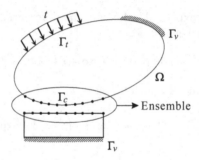

Fig. 1. An elastic body coming into contact with a foundation.

Table 1. The descriptions of entropy in statistical physics and contact mechanics.

Description in statistical physics	Description in contact mechanics
A simple system or a particle	A potential contact node
Microstate of a system	Normalized contact force of a node
A complex system	A number of contact nodes
A microstate of a complex system	A contact state of a number of potential contact nodes
An ensemble of complex systems	All potential contact nodes and their contact states
Probability of a microstate	Normalized contact force of a node
Temperature	Displacement
Non-isolation concerns temperature	Non-isolation concerns displacement
Expectation of energy	Expectation of work by contact forces

In statistical physics, the *microstate* is the distribution of the particles in a space. The figure shows three microstates of the system, they contribute to the macrostate of an ensemble. Assume that we have a system of four particles in a two dimensional space that is subdivided into three equal-volume parts as shown in Fig. 3. The *probabilities* of the system occupying microstates 1, 2, and 3 are p_1, p_2 and p_3, respectively, where $p_1 + p_2 + p_3 = 1$.

In a contact mechanics problem, each contact node (specifically, the normalized contact force) that appears in a separate column is assumed to be a system, see Fig. 3. Each normalized nodal contact force λ_i, for $i = 1, \cdots, 5$, is in the range between 0 and 1, thus representing a "probability", as the sum of the normalized nodal contact forces is 1, that is, $\lambda_1 + \lambda_2 + \lambda_3 + \lambda_4 + \lambda_5 = 1$.

Therefore, both the "microstate of a system" and the "probability of a microstate" are identified with the "normalized nodal contact force".

Analogously to the derivation in Sect. 2, we write the entropy in the form $S = -\rho \boldsymbol{\lambda}^T \ln \boldsymbol{\lambda}$, and the Lagrangian function for the elastic contact problem is

$$L(\boldsymbol{\lambda}, \alpha, \beta) = -\rho \boldsymbol{\lambda}^T \ln \boldsymbol{\lambda} + \alpha \boldsymbol{\lambda}^T d(\mathbf{u}) + \beta(\mathbf{1}^T \boldsymbol{\lambda} - 1), \tag{8}$$

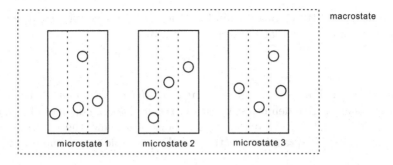

Fig. 2. Four particles in a two dimensional space.

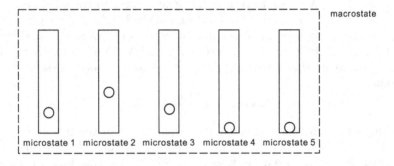

Fig. 3. Each contact node is assumed as a system.

where α and β are Lagrange multipliers, $d(\mathbf{u})$ is the function of displacement related to the potential contact nodes, and $\boldsymbol{\lambda}$ is the vector of normalized contact forces on the potential contact nodes – which therefore satisfies $\mathbf{1}^T \boldsymbol{\lambda} = 1$. According to the stationary condition of the Lagrangian function with respect to $\boldsymbol{\lambda}$, we have

$$- \rho \ln \boldsymbol{\lambda} - \rho \mathbf{1} + \alpha d(\mathbf{u}) + \beta \mathbf{1} = \mathbf{0}, \tag{9}$$

and thus we obtain

$$\boldsymbol{\lambda} = \frac{\exp\left(\frac{\alpha}{\rho} d(\mathbf{u})\right)}{\mathbf{1}^T \exp\left(\frac{\alpha}{\rho} d(\mathbf{u})\right)}. \tag{10}$$

The next step is to find the Lagrange multiplier α. In statistical physics, the Lagrange multiplier γ is eliminated by using the relation $\frac{\partial S}{\partial \langle E \rangle} = \frac{1}{T}$, which relates the energy to the temperature. However, in elastic contact problems, the energy in the ensemble we have just defined is the work done by the contact forces, which are related to the displacement of the elastic body. Since the work here concerns the displacement, we can eliminate the Lagrange multiplier α by using the principle of the minimum potential energy of the structure. In the finite element analysis of the displacement of a structure with a contact condition, the principle of minimum potential energy is given by

$$\min_{\mathbf{u} \in \mathcal{U}} \pi(\mathbf{u}), \tag{11}$$

with $\pi(\mathbf{u}) = \frac{1}{2}\mathbf{u}^T\mathbf{K}\mathbf{u} - \mathbf{c}^T\mathbf{u}$ and $\mathcal{U} = \{\mathbf{u} \in \Re^n | d(\mathbf{u}) \in \Re^m, d(\mathbf{u}) \leq \mathbf{0}\}$, where $\mathbf{K} \in \Re^{n \times n}$ is a positive definite stiffness matrix, n gives the degrees of freedom, and $\mathbf{c} \in \Re^n$ is the vector combining the body force and the external force. The corresponding Lagrangian function for the principle of minimum potential energy is

$$L(\mathbf{u}, \boldsymbol{\lambda}, \alpha) = \pi(\mathbf{u}) + \alpha \boldsymbol{\lambda}^T d(\mathbf{u}). \tag{12}$$

We can see that $\alpha\boldsymbol{\lambda}$ is the Lagrange multiplier vector to the constraint $d(\mathbf{u})$, and $\mathbf{1}^T\alpha\boldsymbol{\lambda} = \alpha\mathbf{1}^T\boldsymbol{\lambda} = \alpha$, so α is the total sum of the Lagrange multipliers. Since $\boldsymbol{\lambda} = \frac{\alpha\boldsymbol{\lambda}}{\alpha}$ is the normalized Lagrange multiplier, α is therefore the total sum of contact forces on the potential contact nodes and $\boldsymbol{\lambda}$ is the normalized contact forces. By differentiating it partially with respect to $\mathbf{u}$, we obtain the stationary condition

$$\mathbf{K}\mathbf{u} - \mathbf{c} + \alpha\frac{\partial d(\mathbf{u})}{\partial \mathbf{u}}\boldsymbol{\lambda} = \mathbf{0}. \tag{13}$$

Here $d(\mathbf{u})$ is the function of the relative displacement of the potential contact nodes. For node-to-node contact mode processing in finite element analysis, it can take the form $d(\mathbf{u}) = \mathbf{A}\mathbf{u} - \mathbf{g}$, where $\mathbf{A} \in \Re^{m \times n}$ is the transformation matrix and $\mathbf{g} \in \Re^m$ is the gap between two contact bodies. Therefore, using Eq. (13), we obtain the following equation for the displacement $\mathbf{u}$ in terms of $\mathbf{c}, \mathbf{A}$, and $\boldsymbol{\lambda}$:

$$\mathbf{u} = \mathbf{K}^{-1}\left(\mathbf{c} - \alpha\mathbf{A}^T\boldsymbol{\lambda}\right). \tag{14}$$

As stated above, in the ensemble constructed by the nodes on the potential contact surface, the expectation of energy is the work done by the contact forces, that is,

$$\langle E \rangle = \alpha\boldsymbol{\lambda}^T d(\mathbf{u}). \tag{15}$$

If there is a non-zero relative displacement between a contact node pair in the normal direction of the contact surface, then the contact force on the contact node pair is zero because there is no contact between the two nodes. If, on the other hand, the displacement is zero, this means that contact occurs between the two nodes and there will be a contact force, and therefore we obtain $\langle E \rangle = 0$. These are referred to as the complementary conditions in contact problems. Taking $d(\mathbf{u}) = \mathbf{A}\mathbf{u} - \mathbf{g}$ in node-to-node contact mode and substituting (14) into (15), we have

$$\alpha = \frac{\boldsymbol{\lambda}^T(\mathbf{A}\mathbf{K}^{-1}\mathbf{c} - \mathbf{g})}{\boldsymbol{\lambda}^T\mathbf{A}\mathbf{K}^{-1}\mathbf{A}^T\boldsymbol{\lambda}}. \tag{16}$$

Then, by substituting the right-hand side into (14), the displacement $\mathbf{u}$ can be expressed in terms of the normalized contact force $\boldsymbol{\lambda}$, that is,

$$\mathbf{u} = \mathbf{K}^{-1}\mathbf{c} - \frac{\boldsymbol{\lambda}^T(\mathbf{A}\mathbf{K}^{-1}\mathbf{c} - \mathbf{g})}{\boldsymbol{\lambda}^T\mathbf{A}\mathbf{K}^{-1}\mathbf{A}^T\boldsymbol{\lambda}}\mathbf{K}^{-1}\mathbf{A}^T\boldsymbol{\lambda}. \tag{17}$$

Comparing the three Eqs. (7), (10), and (17), we see that in the first equation, the microstate probability $\mathbf{p}$ is expressed in terms of the temperature T and particle energy $\mathbf{e}$; in the second equation, the normalized contact force $\boldsymbol{\lambda}$ is expressed in terms of the total contact force α and the displacement $\mathbf{u}$; and in the third equation, the displacement $\mathbf{u}$ is expressed in terms of the normalized contact force $\boldsymbol{\lambda}$.

4   An Example

Example: An elastic structure in the state of plane strain is fixed at the top of the left part, and a force is put on the end of the right part to make the two parts come to contact to each other, as shown in Fig. 4, which shows the post output of stresses of the deformed structure. The Young's modulus of the elastic body is $E = 2900\,\mathrm{N/cm}^2$, and Poisson's ratio is $\nu = 0.4$, the force $f = 10\,\mathrm{N/cm}^2$. There is no gap between the elastic bodies. Figures 5 shows the results of contact forces on the potential contact nodes. The number marker on the lines shows the iteration number, and the square marker shows the reference solution by the commercial code. The iteration numbers are not plotted for the results of node 1 and node 10 for the conciseness of the figures. The initial value of λ is set to $\frac{1}{10}$, 6 iterations are used to calculate the solutions.

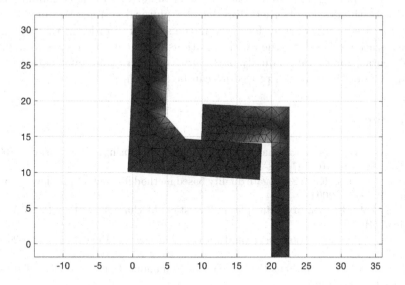

Fig. 4. Two elastic bodies come to contact.

In the iterations, we can find entropy is always decreasing as the iteration numbers increasing. As the maximum of entropy is used as an initial value, so in the iterations followed, the entropy should be smaller. If there is no stop for the iteration, it will decreases to zero monotonously as the parameter ρ becomes

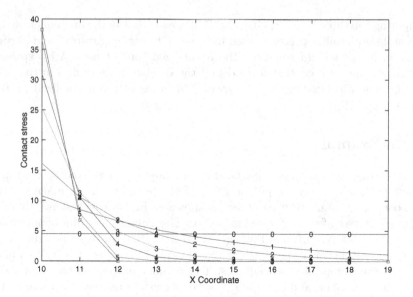

Fig. 5. The results of nodal contact stresses.

smaller and smaller. This shows one character of entropy being commonly associated with the degree of order, disorder of a physical system. We may assume the normalized nodal contact forces be in the state of disorder at the beginning of the iteration procedure, then with iterations being carried on, the normalized nodal contact forces are gradually in the state of order. More information about the analysis of the iteration procedure can be found in [7].

References

1. Wriggers, P.: Computational Contact Mechanics. Springer, New York (2006). https://doi.org/10.1007/978-3-540-32609-0
2. Xuan, Z.C., Lee, K.H.: Surrogate duality based method for contact problems. Optim. Eng. **5**, 59–75 (2004)
3. Gibbs, J.W.: Elementary Principles in Statistical Mechanics. Ox Bow Press, Woodbridge (1981)
4. Jaynes, E.T.: Information theory and statistical mechanics. Phys. Rev. **166**, 620–630 (1957)
5. Shannon, C.: The Mathematical Theory of Communication. University of Illunois Press, Urbana (1964)
6. Lemons, D.S.: A Student's Guide to Entropy. Cambridge University Press, Cambridge (2013)
7. Xuan, Z.C., Papadopoulos, P., Li, J.Y., Zhang, L.L.: An entropy-based evaluation of contact forces in continuum mechanics of elastic structures. Finite Elem. Anal. Des. **114**, 78–84 (2016)

Numerical Methods for Fractional Diffusion Problems

A Second Order Time Accurate SUSHI Method for the Time-Fractional Diffusion Equation

Abdallah Bradji[✉]

Department of Mathematics, Faculty of Sciences, University of Annaba, Annaba, Algeria
abdallah.bradji@gmail.com, abdallah.bradji@etu.univ-amu.fr
https://www.i2m.univ-amu.fr/~bradji/

Abstract. SUSHI (Scheme Using Stabilization and Hybrid Interfaces) is a finite volume method developed at the first time to approximate heterogeneous and anisotropic diffusion problems. It has been applied later to approximate several types of partial differential equations. The formulation of SUSHI involves a consistent and stable Discrete Gradient which is developed for a large class of nonconforming meshes in any space dimension.

In this note, we establish a second order time accurate implicit scheme for the Time Fractional Diffusion Equation. The space discretization is based on the use of SUSHI whereas the time discretization is performed using a uniform mesh. We state and prove a discrete *a priori estimate* from which we derive an optimal convergence order in $L^\infty(L^2)$.

Keywords: Time fractional diffusion equation · SUSHI scheme
Discrete gradient · Second order implicit scheme

1 Problem to Be Solved and Aim of This Paper

We consider the following time fractional diffusion equation:

$$\partial_t^\alpha u(\boldsymbol{x},t) - \Delta u(\boldsymbol{x},t) = f(\boldsymbol{x},t), \ (\boldsymbol{x},t) \in \Omega \times (0,T), \tag{1}$$

where Ω is an open polygonal bounded subset in $\mathbb{R}^d$, $T > 0$, and f is a given function. Here the operator ∂_t^α is the Caputo derivative defined by:

$$\partial_t^\alpha u(\boldsymbol{x},t) = \frac{1}{\Gamma(1-\alpha)} \int_0^t (t-s)^{-\alpha} \frac{\partial u(\boldsymbol{x},s)}{\partial s} ds, \ 0 < \alpha < 1. \tag{2}$$

Initial condition is given by, for a given function u^0 defined on Ω

$$u(\boldsymbol{x},0) = u^0(\boldsymbol{x}), \ \boldsymbol{x} \in \Omega. \tag{3}$$

© Springer Nature Switzerland AG 2019
G. Nikolov et al. (Eds.): NMA 2018, LNCS 11189, pp. 197–206, 2019.
https://doi.org/10.1007/978-3-030-10692-8_22

Homogeneous Dirichlet boundary conditions are given by

$$u(\boldsymbol{x}, t) = 0, \quad (\boldsymbol{x}, t) \in \partial\Omega \times (0, T). \tag{4}$$

Fractional differential equations have been successfully used in theory and they appear in many areas of application, see [6].

In this work, we consider the SUSHI developed in [5] to approximate time fractional diffusion equations. SUSHI uses general nonconforming meshes in which the control volumes can only be assumed to be polyhedral (the boundary of each control volume is a finite union of subsets of hyperplanes). We first establish a finite volume scheme in which the gradient of the unknown solution is replaced by the discrete gradient of the discrete solution. The time discretization is performed using a uniform mesh and the approximation of the time fractional derivative is defined by the so-called $L2 - 1_\sigma$ formula developed in [1,6]. The scheme uses a Crank-Nicolson like method. We prove an *a prior estimate* for the discrete solution in discrete norms of $L^\infty(L^2)$ and $L^\infty(H_0^1)$. From this *a prior estimate*, we prove an optimal convergence order in a discrete norm of $L^\infty(L^2)$, whereas, the stated *a prior estimate* does not serve us to derive an optimal convergence order in discrete norm of $L^\infty(H_0^1)$, see Remark 1 below for more details and for a possible different *a prior estimate* which yields optimal convergence order in a discrete norm of $L^\infty(H_0^1)$. The present note can be viewed as a continuation of our previous paper [3] in which we analyzed a first order finite volume scheme for time fractional diffusion equation. This paper is organized as follows. In Sects. 2 and 3, we recall for the sake of completeness some known results concerning the space and time discretizations. In Sect. 4, we derive the numerical scheme and in Sect. 5, we prove the convergence of the stated scheme.

2   The Spatial Mesh and the Definition of a Discrete Gradient

We consider as discretization in space the mesh of [5]. In brief, such mesh is defined as the triplet $\mathscr{D} = (\mathscr{M}, \mathscr{E}, \mathscr{P})$ where $\mathscr{M}$ is the set of cells, $\mathscr{E}$ is the set of edges, and $\mathscr{P}$ is a set of points $\boldsymbol{x}_K$ in each cell K. We assume that, for all $K \in \mathscr{M}$, there exists a subset $\mathscr{E}_K$ of $\mathscr{E}$ such that $\partial K = \cup_{\sigma \in \mathscr{E}_K} \overline{\sigma}$. For any $\sigma \in \mathscr{E}$, we denote by $\mathscr{M}_\sigma = \{K, \sigma \in \mathscr{E}_K\}$. We then assume that, for any $\sigma \in \mathscr{E}$, either $\mathscr{M}_\sigma$ has exactly one element and then $\sigma \subset \partial\Omega$ (the set of these interfaces, called boundary interfaces, denoted by $\mathscr{E}_{\text{ext}}$) or $\mathscr{M}_\sigma$ has exactly two elements (the set of these interfaces, called interior interfaces, denoted by $\mathscr{E}_{\text{int}}$). For all $\sigma \in \mathscr{E}$, we denote by $\boldsymbol{x}_\sigma$ the barycentre of σ. For all $K \in \mathscr{M}$ and $\sigma \in \mathscr{E}$, we denote by $\mathbf{n}_{K,\sigma}$ the unit vector normal to σ outward to K. Denoting by $d_{K,\sigma}$ the Euclidean distance between $\boldsymbol{x}_K$ and the hyperplane including σ, one assumes that $d_{K,\sigma} > 0$. We then denote by $\mathscr{D}_{K,\sigma}$ the cone with vertex $\boldsymbol{x}_K$ and basis σ. Also, h_K is used to denote the diameter of K. For more details on the mesh, we refer to [5, Definition 2.1, p. 1012].

We define the discrete space $\mathscr{X}_{\mathscr{D},0}$ as the set of all $v = \left((v_K)_{K \in \mathscr{M}}, (v_\sigma)_{\sigma \in \mathscr{E}}\right)$, where $v_K, v_\sigma \in \mathbb{R}$ and $v_\sigma = 0$ for all $\sigma \in \mathscr{E}_{\text{ext}}$. Let $H_\mathscr{M}(\Omega) \subset L^2(\Omega)$ be the space of functions which are constant on each control volume K of the mesh $\mathcal{M}$. For all $v \in \mathcal{X}_{\mathscr{D}}$, we denote by $\Pi_{\mathcal{M}} v \in H_{\mathcal{M}}(\Omega)$ the function defined by $\Pi_{\mathcal{M}} v(\boldsymbol{x}) = v_K$, for a.e. $\boldsymbol{x} \in K$, for all $K \in \mathcal{M}$. In order to analyze the convergence, we need to consider the size of the discretization $\mathscr{D}$ defined by $h_{\mathscr{D}} = \sup \{\text{diam(K)}, \, K \in \mathcal{M}\}$ and the regularity of the mesh given by

$$\theta_{\mathscr{D}} = \max \left(\max_{\sigma \in \mathcal{E}_{\text{int}}, K, L \in \mathcal{M}} \frac{d_{K,\sigma}}{d_{L,\sigma}}, \max_{K \in \mathcal{M}, \sigma \in \mathcal{E}_K} \frac{h_K}{d_{K,\sigma}} \right).$$

The formulation of the scheme we want to consider involves the discrete gradient, denoted by $\nabla_{\mathscr{D}}$, developed in [5]. The value of $\nabla_{\mathscr{D}} u$, where $u \in \mathcal{X}_{\mathscr{D},0}$, is defined by, for all $K \in \mathcal{M}$, for a.e. $\boldsymbol{x} \in \mathscr{D}_{K,\sigma}$

$$\nabla_{\mathscr{D}} u(\boldsymbol{x}) = \nabla_K u + \left(\frac{\sqrt{d}}{d_{K,\sigma}} (u_\sigma - u_K - \nabla_K u \cdot (\boldsymbol{x}_\sigma - \boldsymbol{x}_K)) \right) \mathbf{n}_{K,\sigma}, \qquad (5)$$

with $\nabla_K u = \frac{1}{m(K)} \sum_{\sigma \in \mathcal{E}_K} m(\sigma) (u_\sigma - u_K) \mathbf{n}_{K,\sigma}$. We define now the inner product defined on $\mathcal{X}_{\mathscr{D},0} \times \mathcal{X}_{\mathscr{D},0}$ and given by $\langle u, v \rangle_F = \int_\Omega \nabla_{\mathscr{D}} u(\boldsymbol{x}) \cdot \nabla_{\mathscr{D}} v(\boldsymbol{x}) d\boldsymbol{x}$.

The time discretization is performed with a constant time step $k = \frac{T}{N+1}$, where $N \in \mathbb{N}^*$, and we shall denote $t_n = nk$, for $n \in [\![0, N+1]\!]$. We denote by ∂^1 and ∂^2 the discrete first time derivative and discrete second time derivative given respectively by

$$\partial^1 v^{j+1} = \frac{v^{j+1} - v^j}{k} \quad \text{and} \quad \partial^2 v^{j+1} = \partial^1(\partial^1 v^{j+1}) = \frac{v^{j+1} - 2v^j + v^{j-1}}{k^2} \qquad (6)$$

Throughout this paper, the letter C stands for a positive constant independent of the parameters of the space and time discretizations.

3 A Convenient Approximation for the Caputo Derivative and Some Known Results

Almost of the results presented in this section are given for instance in [1,6]. For completeness, we recall them. Let us consider the "fractional mesh point" $t_{n+\sigma} = (n+\sigma)k$ where

$$\sigma = 1 - \frac{\alpha}{2}. \qquad (7)$$

Consequently (since $0 < \alpha < 1$) $0 < \sigma < 1$. Using (2), the value $\partial_t^\alpha u(t_{n+\sigma})$ is given by

$$\frac{1}{\Gamma(1-\alpha)} \left(\sum_{j=1}^n \int_{t_{j-1}}^{t_j} (t_{n+\sigma} - s)^{-\alpha} u_s(s) ds + \int_{t_n}^{t_{n+\sigma}} (t_{n+\sigma} - s)^{-\alpha} u_s(s) \right) ds. \qquad (8)$$

For each $j \in [\![1, N+1]\!]$, let $\Pi_{2,j} u$ be the quadratic interpolation defined on (t_{j-1}, t_j) on the points t_{j-1}, t_j, t_{j+1} of u. An explicit expansion for $\Pi_{2,j} u$ yields:

$$(\Pi_{2,j} u(s))' = \partial^1 u(t_{j+1}) + \partial^2 u(t_{j+1}) \left(s - t_{j+\frac{1}{2}} \right) = \partial^1 u(t_j) + \partial^2 u(t_{j+1}) \left(s - t_{j-\frac{1}{2}} \right). \qquad (9)$$

When approximating the terms of the sum (resp. the last term) using quadratic interpolations (resp. a linear interpolation) in (8) of $\partial_t^\alpha u(t_{n+\sigma})$, we have to compute the following integrals:

1. **First set of integrals:**

$$\int_{t_{j-1}}^{t_j} \left(s - t_{j-\frac{1}{2}}\right)(t_{n+\sigma} - s)^{-\alpha}ds = \frac{k^{2-\alpha}}{1-\alpha}b_{n-j}^\sigma, \tag{10}$$

where

$$b_l^\sigma = \frac{1}{2-\alpha}\left((l+\sigma+1)^{2-\alpha} - (l+\sigma)^{2-\alpha}\right) - \frac{1}{2}\left((l+\sigma+1)^{1-\alpha} + (l+\sigma)^{1-\alpha}\right). \tag{11}$$

2. **Second set of integrals:**

$$\int_{t_{j-1}}^{t_j}(t_{n+\sigma} - s)^{-\alpha}ds = \frac{k^{1-\alpha}}{1-\alpha}d_{n+\sigma-j,\alpha}, \tag{12}$$

with, for all $s > 0$, $d_{s,\alpha}$ is given by

$$d_{s,\alpha} = (s+1)^{1-\alpha} - s^{1-\alpha}. \tag{13}$$

3. **Third set of integrals:**

$$\int_{t_n}^{t_{n+\sigma}}(t_{n+\sigma} - s)^{-\alpha}ds = \frac{k^{1-\alpha}}{1-\alpha}\sigma^{1-\alpha}. \tag{14}$$

We then obtained approximation for the fractional derivative $\partial_t^\alpha u(t_{n+\sigma})$ using (8)–(14)

$$\frac{1}{\Gamma(1-\alpha)}\left(\sum_{j=1}^n \int_{t_{j-1}}^{t_j}(t_{n+\sigma} - s)^{-\alpha}\left(\Pi_{2,j}u(s)\right)'ds + \frac{k^{1-\alpha}}{1-\alpha}\sigma^{1-\alpha}\partial^1 u(t_{n+1})\right)$$
$$= \frac{k^{1-\alpha}}{\Gamma(2-\alpha)}\left(\sum_{j=1}^n \partial^1 u(t_j)d_{n+\sigma-j,\alpha} + \partial^2 u(t_{j+1})kb_{n-j}^\sigma + \sigma^{1-\alpha}\partial^1 u(t_{n+1})\right). \tag{15}$$

This gives, after re-ordering the sum and using the fact that $k\partial^2 u(t_{j+1}) = \partial^1 u(t_{j+1}) - \partial^1 u(t_j)$ (see [1, (27)–(28), p. 429])

$$\partial_t^\alpha u(t_{n+\sigma}) \approx \frac{k^{1-\alpha}}{\Gamma(2-\alpha)}\sum_{j=0}^n c_{n-j}^{\sigma,n}\partial^1 u(t_{j+1}), \tag{16}$$

where, for all $n \geq 1$

$$c_j^{\sigma,n} = d_{j+\sigma-1,\alpha} + b_j^\sigma - b_{j-1}^\sigma, \quad \forall j \in [\![1,n-1]\!], \quad c_0^{\sigma,n} = \sigma^{1-\alpha} + b_0^\sigma, \quad c_n^{\sigma,n} = d_{n+\sigma-1,\alpha} - b_{n-1}^\sigma, \tag{17}$$

and $n = 0$

$$c_0^{\sigma,0} = \sigma^{1-\alpha}. \tag{18}$$

Let us denote

$$\Lambda_{n+\sigma} u = \sum_{j=0}^{n} k\lambda_j^{n+1} \partial^1 u(t_{j+1}), \tag{19}$$

where

$$\lambda_j^{n+1} = \frac{c_{n-j}^{\sigma,n}}{k^\alpha \Gamma(2-\alpha)}, \tag{20}$$

The following lemma summarizes some properties of the time discretization. All these results are proved in [1] except the second estimate in (21) which can be justified by choosing $\varphi(t) = t$, for all $t \in [0, T]$ in (23) below. This yields $\partial^\alpha \varphi(t_{n+\sigma}) = \Lambda_{n+\sigma}\varphi$. Using now (19) implies $\partial^\alpha \varphi(t_{n+\sigma}) = \sum_{j=0}^{n} k\lambda_j^{n+1}$. Gathering this with definition (2) (or also expansion (8)) of $\partial^\alpha \varphi(t_{n+\sigma})$ yields

$$\sum_{j=0}^{n} k\lambda_j^{n+1} = \frac{1}{\Gamma(1-\alpha)} \int_0^{t_{n+\sigma}} (t_{n+\sigma} - s)^{-\alpha} ds = \frac{t_{n+\sigma}^{1-\alpha}}{\Gamma(2-\alpha)} \leq \frac{T^{1-\alpha}}{\Gamma(2-\alpha)}.$$

Lemma 1 (Some results concerning the time discretization, cf. [1]). *Let $d_{j,\alpha}$, b_l^σ, and $c_j^{\sigma,n}$ be defined respectively by (13), (11), and (17)–(18). For any $n \in [\![0, N]\!]$, for any $j \in [\![0, n]\!]$, let λ_j^{n+1} be given by (20) and $\Lambda_{n+\sigma}u$ be defined by (19). Then the following results hold:*

1. **Properties of the coefficients** λ_j^{n+1}, cf. [1, Lemma 4, p. 431].

$$\lambda_n^{n+1} > \lambda_{n-1}^{n+1} > \ldots > \lambda_0^{n+1} > \lambda_0 = \frac{1}{2T^\alpha \Gamma(1-\alpha)} \quad \text{and} \quad \sum_{j=0}^{n} k\lambda_j^{n+1} \leq \frac{T^{1-\alpha}}{\Gamma(2-\alpha)}. \tag{21}$$

2. **Stability result,** cf. [1, Corollary 1, p. 427]. *For all $(\beta^j)_{j=0}^{N+1} \in \mathbb{R}^{N+2}$, for any $n \in [\![0, N+1]\!]$:*

$$(\sigma\beta^{n+1} + (1-\sigma)\beta^n) \sum_{j=0}^{n} \lambda_j^{n+1}(\beta^{j+1} - \beta^j) \geq \frac{1}{2} \sum_{j=0}^{n} \lambda_j^{n+1} \left((\beta^{j+1})^2 - (\beta^j)^2 \right). \tag{22}$$

3. **Consistency result,** cf. [1, Lemma 2, p. 429]. *For any $\varphi \in \mathscr{C}^3([0, T])$:*

$$|\partial^\alpha \varphi(t_{n+\sigma}) - \Lambda_{n+\sigma}\varphi| \leq Ck^{3-\alpha} |\varphi^{(3)}|_{\mathscr{C}([0,T])}. \tag{23}$$

4   Formulation of a New Implicit Scheme Using the Discrete Gradient (5)

Taking $t = t_{n+\sigma}$ in Eq. (1) and using (16) and (19) (or also (23)) yield

$$\sum_{j=0}^{n} k\lambda_j^{n+1} \partial^1 u(t_{j+1}) - \Delta u(t_{n+\sigma}) \approx f(t_{n+\sigma}). \tag{24}$$

We have, thanks to a convenient Taylor expansion

$$\sigma u(t_{n+1}) + (1 - \sigma)u(t_n) = u(t_{n+\sigma}) + \mathbb{T}^1,$$ (25)

where

$$|\mathbb{T}^1| \leq \frac{k^2}{2}\|u\|_{\mathscr{C}^2([0,T])}.$$ (26)

From (24) and (25), we deduce that $\sum_{j=0}^{n} k\lambda_j^{n+1}\partial^1 u(t_{j+1}) - \Delta u^{n+\sigma} \approx f(t_{n+\sigma})$.
This motivates us to suggest the following finite volume scheme for problem (1)–(4):

1. **Approximation of initial condition (3).** Find $u_{\mathscr{D}}^0 \in \mathscr{X}_{\mathscr{D},0}$ such that

$$\langle u_{\mathscr{D}}^0, v\rangle_F = -\left(\Delta u^0, \Pi_{\mathscr{M}} v\right)_{L^2(\Omega)}, \ \forall v \in \mathscr{X}_{\mathscr{D},0}.$$ (27)

2. **Discretization of the time factional diffusion equation.** For any $n \in [\![0, N]\!]$, find $u_{\mathscr{D}}^{n+1} \in \mathscr{X}_{\mathscr{D},0}$ such that, for all $v \in \mathscr{X}_{\mathscr{D},0}$

$$\sum_{j=0}^{n} \lambda_j^{n+1}\left(\Pi_{\mathscr{M}}\left(u_{\mathscr{D}}^{j+1} - u_{\mathscr{D}}^j\right), \Pi_{\mathscr{M}} v\right)_{L^2(\Omega)} + \langle u_{\mathscr{D}}^{n+\sigma}, v\rangle_F = (f(t_{n+\sigma}), \Pi_{\mathscr{M}} v)_{L^2(\Omega)},$$ (28)

where $(\cdot, \cdot)_{L^2(\Omega)}$ denotes the $L^2(\Omega)$-inner product and $v^{n+\sigma}$ denotes the two-point barycentric element given by

$$v^{n+\sigma} = \sigma v^{n+1} + (1 - \sigma)v^n.$$ (29)

5   Convergence Results for the Finite Volume Scheme (27)–(28)

We now state one of the main results of this note, that is the convergence of scheme (27)–(28).

Theorem 1 *(Error estimate for scheme (27)–(28)). Let Ω be a polyhedral open bounded subset of $\mathbb{R}^d$, where $d \in \mathbb{N}\setminus\{0\}$, and $\partial\Omega = \overline{\Omega}\setminus\Omega$ its boundary. Assume that the solution of (1)–(4) satisfies $u \in \mathscr{C}^3([0, T]; \mathscr{C}^2(\overline{\Omega}))$ and $\theta_{\mathscr{D}}$ satisfies $\theta \geq \theta_{\mathscr{D}}$. Let $\nabla_{\mathscr{D}}$ be the discrete gradient defined as in (5). Let $k = \frac{T}{N+1}$, with $N \in \mathbb{N}^*$, and denote by $t_n = nk$, for $n \in [\![0, N+1]\!]$. Let $d_{j,\alpha}$, b_l^σ, and $c_j^{\sigma,n}$ be defined respectively by (13), (11), and (17)–(18). For any $n \in [\![0, N]\!]$, for any $j \in [\![0, n]\!]$, we define the coefficients λ_j^{n+1} as in (20).*

Then there exists a unique solution $(u_{\mathscr{D}}^n)_{n=0}^{N+1} \in \mathscr{X}_{\mathscr{D},0}^{N+2}$ for scheme (27)–(28) and the following error estimates hold:

– $L^\infty(L^2)$-estimate. For all $n \in [\![0, N+1]\!]$

$$\|u(t_n) - \Pi_{\mathscr{M}} u_{\mathscr{D}}^n\|_{L^2(\Omega)} \leq C(k^2 + h_{\mathscr{D}})\|u\|_{\mathscr{C}^3([0,T]; \mathscr{C}^2(\overline{\Omega}))}.$$ (30)

– $L^\infty(H_0^1)$-estimate. For all $n \in [\![0, N]\!]$

$$\left(\lambda_n^{n+1}\right)^{-\frac{1}{2}} \|\nabla u^{n+\sigma} - \nabla_{\mathscr{D}} u_{\mathscr{D}}^{n+\sigma}\|_{L^2(\Omega)} \leq C(k^2 + h_{\mathscr{D}}) \|u\|_{\mathscr{C}^3([0,T];\mathscr{C}^2(\overline{\Omega}))}, \quad (31)$$

where $v^{n+\sigma}$ is defined by (29).

To prove Theorem 1, we need to use the following discrete *a priori estimate* result.

Theorem 2 (A priori estimate for the discrete problem). *Under the same hypotheses of Theorem 1, assume that there exists* $(\eta^n)_{n=0}^{N+1} \in (\mathscr{X}_{\mathscr{D},0})^{N+2}$ *such that* $\eta_{\mathscr{D}}^0 = 0$ *and for all* $n \in [\![0, N]\!]$

$$\sum_{j=0}^{n} \lambda_j^{n+1} \left(\Pi_{\mathscr{M}}(\eta_{\mathscr{D}}^{j+1} - \eta_{\mathscr{D}}^j), \Pi_{\mathscr{M}} v\right)_{L^2(\Omega)} + \langle \eta_{\mathscr{D}}^{n+\sigma}, v\rangle_F = \left(\mathscr{S}^{n+1}, \Pi_{\mathscr{M}} v\right)_{L^2(\Omega)},$$

$$(32)$$

where $\mathscr{S}^{n+1} \in L^2(\Omega)$, *for all* $n \in [\![0, N]\!]$. *Then, the following estimate holds: For all* $n \in [\![0, N]\!]$:

$$\|\Pi_{\mathscr{M}} \eta^{n+1}\|_{L^2(\Omega)}^2 + \left(\lambda_n^{n+1}\right)^{-1} \|\nabla_{\mathscr{D}} \eta^{n+\sigma}\|_{L^2(\Omega)}^2 \leq \frac{C}{\lambda_0} \mathscr{S}^2, \quad (33)$$

where $\mathscr{S} = \max\limits_{n=0}^{N} \|\mathscr{S}^{n+1}\|_{L^2(\Omega)}$.

Sketch of Proof of Theorem 2

Taking $v = u_{\mathscr{D}}^{n+\sigma}$ in (32), using the Cauchy Schwarz inequality, the Poincaré inequality [5, Lemma 5.4, p. 1038], and the Young inequality $xy \leq x^2/2 + y^2/2$ imply that

$$\sum_{j=0}^{n} \lambda_j^{n+1} \left(\Pi_{\mathscr{M}}(\eta_{\mathscr{D}}^{j+1} - \eta_{\mathscr{D}}^j), \Pi_{\mathscr{M}} v\right)_{L^2(\Omega)} + \frac{1}{2}\|\nabla_{\mathscr{D}} \eta^{n+\sigma}\|_{L^2(\Omega)}^2 \leq C\|\mathscr{S}^{n+1}\|_{L^2(\Omega)}^2.$$

$$(34)$$

We first remark that the terms of the sum of (34) can be written as

$$\int_\Omega \left(\sigma\Pi_{\mathscr{M}} \eta_{\mathscr{D}}^{n+1}(\boldsymbol{x}) + (1-\sigma)\Pi_{\mathscr{M}} \eta_{\mathscr{D}}^n(\boldsymbol{x})\right) \left(\Pi_{\mathscr{M}} \eta_{\mathscr{D}}^{j+1}(\boldsymbol{x}) - \Pi_{\mathscr{M}} \eta_{\mathscr{D}}^j(\boldsymbol{x})\right) d\boldsymbol{x}.$$

Using this form and applying (22) with $\beta^j = \Pi_{\mathscr{M}} \eta_{\mathscr{D}}^j(\boldsymbol{x})$, inequality (34) implies that

$$\frac{1}{2}\sum_{j=0}^{n} \lambda_j^{n+1} \left(\|\Pi_{\mathscr{M}} \eta^{j+1}\|_{L^2(\Omega)}^2 - \|\Pi_{\mathscr{M}} \eta^j\|_{L^2(\Omega)}^2\right) + \frac{1}{2}\|\nabla_{\mathscr{D}} \eta_{\mathscr{D}}^{n+\sigma}\|_{L^2(\Omega)}^2 \leq C(\mathscr{S})^2.$$

$$(35)$$

This implies that, thanks to (21) (which implies that $1 \leq \frac{\lambda_0^{n+1}}{\lambda_0}$)

$$\lambda_n^{n+1}\|\Pi_{\mathscr{M}} \eta^{n+1}\|_{L^2(\Omega)}^2 + \|\nabla_{\mathscr{D}} \eta_{\mathscr{D}}^{n+\sigma}\|_{L^2(\Omega)}^2 \leq \sum_{j=1}^{n} (\lambda_j^{n+1} - \lambda_{j-1}^{n+1})\|\Pi_{\mathscr{M}} \eta^j\|_{L^2(\Omega)}^2 + \frac{C\lambda_0^{n+1}}{\lambda_0}(\mathscr{S})^2. \quad (36)$$

To prove (33), we will use mathematical induction on n. Let us set $n = 0$ in (36). This gives (33) when $n = 0$. Assume that the estimates (33) holds for $n \leq m$ and prove this estimate for $n = m + 1$. Taking $n = m$ in (36) yields that, thanks to (21)

$$\lambda_m^{m+1} \|\Pi_{\mathscr{M}}\, \eta^{m+1}\|_{L^2(\Omega)}^2 + \|\nabla_{\mathscr{D}}\, \eta_{\mathscr{D}}^{m+\sigma}\|_{L^2(\Omega)}^2 \leq (\lambda_m^{m+1} - \lambda_0^{m+1}) \frac{C}{\lambda_0} (\mathscr{S})^2 + \frac{C\lambda_0^{m+1}}{\lambda_0} (\mathscr{S})^2$$

$$\leq \frac{C\lambda_m^{m+1}}{\lambda_0} (\mathscr{S})^2. \tag{37}$$

This completes the proof of Theorem 2. ∎

Sketch of Proof of Theorem 1

The existence and uniqueness for (27)–(28) stem from the fact that $\|\nabla_{\mathscr{D}} \cdot\|_{L^2(\Omega)^d}$ is a norm on $\mathscr{X}_{\mathscr{D},0}$. To prove (30)–(31), we consider the scheme: for all $n \in [\![0, N + 1]\!]$, find $\bar{u}_{\mathscr{D}}^n \in \mathscr{X}_{\mathscr{D},0}$ such that

$$\langle \bar{u}_{\mathscr{D}}^n, v \rangle_F = -\left(\Delta u(t_n), \Pi_{\mathscr{M}}\, v\right)_{L^2(\Omega)}, \quad \forall v \in \mathscr{X}_{\mathscr{D},0}. \tag{38}$$

Taking $n = 0$ in the scheme (38), using (3), and comparing the result with (27) lead to $\eta_{\mathscr{D}}^0 = 0$, where, for all $n \in [\![0, N + 1]\!]$, $\eta_{\mathscr{D}}^n \in \mathscr{X}_{\mathscr{D},0}$ is given by $\eta_{\mathscr{D}}^n = u_{\mathscr{D}}^n - \bar{u}_{\mathscr{D}}^n$.

First step: comparison between $\bar{u}_{\mathscr{D}}^n$ and exact solution u. We have (see [4,5]), for all $n \in [\![1, N + 1]\!]$, for all $j \in \{0, 1\}$:

$$\|\partial^j (\Pi_{\mathscr{M}}\, \bar{u}_{\mathscr{D}}^n - u(t_n))\|_{L^2(\Omega)} + \|\nabla_{\mathscr{D}}\, \bar{u}_{\mathscr{D}}^n - \nabla u(t_n)\|_{L^2(\Omega)} \leq Ch_{\mathscr{D}} \|u\|_{\mathscr{C}^j([0,T];\, \mathscr{C}^2(\overline{\Omega}))}, \tag{39}$$

where $\partial^0(v^n) = v^n$ and ∂^1 is defined in (6).

Second step: comparison between schemes (27)–(28) and (38). From (38), we deduce that

$$\langle \bar{u}_{\mathscr{D}}^{n+\sigma}, v \rangle_F = \langle \sigma \bar{u}_{\mathscr{D}}^{n+1} + (1 - \sigma)\bar{u}_{\mathscr{D}}^n, v \rangle_F = \left(-\Delta u^{n+\sigma}, \Pi_{\mathscr{M}} v\right)_{L^2(\Omega)}, \quad \forall v \in \mathscr{X}_{\mathscr{D},0}. \tag{40}$$

Subtracting (40) from (28) (recall that $\eta_{\mathscr{D}}^n = u_{\mathscr{D}}^n - \bar{u}_{\mathscr{D}}^n$)

$$\sum_{j=0}^n \lambda_j^{n+1} \left(\Pi_{\mathscr{M}} (u_{\mathscr{D}}^{j+1} - u_{\mathscr{D}}^j), \Pi_{\mathscr{M}} v\right)_{L^2(\Omega)} + \langle \eta_{\mathscr{D}}^{n+\sigma}, v \rangle_F = \left(f(t_{n+\sigma}) + \Delta u^{n+\sigma}, \Pi_{\mathscr{M}} v\right)_{L^2(\Omega)}. \tag{41}$$

Subtracting $\sum_{j=0}^n \lambda_j^{n+1} \left(\Pi_{\mathscr{M}} (\bar{u}_{\mathscr{D}}^{j+1} - \bar{u}_{\mathscr{D}}^j), \Pi_{\mathscr{M}} v\right)_{L^2(\Omega)}$ from both sides of (41) and using (25) and (1) (which implies that $\Delta u^{n+\sigma} = \Delta u(t_{n+\sigma}) + \Delta \mathbb{T}^1 = \partial_t^\alpha u(t_{n+\sigma}) - f(t_{n+\sigma}) + \Delta \mathbb{T}^1$) yield

$$\sum_{j=0}^n \lambda_j^{n+1} \left(\Pi_{\mathscr{M}} (\eta_{\mathscr{D}}^{j+1} - \eta_{\mathscr{D}}^j), \Pi_{\mathscr{M}} v\right)_{L^2(\Omega)} + \langle \eta_{\mathscr{D}}^{n+\sigma}, v \rangle_F = \left(\mathscr{S}^{n+1}, \Pi_{\mathscr{M}} v\right)_{L^2(\Omega)}, \tag{42}$$

where $\mathscr{S}^{n+1} = \partial_t^\alpha u(t_{n+\sigma}) - \sum_{j=0}^{n} \lambda_j^{n+1} \Pi_{\mathscr{M}}(\bar{u}_{\mathscr{D}}^{j+1} - \bar{u}_{\mathscr{D}}^j) + \Delta\mathbb{T}^1$. Using the trian-

gle inequality, the definition (19) of $\Lambda_{n+\sigma}$, the consistency result (23), second

estimate in (21), and (39) yields (recall that $\mathscr{S} = \max_{n=0}^{N} \|\mathscr{S}^{n+1}\|_{L^2(\Omega)}$)

$$\mathscr{S} = \|\partial_t^\alpha u(t_{n+\sigma}) - \Lambda_{n+\sigma} + \sum_{j=0}^{n} k\lambda_j^{n+1} \partial^1 \left((u(t_{j+1}) - \Pi_{\mathscr{M}} \bar{u}_{\mathscr{D}}^{j+1} \right) + \Delta\mathbb{T}^1 \|_{L^2(\Omega)}$$

$$\leq C(k^2 + h_{\mathscr{D}}) \|u\|_{\mathscr{C}^3([0,T];\,\mathscr{C}^2(\overline{\Omega}))}. \tag{43}$$

Since $(\eta^n)_{n=0}^{N+1} \in (\mathscr{X}_{\mathscr{D},0})^{N+2}$ is satisfying (42) and $\eta_{\mathscr{D}}^0 = 0$, then it satisfies hypotheses of Theorem 2. Applying now estimate (33) of Theorem 2 and using (43) to get

$$\|\Pi_{\mathscr{M}} \eta^{n+1}\|_{L^2(\Omega)}^2 + (\lambda_n^{n+1})^{-1} \|\nabla_{\mathscr{D}} \eta^{n+\sigma}\|_{L^2(\Omega)}^2 \leq C(k^2 + h_{\mathscr{D}})^2 \|u\|_{\mathscr{C}^3([0,T];\,\mathscr{C}^2(\overline{\Omega}))}^2. \tag{44}$$

Gathering this with the triangle inequality, estimate (39), and the first estimate in (21) (which yields that $(\lambda_n^{n+1})^{-1} \leq (\lambda_0)^{-1}$) yields the desired estimates (30)–(31) of Theorem 1. ∎

Remark 1 (On the convergence order stated in Theorem 1). The $L^\infty(L^2)$–error estimate (30) is optimal in the sense that it is the same one proved in [4] for a Crank-Nicolson finite volume scheme approximating the heat equation. Since $\lambda_n^{n+1} = \frac{\sigma^{1-\alpha}+b_0^\sigma}{k^\alpha \Gamma(2-\alpha)}$, hence λ_n^{n+1} is of order $k^{-\alpha}$. Therefore, the $L^\infty(H_0^1)$–error estimate (31) implies that $\|\nabla_{\mathscr{D}} u_{\mathscr{D}}^{n+\sigma} - \nabla u^{n+\sigma}\|_{L^2(\Omega)}$ is of order $k^{-\frac{\alpha}{2}}(k^2 + h_{\mathscr{D}})$ which is a conditional convergence and consequently is not optimal. Seems that the *a priori estimate* (33) of Theorem 2 serves us to get only optimal error estimate in $L^\infty(L^2)$. We expect that another improved *a priori estimate* (see the case of first order schemes for time fractional diffusion equations in [2, Remark 1, p. 443]) can be proved and yields an optimal order in $L^\infty(H_0^1)$. This task will be addressed thoroughly in a future paper.

Remark 2 (On the regularity assumption of Theorem 1). The regularity assumption $\mathscr{C}^3(\mathscr{C}^2)$ of Theorem 1 stems from two essential facts. The first one is the regularity $\mathscr{C}^3$ w.r.t. time required in (23) whereas the second fact is the regularity $\mathscr{C}^2$ w.r.t. space required in (39). In fact, the regularity $\mathscr{C}^2$ in space is assumed in [5, Theorem 4.8, p. 1033] where a first convergence order, in discrete L^2 and H^1 norms, is proved for SUSHI (as approximation of elliptic equations). The regularity $\mathscr{C}^3(\mathscr{C}^2)$ of Theorem 1 can be weakened to $\mathscr{C}^3(H^2)$ in the particular cases when $d = 2$ or $d = 3$, see [5, Remark 4.9, pp. 1033–1034]. The regularity $\mathscr{C}^3(\mathscr{C}^4)$ (which is stronger than that of Theorem 1) is assumed for instance in [1, Lemma 5, p. 433] to deal with a second order finite difference scheme approximating the one dimensional case of problem (1)–(4).

6   Conclusion and Perspectives

We established a second order time accurate finite volume scheme for a time fractional diffusion equation in any space dimension. One of the main perspectives is to work on a new *a priori estimate* which serves to derive optimal error estimate in $L^\infty(H_0^1)$, see Remark 1.

References

1. Alikhanov, A.-A.: A new difference scheme for the fractional diffusion equation. J. Comput. Phys. **280**, 424–438 (2015)
2. Bradji, A.: Convergence order of gradient schemes for time-fractional partial differential equations. C. R. Math. Acad. Sci. Paris **356**(4), 439–448 (2018)
3. Bradji, A., Fuhrmann, J.: Convergence order of a finite volume scheme for the time-fractional diffusion equation. In: Dimov, I., Faragó, I., Vulkov, L. (eds.) NAA 2016. LNCS, vol. 10187, pp. 33–45. Springer, Cham (2017). https://doi.org/10.1007/978-3-319-57099-0_4
4. Bradji, A.: An analysis of a second-order time accurate scheme for a finite volume method for parabolic equations on general nonconforming multidimensional spatial meshes. Appl. Math. Comput. **219**(11), 6354–6371 (2013)
5. Eymard, R., Gallouët, T., Herbin, R.: Discretization of heterogeneous and anisotropic diffusion problems on general nonconforming meshes. IMA J. Numer. Anal. **30**(4), 1009–1043 (2010)
6. Gao, G.-H., Sun, Z.-Z., Zhang, H.-W.: A new fractional numerical differentiation formula to approximate the Caputo fractional derivative and its applications. J. Comput. Phys. **259**, 33–50 (2014)

A High Order Numerical Method for Solving Nonlinear Fractional Differential Equation with Non-uniform Meshes

Lili Fan[1] and Yubin Yan[2]([✉]) [iD]

[1] Department of Mathematics, Lvliang University,
Lvliang, People's Republic of China
634936564@qq.com
[2] Department of Mathematics, University of Chester,
Thornton Science Park, Pool Lane, Ince CH2 4NU, UK
y.yan@chester.ac.uk

Abstract. We introduce a high-order numerical method for solving nonlinear fractional differential equation with non-uniform meshes. We first transform the fractional nonlinear differential equation into the equivalent Volterra integral equation. Then we approximate the integral by using the quadratic interpolation polynomials. On the first subinterval $[t_0, t_1]$, we approximate the integral with the quadratic interpolation polynomials defined on the nodes t_0, t_1, t_2 and in the other subinterval $[t_j, t_{j+1}], j = 1, 2, \ldots N - 1$, we approximate the integral with the quadratic interpolation polynomials defined on the nodes t_{j-1}, t_j, t_{j+1}. A high-order numerical method is obtained. Then we apply this numerical method with the non-uniform meshes with the step size $\tau_j = t_{j+1} - t_j = (j + 1)\mu$ where $\mu = \frac{2T}{N(N+1)}$. Numerical results show that this method with the non-uniform meshes has the higher convergence order than the standard numerical methods obtained by using the rectangle and the trapzoid rules with the same non-uniform meshes.

Keywords: Nonlinear fractional differential equation
Numerical method · Non-uniform meshes

1 Introduction

Consider the following nonlinear fractional differential equation, with $\alpha > 0$,

$$\begin{array}{cc} {}^{C}_{0}D^{\alpha}_t y(t) = f(t, y(t)), \; t > 0, \; y^{(k)}(0) = y^{(k)}_0, \; k = 0, 1, \ldots, \lceil \alpha \rceil - 1, & (1) \end{array}$$

where ${}^{C}_{0}D^{\alpha}_t y(t)$ denotes the Caputo fractional derivative and $\lceil \alpha \rceil$ is the smallest integer $\geq \alpha$. Here $y^{(k)}_0$ are the initial values.

© Springer Nature Switzerland AG 2019
G. Nikolov et al. (Eds.): NMA 2018, LNCS 11189, pp. 207–215, 2019.
https://doi.org/10.1007/978-3-030-10692-8_23

It is well-known that (1) is equivalent to, [4, Lemma 2.3]

$$y(t) = \sum_{\nu=0}^{\lceil \alpha \rceil - 1} y_0^{(\nu)} \frac{t^\nu}{\nu!} + \frac{1}{\Gamma(\alpha)} \int_0^t (t-s)^{\alpha-1} f(s, y(s)) \, ds. \tag{2}$$

For the existence and uniqueness of the solution of (1) and the application of the Newton iteration method for solving the nonlinear equation of the proposed numerical method, we demand that the function f is continuous on a suitable set $(0, T) \times (c, d)$ and $f(t, \cdot) \in C^2[c, d]$ for some $c, d \in \mathbb{R}^+$ and any fixed $t \in [0, T]$. Under these assumptions, Diethelm et al. [4, Theorems 2.1, 2.2] showed that (1) has a unique solution y on some interval $[0, T]$.

There are many works in the literature to consider the numerical methods for solving (1), see, e.g., [1,2,4,8,13,15,16]. Most numerical methods for solving (1) are designed and analyzed with the uniform meshes, see, e.g., [4–6,8,16]. Since the fractional differential equation is a nonlocal problem and the derivative of the solution of (1) has the singularity at $t = 0$, it is not possible to obtain the high order numerical methods with uniform meshes. Therefore it is natural to use the non-uniform meshes to capture the singularity near $t = 0$. Diethelm [3, Theorem 3.1] used the graded meshes to recover the optimal convergence order for the approximation of the Hadamard finite-part integral. Recently Stynes et al. [11,12] applied the graded meshes to recover the convergence order of the finite difference method for solving time-fractional diffusion equation when the solution is not sufficiently smooth. Li et al. [7] considered the error estimates of the rectangle formula, trapezoid formula and the predictor-corrector scheme with non-uniform meshes for solving (1) under the assumption that the solution is sufficiently smooth. Other works for solving fractional differential equations with non-uniform meshes may be found in, for example, [7,14,17,18].

Recently, Liu et al. [9] designed a predictor-corrector numerical method for solving (1) with graded meshes and the detailed error estimates are provided. Liu et al. [10] also introduced a numerical method with non-uniform meshes for solving (1) and the detailed error estimates are considered. This paper is the continuation of the works in [9,10] and we will introduce a high order numerical method for solving (1) with non-uniform meshes. More precisely, we first approximate the integral in (2) with the piecewise quadratic interpolation polynomials with non-uniform meshes. We then use the Newton iteration for solving the nonlinear equation. Numerical examples show that this method has the high order convergence for solving nonlinear fractional differential equation with non-uniform meshes, where the solution of the fractional differential equation has the low regularity at $t = 0$.

The novelties of this work are as follows:

1. A new way to approximate the integral in the Volterra integral Eq. (2) by using the piecewise quadratic polynomials is introduced.
2. A high order numerical method for solving nonlinear fractional differential Eq. (1) with non-uniform meshes is obtained which is particular useful when f is not sufficiently smooth with respect to time t. The convergence order of the

proposed numerical method is higher than the available numerical methods in [7,9,10] for solving (1) with non-uniform meshes.

The paper is organized as follows. In Sect. 2, we introduce a high-order numerical method for solving (1). In Sect. 3, we give a numerical example which shows that the numerical results are consistent with the theoretical results.

2   A High Order Numerical Method with Non-uniform Meshes

For simplicity, we only consider the case with $\alpha \in (0,1)$ below. Similarly one may consider the general case with $\alpha > 1$. More precisely, we shall consider the numerical algorithm for solving, with $\alpha \in (0,1)$,

$$y(t) - y(0) = \frac{1}{\Gamma(\alpha)} \int_0^t (t-s)^{\alpha-1} f(s, y(s)) \, ds. \tag{3}$$

Let N be a positive integer. Let $0 = t_0 < t_1 < \cdots < t_n < t_{n+1} = T$, $n = 0, 1, 2, \ldots, N-1$ be the time partition of $[0, T]$. We want to find the approximate value y_{n+1} of $y(t_{n+1})$ at $t = t_{n+1}$. We shall consider the approximation of the following integral, with $n = 0, 1, 2, \ldots, N-1$,

$$I_{n+1} = \int_0^{t_{n+1}} (t_{n+1} - s)^{\alpha-1} f(s, y(s)) \, ds = \sum_{j=0}^n \int_{t_j}^{t_{j+1}} (t_{n+1} - s)^{\alpha-1} f(s, y(s)) \, ds.$$

It can be approximated by the following approach

$$I_{n+1} \approx \sum_{j=0}^n \int_{t_j}^{t_{j+1}} (t_{n+1} - s)^{\alpha-1} \tilde{f}_j(s, y(s)) \, ds,$$

where $\tilde{f}_j(s, y(s)), j = 0, 1, 2, \ldots, n$ is the approximation of $f(s, y(s))$ on the interval $[t_j, t_{j+1}]$.

It will lead to different scheme by choosing different $\tilde{f}_j(s, y(s))$. In [7], Li et al. introduced the fractional rectangle, trapezoid, and predictor-corrector methods respectively. In this paper, we shall construct a high order numerical method for solving (2) by approximating the integral in (2) with the piecewise quadratic polynomials. Numerical examples in Sect. 3 show that the convergence order of the proposed numerical method is almost 3 for the sufficiently smooth function $f(t, y(t))$ and the suitable chosen non-uniform meshes as expected.

Let $P_2^{(j)}(s)$ denote the quadratic interpolation polynomial approximation of $f(s, y(s))$ on the interval $[t_j, t_{j+1}], j = 0, 1, 2, \ldots, n$, where $P_2^{(0)}(s)$ is the quadratic interpolation polynomial of $f(t, y(t))$ on the nodes t_0, t_1, t_2 and

$P_2^{(j)}(s), j = 1, 2, \ldots, n$ are the quadratic interpolation polynomial of $f(t, y(t))$ on the nodes t_{j-1}, t_j, t_j. More precisely, we have

$$
\begin{aligned}
P_2^{(0)}(s) = {}& \frac{(s - t_1)(s - t_2)}{(t_0 - t_1)(t_0 - t_2)} f(t_0, y(t_0)) + \frac{(s - t_0)(s - t_2)}{(t_1 - t_0)(t_1 - t_2)} f(t_1, y(t_1)) \\
& + \frac{(s - t_0)(s - t_1)}{(t_2 - t_0)(t_2 - t_1)} f(t_2, y(t_2)),
\end{aligned}
$$

and, with $j = 1, 2, \ldots, n$,

$$
\begin{aligned}
P_2^{(j)}(s) = {}& \frac{(s - t_j)(s - t_{j+1})}{(t_{j-1} - t_j)(t_{j-1} - t_{j+1})} f(t_{j-1}, y(t_{j-1})) + \frac{(s - t_{j-1})(s - t_{j+1})}{(t_j - t_{j-1})(t_j - t_{j+1})} f(t_j, y(t_j)) \\
& + \frac{(s - t_{j-1})(s - t_j)}{(t_{j+1} - t_{j-1})(t_{j+1} - t_j)} f(t_{j+1}, y(t_{j+1})).
\end{aligned}
$$

We then get the following numerical approximate scheme for approximating (2),

$$
y_{n+1} - y_0 = \sum_{j=0}^{n+1} \tilde{w}_{j,n+1} f(t_j, y_j), \tag{4}
$$

where, after some simple but tedious calculations,

$$
\frac{1}{\Gamma(\alpha + 3)} \tilde{w}_{j,n+1} \tag{5}
$$

$$
= \begin{cases}
\frac{2}{(t_0 - t_1)(t_0 - t_1)} C_0 & j = 0, \\
\frac{2}{(t_1 - t_0)(t_1 - t_2)} C_1 + \frac{1}{(t_1 - t_2)(t_1 - t_3)} D_1 & j = 1, \\
\frac{2}{(t_2 - t_0)(t_2 - t_1)} C_2 + \frac{1}{(t_2 - t_1)(t_2 - t_3)} D_2 + \frac{1}{(t_2 - t_3)(t_2 - t_4)} E_2 & j = 2, \\
\frac{1}{(t_j - t_{j-2})(t_j - t_{j-1})} C_j + \frac{1}{(t_j - t_{j-1})(t_j - t_{j+1})} D_j + \frac{1}{(t_j - t_{j+1})(t_j - t_{j+2})} E_j & j = 3, 4, \ldots, n-1, \\
\frac{1}{(t_n - t_{n-2})(t_n - t_{n-1})} D_n + \frac{1}{(t_n - t_{n-1})(t_n - t_{n+1})} E_n & j = n, \\
\frac{1}{(t_{n+1} - t_{n-1})(t_{n+1} - t_n)} E_{n+1} & j = n+1.
\end{cases}
$$

Here

$$
\begin{aligned}
C_0 = {}& \alpha(\alpha + 1)\left[(t_{n+1} - t_0)^{\alpha+2} - (t_{n+1} - t_1)^{\alpha+2}\right] \\
& - \alpha(\alpha + 2)(2t_{n+1} - t_1 - t_2)\left[(t_{n+1} - t_0)^{\alpha+1} - (t_{n+1} - t_1)^{\alpha+1}\right] \\
& + (\alpha + 1)(\alpha + 2)(t_{n+1} - t_1)(t_{n+1} - t_2)\left[(t_{n+1} - t_0)^\alpha - (t_{n+1} - t_1)^\alpha\right],
\end{aligned}
$$

$$
\begin{aligned}
C_1 = {}& \alpha(\alpha + 1)\left[(t_{n+1} - t_0)^{\alpha+2} - (t_{n+1} - t_1)^{\alpha+2}\right] \\
& - \alpha(\alpha + 2)(2t_{n+1} - t_0 - t_2)\left[(t_{n+1} - t_0)^{\alpha+1} - (t_{n+1} - t_1)^{\alpha+1}\right] \\
& + (\alpha + 1)(\alpha + 2)(t_{n+1} - t_0)(t_{n+1} - t_2)\left[(t_{n+1} - t_0)^\alpha - (t_{n+1} - t_1)^\alpha\right]
\end{aligned}
$$

$$
\begin{aligned}
D_1 = {}& \alpha(\alpha + 1)\left[(t_{n+1} - t_2)^{\alpha+2} - (t_{n+1} - t_3)^{\alpha+2}\right] \\
& - \alpha(\alpha + 2)(2t_{n+1} - t_2 - t_3)\left[(t_{n+1} - t_2)^{\alpha+1} - (t_{n+1} - t_3)^{\alpha+1}\right] \\
& + (\alpha + 1)(\alpha + 2)(t_{n+1} - t_2)(t_{n+1} - t_3)\left[(t_{n+1} - t_2)^\alpha - (t_{n+1} - t_3)^\alpha\right],
\end{aligned}
$$

$$C_2 = \alpha(\alpha + 1)\left[(t_{n+1} - t_0)^{\alpha+2} - (t_{n+1} - t_1)^{\alpha+2}\right]$$
$$- \alpha(\alpha + 2)(2t_{n+1} - t_0 - t_1)\left[(t_{n+1} - t_0)^{\alpha+1} - (t_{n+1} - t_1)^{\alpha+1}\right]$$
$$+ (\alpha + 1)(\alpha + 2)(t_{n+1} - t_0)(t_{n+1} - t_1)\left[(t_{n+1} - t_0)^{\alpha} - (t_{n+1} - t_1)^{\alpha}\right],$$

$$D_2 = \alpha(\alpha + 1)\left[(t_{n+1} - t_2)^{\alpha+2} - (t_{n+1} - t_3)^{\alpha+2}\right]$$
$$- \alpha(\alpha + 2)(2t_{n+1} - t_1 - t_3)\left[(t_{n+1} - t_2)^{\alpha+1} - (t_{n+1} - t_3)^{\alpha+1}\right]$$
$$+ (\alpha + 1)(\alpha + 2)(t_{n+1} - t_1)(t_{n+1} - t_3)\left[(t_{n+1} - t_2)^{\alpha} - (t_{n+1} - t_3)^{\alpha}\right],$$

$$E_2 = \alpha(\alpha + 1)\left[(t_{n+1} - t_3)^{\alpha+2} - (t_{n+1} - t_4)^{\alpha+2}\right]$$
$$- \alpha(\alpha + 2)(2t_{n+1} - t_3 - t_4)\left[(t_{n+1} - t_3)^{\alpha+1} - (t_{n+1} - t_4)^{\alpha+1}\right]$$
$$+ (\alpha + 1)(\alpha + 2)(t_{n+1} - t_3)(t_{n+1} - t_4)\left[(t_{n+1} - t_3)^{\alpha}(t_{n+1} - t_4)^{\alpha}\right],$$

$$C_j = \alpha(\alpha + 1)\left[(t_{n+1} - t_j)^{\alpha+2} - (t_{n+1} - t_{j+1})^{\alpha+2}\right]$$
$$- \alpha(\alpha + 2)(2t_{n+1} - t_{j-2} - t_{j-1})\left[(t_{n+1} - t_j)^{\alpha+1} - (t_{n+1} - t_{j+1})^{\alpha+1}\right]$$
$$+ (\alpha + 1)(\alpha + 2)(t_{n+1} - t_{j-2})(t_{n+1} - t_{j-1})\left[(t_{n+1} - t_j)^{\alpha} - (t_{n+1} - t_{j+1})^{\alpha}\right],$$

$$D_j = \alpha(\alpha + 1)\left[(t_{n+1} - t_j)^{\alpha+2} - (t_{n+1} - t_{j+1})^{\alpha+2}\right]$$
$$- \alpha(\alpha + 2)(2t_{n+1} - t_{j-1} - t_{j+1})\left[(t_{n+1} - t_j)^{\alpha+1} - (t_{n+1} - t_{j+1})^{\alpha+1}\right]$$
$$+ (\alpha + 1)(\alpha + 2)(t_{n+1} - t_{j-1})(t_{n+1} - t_{j+1})\left[(t_{n+1} - t_j)^{\alpha} - (t_{n+1} - t_{j+1})^{\alpha}\right],$$

$$E_j = \alpha(\alpha + 1)\left[(t_{n+1} - t_j)^{\alpha+2} - (t_{n+1} - t_{j+1})^{\alpha+2}\right]$$
$$- \alpha(\alpha + 2)(2t_{n+1} - t_{j+1} - t_{j+2})\left[(t_{n+1} - t_j)^{\alpha+1} - (t_{n+1} - t_{j+1})^{\alpha+1}\right]$$
$$+ (\alpha + 1)(\alpha + 2)(t_{n+1} - t_{j+1})(t_{n+1} - t_{j+2})\left[(t_{n+1} - t_j)^{\alpha} - (t_{n+1} - t_{j+1})^{\alpha}\right],$$

$$D_n = \alpha(\alpha + 1)\left[(t_{n+1} - t_{n-1})^{\alpha+2} - (t_{n+1} - t_n)^{\alpha+2}\right]$$
$$- \alpha(\alpha + 2)(2t_{n+1} - t_{n-2} - t_{n-1})\left[(t_{n+1} - t_{n-1})^{\alpha+1} - (t_{n+1} - t_n)^{\alpha+1}\right]$$
$$+ (\alpha + 1)(\alpha + 2)(t_{n+1} - t_{n-2})(t_{n+1} - t_{n-1})\left[(t_{n+1} - t_{n-1})^{\alpha} - (t_{n+1} - t_n)^{\alpha}\right],$$

$$E_n = \alpha(\alpha + 1)(t_{n+1} - t_n)^{\alpha+2} - \alpha(\alpha + 2)(t_{n+1} - t_{n-1})(t_{n+1} - t_n)^{\alpha+1}$$
$$+ (\alpha + 1)(\alpha + 2)(t_{n+1} - t_{n-1})(t_{n+1} - t_n)^{\alpha},$$

$$E_{n+1} = \alpha(\alpha + 1)(t_{n+1} - t_n)^{\alpha+2} - \alpha(\alpha + 2)(2t_{n+1} - t_{n-1} - t_n)(t_{n+1} - t_n)^{\alpha+1}$$
$$+ (\alpha + 1)(\alpha + 2)(t_{n+1} - t_{n-1})(t_{n+1} - t_n)(t_{n+1} - t_n)^{\alpha}.$$

Now we need to solve y_{n+1} in (4) with the weights defined in (5). Let us here only consider how to calculate y_1. The calculation of $y_l, l \geq 2$ is similar. Note that, by (4),

$$y_1 = y_0 + \tilde{w}_{0,1} f(t_0, y_0) + \tilde{w}_{1,1} f(t_1, y_1).$$

Denote

$$g(y_1) = y_1 - [y_0 + \tilde{w}_{0,1} f(t_0, y_0) + \tilde{w}_{1,1} f(t_1, y_1)].$$

We then need to solve $g(y_1) = 0$ which is a nonlinear equation with respect to the variable y_1. Let $z_0 = y_0$ be the initial guess, then y_1 can be approximated by $z_M \approx y_1$ which is obtained by the following Newton iteration formula

$$z_{l+1} = z_l - \frac{g(z_l)}{g'(z_l)}, \ l = 0, 1, 2, \ldots, M.$$

Here g' denotes the derivative of g and the positive integer $M \in \mathbb{N}$ can be determined by using the error control quantity $|z_M - z_{M-1}| < 10^{-10}$. The assumption for f in our paper guarantees that the sequence $z_l, l = 0, 1, 2, \ldots$ is convergent.

Remark 1. The work in this paper is the extension of the work in [7] where the authors introduced the fractional rectangle, trapezoid and predictor-corrector methods with non-uniform meshes for solving (1). The stability and error estimates are discussed in [7]. One may use the similar approach to discuss the stability and error estimates of the proposed numerical method (4) in this work.

3 Numerical Results

We will now look at some numerical results for the numerical method defined in (4) with non-uniform mesh with the time step size

$$\tau_j = t_{j+1} - t_j = (j+1)\mu, \ j = 0, 1, 2, \ldots, N-1, \tag{6}$$

where $\mu = \frac{2T}{N(N+1)}$.

Remark 2. Following the analysis in [7, Sect. 4], if $f(t, y(t))$ is sufficiently smooth, the convergence orders of the proposed numerical methods in [7] for both uniform and non-uniform meshes are highly possible the same. But for non-smooth function $f(t, y(t))$, non-uniform meshes are much suitable than uniform meshes. For the non-smooth function $f(t, y(t))$, Li et al. [7] proved the error estimates for the fractional rectangle, trapezoid and predictor-corrector methods with the concrete non-uniform meshes (6). The mesh (6) is not the unique non-uniform meshes in literature. In general one may consider the graded meshes with $t_j = T(j/N)^r, r > 0$, see, e.g., [9,10]. In fact, after a simple calculation, one may see that the mesh (6) is equivalent to the graded mesh with $r = 2$. In some cases, one may get better convergence order when choosing the graded mesh with $r > 2$ for some non-smooth function $f(t, y(t))$.

In this section, we shall consider two examples. In the first example, we consider the case where the solution of (1) is very smooth and in the second example we consider the case where the solution is less regular.

Example 1. Consider, with $\alpha \in (0,1)$ and $\beta > 0$,

$$\,_0^C D_t^\alpha y(t) = \frac{\Gamma(1+\beta)}{\Gamma(1+\beta-\alpha)} t^{\beta-\alpha} + t^{2\beta} - y^2, \qquad (7)$$

where $f(t,y) = \frac{\Gamma(1+\beta)}{\Gamma(1+\beta-\alpha)} t^{\beta-\alpha} + t^{2\beta} - y^2$ and the exact solution is $y(t) = t^\beta$. We choose $\beta = 2$ and the exact solution is now very smooth $y(t) = t^2$.

In Table 1, we list the experimentally determined convergence orders for the quadratic method (4) with respect to the different $\alpha = 0.4, 0.6, 0.8$. We observe that the quadratic method with the non-uniform meshes has the convergence order almost 3 as we expected.

Table 1. Errors at $T = 1$ by using quadrature method (4) in Example 1

Meshes	N	$\alpha = 0.4$	EOC	$\alpha = 0.6$	EOC	$\alpha = 0.8$	EOC
Uniform	40	2.61E−06		6.85EE−07		3.70E−06	
	80	6.56E−07	1.99	1.75E−07	1.97	9.26E−07	1.99
	160	1.64E−07	1.99	4.39E−08	1.99	2.32E−07	1.99
	320	4.10E−08	1.99	1.10E−08	1.99	5.79E−08	2.00
	640	1.03E−08	2.00	2.74E−09	2.00	1.45E−08	2.00
Non-Uniform	40	1.58E−06		2.09E−06		1.80E−06	
	80	2.01E−07	2.98	2.59E−07	3.00	2.21E−07	3.02
	160	2.54E−08	2.98	3.23E−08	3.00	2.73E−08	3.01
	320	3.21E−09	2.99	4.03E−09	3.00	3.39E−09	3.00
	640	4.04E−10	2.99	5.04E−10	3.00	4.24E−10	3.00

Example 2. We consider the same equation as in Example 1 with $\beta = 0.9$. In this case the exact solution is $y(t) = t^{0.9}$ which is not so regular. In Table 2, we observe that the convergence orders are much lower than the smooth case in Example 1 for both uniform and non-uniform meshes. This is because $f(t, y(t))$ behaves as $t^{\beta-\alpha}, \beta = 0.9$ in Example 2 which is less smoother than $t^{2-\alpha}$ in Example 1. The convergence order depends on the smoothness of the regularity of $f(t, y(t))$, see [7,9,10].

Table 2. Errors at T = 1 by using quadrature method (4) in Example 2

Meshes	N	$\alpha = 0.4$	EOC	$\alpha = 0.6$	EOC	$\alpha = 0.8$	EOC
Uniform	40	4.42E−04		1.91E−03		5.76E−03	
	80	2.44E−04	0.86	1.04E−03	0.88	3.11E−03	0.89
	160	1.32E−04	0.88	5.59E−04	0.89	1.67E−03	0.89
	320	7.13E−04	0.89	2.99E−04	0.90	8.94E−04	0.90
	640	3.83E−05	0.90	1.61E−04	0.90	4.79E−04	0.90
Non-Uniform	40	1.03E−04		2.43E−04		4.58E−04	
	80	2.97E−05	1.79	6.98E−05	1.79	1.31E−04	1.79
	160	8.54E−06	1.80	2.01E−05	1.79	3.78E−05	1.80
	320	2.45E−06	1.80	5.76E−06	1.80	1.08E−05	1.80
	640	7.04E−07	1.80	1.65E−06	1.80	3.11E−06	1.80

References

1. Cao, J., Xu, C.: A high order schema for the numerical solution of the fractional ordinary differential equations. J. Comp. Phys. **238**, 154–168 (2013)
2. Deng, W.H.: Short memory principle and a predict-corrector approach for fractional differential equations. J. Comput. Appl. Math. **206**, 1768–1777 (2007)
3. Diethelm, K.: Generalized compound quadrature formulae for finite-part integral. IMA J. Numer. Anal. **17**, 479–493 (1997)
4. Diethelm, K., Ford, N.J.: Analysis of fractional differential equations. J. Math. Anal. Appl. **265**, 229–248 (2002)
5. Diethelm, K., Ford, N.J., Freed, A.D.: Detailed error analysis for a fractional Adams method. Numer. Algorithms **36**, 31–52 (2004)
6. Diethelm, K., Ford, N.J., Freed, A.D.: A predictor-corrector approach for the numerical solution of fractional differential equations. Nonlinear Dyn. **29**, 3–22 (2002)
7. Li, C., Yi, Q., Chen, A.: Finite difference methods with non-uniform meshes for nonlinear fractional differential equations. J. Comput. Phys. **316**, 614–631 (2016)
8. Li, C., Zeng, F.: The finite difference methods for fractional ordinary differential equations. Numer. Funct. Anal. Optim. **34**, 149–179 (2013)
9. Liu, Y., Roberts, J., Yan, Y.: A note on finite difference methods for nonlinear fractional differential equations with non-uniform meshes. Int. J. Comput. Math. **95**, 1151–1169 (2018)
10. Liu, Y., Roberts, J., Yan, Y.: Detailed error analysis for a fractional Adams method with graded meshes. Numer. Algor. **78**(2018), 1195–1216 (2017). https://doi.org/10.1007/s11075-017-0419-5
11. Stynes, M.: Too much regularity may force too much uniqueness. Fractional Calc. Appl. Anal. **19**, 1554–1562 (2016)
12. Stynes, M., O'riordan, E., Gracia, J.L.: Error analysis of a finite difference method on graded meshes for a time-fractional diffusion equation. SIAM J. Numer. Anal. **55**, 1057–1079 (2017)

13. Pal, K., Liu, F., Yan, Y.: Numerical solutions of fractional differential equations by extrapolation. In: Dimov, I., Faragó, I., Vulkov, L. (eds.) FDM 2014. LNCS, vol. 9045, pp. 299–306. Springer, Cham (2015). https://doi.org/10.1007/978-3-319-20239-6_32

14. Quintana-Murillo, J., Yuste, S.B.: A finite difference method with non-uniform timesteps for fractional diffusion and diffusion-wave equations. Eur. Phys. J. Spec. Top. **222**, 1987–1998 (2013)

15. Yan, Y., Pal, K., Ford, N.J.: Higher order numerical methods for solving fractional differential equations. BIT Numer. Math. **54**, 555–584 (2014)

16. Zhao, L., Deng, W.H.: Jacobi-predictor-corrector approach for the fractional ordinary differential equations. Adv. Comput. Math. **40**, 137–165 (2014)

17. Yuste, S.B., Quintana-Murillo, J.: Fast, accurate and robust adaptive finite difference methods for fractional diffusion equations. Numer. Algor. **71**, 207–228 (2016)

18. Zhang, Y., Sun, Z., Liao, H.: Finite difference methods for the time fractional diffusion equation on non-uniform meshes. J. Comput. Phys. **265**, 195–210 (2014)

Identification of the Right-Hand Side of an Equation with a Fractional Power of an Elliptic Operator

Petr N. Vabishchevich[1,2]([✉])[iD]

[1] Nuclear Safety Institute, Russian Academy of Sciences,
52, B. Tulskaya, 115191 Moscow, Russia
vabishchevich@gmail.com
[2] Akhmet Yasawi International Kazakh-Turkish University,
29, str. Sattarkhanov, 161200 Turkistan, Kazakhstan

Abstract. An inverse problem of identifying the right-hand side of an equation with a fractional power of an elliptic operator by the solution is considered. The direct problem is solved via solving a Cauchy problem for a pseudo-parabolic equation. The problem of identifying the right-hand side is reduced to a retrospective problem for this pseudo-parabolic equation. An iterative method is employed to adjust the initial condition. The results of numerical experiments for a 2D inverse problem are presented.

Keywords: Fractional power of an elliptic operator
Identification of the right-hand side · Retrospective inverse problem
Iterative method for solving an ill-posed problem

1 Introduction

An important class of inverse problems for stationary partial differential equations is connected with the evaluation of an unknown right-hand side of an equation. An additional information about the solution is given in the entire computational domain or in some part of it, in particular, on the boundary. If the solution is known in the whole domain with some error, then the problem of reconstructing the right-hand side actually consists in calculating the differential operator for a function given approximately. Problems of calculating values of an unbounded operator belong to the class of ill-posed ones and various regularizing numerical algorithms [4,9] have been developed for their approximate solution. As for elliptic equations of second order, the main approaches for them are considered in the book [6].

This research was supported by the Russian Foundation for Basic Research (project 17-01-00689), Bulgarian NSF Grant DN 12/1 and the Ministry of Education and Science of the Republic of Kazakhstan (project AP05133873).

G. Nikolov et al. (Eds.): NMA 2018, LNCS 11189, pp. 216–223, 2019.
https://doi.org/10.1007/978-3-030-10692-8_24

Many new non-local mathematical models are associated with fractional powers of elliptic operators [10]. An example is the problem of super-diffusion (fractional in space). At present, for numerical solving such non-classical problems, there are actively developed approaches with rational approximation a fractional power of an elliptic operator [1,2]. We have proposed [11] a numerical algorithm to solve an equation with fractional power elliptic operators that is based on a transition to a pseudo-parabolic equation. For an auxiliary Cauchy problem, the standard two-level schemes are applied.

In the present work, we consider an inverse problem of identifying the right-hand side of an equation with a fractional power of an elliptic second-order operator. It reduces to a problem with inverse time for a pseudo-parabolic equation. This retrospective problem is solved iteratively with sequential adjusting the initial condition. Earlier we have used this approach (see [6,7]) when considering the retrospective inverse problem for a parabolic equation. The standard finite-element approximation in space is used with fully implicit time-stepping.

2 Problem Formulation

Let Ω be a bounded domain ($\Omega \subset \mathbb{R}^d, d = 1, 2, \ldots$) with a piecewise smooth boundary $\partial\Omega$. Let $(\cdot,\cdot), \|\cdot\|$ be the scalar product and norm in $H = L_2(\Omega)$, respectively: $(u,v) = \int_\Omega u(\boldsymbol{x})v(\boldsymbol{x})d\boldsymbol{x}$, $\|u\| = (u,u)^{1/2}$. For functions $u(\boldsymbol{x}), v(\boldsymbol{x}) \in H^1(\Omega)$, we define the bilinear form $a(\cdot,\cdot)$ in the following way:

$$a(u,v) = \int_\Omega (k\nabla u \cdot \nabla v + c\,u\,v)d\boldsymbol{x} + \int_{\partial\Omega} \mu\,u\,v\,d\boldsymbol{x}.$$

The coefficients $k(\boldsymbol{x})$, $c(\boldsymbol{x})$ and $\mu(\boldsymbol{x})$ are smooth functions in $\overline{\Omega}$ and $k(\boldsymbol{x}) \geq \kappa > 0$, $\mu(\boldsymbol{x}) > 0$, $\boldsymbol{x} \in \Omega$. We have $a(u,v) = a(v,u)$, $a(u,u) \geq \delta\|u\|^2$, with a constant $\delta > 0$.

For the bilinear form $a(\cdot,\cdot)$, we put $(a(u,v) = (\mathcal{A}u,v))$ into the correspondence an elliptic operator $\mathcal{A}$ such that

$$\mathcal{A}u = -\nabla(k(\boldsymbol{x})\nabla u) + c(\boldsymbol{x})u, \quad \boldsymbol{x} \in \Omega. \tag{1}$$

We use calligraphic letters for denoting operators in infinite dimensional spaces and standard capital letters for their finite dimensional approximations. The operator $\mathcal{A}$ is defined on the set of functions $u(\boldsymbol{x})$ that satisfy on the boundary $\partial\Omega$ the following conditions:

$$k(\boldsymbol{x})\frac{\partial u}{\partial n} + \mu(\boldsymbol{x})u = 0, \quad \boldsymbol{x} \in \partial\Omega. \tag{2}$$

For the spectral problem

$$\mathcal{A}\varphi_k = \lambda_k\varphi_k, \quad \boldsymbol{x} \in \Omega, \quad k(\boldsymbol{x})\frac{\partial\varphi_k}{\partial n} + g(\boldsymbol{x})\varphi_k = 0, \quad \boldsymbol{x} \in \partial\Omega,$$

we have $0 < \lambda_1 \leq \lambda_2 \leq \ldots$, and the eigenfunctions φ_k, $\|\varphi_k\| = 1$, $k = 1, 2, \ldots$ form a basis in $L_2(\Omega)$. Therefore,

$$u = \sum_{k=1}^{\infty} (u, \varphi_k)\varphi_k.$$

Let the operator $\mathcal{A}$ be defined in the following domain:

$$D(\mathcal{A}) = \{u \mid u(\boldsymbol{x}) \in L_2(\Omega), \ \sum_{k=0}^{\infty} |(u, \varphi_k)|^2 \lambda_k < \infty\}.$$

The operator $\mathcal{A}$ is self-adjoint and positive definite:

$$\mathcal{A} = \mathcal{A}^* \geq \delta\mathcal{I}, \tag{3}$$

where $\mathcal{I}$ is the identity operator in H. For δ, we have $\delta = \lambda_1$. In applications, the value of λ_1 is unknown (the spectral problem must be solved). Therefore, we assume that $\delta \leq \lambda_1$ in (3). When we use the spectral definition of the fractional power of the operator, we have

$$\mathcal{A}^{\alpha}u = \sum_{k=0}^{\infty} (u, \varphi_k)\lambda_k^{\alpha}\varphi_k, \quad 0 < \alpha < 1.$$

More general and mathematically complete definition of fractional powers of elliptic operators is given in [3].

The direct problem consists in evaluating $u(\boldsymbol{x})$ from the equation

$$\mathcal{A}^{\alpha}u = f, \quad 0 < \alpha < 1, \tag{4}$$

with a given $f(\boldsymbol{x})$, $\boldsymbol{x} \in \Omega$. In the inverse problem, we search the right-hand side $f(\boldsymbol{x})$ of Eq. (4) using the given $u(\boldsymbol{x})$.

The solution of the inverse problem of identifying the right-hand side of Eq. (4) can be found via solving the well-posed problem for the equation with a fractional power of an elliptic operator and an ill-posed problem of identifying the right-hand side of an elliptic equation.

From Eq. (4), we have $\mathcal{A}^{-1+\alpha}u = \mathcal{A}^{-1}f$. Assume that

$$f = \mathcal{A}\varphi. \tag{5}$$

For φ with $\beta = 1 - \alpha$, we have

$$\mathcal{A}^{\beta}\varphi = u, \quad 0 < \beta < 1. \tag{6}$$

Thus, we can firstly solve the problem (6) for φ using some approaches to solving problems with a fractional power of an elliptic operator. After that, we can find $f(\boldsymbol{x})$ from (5) (ill-posed problem of numerical differentiation).

3   Retrospective Problem

Our approach is based on reducing the inverse problem of identifying the right-hand side of Eq. (4) to the problem with inverse time for the auxiliary evolution equation. We apply the algorithm for solving approximately direct problem for Eq. (4) that is based on its equivalence to solution of an auxiliary pseudo-time dependent problem [11].

Let $w(x,t)$ be a function

$$w(t) = \delta_h^\alpha (t\mathcal{D} + \delta_h \mathcal{I})^{-\alpha} w(0), \quad \mathcal{D} = \mathcal{A} - \delta_h \mathcal{I}.$$

Due to (3), we have $\mathcal{D} = \mathcal{D}^* \geq 0$. By this construction $w(1) = \delta_h^\alpha \mathcal{A}^{-\alpha} w(0)$ and comparing it with the solution of (4) we see that if we take $w(0) = \delta_h^{-\alpha} f$ then $w(1) = \mathcal{A}^{-\alpha} f = u$, i.e. this is the solution of the direct problem for Eq. (4).

It is also easy to see that $w(t)$ satisfies the following pseudo-parabolic initial value problem

$$(t(\mathcal{A} - \delta_h \mathcal{I}) + \delta_h \mathcal{I})\frac{dw}{dt} + \alpha(\mathcal{A} - \delta_h \mathcal{I})w = 0, \quad 0 < t \leq 1, \tag{7}$$

$$w(0) = \delta_h^{-\alpha} f. \tag{8}$$

Therefore, the solution of Eq. (4) coincides with the solution of the Cauchy problem (7), (8) at the pseudo-time moment $t = 1$: $u = w(1)$. Thus, when solving the direct problem for Eq. (4), from the known right-hand side $f(x)$, we evaluate $u(x)$ and so, we have the well-posed Cauchy problem (7), (8).

A quite different situation occurs, when we consider the inverse problem for Eq. (4), where the right-hand side $f(x)$ is evaluated using the given $u(x)$. In this case, we search the solution of Eq. (7) with the given

$$w(1) = u. \tag{9}$$

For the right-hand side of Eq. (4), we have $f = \delta_h^\alpha w(0)$. Thus, we arrive at the retrospective inverse problem (the problem with inverse time) for the pseudo-parabolic equation (7).

4   Numerical Algorithm

To solve numerically the retrospective inverse problem, different approaches [6] are available. They can be divided into two main classes. The first class (different variants of the quasi-inversion method) employs regularizing algorithms that are based on a perturbation of Eq. (7). The second class (the Tikhonov regularization method, non-local boundary conditions) is related to a perturbation of the boundary condition (9). In the present paper, we start from the iterative method for adjusting the initial condition for Eq. (7), which was applied to the retrospective problem for the parabolic equation in the work [7].

For numerical solving the initial-boundary value problem (7), (9), discretization in space is constructed using the finite element method [8]. Define the subspace of finite elements $V_h \subset H^1(\Omega)$ and the discrete elliptic operator A as $(Ay, v) = a(y, v)$, $\forall\, y, v \in V_h$. The operator A acts on the finite dimensional space V_h ($A : V_h \to V_h$) and

$$A = A^* \geq \delta_h I, \tag{10}$$

where I is the identity operator in V_h and $\delta_h > 0$.

After discretization in space, we arrive at the equation

$$(t(A - \delta_h I) + \delta_h I)\frac{dy}{dt} + \alpha(A - \delta_h I)y = 0, \quad 0 < t \leq 1, \tag{11}$$

for $y(t) \in V_h$. From (9), we get

$$y(1) = \psi, \tag{12}$$

where $\psi = Pu$ with P denoting L_2-projection onto V_h.

To solve numerically the problem (11), (12) we apply implicit two-level scheme, see, e.g. [5]. Let τ be the step-size of a uniform grid in time such that $y_n = y(t_n)$, $t_n = n\tau$, $n = 0, 1, \dots, N$, $N\tau = 1$. We approximate Eq. (11) by the following fully implicit two-level scheme

$$(t_{n+1} D + \delta_h I)\frac{y_{n+1} - y_n}{\tau} + \alpha D y_{n+1} = 0, \quad n = 0, 1, \dots, N - 1, \tag{13}$$

where $D = A - \delta_h I$. Tn accordance with (12), we put

$$y_N = \psi. \tag{14}$$

We search y_0 that provides the approximate solution $f \approx \delta_h^\alpha y_0$ of the inverse problem (4).

For solving the inverse problem (13), (14), we apply the simplest iterative process that is based on sequential adjusting the initial condition, where at each iteration, we solve the direct problem. The direct problem corresponds to the specification of the initial condition

$$y_0 = v. \tag{15}$$

The initial condition is adjusted iteratively in order to satisfy condition (14) at the final time moment $t = 1$.

From (13), (15), for a given y_0, at the final time moment, we get

$$y_N = \sum_{n=1}^{N} S_n v, \tag{16}$$

where S_{n+1} is the operator of the transition from the time level t_n to the next level t_{n+1}:

$$S_{n+1} = ((t_{n+1} + \tau\alpha)D + \delta_h I)^{-1}(t_{n+1} D + \delta_h I), \quad n = 0, 1, \dots, N - 1. \tag{17}$$

In view of (14)–(16), for the approximate solution of the inverse problem (13), (14), it seems natural to put into the correspondence the solution of the following operator equation:

$$Bv = \psi, \quad B = \sum_{n=1}^{N} S_n. \tag{18}$$

In view of the self-adjointness of the operator A, the transition operators S_{n+1}, $n = 0, 1, \ldots, N - 1$ and the operator B in (18) are self-adjoint, too. Single-valued solvability of Eq. (18) holds due to the positivity of the operator D. In our case, for the operator B, defined according to (18), we have

$$0 < B = B^* < I. \tag{19}$$

To solve Eqs. (18), (19), we can use the explicit two-level iterative method in the form

$$\frac{v_{k+1} - v_k}{s_{k+1}} + Bv_k = \psi, \quad k = 0, 1, \ldots, \tag{20}$$

with a given v_0, where s_{k+1} are iterative parameters. The solution of the evolutionary problem that corresponds the initial condition v_k, we denote as $y^{(k)}$.

The above iterative method corresponds to the following implementation of calculations for solving the retrospective inverse problem (13), (14). First, for a given v_k, we solve the direct problem

$$(t_{n+1}D + \delta_h I)\frac{y_{n+1}^{(k)} - y_n^{(k)}}{\tau} + \alpha Dy_{n+1}^{(k)} = 0, \; n = 0, 1, \ldots, N - 1, \quad y_0^{(k)} = v_k,$$

in order to evaluate $y_N^{(k)}$. After obtaining the solution of the direct problem at the final time moment, we adjust in accordance with (14) the initial condition in the following way: $v_{k+1} = v_k - s_{k+1}(y_N^{(k)} - \psi)$.

As it follows from the general theory of iterative methods, the rate of convergence of the method (20) when solving Eq. (18) is determined by the constants of energy equivalence γ_j, $j = 1, 2$: $\gamma_1 I \leq B \leq \gamma_2 I$, $\gamma_1 > 0$. By (19), we can put $\gamma_2 = 1$. The positive constant γ_1 depends on the grid and has a value in the vicinity of zero.

For the stationary iterative method ($s_k = s_0 = $ const in (20)), the convergence conditions for our case have the form $s_0 \leq 2$. For the optimal constant value of the iterative parameter, we have $s_0 \approx 1$. To accelerate convergence, it is necessary to focus on the application of iterative methods of variational type. Using the iterative method of minimal residuals, for iterative parameters, we have

$$s_{k+1} = \frac{(Br_k, r_k)}{(Br_k, Br_k)}, \quad r_k = Bv_k - \psi. \tag{21}$$

In this case, two direct problems are solved at each iteration.

5   Numerical Example

The proposed computational algorithm of identification has been verified in the framework of a quasi-real numerical experiment. The problem is considered in the unit square: $\Omega = \{x \mid x = (x_1, x_2),\ 0 < x_i < 1,\ i = 1, 2\}$ with $k(x) = 1$, $c(x) = 0$, $\mu(x) = 1$ and $\alpha = 0.5$. First, for the given right-hand side $f(x) = \exp(-10((x_1 - 0.4)^2 + (x_2 - 0.3)^2))$, we solve the direct problem (4). Next, the obtained numerical solution $u(x)$ is used to solve the inverse problem.

Below, we will present typical results. The problem was solved on a uniform grid with 50 intervals in each direction using finite elements P_2 on triangles. The parameter $\delta_h = 3.4141$ was obtained numerically from the solution of the spectral problem. The right-hand side and the solution of the direct problem are shown in Fig. 1.

For solving the inverse problem, we applied the iterative method with step selection according to (21). The error of evaluation of the right-hand side is $e_k = f_k - f$, $f_k = \delta_h^\alpha v_k$ with the initial approximation $v_0 = 0$. The error for the first four iterations is shown in Figs. 2 and 3. A high convergence of the iterative method is observed here. Similarly, problems with noisy data are considered when iterations break off when reaching the residual level.

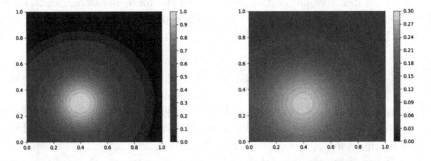

Fig. 1. The direct problem: the RHS $f(x)$ (left) and solution $u(x)$ (right).

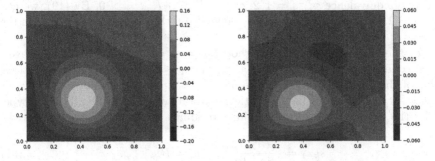

Fig. 2. The inverse problem: errors $e_1(x)$ (left) and $e_2(x)$ (right).

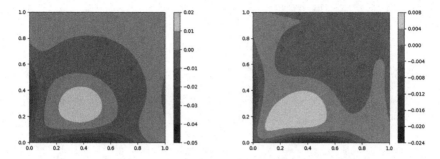

Fig. 3. The inverse problem: errors $e_3(x)$ (left) and $e_4(x)$ (right).

References

1. Aceto, L., Novati, P.: Rational approximation to the fractional Laplacian operator in reaction-diffusion problems. J. Sci. Comput. **39**(1), A214–A228 (2017)
2. Bonito, A., Pasciak, J.: Numerical approximation of fractional powers of elliptic operators. Math. Comput. **84**(295), 2083–2110 (2015)
3. Carracedo, C.M., Alix, M.S., Sanz, M.: The Theory of Fractional Powers of Operators. Elsevier, Amsterdam (2001)
4. Morozov, V.A.: Methods for Solving Incorrectly Posed Problems. Springer, New York (2012). https://doi.org/10.1007/978-1-4612-5280-1
5. Samarskii, A.A.: The Theory of Difference Schemes. Marcel Dekker, New York (2001)
6. Samarskii, A.A., Vabishchevich, P.N.: Numerical Methods for Solving Inverse Problems of Mathematical Physics. Walter de Gruyter (2007)
7. Samarskii, A.A., Vabishchevich, P.N., Vasil'ev, V.I.: Iterative solution of the retrospective inverse heat conduction problem. Math. Model. **9**(5), 119–127 (1997). in Russian
8. Thomée, V.: Galerkin Finite Element Methods for Parabolic Problems. Springer, Berlin (2006). https://doi.org/10.1007/BFb0071790
9. Tikhonov, A.N., Arsenin, V.Y.: Methods for Solving Ill-Posed Problems. Wiley, New York (1977)
10. Uchaikin, V.V.: Fractional Derivatives for Physicists and Engineers. Higher Education Press, Beijing (2013)
11. Vabishchevich, P.N.: Numerically solving an equation for fractional powers of elliptic operators. J. Comp. Phys. **282**(1), 289–302 (2015)

References

Orthogonal Polynomials and Numerical Quadratures

Definite Quadrature Formulae of Order Three Based on the Compound Midpoint Rule

Ana Avdzhieva⬤, Vesselin Gushev⬤, and Geno Nikolov$^{(\boxtimes)}$⬤

Faculty of Mathematics and Informatics, Sofia University "St. Kliment Ohridski",
5, James Bourchier blvd., 1164 Sofia, Bulgaria
{aavdzhieva,v_gushev,geno}@fmi.uni-sofia.bg

Abstract. A sequence of definite quadrature formulae of order three based on the compound midpoint rule is constructed. Their error constants are evaluated and simple a posteriori error estimates are derived.

Keywords: Definite quadrature formulae
Euler-Maclaurin summation formula · Peano kernel
A posteriori error estimate

1 Introduction and Statement of the Results

The definite integral

$$I[f] := \int\limits_0^1 f(x)\,dx$$

is approximated by quadrature formulae, which are linear functionals of the form

$$Q[f] := \sum_{i=0}^n a_i f(x_i), \quad 0 \le x_0 < x_1 < \cdots < x_n \le 1. \tag{1}$$

Throughout this paper, π_m stands for the set of algebraic polynomials of degree at most m. Quadrature formula (1) is said to have algebraic degree of precision m (in short, $ADP(Q) = m$), if its remainder functional

$$R[Q; f] := I[f] - Q[f]$$

vanishes whenever $f \in \pi_m$, and $R[Q; f] \ne 0$ when $f \in \pi_{m+1} \setminus \pi_m$.

Definition 1. *Quadrature formula (1) is said to be definite of order r, $r \in \mathbb{N}$, if there exists a real non-zero constant $c_r(Q)$ such that*

$$R[Q; f] = c_r(Q) f^{(r)}(\xi)$$

for every $f \in C^r[0,1]$, with some $\xi \in [0,1]$ depending on f. Furthermore, Q is called positive (resp., negative) definite of order r, if $c_r(Q) > 0$ $(c_r(Q) < 0)$.

Supported by the Sofia University Research Fund under Contract 80-10-139/2018.

G. Nikolov et al. (Eds.): NMA 2018, LNCS 11189, pp. 227–234, 2019.
https://doi.org/10.1007/978-3-030-10692-8_25

Definition 2. *A function* $f \in C^r[0,1]$ *is called* r-*positive (resp.,* r-*negative) if* $f^{(r)}(x) \geq 0$ *(resp.* $f^{(r)}(x) \leq 0$) *for every* $x \in [0,1]$.

If $\{Q^+, Q^-\}$ is a pair of a positive and a negative definite quadrature formula of order r and f is an r-positive function, then $Q^+[f] \leq I[f] \leq Q^-[f]$. This simple observation is a base for derivation of a posteriori error estimates and rules for termination of calculations in the algorithms for numerical integration (see [4] for a survey). Most of quadratures used in practice (e.g., quadrature formulae of Gauss, Radau, Lobatto, Newton-Cotes) are definite of certain order. The best known definite quadratures are the compound midpoint and trapezium rules

$$Q_n^{Mi}[f] = \frac{1}{n} \sum_{k=1}^{n} f\left(\frac{2k-1}{2n}\right), \quad Q_{n+1}^{Tr}[f] = \frac{1}{2n}\big(f(0) + f(1)\big) + \frac{1}{n} \sum_{k=1}^{n-1} f\left(\frac{k}{n}\right),$$

which are respectively positive and negative definite of order two with error constants $c_2(Q_n^{Mi}) = 1/(24n^2)$ and $c_2(Q_{n+1}^{Tr}) = -1/(12n^2)$. Moreover, Q_{n+1}^{Tr} and Q_n^{Mi} are respectively *optimal negative definite* and *asymptotically optimal positive definite* quadrature formulae of order two (we refer to [2,5–9,11] for more details about optimal and asymptotically optimal definite quadratures).

The simplest pair of definite quadrature formulae of odd order is the left and the right rectangles quadratures,

$$Q_+[f] = \frac{1}{n} \sum_{k=0}^{n-1} f\left(\frac{k}{n}\right), \qquad Q_-[f] = \frac{1}{n} \sum_{k=1}^{n} f\left(\frac{k}{n}\right).$$

If f is 1-positive (or non-decreasing), then $R[Q_+; f] \geq 0, R[Q_-; f] \leq 0$, and

$$|R[Q_\pm; f]| \leq Q_-[f] - Q_+[f] = \frac{1}{n}\big(f(1) - f(0)\big).$$

Definition 3. *Quadrature formula (1) is called symmetrical, if*

$$a_k = a_{n-k}, \quad k = 0, \ldots, n; \tag{2}$$

$$x_k = 1 - x_{n-k}, \quad k = 0, \ldots, n. \tag{3}$$

Quadrature formula (1) is nodes-symmetrical, if only condition (3) is satisfied. The reflected quadrature formula to (1) is defined by

$$\tilde{Q}[f] = \tilde{Q}[Q; f] := \sum_{k=0}^{n} a_k f(x_{n-k}).$$

While, most often, definite quadrature formulae of even order are symmetrical, $Q_\pm$ are non-symmetrical and are obtained from each other by a reflection.

By adding (if necessary) nodes with weights equal to zero, every quadrature formula may be viewed as nodes-symmetrical. The next statement shows that our observations about the left and right rectangles quadratures apply to a more general situation.

Proposition 1 ([1]).

(i) If Q is a positive definite quadrature formula of order r (r – odd), then its reflected quadrature formula $\widetilde{Q}$ is negative definite of order r and vice versa. Moreover, $c_r(\widetilde{Q}) = -c_r(Q)$.

(ii) If quadrature formula Q in (1) is nodes-symmetrical and definite of order r (r – odd), and f is an r-positive or r-negative function, then, with Q^* standing for either Q or $\widetilde{Q}$ we have

$$\left| R[Q^*; f] \right| \le B[Q; f] := \left| \sum_{k=0}^{\lfloor \frac{n}{2} \rfloor} \left(a_k - a_{n-k}\right)\left(f(x_{n-k}) - f(x_k)\right) \right|. \qquad (4)$$

(iii) Under the same assumptions for Q and f as in (ii), for $\hat{Q} = (Q + \widetilde{Q})/2$ we have

$$\left| R[\hat{Q}; f] \right| \le \frac{1}{2} B[Q; f].$$

Proposition 1 implies, in particular, that definite quadrature formulae of odd order are never symmetrical. The error estimate (4) is especially simple when Q is of almost Chebyshev type, i.e. almost all weights of Q are equal to each other. In [1] we constructed definite quadrature formulae of order three with equidistant nodes which are of almost Chebyshev type. Here we apply a similar approach to construct definite quadrature formulae of order three with nodes

$$y_{0,n} = 0, \qquad y_{k,n} = \frac{2k-1}{2n}, \qquad k = 1, \ldots, n, \qquad y_{n+1,n} = 1.$$

Theorem 1. *For every $n \in \mathbb{N}$, $n \ge 8$, the (nodes-symmetrical) quadrature formula $Q_n[f] = \sum_{k=0}^{n+1} A_{k,n} f(y_{k,n})$ with coefficients $A_{k,n} = \dfrac{1}{n}$, $4 \le k \le n-3$,*

$$A_{0,n} = \frac{-42+41\sqrt{3}}{162n}, \ A_{1,n} = \frac{678-203\sqrt{3}}{432n}, \ A_{2,n} = \frac{357+199\sqrt{3}}{648n}, \ A_{3,n} = \frac{164-13\sqrt{3}}{144n},$$

$$A_{n-2,n} = \frac{225-\sqrt{3}}{216n}, \ A_{n-1,n} = \frac{189+2\sqrt{3}}{216n}, \ A_{n,n} = \frac{234-\sqrt{3}}{216n}, \quad A_{n+1,n} = 0,$$

is positive definite of order three with the error constant

$$c_3(Q_n) = \frac{\sqrt{3}}{216n^3} + \frac{169\sqrt{3} - 210}{2592n^4}.$$

The reflected quadrature formula $\widetilde{Q}_n$ is negative definite of order three with the error constant $c_3(\widetilde{Q}_n) = -c_3(Q_n)$. If f is a 3-positive or 3-negative function, then with Q_n^* standing for either Q_n or $\widetilde{Q}_n$ we have the a posteriori error estimate

$$\left| R[Q_n^*; f] \right| \le \left| \frac{41\sqrt{3} - 42}{162n}\left(f(y_{0,n}) - f(y_{n+1,n})\right) - \frac{67\sqrt{3} - 70}{144n}\left(f(y_{1,n}) - f(y_{n,n})\right) \right.$$

$$\left. + \frac{193\sqrt{3} - 210}{648n}\left(f(y_{2,n}) - f(y_{n-1,n})\right) - \frac{37\sqrt{3} - 42}{432n}\left(f(y_{3,n}) - f(y_{n-2,n})\right) \right|,$$

and for $\hat{Q}_n = (Q_n + \tilde{Q}_n)/2$ the halved a posteriori error estimate holds:

$$\left| R[\hat{Q}_n; f] \right| \leq \left| \frac{41\sqrt{3}-42}{324n}\left(f(y_{0,n})-f(y_{n+1,n})\right) - \frac{67\sqrt{3}-70}{288n}\left(f(y_{1,n})-f(y_{n,n})\right) \right.$$
$$\left. + \frac{193\sqrt{3}-210}{1296n}\left(f(y_{2,n})-f(y_{n-1,n})\right) - \frac{37\sqrt{3}-42}{864n}\left(f(y_{3,n})-f(y_{n-2,n})\right) \right|.$$

The rest of the paper is organized as follows. Section 2 provides some preliminaries. In Sect. 3 we present the proof of Theorem 1. In the construction of our quadrature formulae we perform some optimization, minimizing their error constants and trying at the same time to keep almost Chebyshev structure.

2   Preliminaries

By $W_1^r = W_1^r[0,1]$, $r \in \mathbb{N}$, we denote the Sobolev class of functions

$$W_1^r[0,1] := \{f \in C^{r-1}[0,1] : f^{(r-1)} \text{ abs. continuous}, \int_0^1 |f^{(r)}(t)| \, dt < \infty\}.$$

If $\mathcal{L}$ is a linear functional defined in $W_1^r[0,1]$ which vanishes on π_{r-1}, then, by a classical result of Peano [10], $\mathcal{L}$ admits the integral representation

$$\mathcal{L}[f] = \int_0^1 K_r(t) f^{(r)}(t) \, dt, \qquad K_r(t) = \mathcal{L}\left[\frac{(\cdot - t)_+^{r-1}}{(r-1)!}\right], \quad t \in [0,1],$$

where $u_+(t) = \max\{t, 0\}$, $t \in \mathbb{R}$.

In the case $\mathcal{L}$ is the remainder $R[Q; \cdot]$ of a quadrature formula Q with $ADP(Q) \geq r-1$, the function $K_r(t) = K_r(Q; t)$ is referred to as the r-th Peano kernel of Q. For Q as in (1), explicit representations for $K_r(Q; t)$, $t \in [0,1]$, are

$$K_r(Q; t) = \frac{(1-t)^r}{r!} - \frac{1}{(r-1)!}\sum_{i=0}^{n} a_i(x_i - t)_+^{r-1}, \tag{5}$$

$$K_r(Q; t) = (-1)^r\left[\frac{t^r}{r!} - \frac{1}{(r-1)!}\sum_{i=0}^{n} a_i(t - x_i)_+^{r-1}\right]. \tag{6}$$

Since for $f \in C^r[0,1]$ we have $R[Q; f] = \int_0^1 K_r(Q; t) f^{(r)}(t) \, dt$, it is clear that Q is a positive (negative) definite quadrature formula of order r if and only if $ADP(Q) = r-1$ and $K_r(Q; t) \geq 0$ (resp. $K_r(Q; t) \leq 0$) for all $t \in [0,1]$.

The following is a particular case of the Euler-Maclaurin formula (cf. [3]):

Lemma 1. Assume that $f \in W_1^3[0,1]$. Then

$$I[f] = Q_n^{Mi}[f] + \frac{1}{24n^2}\left[f'(1) - f'(0)\right] - \frac{1}{n^3}\int_0^1 \tilde{B}_3\left(nx - \frac{1}{2}\right)f^{(3)}(x) \, dx, \tag{7}$$

where $\tilde{B}_3$ is the 1-periodic extension of $B_3(x) = \dfrac{x^3}{6} - \dfrac{x^2}{4} + \dfrac{x}{12}$, the Bernoulli polynomial of degree three. Moreover, $|\tilde{B}_3(x)| \leq \dfrac{\sqrt{3}}{216}$ for every $x \in \mathbb{R}$.

3   Proof of Theorem 1

We rewrite formula (7) in Lemma 1 in the following form:

$$I[f] = Q_n^{Mi}[f] + \frac{1}{24\,n^2}\left[f'(1) - f'(0)\right] - \frac{\sqrt{3}}{216\,n^3}\left[f''(1) - f''(0)\right]$$

$$+\frac{1}{n^3}\int_0^1\left(\frac{\sqrt{3}}{216} - \tilde{B}_3\left(nx - \frac{1}{2}\right)\right)f^{(3)}(x)\,dx =: \bar{Q}[f] + R[\bar{Q}; f]\,,$$

$$\bar{Q}[f] = Q_n^{Mi}[f] - \frac{1}{24n^2}f'(0) + \frac{\sqrt{3}}{216n^3}f''(0) + \frac{1}{24n^2}f'(1) - \frac{\sqrt{3}}{216n^3}f''(1)\,. \quad (8)$$

According to Lemma 1, $\bar{Q}$ is a positive definite quadrature formula, though not of the desired type as it involves values of integrand's derivatives. We approximate the derivatives values at the end-points appearing in $\bar{Q}$ by pairs of formulae for numerical differentiation involving values at the closest nodes. The reason for not using single formulae for numerical differentiation is that it is not a priory clear whether they will result in a positive definite quadrature formula, so we need some flexibility to achieve definiteness.

Thus, $f'(0)$ and $f''(0)$ are approximated as follows:

$$f'(0) \approx \frac{n}{3}\left[-8f(y_{0,n}) + 9f(y_{1,n}) - f(y_{2,n})\right] =: D_{1,1}[f]\,,$$

$$f'(0) \approx n\left[-2f(y_{1,n}) + 3f(y_{2,n}) - f(y_{3,n})\right] =: D_{1,2}[f]\,,$$

$$f''(0) \approx \frac{n^2}{3}\left[8f(y_{0,n}) - 12f(y_{1,n}) + 4f(y_{2,n})\right] =: D_{2,1}[f]\,,$$

$$f''(0) \approx n^2\left[f(y_{1,n}) - 2f(y_{2,n}) + f(y_{3,n})\right] =: D_{2,2}[f]\,.$$

For $\alpha, \beta \in \mathbb{R}$, we set

$$D_1^\alpha[f] := \alpha D_{1,1}[f] + (1-\alpha)D_{1,2}[f]\,, \quad D_2^\beta[f] := \beta D_{2,1}[f] + (1-\beta)D_{2,2}[f]\,.$$

We approximate $f'(1)$ and $f''(1)$ by $\tilde{D}_{1,i}[f]$ and $\tilde{D}_{2,i}[f]$, which are the reflected formulae for numerical differentiation $D_{1,i}$ and $D_{2,i}$, $i = 1, 2$, precisely,

$$\tilde{D}_{1,i}[f] = D_{1,i}[g]\,, \quad g(x) = -f(1-x)\,, \qquad \tilde{D}_{2,i}[f] = D_{2,i}[h]\,, \quad h(x) = f(1-x)\,.$$

For $\gamma, \delta \in \mathbb{R}$, we set

$$\tilde{D}_1^\gamma[f] := \gamma\tilde{D}_{1,1}[f] + (1-\gamma)\tilde{D}_{1,2}[f]\,, \quad \tilde{D}_2^\delta[f] := \delta\tilde{D}_{2,1}[f] + (1-\delta)\tilde{D}_{2,2}[f]\,.$$

Note that the functionals $L_1[f] := f'(0) - D_1^\alpha[f]$, $L_2[f] := f''(0) - D_2^\beta[f]$, $\tilde{L}_1[f] := f'(1) - \tilde{D}_1^\gamma[f]$ and $\tilde{L}_2[f] := f''(1) - \tilde{D}_2^\delta[f]$ vanish on π_2, hence the replacement of $f'(0)$, $f''(0)$, $f'(1)$ and $f''(1)$ in (8) with $D_1^\alpha[f]$, $D_2^\beta[f]$, $\tilde{D}_1^\gamma[f]$ and $\tilde{D}_2^\delta[f]$, respectively, yields a quadrature formula

$$Q[f] := Q_n[\alpha, \beta, \gamma, \delta; f] = \sum_{k=0}^{n+1} A_{k,n}f(y_{k,n})\,, \quad (9)$$

which evaluates $I[f]$ to the exact value whenever $f \in \pi_2$. Assuming that $n \geq 8$, we have $A_{k,n} = 1/n$ for $4 \leq k \leq n - 3$. It is not difficult to see that $\{A_{k,n}\}_{k=0}^3$ depend on a single parameter, say θ, while $\{A_{k,n}\}_{k=n-2}^{n+1}$ depend on another single parameter, say ϱ, where $\theta := 9\alpha + \sqrt{3}\beta$, $\varrho := 9\gamma - \sqrt{3}\delta$. Specifically,

$$A_{0,n} = \frac{\theta}{81n}, \ A_{1,n} = \frac{234 + \sqrt{3} - 5\theta}{216n}, \ A_{2,n} = \frac{567 - 6\sqrt{3} + 10\theta}{648n}, \ A_{3,n} = \frac{225 + \sqrt{3} - \theta}{216n},$$

$$A_{n-2,n} = \frac{225 - \sqrt{3} - \varrho}{216n}, \ A_{n-1,n} = \frac{567 + 6\sqrt{3} + 10\varrho}{648n}, \ A_{n,n} = \frac{234 - \sqrt{3} - 5\varrho}{216n}, \ A_{n+1,n} = \frac{\varrho}{81n}.$$

A closer look at $Q = Q_n[\alpha, \beta, \gamma, \delta]$ and its third Peano kernel shows that

$$R[Q; f] = R[\bar{Q}; f] - \frac{1}{24n^2} L_1[f] + \frac{\sqrt{3}}{216n^3} L_2[f] + \frac{1}{24n^2} \tilde{L}_1[f] - \frac{\sqrt{3}}{216n^3} \tilde{L}_2[f],$$

$$K_3(Q; t) = K_3(\bar{Q}; t) - \frac{1}{24n^2} \left[K_3(L_1; t) - K_3(\tilde{L}_1; t) \right] + \frac{\sqrt{3}}{216n^3} \left[K_3(L_2; t) - K_3(\tilde{L}_2; t) \right].$$

From the definition of the Peano kernels, we see that $K_3(L_1; \cdot)$ and $K_3(L_2; \cdot)$ vanish identically on the interval $[y_{3,n}, 1]$ while $K_3(\tilde{L}_1; \cdot)$ and $K_3(\tilde{L}_2; \cdot)$ vanish identically on the interval $[0, y_{n-2,n}]$. Hence,

$$K_3(Q; t) = K_3(\bar{Q}; t) = n^{-3} \left[\frac{\sqrt{3}}{216} - \tilde{B}_3 \left(nt - \frac{1}{2} \right) \right] \geq 0, \quad t \in [y_{3,n}, y_{n-2,n}]$$

and we need to verify condition $K_3(Q; t) \geq 0$ only on $[0, y_{3,n}]$ and $[y_{n-2,n}, 1]$. Assuming this condition is satisfied, for the error constant of Q we have

$$c_3(Q) = \int_0^{y_{3,n}} K_3(Q; t) \, dt + \int_{y_{n-2,n}}^1 K_3(Q; t) \, dt + \frac{\sqrt{3}(n-5)}{216n^4}, \qquad (10)$$

the last summand being the integral of $K_3(Q; \cdot)$ on the interval $[y_{3,n}, y_{n-2,n}]$.

We aim to minimize the integrals in (10) with respect to parameters θ and ϱ subject to the requirement $K_3(Q; t) \geq 0$ on $[0, y_{3,n}]$ and $[y_{n-2,n}, 1]$.

3.1 Positivity of $K_3(Q; t)$ on $[0, y_{3,n}]$

Applying formula (6) for Peano kernels with $r = 3$, after the change of variable $t = u/n$ we arrive at the following representation of $K_3(Q; t)$ for $t \in [0, y_{3,n}]$:

$$K_3(Q; t) = -\frac{1}{6n^4} \left[u^3 - \frac{\theta}{27} u^2 - \frac{234 + \sqrt{3} - 5\theta}{72} \left(u - \frac{1}{2} \right)_+^2 - \frac{576 - 6\sqrt{3} + 10\theta}{216} \left(u - \frac{3}{2} \right)_+^2 \right.$$
$$\left. - \frac{225 + \sqrt{3} - \theta}{72} \left(u - \frac{5}{2} \right)_+^2 \right] =: -\frac{1}{6n^4} \varphi(\theta; u) = -\frac{1}{6n^4} \varphi(u).$$

We have the equivalence $K_3(Q; t) \geq 0$, $t \in [0, y_{3,n}] \Leftrightarrow \varphi(u) \leq 0$, $u \in [0, 5/2]$. Assuming $\varphi(u) \leq 0$, $u \in [0, 5/2]$, we find

$$\int_0^{y_{3,n}} K_3(Q; t) \, dt = -\frac{1}{6n^4} \int_0^{\frac{5}{2}} \varphi(u) \, du = \frac{-387 + 48\sqrt{3} + 40\theta}{10368n^4}. \qquad (11)$$

To minimize $c_3(Q)$, in view of (10), we need to find the smallest value of θ ensuring that $\varphi(u) \leq 0$, $u \in [0, 5/2]$. We consider separately three cases.

Case 1: $u \in [0, 1/2]$. In this case $\varphi(u) = \dfrac{1}{27} u^2(27u - \theta)$ and condition $\varphi(u) \leq 0$, $u \in [0, 1/2]$ is equivalent to $\theta \geq \dfrac{27}{2}$.

Case 2: $u \in [1/2, 3/2]$. We set $v = u - \frac{1}{2}$, $v \in [0, 1]$, then

$$\varphi(u) = u^3 - \frac{\theta}{27} u^2 - \frac{234 + \sqrt{3} - 50\theta}{72} \left(u - \frac{1}{2}\right)^2$$
$$= v^3 - \frac{126 + \sqrt{3}}{72} v^2 + \frac{3}{4} v + \frac{1}{8} + \frac{\theta}{216} (7v^2 - 8v - 2) =: \varphi_1(v).$$

Since $7v^2 - 8v - 2 < 0$ for all $v \in [0, 1]$, it follows that $\varphi(u) < 0$ for every $u \in [1/2, 3/2]$ provided θ is big enough; in addition, if the latter condition holds for some θ_0, it will hold also for all $\theta > \theta_0$. The smallest value of θ such that $\varphi_1(v) \leq 0$ for all $v \in [0, 1]$ should be such that φ_1 has a double zero in $(0, 1)$, i.e. θ is a zero of $D(\varphi_1)$, the discriminant of φ_1. Using Wolfram's *Mathematica*, we find $D(\varphi_1)$, which is a quintic polynomial of θ with four distinct real zeros: $\theta_1 = -53.4866...$, $\theta_2 = 5.4754...$, $\theta_3 = 14.5060...$ and $\theta_4 = 47.1464...$ For $\theta = \theta_3$ the polynomial φ_1 has the maximum value in $[0, 1]$, which is equal to zero. Therefore, this case imposes the restriction $\theta \geq \theta_3$.

Case 3: $u \in [1/2, 3/2]$. We set $v = u - \frac{3}{2}$, $v \in [0, 1]$, and obtain

$$\varphi(u) = u^3 - \frac{\theta}{27} u^2 - \frac{234 + \sqrt{3} - 50\theta}{72} \left(u - \frac{1}{2}\right)^2 - \frac{576 - 6\sqrt{3} + 100\theta}{216} \left(u - \frac{3}{2}\right)^2$$
$$= \frac{1}{72} \left[72v^3 + (\sqrt{3} - 99) v^2 + 2(9 - \sqrt{3}) v + 9 - \sqrt{3} - \theta(v + 1)^2\right] =: \varphi_2(v).$$

Similarly to *Case 2*, with Wolfram's *Mathematica* we find that $\varphi_2(v) \leq 0$ for $\theta \geq \tilde{\theta}_1 = 10.7321$, where $\tilde{\theta}_1$ is the smallest zero of the discriminant $D(\varphi_2)$ of φ_2, i.e. the optimal value of θ in *Case 3* is $\tilde{\theta}_1 = 10.7321$.

Summarizing the three cases considered above, we see that the optimal value of θ ensuring that $\varphi(\theta; u) \leq 0$ for all $u \in [0, 5/2]$, is $\theta = \theta_3$. Our choice is a slightly greater value, $\theta^* = \dfrac{41\sqrt{3} - 42}{2} \approx 14.5070$, allowing "exact" expressions for the first four coefficients of the quadrature formula in Theorem 1.

3.2 Positivity of $K_3(Q; t)$ on $[y_{n-2,n}, 1]$

We apply (5) with $r = 3$ and Q being quadrature formula (9) to obtain:

$$K_3(Q; t) = \frac{(1-t)^3}{6} - \frac{1}{2} \sum_{k=0}^{n+1} A_{k,n}(y_{k,n} - t)_+^2 = \frac{(1-t)^3}{6} - \frac{1}{2} \sum_{k=0}^{n+1} A_{k,n}(1 - t - y_{n+1-k,n})_+^2$$

$$\underset{=}{\{x = 1 - t\}} \frac{x^3}{6} - \frac{1}{2} \sum_{k=0}^{n+1} A_{n+1-k,n}(x - y_{k,n})_+^2 := \tilde{K}_3(Q; x).$$

We have, with $\psi(u) = \psi(\varrho; u)$ given by

$$\psi(u) = u^3 - \frac{\varrho}{27}\,u^2 - \frac{234 - \sqrt{3} - 5\varrho}{72}\left(u - \frac{1}{2}\right)_+^2 - \frac{576 + 6\sqrt{3} + 10\varrho}{216}\left(u - \frac{3}{2}\right)_+^2,$$

$$\int_{y_{n-3,n}}^1 K_3(Q;t)\,dt = \int_0^{y_{3,n}} \widetilde{K}_3(Q;x)\,dx \overset{\{x=u/n\}}{=} \frac{1}{6n^4}\int_0^{\frac{5}{2}} \psi(\varrho;u)\,du.$$

By a straightforward calculation we obtain

$$\int_{y_{n-2,n}}^1 K_3(Q;t)\,dt = \frac{387 + 48\sqrt{3} - 40\varrho}{10368 n^4}. \tag{12}$$

From the equivalence $K_3(Q;t) \geq 0$, $t \in [y_{n-2,n}, 1] \Leftrightarrow \psi(\varrho; u) \geq 0$, $u \in [0, 5/2]$, we see that, to minimize the error constant $c_3(Q)$, we need to find the largest ϱ such that $\psi(\varrho; u) \geq 0$ for all $u \in [0, 5/2]$. Since $\psi(\varrho; u) = u^2\left(u - \frac{\varrho}{27}\right)$ for $u \in [0, 1/2]$, a necessary condition is $\varrho \leq 0$. It is easily verified that the choice $\varrho = \varrho^* = 0$ is optimal, as it secures the non-negativity of ψ on $[0, 5/2]$.

We observe that, with the (almost) optimal choice $(\theta, \varrho) = (\theta^*, \varrho^*)$, quadrature formula (9) coincides with the one in Theorem 1. Its error constant is easily evaluated from (10)–(12). This proves the first part of Theorem 1. The second part of Theorem 1 is a direct consequence of Proposition 1.

References

1. Avdzhieva, A., Gushev, V., Nikolov, G.: Definite quadrature formulae of order three with equidistant nodes. Ann. Univ. Sofia, Fac. Math. Inf. **104**, 155–170 (2017)
2. Avdzhieva, A., Nikolov, G.: Asymptotically optimal definite quadrature formulae of 4th order. J. Comput. Appl. Math. **311**, 565–582 (2017)
3. Braß, H.: Quadraturverfahren. Vandenhoeck & Ruprecht, Göttingen (1977)
4. Förster, K.-J.: Survey on stopping rules in quadrature based on Peano kernel methods. Suppl. Rend. Circ. Math. Palermo, Ser. II **33**, 311–330 (1993)
5. Jetter, K.: Optimale Quadraturformeln mit semidefiniten Peano-Kernen. Numer. Math. **25**, 239–249 (1976)
6. Köhler, P., Nikolov, G.: Error bounds for optimal definite quadrature formulae. J. Approx. Theory **81**, 397–405 (1995)
7. Lange, G.: Beste und optimale definite Quadraturformel. Ph.D. Thesis, Technical University Clausthal, Germany (1977)
8. Lange, G.: Optimale definite Quadraturformel. In: Hämmerlin, G., (Ed.), Numerische Integration, ISNM vol. 45, Birkhäuser, Basel, Boston, Stuttgart, pp. 187–197 (1979)
9. Nikolov, G.: On certain definite quadrature formulae. J. Comput. Appl. Math. **75**, 329–343 (1996)
10. Peano, G.: Resto nelle formule di quadratura espresso con un integrale definito. Atti della Reale Accademia dei Lincei Rendiconti (Ser. 5) **22**, 562–569 (1913)
11. Schmeisser, G.: Optimale Quadraturformeln mit semidefiniten Kernen. Numer. Math. **20**, 32–53 (1972)

Finite Difference Schemes and Classical Transcendental Functions

Edik A. Ayryan[1], Mikhail D. Malykh[2]($\boxtimes$) (iD), Leonid A. Sevastianov[1,2] (iD),
and Yu Ying[2,3]

[1] Joint Institute for Nuclear Research (Dubna),
Joliot-Curie, 6, Dubna, Moscow Region 141980, Russia
ayrjan@jinr.ru
[2] Department of Applied Probability and Informatics,
Peoples' Friendship University of Russia (RUDN University),
6 Miklukho-Maklaya Street, Moscow 117198, Russia
{malykh_md,sevastianov_la}@rudn.university
[3] Department of Algebra and Geometry, Kaili University,
3 Kaiyuan Road, Kaili 556011, China
yingy6165@gmail.com

Abstract. In the first part of the article we give a brief review of various approaches to symbolic integration of ordinary differential equations (Liouvillian approach, power series method) from the point of view of numerical methods. We aim to show that all higher transcendental functions were considered in the past centuries as solutions of such differential equations, for which the application of the computational techniques of that time was particularly efficient. Nowadays the finite differences method is a standard method for integration of differential equations. Our main idea is that now all transcendental functions can be considered as solutions of such differential equations, for which the application of this method is particularly efficient.

In the second part of the article we consider an autonomous system of differential equations with algebraic integrals of motion and try to find a totally conservative difference scheme. There are only two cases when the system can be discretized by explicit totally conservative scheme: integrals specify an elliptic curve or unicursal curve. For autonomous systems describing the Jacobi elliptic functions we construct the finite differences scheme, which conserves all algebraic integrals and defines one-to-one correspondence between the layers. We can see that this scheme truly describes the periodicity of the motion.

Keywords: Conservative difference scheme · Elliptic function
Symbolic integration · Algebraic curve · Algebraic correspondence

The publication has been prepared with the support of the "RUDN University Program 5-100" and funded by RFBR according to the research projects No. 18-07-00567 and No. 18-51-18005.

© Springer Nature Switzerland AG 2019
G. Nikolov et al. (Eds.): NMA 2018, LNCS 11189, pp. 235–242, 2019.
https://doi.org/10.1007/978-3-030-10692-8_26

1   Introduction: Symbolic Integration of Differential Equation and the Progress in Numerical Methods

Many problems which we try to solve in finite terms have arisen many centuries ago. The numerical methods of the last centuries dictated their formulations. For example, we study the compass-and-straightedge constructions but we have not used these devices for a long time.

In the studies of Galois differential theory [1], we use the concept of elementary functions. In the time of Liouville, when these studies were initiated, the functions have been considered elementary if their tables of values were available for common use. But at present, this class of functions is much narrower than the set of functions, for which the computation algorithms are implemented in all systems of computer algebra. Thus, now Galois differential theory seems to be mostly of historical interest.

In XIX century the power series were really used for the numerical integration of the differential equations. Therefore mathematicians of the XIX century have allocated those functions which can be presented by power series [2]. The main idea can be stated as follows: the elementary functions and the higher transcendental functions are such solutions of autonomous system of the differential equations

$$\dot{\boldsymbol{x}} = F(\boldsymbol{x}), \quad F \in \mathbb{Q}[\boldsymbol{x}], \tag{1}$$

which can be calculated by means of power series for all values of variable t.

The elementary functions (like $\exp t$, $\sin t$, $\cos t$) are such solutions which can be represented by power series everywhere. The solutions of any system of the linear differential equations can be represented by power series which converge everywhere, but all these solutions are expressed with the help of the solution of the scalar equation $\dot{x} = x$, known as $\exp t$.

The higher transcendental functions are such solutions of autonomous systems of the differential equations, which can be represented as a ratio of two power series

$$x_1 = \frac{a_0 + a_1 t + a_2 t^2 + \ldots}{b_0 + b_1 t + b_2 t^2 + \ldots},$$

which converge for all values of t. This means that $\boldsymbol{x}$ is a meromorphic function of t and the system has the Painlevé property.

Example 1. The Jacobi elliptic functions

$$p = \text{sn } t, \quad q = \text{cn } t, \quad r = \text{dn } t$$

are the solution of the nonlinear system

$$\begin{cases} \dot{p} = qr \\ \dot{q} = -pr \\ \dot{r} = -k^2 pq \end{cases} \tag{2}$$

with initial conditions

$$p = 0, \; q = r = 1 \text{ at } t = 0.$$

These functions can be represented everywhere as ratios of two power series.

In analytical theory of ODEs the notion of integration in finite terms means:

- constructing the algebraic substitution which reduces the initial system to a system with the Painlevé property;
- searching the symbolic expression for the solution in elementary or higher transcendental functions.

Anyway we can see the reference to the computational techniques of the past centuries.

A modern method for integration of an autonomous system of differential equations is the finite differences method (FDM). We believe that all transcendental functions can be considered as solutions of such differential equations, for which the application of the finite difference method is particularly efficient [3,4]. In the present report, we would like to consider one of the most important classes of such functions, namely, the elliptic functions.

2 Notations

Consider the autonomous system of the differential equations (1) on the interval $0 \le t \le T$ with the initial condition $x|_{t=0} = x_0$. We divide the interval $[0, T]$ into parts with step Δt by the points $t_1, \ldots t_{N-1}$ and take $t_0 = 0$ and $t_N = T$. The value of the approximate solution at $t = t_n$ is designated as x_n and the value of the exact solution is designated as $x(t_n)$. FDM suggests replacing the original system of differential equations with algebraic equations (scheme) of the form

$$F(x, \hat{x}; \Delta t) = 0$$

in the commonly used notations [5]. This equation defines algebraical correspondence between the neighboring layers x and $\hat{x}$, which are usually investigated as points on two affine or projective spaces.

3 Totally Conservative Difference Schemes

Since the difference scheme is a system of algebraical equations, we can conserve algebraical properties of the exact solution. However standard explicit schemes don't conserve algebraic integrals of motion.

Example 2. The system (2) has two quadratic integrals

$$p^2 + q^2 = \text{const}, \quad \text{and} \quad k^2 p^2 + r^2 = \text{const}.$$

The standard scheme of Runge-Kutta does not conserve them.

Definition 1. *The differential scheme*

$$F(\boldsymbol{x}, \hat{\boldsymbol{x}}; \varDelta t) = 0 \tag{3}$$

is called totally conservative iff for any algebraical integral $u(\boldsymbol{x})$ *the equation*

$$u(\hat{\boldsymbol{x}}) = u(\boldsymbol{x})$$

is a consequence of the system (3).

Note 1. The equality is conserved precisely if transition from one layer to another layer becomes precise, without rounding errors.

Note 2. Modern systems of computer algebra can't find all algebraic integrals for a given system. All realized algorithms, for example, our package Lagutinski for Sage [6], can find all rational integrals if the user indicates a bound for the degree of required integrals.

By the theorem of Cooper, the implicit midpoint rule

$$\frac{\hat{\boldsymbol{x}} - \boldsymbol{x}}{\varDelta t} = \frac{F(\hat{\boldsymbol{x}}) + F(\boldsymbol{x})}{2}$$

automatically inherits each quadratic conservation law [7]. If the field of algebraical integrals of a dynamic system is generated by quadratic forms, then the implicit midpoint rule is totally conservative.

Example 3. All algebraic integrals of motion for the system (2) of Example 2 are quadratic forms, thus, the implicit midpoint rule gives totally conservative scheme for Jacobi elliptic functions. Although this scheme is implicit, and for transition to the next layer we have to solve a system of algebraic equations, calculations are easily feasible on the modern computer. Furthermore, we have solved this system of algebraic equations without uncontrollable error of numerical methods by Gröbner basis technique. For transition we have solved numerically univariable equations of the 5th degree.

The implicit nature of the scheme is the main difficulty for theoretical investigations and also for practical computations.

Problem 1. Given a system of differential equations (3) and algebraic integrals of motions, construct an explicit difference scheme, exactly conserving the integrals of motion.

Here we give the solution for the case when the integrals of motion specify a curve in the space $\mathbb{V}$, where $\boldsymbol{x}$ varies.

Example 4. The two quadratic integrals for the system (2) specify an elliptic curve in the space $Opqr$. All layers coincide with this curve, exact solution define an automorphism of this curve and totally conservative scheme also defines an algebraic correspondence on this curve.

Theorem 1. *If the integrals of motion specify a curve of genus $\rho > 1$ in the space $\mathbb{V}$, then totally conservative explicit schemes do not exist.*

Proof. For the proof we can use Zeuthen theorem [8, Sect. 65]. Let there be an algebraic correspondence between curves C_1 and C_2,

- a general point P_1 at the curve C_1 corresponds to α_2 various points at the curve C_2, and a general point P_2 at the curve C_2 corresponds to α_1 various points at the curve C_1;
- η_1 is the number of the coincidences of two points P_1 corresponding to a point P_2 at the curve C_2, and similarly η_2 is the number of the coincidences of two points P_2;
- ρ_i is the genus of the curve C_i, $i = 1, 2$.

These numbers are connected by Zeuthen formula

$$\eta_2 - \eta_1 = 2\alpha_1(\rho_2 - 1) - 2\alpha_2(\rho_1 - 1).$$

If the integrals of motion specify a curve C in the space $\mathbb{V}$, any totally conservative difference scheme defines an algebraic correspondence between layers $C_1 = C$ and $C_2 = C$. This scheme is explicit iff a general point P_1 in the first layer corresponds to one point in the second layer, thus

$$\alpha_2 = 1, \quad \eta_2 = 0.$$

By Zuethen formula

$$-\eta_1 = 2(\alpha_1 - 1)(\rho - 1),$$

which is possible iff $\alpha_1 = 1$. This means that the correspondence is birational automorphism.

Difference scheme has natural parameter Δt, thus it defines an one-parametric family of birational automorphisms. This is imposible, because by Hurwitz theorem the number N of all birational automorphisms of a curve with the genus $\rho > 1$ is finite [9].

In general, a curve of degree equal or higher than 4 has a genus $\rho > 1$, thus there aren't totally conservative explicit schemes by purely geometric reasons. There are only two cases when the system (1) can be discretized by explicit totally conservative scheme:

- if the integrals specify an elliptic curve (the genus is equal to 1),
- if the integrals specify an unicursal curve (the genus is equal to 0).

Our example with Jacobi elliptic functions belongs to the first case.

Theorem 2. *Any explicit totally conservative difference scheme with elliptic layers defines birational automorphism and can be written as*

$$\int\limits_{x}^{\hat{x}} H dx_1 = \lambda(\Delta t),$$

where $H dx_1$ is the differential form of the first kind.

Sketch of proof. For explicit scheme, $\hat{x}$ is a rational function of x,

$$\int_0^{\hat{x}} H dx_1 = \alpha(\Delta t) \cdot \int_0^x H dx_1 + \beta(\Delta t).$$

Here α is an algebraic function without singularities, that is a constant. At $\Delta t = 0$ the automorphism is identity, thus $\alpha = 1$.

Example 5. The differential form of the first kind on the curve

$$p^2 + q^2 = \text{const} \quad \text{and} \quad k^2 p^2 + r^2 = \text{const}$$

is equal to $\frac{dp}{qr}$, thus the explicit totally conservative scheme (if it exists) can be written as

$$\int_{(p,q,r)}^{(\hat{p},\hat{q},\hat{r})} \frac{dp}{qr} = \lambda(\Delta t). \tag{4}$$

The exact solution is described also as

$$\int_{(p,q,r)}^{(\hat{p},\hat{q},\hat{r})} \frac{dp}{qr} = \Delta t.$$

By additions theorem for Jacobi functions we can write Eq. (4) in the algebraical form as

$$\begin{cases} \hat{p} = \dfrac{p \operatorname{cn} \lambda \operatorname{dn} \lambda - \operatorname{sn} \lambda qr}{1 - k^2 p^2 \operatorname{sn}^2 \lambda} \\[2mm] \hat{q} = \dfrac{q \operatorname{cn} \lambda - \operatorname{sn} \lambda \operatorname{dn} \lambda pr}{1 - k^2 p^2 \operatorname{sn}^2 \lambda} \\[2mm] \hat{r} = \dfrac{r \operatorname{dn} \lambda - k^2 \operatorname{sn} \lambda \operatorname{cn} \lambda pq}{1 - k^2 p^2 \operatorname{sn}^2 \lambda}. \end{cases}$$

Thus, $\operatorname{sn} \lambda$ has to be an algebraical function of Δt and hasn't to be equal to $\operatorname{sn} \Delta t$. The difference scheme approximates the differential equations iff

$$\lambda = \Delta t + \mathcal{O}(\Delta t^2),$$

and thus

$$\operatorname{sn} \lambda = \Delta t, \quad \operatorname{cn} \lambda = \sqrt{1 - \Delta t^2}, \quad \operatorname{dn} \lambda = \sqrt{1 - k^2 \Delta t^2}.$$

This difference scheme gives us exactly Gudermann's method for calculation of the elliptic functions [10, Abh. 1].

4   Totally Periodic Schemes

As is known, the elliptic functions are periodic and the scheme from Example 5 is also periodic in some sense.

Definition 2. *A difference scheme is called totally periodic if there is a sequence* $\{\Delta t_n \in \overline{\mathbb{Q}}\}$ *such that* $x_n = x_0$, *where* $\{x_m\}_{m=0}^{N}$ *is the approximate solution at* $\Delta t = \Delta t_n$.

For explicit totally conservative scheme we have

$$x_n = x_0 \quad \Rightarrow \quad n\lambda = \int\limits_{x_0}^{x_0} H \, dx_1 = 4K.$$

For the scheme from Example 5 sn $\lambda = \Delta t$, thus $\Delta t_n = $ sn $\frac{4K}{n}$ and therefore our scheme is totally periodic. Furthermore, we have calculated Δt_n at $n = 2^s$ by formulas of a half corner (Fig. 1).

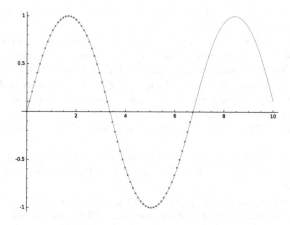

Fig. 1. Graph of sn$(t, \frac{1}{2})$ and points found by the totally periodic scheme at $n = 2^6$.

In Definition 2 n is the number of points per period and $n\Delta t_n$ is the period of the approximate solution. Obviously, if $n\Delta t_n \to T$, then the number T is the period of the exact solution. In our example the approximate period is equal to

$$n\Delta t_n = n \text{ sn } \frac{4K}{n} = 4K - \frac{k^2 + 1}{6} \frac{4^3 K^3}{n^2} + \mathcal{O}\left(\frac{1}{n^4}\right).$$

Thus our difference scheme conserves exact the periodical nature of motion, but we calculate the value of the period with a small error.

5   Conclusion: What Are Elliptic Functions?

The elliptic functions can be considered as solutions of such differential equations, for which we can write a very good difference scheme. This scheme is

- explicit, that is, calculations don't require the solution of nonlinear equations,
- totally conservative, that is, all algebraical integrals of motion are conserved exactly,
- totally periodic, that is, the periodical nature of the motion is conserved exactly and value of the period is conserved with small error.

In general, an autonomous system with algebraical integrals can't be approximated by explicit totally conservative difference scheme.

References

1. Goriely, A.: Integrability and Nonintegrability of Dynamical Systems. World Scientific, Singapore (2001)
2. Schlesinger, L.: Einführung in die Theorie der gewöhnlichen Differentialgleichungen auf funktionentheoretischer Grundlage. Walter de Gruyter & Co., Berlin und Leipzig (1922)
3. Ayryan, E.A., Malykh, M.D., Sevastianov, L.A.: Finite differences method and integration of the differential equations in finite terms. In: Preprints of the Joint Institute for Nuclear Research (Dubna), no. P11-2018-17 (2018)
4. Malykh, M.D., Sevastianov, L.A., Ying, Y.: Elliptic functions and finite difference method. In: International Conference Polynomial Computer Algebra 2018, St. Petersburg, pp. 66–68 (2018)
5. Samarskii, A.A.: The Theory of Difference Schemes. Marcel Dekker, New York (2001)
6. Malykh, M.D.: On integration of the first order differential equations in finite terms. In: IOP Conference Series: Journal of Physics: Conference Series, vol. 788, p. 012026 (2017)
7. Sanz-Serna, J.M.: Symplectic Runge-Kutta schemes for adjoint equations, automatic differentiation, optimal control, and more. SIAM Rev. **58**(1), 3–33 (2016)
8. Zeuthen, H.G.: Lehrbuch der abzählenden Methoden der Geometrie. Teubner, Leipzig (1914)
9. Badr, E.E., Saleem, M.A.: Cyclic Automorphisms groups of genus 10 nonhyperelliptic curves. arXiv:1307.1254 (2013)
10. Weierstrass, K.: Math. Werke. Bd. 1. Mayer&Müller, Berlin (1894)

Bounds for the Extreme Zeros
of Laguerre Polynomials

Geno Nikolov[ID] and Rumen Uluchev[✉][ID]

Faculty of Mathematics and Informatics, Sofia University "St. Kliment Ohridski",
5, James Bourchier blvd., 1164 Sofia, Bulgaria
{geno,rumenu}@fmi.uni-sofia.bg

Abstract. By applying well-known techniques such as the Gershgorin Circle Theorem and the Euler-Rayleigh method (the latter assisted by some computer algebra), we obtain new bounds for the extreme zeroes of the n-th Laguerre polynomial. It turns out that these bounds are competitive to some of the known best bounds.

Keywords: Extreme zeros of Laguerre polynomials
Gershgorin circle theorem · Euler-Rayleigh method

1 Introduction and Statement of the Results

The n-th degree Laguerre polynomial $L_n^{(\alpha)}$ is given by the explicit formula

$$L_n^{(\alpha)}(x) = \sum_{\nu=0}^{n} (-1)^\nu \binom{n+\alpha}{n-\nu} \frac{x^\nu}{\nu!}. \tag{1}$$

Since $L_n^{(\alpha)}$ is orthogonal on $(0, \infty)$ with respect to the Laguerre weight function $w_\alpha(x) = x^\alpha e^{-x}$, $\alpha > -1$, its zeros are positive and simple. Throughout this paper,

$$x_{1n}(\alpha) < x_{2n}(\alpha) < \cdots < x_{nn}(\alpha)$$

will stand for the zeros of $L_n^{(\alpha)}$. We are interested in bounds for $x_{1n}(\alpha)$ and $x_{nn}(\alpha)$, the extreme zeros of $L_n^{(\alpha)}$. Following the terminology from [5], we call *inside bounds* the upper bounds for $x_{1n}(\alpha)$ and the lower bounds for $x_{nn}(\alpha)$, while the lower bounds for $x_{1n}(\alpha)$ and the upper bounds for $x_{nn}(\alpha)$ are called *outside bounds*.

Let us give a brief account on the known bounds for the extreme zeros of $L_n^{(\alpha)}$, which do not involve zeros of the Bessel function. We start with an upper bound for $x_{nn}(\alpha)$ due to Szegő [12, Theorem 6.31.2]:

$$x_{nn}(\alpha) < 2n + \alpha + 1 + \sqrt{(2n+\alpha+1)^2 + 1/4 - \alpha^2}, \quad \alpha > -1. \tag{2}$$

Supported by the Bulgarian National Research Fund under Contract DN 02/14 and by the Sofia University Research Fund under Contract 80-10-139/2018.

G. Nikolov et al. (Eds.): NMA 2018, LNCS 11189, pp. 243–250, 2019.
https://doi.org/10.1007/978-3-030-10692-8_27

In [9] Ismail and Li developed a powerful tool for estimation of the extreme zeros of orthogonal polynomials, in particular, they proved that all the zeros of $L_n^{(\alpha)}$ are in the interval with endpoints

$$2n + \alpha - 2 \pm \sqrt{1 + 4(n-1)(n+\alpha-1)\cos^2 \frac{\pi}{n+1}} \, .$$

Krasikov [10] proved the following outside bounds, valid for $\alpha \geq 8$ and $n \geq 7$:

$$x_{1n}(\alpha) > s - r + \frac{(s-r)^{2/3}}{2r^{1/3}}, \tag{3}$$

$$x_{nn}(\alpha) < s + r + \frac{(s+r)^{2/3}}{2r^{1/3}}, \tag{4}$$

where $s = 2n + \alpha + 1$ and $r = \sqrt{4n^2 + 2(\alpha+1)(2n-1)}$.

Dimitrov and Nikolov [3] proved further outside bounds, valid for all $\alpha > -1$:

$$x_{1n}(\alpha) > \frac{2n^2 + (\alpha-1)n + 2(\alpha+1) - 2(n-1)\sqrt{n^2 + (\alpha+1)(n+2)}}{n+2},$$

$$x_{nn}(\alpha) < \frac{2n^2 + (\alpha-1)n + 2(\alpha+1) + 2(n-1)\sqrt{n^2 + (\alpha+1)(n+2)}}{n+2}.$$

Turning towards the inside bounds, let us mention the upper bounds

$$x_{1n}(\alpha) < \frac{(\alpha+1)(\alpha+2)}{n+\alpha+1}, \tag{5}$$

$$x_{1n}(\alpha) < \frac{(\alpha+1)(\alpha+3)}{2n+\alpha+1}, \tag{6}$$

$$x_{1n}(\alpha) < \frac{(\alpha+1)(\alpha+2)(\alpha+4)(2n+\alpha+1)}{(5\alpha+11)n(n+\alpha+1) + (\alpha+1)^2(\alpha+2)}, \tag{7}$$

obtained respectively by Hahn [7], Szegő [12, Eq. 6.31.12] and Gupta and Muldoon [6], as well as the upper bound for $x_{1n}(\alpha)$:

$$\frac{(\alpha+2)_2(3n+2\alpha+2) - \sqrt{(\alpha+2)_2 \left[9(\alpha+2)_2 + 2(2\alpha+5)(\alpha^2+5\alpha+10)(n-1) + (5\alpha^2+25\alpha+38)(n-1)^2\right]}}{2(n+\alpha+1)_2} \tag{8}$$

proved by Driver and Jordaan [4] (as usual, $(\beta)_k := \beta(\beta+1)\cdots(\beta+k-1)$).

Finally, we quote the following lower bounds for $x_{nn}(\alpha)$, $\alpha > -1$:

$$x_{nn}(\alpha) > 4n + \alpha - 16\sqrt{2n}, \tag{9}$$

$$x_{nn}(\alpha) > 3n - 4, \tag{10}$$

$$x_{nn}(\alpha) > 2n + \alpha - 1, \tag{11}$$

$$x_{nn}(\alpha) > 2n + \alpha - 2 + \sqrt{n^2 - 2n + \alpha n + 2}, \tag{12}$$

due to Bottemma [1], Neumann [11], Szegő [12, Eq. 6.2.14], and Driver and Jordaan [4], respectively.

Here we prove the following bounds for the extreme zeros of $L_n^{(\alpha)}$:

Theorem 1. *For all $\alpha > -1$ and $n \geq 6$, the smallest zero $x_{1n}(\alpha)$ of the Laguerre polynomial $L_n^{(\alpha)}$ satisfies the inequalities:*

(i) $x_{1n}(\alpha) < \dfrac{(\alpha+1)(\alpha+5)\left[(5\alpha+11)n(n+\alpha+1)+(\alpha+1)^2(\alpha+2)\right]}{(2n+\alpha+1)\left[(7\alpha+19)n(n+\alpha+1)+(\alpha+1)^2(\alpha+2)\right]}$;

(ii) $x_{1n}(\alpha) > \left(\dfrac{(\alpha+1)^5(\alpha+2)^2(\alpha+3)(\alpha+4)(\alpha+5)}{n(n+\alpha+1)(2n+\alpha+1)\left[(7\alpha+19)n(n+\alpha+1)+(\alpha+1)^2(\alpha+2)\right]}\right)^{1/5}$.

Theorem 2. *For all $\alpha > -1$, the following are lower bounds for the largest zero $x_{nn}(\alpha)$ of the Laguerre polynomial $L_n^{(\alpha)}$:*

(i) $3n+\alpha-2-\dfrac{(n-1)n}{2n+\alpha+1}$, $n \geq 3$;

(ii) $4n+\alpha-3-\dfrac{2(n-1)n(3n+2\alpha-2)}{5n^2+(5\alpha-6)n+(\alpha-1)(\alpha-2)}$, $n \geq 4$;

(iii) $5n+\alpha-4-\dfrac{(n-1)n\left[28n^2+(35\alpha-43)n+10\alpha^2-28\alpha+18\right]}{14n^3+(21\alpha-29)n^2+(9\alpha^2-29\alpha+22)n+(\alpha-1)(\alpha-2)(\alpha-3)}$, $n \geq 5$;

(iv) $6n+\alpha-5-r(n,\alpha)$, $n \geq 6$, *where*

$$r(n,\alpha) = \frac{4(n-1)n\left[30n^3+(54\alpha-77)n^2+(30\alpha^2-95\alpha+72)n+(\alpha-1)(\alpha-2)(5\alpha-12)\right]}{42n^4+(84\alpha-130)n^3+(56\alpha^2-195\alpha+165)n^2+(14\alpha^3-85\alpha^2+165\alpha-100)n+(\alpha-1)(\alpha-2)(\alpha-3)(\alpha-4)}.$$

Theorem 3. *For all $\alpha \geq 0$ and $n \geq 4$, the extreme zeros of the Laguerre polynomial $L_n^{(\alpha)}$ satisfy the inequalities:*

$$x_{1n}(\alpha) \geq \frac{1}{2}\left[\left(\sqrt{n-2+\alpha}-\sqrt{n-2}\right)^2+\left(\sqrt{n-1+\alpha}-\sqrt{n-1}\right)^2\right]; \quad (13)$$

$$x_{nn}(\alpha) \leq \frac{1}{2}\left[\left(\sqrt{n-2+\alpha}+\sqrt{n-2}\right)^2+\left(\sqrt{n-1+\alpha}+\sqrt{n-1}\right)^2\right]. \quad (14)$$

Let us make a short comment on how our results compare with some of the bounds quoted above. Theorem 1(i) certainly improves inequalities (5)–(8). The two-sided bounds for $x_{1n}(\alpha)$ in Theorem 1 are especially sharp for small α, in particular they imply $\lim_{\alpha \to -1} \frac{x_{1n}(\alpha)}{\alpha+1} = \frac{1}{n}$. The bounds for $x_{nn}(\alpha)$ in Theorem 2 are listed with increasing sharpness from (i) to (iv). Since $\lim_{\alpha \to \infty} \frac{x_{nn}(\alpha)}{\alpha} = 1$ (compare, e.g., (2) and (9)), all these bounds are asymptotically sharp when n is fixed and α grows. For such n and α, (iv) compares favorably with the bounds (10)–(12); moreover, (iv) compares favorably with (11) and (12) when α is fixed and n is large, see Corollary 1 in the next section. Regarding the outside bounds in Theorem 3, (14) both improves and extends for a wider range of n and α the Krasikov upper bound (4). Estimate (13) is weaker than Krasikov's lower bound (3) but holds for a wider range of n and α.

The rest of this paper is organized as follows. In Sect. 2 a short description of the Euler-Rayleigh method is given, and then this method is applied for proving Theorems 1 and 2. The proof of Theorem 3 is based on the Gershgorin Circle Theorem, and is given in Sect. 3.

2   Proof of Theorems 1 and 2

As was already mentioned, our proof of Theorems 1 and 2 exploits the so-called Euler-Rayleigh method. This method has been described in details by Ismail and Muldoon [8], where the interested reader may find also some facts about its history. This method has been originally designed for the estimation of the (modulus of) smallest zero of entire functions having only real roots $\{x_i\}_{i=1}^\infty$ such as the Bessel functions and alike, therefore it deals with sums of powers of the reciprocal of these roots, $S_k = \sum_{i=1}^\infty x_i^{-k}$, $k \in \mathbb{N}$. Here Euler-Rayleigh method is applied to real-root polynomials, and we prefer to work with power sums of the roots rather than their reciprocals. The necessary details are given below.

Let P be a monic polynomial of degree n with zeros $\{x_i\}_{i=1}^n$,

$$P(x) = x^n - b_1\, x^{n-1} + b_2\, x^{n-2} - \cdots + (-1)^n b_n = \prod_{i=1}^n (x - x_i)\,. \qquad (15)$$

For $k \in \mathbb{N}_0$, the power sums $p_k = p_k(P) := \sum_{i=1}^n x_i^k$, with $p_0 = n = \deg P$, and the coefficients $\{b_i\}_{i=1}^n$ of P are connected by the Newton identities (cf. [13])

$$p_r + \sum_{i=1}^{\min\{r-1,n\}} (-1)^i p_{r-i}\, b_i + (-1)^r r\, b_r = 0\,, \qquad (b_i = 0\,, \ \ i > n)\,.$$

On using Newton's identities, one easily obtains the following lemma:

Lemma 1. *The following formulae hold for p_r, $1 \le r \le 6$, assuming $n \ge r$:*

$p_1 = b_1$;

$p_2 = b_1^2 - 2b_2$;

$p_3 = b_1^3 - 3b_1 b_2 + 3b_3$;

$p_4 = b_1^4 - 4b_1^2 b_2 + 4b_1 b_3 + 2b_2^2 - 4b_4$;

$p_5 = b_1^5 - 5b_1^3 b_2 + 5b_1^2 b_3 + 5b_1 b_2^2 - 5b_1 b_4 - 5b_2 b_3 + 5b_5$;

$p_6 = b_1^6 - 6b_1^4 b_2 + 6b_1^3 b_3 + 9b_1^2 b_2^2 - 6b_1^2 b_4 - 12b_1 b_2 b_3 + 6b_1 b_5 - 2b_2^3 + 3b_3^2 + 6b_2 b_4 - 6b_6$.

Let us set

$$\ell_k(P) := \frac{p_k(P)}{p_{k-1}(P)}\,, \qquad u_k(P) := \left[p_k(P)\right]^{1/k}\,, \qquad k \in \mathbb{N}\,.$$

The following statement is a slight modification of Lemma 3.2 in [8].

Proposition 1. *Let P be a monic polynomial of degree n with positive zeros $x_1 < x_2 < \cdots < x_n$. Then the largest zero x_n of P satisfies the inequalities*

$$\ell_k(P) < x_n < u_k(P)\,, \qquad k \in \mathbb{N}\,. \qquad (16)$$

Moreover, $\{\ell_k(P)\}_{k=1}^\infty$ is monotonically increasing, $\{u_k(P)\}_{k=1}^\infty$ is monotonically decreasing, and

$$\lim_{k \to \infty} \ell_k(P) = \lim_{k \to \infty} u_k(P) = x_n\,. \qquad (17)$$

For the reader convenience, we sketch the proof.

Proof. For $i = 1, \ldots, n-1$, we set $t_i := \frac{x_i}{x_n}$, then $0 < t_i < 1$. Inequalities (16) and the limit relations (17) now readily follow from the representations

$$\ell_k(P) = \frac{t_1^k + \cdots + t_{n-1}^k + 1}{t_1^{k-1} + \cdots + t_{n-1}^{k-1} + 1} \, x_n, \qquad u_k(P) = \left(t_1^k + \cdots + t_{n-1}^k + 1\right)^{1/k} x_n.$$

The latter representation of $u_k(P)$ together with $0 < t_i < 1$ implies the monotonicity of $\{u_k(P)\}_{k=1}^\infty$. The monotonicity of $\{\ell_k(P)\}_{k=1}^\infty$ is a consequence of the inequality $p_k^2(P) \leq p_{k-1}(P) \, p_{k+1}(P)$, which follows from the Cauchy-Schwarz inequality. $\square$

2.1   Proof of Theorem 1

Let P be the reciprocal polynomial of $L_n^{(\alpha)}$ with leading coefficient 1, then P has zeroes $x_i = 1/x_{n+1-i,n}(\alpha)$, $i = 1, \ldots, n$. Writing $P(x)$ in the form,

$$P(x) = x^n L_n^{(\alpha)}(x^{-1}) / \binom{n+\alpha}{n} = x^n + \sum_{i=1}^n (-1)^{n-i} b_i x^{n-i},$$

we deduce from (1) that the coefficients $\{b_i\}$ are given by

$$b_i = \binom{n}{i} \frac{\Gamma(\alpha+1)}{\Gamma(n+1-i+\alpha)}, \qquad i = 1, \ldots, n.$$

We apply Lemma 1 to evaluate the power sums $p_k(P)$, $k = 1, \ldots, 5$, and find with the help of the Wolfram *Mathematica*:

$$p_1 = \frac{n}{\alpha+1};$$

$$p_2 = \frac{n(n+\alpha+1)}{(\alpha+1)^2(\alpha+2)};$$

$$p_3 = \frac{n(n+\alpha+1)(2n+\alpha+1)}{(\alpha+1)^3(\alpha+2)(\alpha+3)};$$

$$p_4 = \frac{n(n+\alpha+1)\left[(5\alpha+11)n(n+\alpha+1) + (\alpha+1)^2(\alpha+2)\right]}{(\alpha+1)^4(\alpha+2)^2(\alpha+3)(\alpha+4)};$$

$$p_5 = \frac{n(n+\alpha+1)(2n+\alpha+1)\left[(7\alpha+19)n(n+\alpha+1) + (\alpha+1)^2(\alpha+2)\right]}{(\alpha+1)^5(\alpha+2)^2(\alpha+3)(\alpha+4)(\alpha+5)}.$$

From Proposition 1 we infer

$$x_n = \frac{1}{x_{1n}(\alpha)} > \ell_k(P) = \frac{p_k(P)}{p_{k-1}(P)} \;\Rightarrow\; x_{1n}(\alpha) < \frac{p_{k-1}(P)}{p_k(P)}, \qquad k = 2, \ldots, 5.$$

The case $k = 5$ is the claim of Theorem 1(i), while the cases $k = 2, 3$ and 4 reproduce correspondingly inequalities (5), (6) and (7), i.e., the results of Hahn [7], Szegő [12, Eq. 6.31.12], and Gupta and Muldoon [6]. Theorem 1(i) improves all these results, since, according to the last claim of Proposition 1, the bounds obtained with larger k in the Euler-Rayleigh method are sharper; it improves also the result of Driver and Jordaan (8), as in [5] these authors observe that (7) compares favorably with (8).

Proposition 1 implies also the inequalities

$$x_n = \frac{1}{x_{1n}(\alpha)} < u_k(P) = [p_k(P)]^{1/k} \Rightarrow x_{1n}(\alpha) > \frac{1}{[p_k(P)]^{1/k}}, \qquad k = 1, \ldots, 5.$$

The claim of Theorem 1(ii) corresponds to the case $k = 5$. Note that, according to Proposition 1, the inequalities obtained with $k = 1, \ldots, 4$ are less precise. □

2.2   Proof of Theorem 2

We rewrite (1) in the form $L_n^{(\alpha)}(x) = \frac{(-1)^n}{n!} P(x)$, where

$$P(x) = x^n + \sum_{i=1}^{n} (-1)^{n-i} b_i \, x^{n-i}, \qquad b_i = \binom{n}{i} \frac{\Gamma(n+\alpha+1)}{\Gamma(i+\alpha+1)}, \qquad i = 1, \ldots, n.$$

We apply Lemma 1 with the above coefficients $\{b_i\}$ to evaluate the power sums $p_k(P)$, $k = 1, \ldots, 6$. With the assistance of Wolfram *Mathematica* we find:

$p_1 = n(n+\alpha)$;

$p_2 = n(n+\alpha)(2n+\alpha-1)$;

$p_3 = n(n+\alpha)\left[5n^2 + (5\alpha-6)n + (\alpha-1)(\alpha-2)\right]$;

$p_4 = n(n+\alpha)\left[14n^3 + (21\alpha-29)n^2 + (9\alpha^2 - 29\alpha + 22)n + (\alpha-1)(\alpha-2)(\alpha-3)\right]$;

$p_5 = n(n+\alpha)\left[42n^4 + (84\alpha-130)n^3 + (56\alpha^2 - 195\alpha + 165)n^2 \right.$

$\qquad\qquad \left. + (14\alpha^3 - 85\alpha^2 + 165\alpha - 100)n + (\alpha-1)(\alpha-2)(\alpha-3)(\alpha-4)\right]$;

$p_6 = n(n+\alpha)\left[132n^5 + (330\alpha-562)n^4 + (300\alpha^2 - 1124\alpha + 1044)n^3 \right.$

$\qquad\qquad + (120\alpha^3 - 757\alpha^2 + 1566\alpha - 1041)n^2$

$\qquad\qquad + (20\alpha^4 - 195\alpha^3 + 692\alpha^2 - 1041\alpha + 548)n$

$\qquad\qquad \left. + (\alpha-1)(\alpha-2)(\alpha-3)(\alpha-4)(\alpha-5)\right]$.

Proposition 1 implies that $x_n = x_{nn}(\alpha)$, the largest zero of P, satisfies

$$x_{nn}(\alpha) > \ell_k(P) = \frac{p_k(P)}{p_{k-1}(P)}, \qquad k = 2, 3, 4, 5, 6,$$

with $\{p_k\}_{k=1}^{6}$ as given above, and the inequalities with larger k are stronger. Employing again *Mathematica*, we find that claims (i)-(iv) of Theorem 2 correspond to the cases $k = 3, 4, 5, 6$ of the above inequality. Note that the case $k = 2$ implies Szegő's bound (11). □

The following is a consequence of Theorem 2, verified with *Mathematica*:

Corollary 1. *For every $\alpha > 1$ and for $k = 2, 3, \ldots, 6$, the largest zero $x_{nn}(\alpha)$ of the Laguerre polynomial $L_n^{(\alpha)}$ satisfies the inequalities*

$$x_{nn}(\alpha) > \frac{4k - 2}{k + 1} n + \alpha - k + 1, \qquad n \geq k. \tag{18}$$

Clearly, (18) with $k = 6$ compares favorably with the estimates (10)–(12). We conjecture that inequality (18) holds true for every $k \in \mathbb{N}$, $k \geq 2$.

Having $\{p_k(P)\}_{k=1}^6$ at our disposal, we may apply Proposition 1 for deriving upper bounds for $x_{nn}(\alpha)$ through the inequalities $x_{nn}(\alpha) \leq [p_k(P)]^{1/k}$. However, since $p_k(P) = \mathcal{O}(n^{k+1})$, the rates of the resulting upper bounds are $\mathcal{O}(n^{1+1/k})$ while $x_{nn}(\alpha) = \mathcal{O}(n)$ as $n \to \infty$.

3   Proof of Theorem 3

The monic Laguerre polynomials $\{\widehat{L}_n^{(\alpha)}\}$ satisfy the three term recurrence relation (cf. [2, Eq. (2.30)])

$$\widehat{L}_n^{(\alpha)}(x) = (x - 2n - \alpha + 1)\widehat{L}_{n-1}^{(\alpha)}(x) - (n-1)(n+\alpha-1)\widehat{L}_{n-2}^{(\alpha)}(x), \quad n \geq 1, \tag{19}$$

with $\widehat{L}_{-1}^{(\alpha)} := 0$ and $\widehat{L}_0^{(\alpha)} := 1$. Assuming $\alpha \geq 0$, we can rewrite the recurrence formula (19) in the form $\det(x\,\mathbf{E}_n - \mathbf{A}_n) = 0$, where $\mathbf{E}_n$ is the $n \times n$ identity matrix, and $\mathbf{A}_n$ is a symmetric tri-diagonal matrix,

$$\mathbf{A}_n = \begin{pmatrix} a_1 & c_2 & 0 & \cdots & 0 & 0 \\ c_2 & a_2 & c_3 & \cdots & 0 & 0 \\ 0 & c_3 & a_3 & \cdots & 0 & 0 \\ \vdots & \vdots & \vdots & \ddots & \vdots & \vdots \\ 0 & 0 & 0 & \cdots & a_{n-1} & c_n \\ 0 & 0 & 0 & \cdots & c_n & a_n \end{pmatrix}$$

with non-zero entries

$$a_k = \alpha + 2k - 1, \qquad\qquad k = 1, \ldots, n,$$
$$c_k = \sqrt{(k - 1)(k - 1 + \alpha)}, \qquad k = 2, \ldots, n.$$

In other words, the zeroes $\{x_{kn}(\alpha)\}_{k=1}^n$ of $L_n^{(\alpha)}$ are the eigenvalues of $\mathbf{A}_n$.

As a consequence from the Gershgorin Circle Theorem (or, equivalently, from the fact that every diagonally dominant matrix is non-singular), see, e.g. [14], we have

$$x_{nn}(\alpha) \leq \max_{1 \leq k \leq n} \{a_k + c_k + c_{k+1}\}, \tag{20}$$

$$x_{1n}(\alpha) \geq \min_{1 \leq k \leq n} \{a_k - c_k - c_{k+1}\}, \tag{21}$$

with the convention that $c_1 = c_{n+1} = 0$. It is easily verified that for $\alpha \geq 0$ and $n \geq 4$ we have $\max_{1 \leq k \leq n}\{a_k + c_k + c_{k+1}\} = a_{n-1} + c_{n-1} + c_n$, whence, by (20), we obtain $x_{nn}(\alpha) \leq a_{n-1} + c_{n-1} + c_n$, which is equivalent to inequality (14).

Similarly, we verify that $\min_{1 \leq k \leq n}\{a_k - c_k - c_{k+1}\} = a_{n-1} - c_{n-1} - c_n$ for $\alpha \geq 0$ and $n \geq 4$, which by (21) implies $x_{1n}(\alpha) \geq a_{n-1} - c_{n-1} - c_n$. The latter is nothing but inequality (13). Theorem 3 is proved.   □

We conclude with showing that (14) improves a strengthened Krasikov bound (4), namely, that $s+r > a_{n-1}+c_{n-1}+c_n$ for $\alpha \geq 0$. Indeed, the latter is equivalent to the inequality

$$4 + \sqrt{4n^2 + 2(\alpha + 1)(2n - 1)} > \sqrt{(n - 2)(n - 2 + \alpha)} + \sqrt{(n - 1)(n - 1 + \alpha)},$$

which is obviously true, since for its left-hand side we have

$$4 + \sqrt{4n^2 + 2(\alpha + 1)(2n - 1)} > \sqrt{2n(2n - 1) + 2(\alpha + 1)(2n - 1)}$$

$$= 2\sqrt{\left(n - \frac{1}{2}\right)(n + \alpha + 1)}.$$

References

1. Bottema, Q.: Die Nullstellen gewisser durch Rekursionsformeln definierter Polynome. Proc. Amsterdam **34**(5), 681–691 (1931)
2. Chihara, T.: An Introduction to Orthogonal Polynomials. Gorn and Breach, New York (1978)
3. Dimitrov, D.K., Nikolov, G.P.: Sharp bounds for the extreme zeros of classical orthogonal polynomials. J. Approx. Theory **162**, 1793–1804 (2010)
4. Driver, K., Jordaan, K.: Bounds for extreme zeros of some classical orthogonal polynomials. J. Approx. Theory **164**, 1200–1204 (2012)
5. Driver, K., Jordaan, K.: Inequalities for extreme zeros of some classical orthogonal and q-orthogonal polynomials. Math. Model. Nat. Phenom. **8**(1), 48–59 (2013)
6. Gupta, D.P., Muldoon, M.E.: Inequalities for the smallest zeros of Laguerre polynomials and their q-analogues. J. Ineq. Pure Appl. Math. 8(1) (2007). Article 24
7. Hahn, W.: Bericht über die Nullstellen der Laguerreschen und der Hermiteschen Polynome. Jahresber. Deutsch. Math.-Ferein. **44**, 215–236 (1933)
8. Ismail, M.E.H., Muldoon, M.E.: Bounds for the small real and purelyimaginary zeros of Bessel and related functions. Met. Appl. Math. Appl. **2**, 1–21 (1995)
9. Ismail, M.E.H., Li, X.: Bounds on the extreme zeros of orthogonal polynomials. Proc. Amer. Math. Soc. **115**, 131–140 (1992)
10. Krasikov, I.: Bounds for zeros of the Laguerre polynomials. J. Approx. Theory **121**, 287–291 (2003)
11. Neumann, E.R.: Beiträge zur Kenntnis der Laguerreschen Polynome. Jahresber. Deutsch. Math.-Ferein **30**, 15–35 (1921)
12. Szegő, G.: Orthogonal Polynomials, 4th edn. American Mathematical Society Colloquium Publications, Providence (1975)
13. Van der Waerden, B.L.: Modern Algebra, vol. 1. Frederick Ungar Publishing Co., New York (1949)
14. van Dorn, E.: Representations and bounds for zeros of orthogonal polynomials and eigenvalues of sign-symmetric tri-diagonal matrices. J. Approx. Theory **51**, 254–266 (1987)

Discrete Fourier Analysis on Lattice Grids

Morten A. Nome[1] and Tor Sørevik[2(✉)]

[1] Department of Mathematics, NTNU, Trondheim, Norway
morten.nome@ntnu.no
[2] Department of Mathematics, University of Bergen, Bergen, Norway
tor.sorevik@math.uib.no

Abstract. Using group theory we describe the relation between lattice sampling grids and the corresponding non-aliasing Fourier basis sets, valid for all 1-periodic lattices. This technique enable us to extend the results established in [16]. We also provide explicit formula for the Lagrange functions and show how the FFT algorithm may be used to compute the expansion coefficients.

Keywords: Trigonometric interpolation · Fourier coefficients

1 Introduction

We are interested in interpolating a periodic function f on $[0,1)^s$ by an $s-$dimensional trigonometric polynomial

$$f(\mathbf{x}) \approx \sum_{\mathbf{k} \in S} c_{\mathbf{k}} e^{2\pi i \mathbf{k} \cdot \mathbf{x}},$$

We do so by sampling f in N grid points $\mathbf{x}_j \in [0,1)^s$ such that

$$f(\mathbf{x}_j) = \sum_{\mathbf{k} \in S} c_{\mathbf{k}} e^{2\pi i \mathbf{k} \cdot \mathbf{x}_j} \quad \forall \, \mathbf{x}_j \in \Omega. \tag{1}$$

Ω is called the *sampling grid*, the vector $\mathbf{k} = (k_1, k_2, \cdots, k_s)$ is a multi-index, commonly called wave numbers or Fourier indices, $\mathbf{k} \cdot \mathbf{x} = \mathbf{k}^T \mathbf{x}$ denotes the innerproduct of $\mathbf{k} \in S$ and $\mathbf{x} \in R^s$, and $S \subset \mathbb{Z}^s$ is a finite set with $|S| = N$. S defines the approximation space $\mathscr{H}_S = \{e^{2\pi i \mathbf{k} \cdot \mathbf{x}} \mid \mathbf{k} \in S\}$. We write

$$If = \sum_{\mathbf{k} \in S} c_{\mathbf{k}} e^{2\pi i \mathbf{k} \cdot \mathbf{x}},$$

where I denotes the interpolation operator. For fixed Ω, S and f, (1) defines a linear system of equations for the coefficients $c_{\mathbf{k}}$. If the system is non-singular, the grid is said to be unisolvent with respect to $\mathscr{H}_S$. If sufficient structure is present in the point set, Ω, the FFT-algorithm may be used to solve (1), offering huge savings in computational cost. Unisolvency also ensures that a set of N Lagrange

© Springer Nature Switzerland AG 2019
G. Nikolov et al. (Eds.): NMA 2018, LNCS 11189, pp. 251–260, 2019.
https://doi.org/10.1007/978-3-030-10692-8_28

functions satisfying $L_\ell(\mathbf{x}_j) = \delta_{\ell,j}$ exists. If these can be described explicitly, the interpolation may be written: $If = \sum_{\ell=1}^{N} f(\mathbf{x}_\ell) L_\ell(\mathbf{x})$.

The obvious extension to multi-dimensional interpolation is done by taking the tensor product of your favorite one-dimensional interpolation grid. Then the well-known one-dimensional theory can be straightforwardly extended. However, the exponential increase in cost severely limits this approach and for that reason using non-tensorial sampling grids such as sparse grids [1,3,18] and lattice grids [4,16] have been suggested.

In this paper we will focus on the lattice grid approach. In particular we will establish the relation between a given lattice grid and its corresponding approximation space, and show how to construct the associated Lagrange functions and efficient computation of the expansion coefficients, $c_{\mathbf{k}}$ by the FFT. This was also an issue in [16]. In that paper our proofs were restricted to rank-1 lattices of prime order. In this paper we generalize this result to all 1-periodic lattices. Framing these problems in terms of group theory gives access to the full arsenal of group theoretical tools. This allows us to more precisely describe the decomposition of higher rank lattices into rank-1 lattices, which among other things are the bases for a variable transformation permitting the FFT to be used for fast computation of the interpolation coefficients. Again this has been done before in [14], but there it remains unclear exactly how to relate the computed coefficients with the corresponding basis functions.

For a thorough understanding of the basic properties of 1-periodic integration lattices we recommend [11,12] and Sects. 1–4 of [17]. In general, a good understanding of Fourier analysis on lattice grids requires basic knowledge of group theory, especially Abelian and quotient groups. Good references are [2,13,14].

Similar work has been pursued by Li, Sun and Xu, and reported in a series of papers [5–9]. Their work is targeting other physical domains in 2 or 3 dimensions such as triangles, hexagons, etc. On the other hand they are not limiting themselves to trigonometric interpolation as they employ variable transformations to obtain Chebyshev's polynomials for algebraic polynomial interpolations. These are then used to develop interpolating quadrature rules.

We first establish the correct correspondence between the interpolating lattice points and the corresponding approximation space. In Sect. 3 we show how to construct a full set of trigonometric Lagrange functions. In Sect. 4 we establish the proper variable transformations allowing us to compute the interpolation coefficients by the FFT.

2   The Correspondence Between the Sampling Grid and the Index Set

An s-dimensional lattice Λ is a finitely generated Abelian group under vector addition. Alternatively it may be viewed as a linear integer combination of s linearly independent basis vectors. When arranging the basis vectors as rows in a matrix, the matrix is said to be an generator matrix for the lattice.

In this paper we will consider only $[0,1)^s$-periodic integration lattices, and thus all lattice points may be written $\mathbf{x} = \frac{\mathbf{z}}{N}$, with $\mathbf{z} \in \mathbb{Z}^s$ [10].

Definition 1. *The lattice (sampling) grid T of Λ is*

$$T = \Lambda \cap [0,1)^s.$$

If Λ is periodic on $[0,1)^s$, then T is a group under addition mod 1, and

$$T \simeq \Lambda/\mathbb{Z}^s.$$

See [11,17] for details. From here on, any addition of lattice points in T will be tacitly understood to be mod 1. We shall write $N = |T|$ throughout this paper.

Definition 2. *Let $\mathbf{x} \in T$. The order d of $\mathbf{x}$ is the least natural number such that $d\mathbf{x} = 0$. The subgroup of T generated by $\mathbf{x}$ is the set*

$$\{\mathbf{x}\} = \{j\mathbf{x} \mid \quad 0 \leq j \leq d-1\}.$$

It is clear that $\{\mathbf{x}\}$ is a cyclic group of order d and its periodic extension is a lattice, $\Lambda_{\mathbf{x}}$ and $\Lambda_{\mathbf{x}} \subseteq \Lambda$.

Definition 3. *Let $\mathbf{x}_1, \mathbf{x}_2, ..., \mathbf{x}_t \in T$. We say that T has rank t, and is generated by $\mathbf{x}_1, \mathbf{x}_2, ..., \mathbf{x}_t$ if*

$$T \simeq \{\mathbf{x}_1\} \oplus \{\mathbf{x}_2\} \oplus \cdots \oplus \{\mathbf{x}_t\}.$$

All finite Abelian groups are direct products of cyclic groups. The orders of the generators of T are called invariants, denoted $d_1, d_2,...,d_t$. The generators are in general not unique, but the invariants are, under the condition that $d_l | d_{l+1}$ for all l. Elementary group theory implies that $N = \prod_{l=1}^{t} d_l$.

Lemma 1. *Let $x = \frac{z}{N} \in T$, with $\mathbf{z} \in \mathbb{Z}^s$, and $a = \gcd(\mathbf{z}, N)$, then the order of $\mathbf{x}$ is N/a.*

Proof. Since a divides $\mathbf{z}$, we have

$$\frac{N}{a} \frac{\mathbf{z}}{N} = \mathbf{0}.$$

Let $\mathbf{z}' = \mathbf{z}/a$. If there exist a natural number $b < \frac{N}{a}$ such that b is the order of $\mathbf{x}$, then elementary group theory dictates that we must have $bc = \frac{N}{a}$ for some natural number $c > 1$. Then

$$\frac{\mathbf{z}'}{c} = b\frac{\mathbf{z}}{N} = \mathbf{0},$$

and this implies that c divides all components of $\mathbf{z}'$. But then ac divides $\mathbf{z}$, hence cannot divide N, so $b = \frac{N}{ac}$ is not an integer, and arriving at a contradiction, we conclude that no $b < \frac{N}{a}$ exists. $\square$

It follows that we may always write $\mathbf{x} = \frac{\mathbf{z}}{d}$, with $\mathbf{z} \in \mathbb{Z}^s$, where d is the order of $\mathbf{x}$. If $\mathbf{x} \in T$ has order N, then $\mathbf{x}$ generates T.

We now turn our attention to approximation spaces corresponding to a particular sampling grid T. As stated above this is equivalent to finding index sets $S \in \mathbb{Z}^s$ or $S_{\mathbf{x}}$ for sampling sets T or $\{\mathbf{x}\}$, respectively. Associated with T (or Λ) is the dual lattice.

Definition 4. *The dual lattice of T is*

$$\Lambda^{\perp} = \{\mathbf{k} \ : \ \mathbf{k} \cdot \mathbf{x} \in \mathbb{Z} \quad \forall \mathbf{x} \in T\}.$$

As $\mathbf{k}, \mathbf{h} \in \Lambda^{\perp} \Rightarrow \mathbf{k} + \mathbf{h} \in \Lambda^{\perp}$, $\Lambda^{\perp}$ is itself a lattice and whenever Λ is 1-periodic, $\Lambda^{\perp}$ is an integer lattice. We may also define

Definition 5. *The dual lattice of $\{\mathbf{x}\}$ is*

$$\Lambda_{\mathbf{x}}^{\perp} = \{\mathbf{k} \ : \ \mathbf{k} \cdot \mathbf{x} \in \mathbb{Z} \quad \forall \mathbf{x} \in \{\mathbf{x}\}\}.$$

A key observation is that two Fourier modes $e^{2\pi i \mathbf{k} \cdot \mathbf{x}}$ and $e^{2\pi i \mathbf{h} \cdot \mathbf{x}}$ are indistinguishable for $\mathbf{x} \in T$ if $\mathbf{k} - \mathbf{h} \in \Lambda^{\perp}$. $\mathbf{k}$ and $\mathbf{h}$ are said to be aliasing. To be useful for us S must contain only non-aliasing indecies.

Note that $\Lambda_x \subseteq \Lambda$ implies $\Lambda^{\perp} \subseteq \Lambda_{\mathbf{x}}^{\perp}$, and that $\Lambda_{\mathbf{x}}^{\perp} = \Lambda^{\perp}$ if the order of $\mathbf{x} \in T$ is N. The subgroup $\{\mathbf{x}\}$ is treated specifically with respect to aliasing.

Lemma 2. *Let d be the order of $\{\mathbf{x}\}$, and let $\mathbf{z}, \mathbf{k}, \mathbf{h} \in \mathbb{Z}^s / \{\mathbf{0}\}$, with $\mathbf{x} = \frac{\mathbf{z}}{d} \in T$. Then*

$$\mathbf{k} \cdot \mathbf{z} \equiv \mathbf{h} \cdot \mathbf{z} \pmod{d} \quad \Leftrightarrow \quad \mathbf{k} - \mathbf{h} \in \Lambda_{\mathbf{x}}^{\perp}. \tag{2}$$

Proof. A simple computation yields

$$\mathbf{k} \cdot \mathbf{z} \equiv \mathbf{h} \cdot \mathbf{z} \pmod{d}$$
$$\Updownarrow$$
$$(\mathbf{k} - \mathbf{h}) \cdot \mathbf{z} = m; \ m \in \mathbb{Z},$$

and the lemma follows from Definition 5. □

Lemma 3. *Let d be the order of $\mathbf{x} \in T$. Then $|S_{\mathbf{x}}| = d$.*

Proof. Let $\mathbf{x} = \frac{\mathbf{z}}{d}$ with $\mathbf{z} \in \mathbb{Z}^s$. The equivalence relation

$$\mathbf{k} = \mathbf{h} \quad \text{if} \quad \mathbf{k} \cdot \mathbf{z} \equiv \mathbf{h} \cdot \mathbf{z} \pmod{d},$$

clearly partitions $\mathbb{Z}^s$ into d equivalence classes. From Lemma 2 we know that this equivalence relation is equivalent to the equivalence relation

$$\mathbf{k} = \mathbf{h} \quad \text{if} \quad \mathbf{k} - \mathbf{h} \in \Lambda_{\mathbf{x}}^{\perp}.$$

Accordingly, these two equivalence relations partition $\mathbb{Z}^s$ in the same number of equivalence classes, and thus $|S_{\mathbf{x}}| = d$. □

We now turn to the construction of unisolvent approximation spaces on T. Since $\Lambda^\perp$ is a normal subgroup of $\mathbb{Z}^s$, we may construct $\mathbb{Z}^s/\Lambda^\perp$, which is a group of equivalence classes under the equivalence relation

$$\mathbf{k} = \mathbf{h} \quad \text{if} \quad \mathbf{k} - \mathbf{h} \in \Lambda^\perp.$$

Theorem 1. *A non-aliasing Fourier index set for T is*

$$S = \mathbb{Z}^s/\Lambda^\perp.$$

Proof. For computations we choose a representative from each equivalence class; it is then evident that no two representatives will alias. Moreover, the set of representatives will be isomorphic to $\mathbb{Z}^s/\Lambda^\perp$ under the addition inherited from $\mathbb{Z}^s/\Lambda^\perp$: if $\mathbf{k}, \mathbf{h}, \mathbf{l} \in \mathbb{Z}^s$ are representatives for $[\mathbf{k}], [\mathbf{h}], [\mathbf{l}] \in \mathbb{Z}^s/\Lambda^\perp$, then $\mathbf{k} + \mathbf{h} = \mathbf{l}$ if $[\mathbf{k}] + [\mathbf{h}] = [\mathbf{l}]$. □

We write $S_\mathbf{x} = \mathbb{Z}^s/\Lambda_\mathbf{x}^\perp$.

Lemma 4. *The cosets of $S_\mathbf{x}$ partition S.*

Proof. Since $\Lambda^\perp$ is a normal subgroup of $\Lambda_\mathbf{x}^\perp \subset \mathbb{Z}^s$, we know from the fundamental theorem of quotient groups [2] that

$$\frac{\mathbb{Z}^s/\Lambda^\perp}{\Lambda_\mathbf{x}^\perp/\Lambda^\perp} \simeq \mathbb{Z}^s/\Lambda_\mathbf{x}^\perp,$$

which says that the cosets of $\mathbb{Z}^s/\Lambda_\mathbf{x}^\perp$ partition $\mathbb{Z}^s/\Lambda^\perp$. Since $S = \mathbb{Z}^s/\Lambda^\perp$ and $S_\mathbf{x} = \mathbb{Z}^s/\Lambda_\mathbf{x}^\perp$, the result follows. □

3   Trigonometric Lagrange Functions for Lattice Grids

Definition 6. *The Dirichlet kernel of S on $[0,1)^s$ is*

$$D_S(\mathbf{x}) = \sum_{\mathbf{k} \in S} e^{2\pi i \mathbf{k} \cdot \mathbf{x}}. \tag{3}$$

We proceed to prove that the Dirichlet kernel is zero on all $\mathbf{x} \in T$ except at the origin.

Lemma 5. *For $\mathbf{x} \in T$.*

$$D_{S_\mathbf{x}}(\mathbf{x}) = \begin{cases} 0; \mathbf{x} \in T \setminus \{\mathbf{0}\} \\ |S_\mathbf{x}|; \mathbf{x} = \mathbf{0} \end{cases}.$$

Proof. Let $\mathbf{x} = \frac{\mathbf{z}}{d}$, with $\mathbf{z} \in \mathbb{Z}^s$. We have $\mathbf{k} \cdot \mathbf{x} = \mathbf{k} \cdot \mathbf{z}/d = \frac{m}{d}$ for all $\mathbf{k} \in S_{\mathbf{x}}$. Now m takes at most d different values, and due to Lemma 2, none of them are equal mod d. This implies that $D_{S_{\mathbf{x}}}(\mathbf{x})$ is a geometric series, and for $\mathbf{x} \neq \mathbf{0}$ we may compute

$$D_{S_{\mathbf{x}}}(\mathbf{x}) = \sum_{\mathbf{k} \in S_{\mathbf{x}}} f_{\mathbf{k}}(\mathbf{x}) = \sum_{\mathbf{k} \in S_{\mathbf{x}}} e^{2\pi i \mathbf{k} \cdot \mathbf{x}} = \sum_{m=0}^{d-1} e^{\frac{2\pi i m}{d}} = 0.$$

The case $\mathbf{x} = \mathbf{0}$ is trivial. □

Theorem 2. *Let* $\mathbf{x} \in T$. *Then*

$$D_S(\mathbf{x}) = \begin{cases} 0; & \mathbf{x} \in T \setminus \{\mathbf{0}\} \\ N; & \mathbf{x} = \mathbf{0} \end{cases}.$$

Proof. Since $S_{\mathbf{x}}$ partitions S, we just rearrange the sum in (3) and apply Lemma 5

$$D_S(\mathbf{x}) = \sum_{\mathbf{k} \in S} e^{2\pi i \mathbf{k} \cdot \mathbf{x}} = \sum_{\mathbf{h} \in S/S_{\mathbf{x}}} \sum_{\mathbf{l} \in S_{\mathbf{x}}} e^{2\pi i (\mathbf{h} + \mathbf{l}) \cdot \mathbf{x}} = \sum_{\mathbf{h} \in S/S_{\mathbf{x}}} e^{2\pi i \mathbf{h} \cdot \mathbf{x}} \sum_{\mathbf{l} \in S_{\mathbf{x}}} e^{2\pi i \mathbf{l} \cdot \mathbf{x}} = \begin{cases} 0; & \mathbf{x} \neq \mathbf{0} \\ N; & \mathbf{x} = \mathbf{0} \end{cases}$$

□

We can now construct a complete set of Lagrange functions.

Corollary 1. *For any* $\mathbf{y} \in T$ *a trigonometric Lagrange function is given as*

$$L_{\mathbf{y}}(\mathbf{x}) = \frac{1}{N} D_S(\mathbf{x} - \mathbf{y}).$$

4   Fast Fourier Transform on Lattice Grids

By utilizing the Smith normal form, the standard FFT-algorithm may be extended in a natural way to lattice grids of any rank. In the following, we shall write $T(A)$ for the sampling grid generated by the generator matrix $A \in \mathbb{R}^{s \times s}$.

Definition 7. *The index set of* $T(A)$ *is*

$$S(A) = \{\mathbf{k}^T = \mathbf{x}^T A^{-1} | \ \mathbf{x} \in T(A)\} \tag{4}$$

For integration lattices, A^{-1} is an integer matrix, and $N = \det A^{-1}$, see [10]. If A is non-singular, we have the following trivial lemma.

Lemma 6. $S(A)$ *is a group, with* $T(A) \simeq S(A)$.

Proof. Let $\mathbf{x}, \mathbf{y}, \mathbf{z} \in T(A)$, let $\mathbf{k}, \mathbf{h}, \mathbf{l} \in S(A)$, with $\mathbf{k} = A^{-1}\mathbf{x}$, $\mathbf{h} = A^{-1}\mathbf{y}$, $\mathbf{l} = A^{-1}\mathbf{z}$ and $\mathbf{x} + \mathbf{y} = \mathbf{z}$. If we agree that $\mathbf{k} + \mathbf{h} = \mathbf{l}$, then under this addition, $S(A)$ is a group, with $T(A) \simeq S(A)$. □

In the preceding section we used the letter S to denote non-aliasing Fourier index sets, and in this subsection we use the same letter to denote lattice index sets. This makes sense because of the following theorem, proved in [16].

Theorem 3. $S(A^T)$ *is a non-aliasing index set for* $T(A)$.

Note that the generator matrix A is not unique. Any matrix UA where U is unimodular (integer matrices with determinant equal ± 1), generates the same lattice and consequently $T(A) = T(UA)$. However, using Definition 7 we see that $S(UA) = \{\mathbf{k}^T U | k \in S(A)\}$ and consequently $S(A) \neq S(UA)$. In [16] we give algorithms for computing good index sets for typical function classes.

To solve (1) efficiently by the multi-dimensional FFT-algorithm the sampling points as well as the index-set must form a hyper-rectangular equidistant grid. This is not the case for $T(A)$ and $S(A^T)$, respectively. However, an appropriate grid/index-set may be obtained by a simple structure preserving variable transformation. The inverse generator matrix A^{-1} may be decomposed by the Smith normal form [15] as

$$\tilde{D} = \tilde{U} A^{-1} \tilde{V}; \qquad \tilde{D} = \operatorname{diag}(d_1, \dots, d_s); \qquad d_\ell | d_{\ell+1},$$

where $\tilde{U}$ and $\tilde{V}$ are $s \times s$ unimodular matrices, and $\tilde{D} \in \mathbb{Z}^{s \times s}$ is a unique diagonal matrix with $d_\ell | d_{\ell+1}$ for $1 \leq \ell < s$, the invariants of $T(A)$. If $t < s$ we may omit the upper $s - t$ rows of U and the leftmost $s - t$ columns of V, writing

$$D = U A^{-1} V, \tag{5}$$

where $U \in \mathbb{Z}^{t \times s}$, $V \in \mathbb{Z}^{s \times t}$, and $D \in \mathbb{Z}^{t \times t}$. Transposition of (5) proves that $T(A)$ and $T(A^T)$ have the same invariants, hence

$$T(A) \simeq T(A^T),$$

and Lemma 6 then implies

$$T(A) \simeq S(A^T).$$

In [12] Lyness and Keast showed that the rows of $D^{-1}U$ generate $T(A)$. Transposition shows that the columns of VD^{-1} similarly generate $T(A^T)$, and hence the columns of $A^{-1}VD^{-1}$ generate $S(A^T)$. The Cartesian grid with d_ℓ points in the ℓ-th coordinate direction may be written $T(D^{-1})$, and its non-aliasing Fourier index set is

$$S(D^{-1}) = \{\mathbf{h}: \quad 0 \leq h_\ell < d_\ell \quad 1 \leq \ell \leq t\}.$$

More formally we may write

$$T(D^{-1}) = \{\mathbf{y} : \mathbf{y} = D^{-1}\mathbf{h}; \ \mathbf{h} \in S(D^{-1})\}.$$

The sampling grid $T(D^{-1})$ and the index space $S(D^{-1})$ are standard regular equidistant t-dimensional grids which allows straightforward use of the FFT for solving Eq. (1). Now let $\mathbf{x} \in T(A)$, $\mathbf{k} \in S(A^T)$, and $\mathbf{h}, \mathbf{l} \in S(D^{-1})$, with

$$\mathbf{x}^T = \mathbf{h}^T D^{-1} U = \mathbf{y}^T U \tag{6}$$

and

$$\mathbf{k} = A^{-1} V D^{-1} \mathbf{l}. \tag{7}$$

The matrix-vector form for computing all $f(\mathbf{x}_j)$ of (1) becomes

$$\mathbf{f} = F\mathbf{c} \quad \text{where} \quad (F)_{j,l} = e^{2\pi i \mathbf{k}_l \mathbf{x}_j}; \ \mathbf{k}_l \in S(A^T); \ \mathbf{x}_j \in T(A). \quad (8)$$

If $\mathbf{x}$, $\mathbf{y}$ are related by (6) and $\mathbf{k}$, $\mathbf{l}$ by (7), then by (5) it follows that

$$\exp(2\pi i \mathbf{x} \cdot \mathbf{k}) = \exp(2\pi i \mathbf{y} \cdot \mathbf{l}).$$

Thus the matrix F in (8) is just a permutation of the matrix produced by the $T(D^{-1})$ grid. The matrix obtained by the $\mathbf{y}$, $\mathbf{l}$ entries correspond to the standard t-dimensional inverse Fourier transform, efficiently carried out by the FFT-algorithm. The permutation is implicitly defined by (6) and (7), and care needs to be taken when matching coefficients with function values in (1).

The computations in (6) and (7) are linear in N. Thus the total complexity is dominated by the FFT which is of order $O(N \log N)$, with a constant factor weekly depending on how N factorize. This complexity stays the same regardless of whether we do a one dimensional FFT of length N (for a rank-1 lattice) or we do a t-dimensional FFT on a $d_1 \times d_2 \times \cdots \times d_t$ array for $N = \prod_{j=1}^{t} d_j$ in the case of a rank-t lattice with N points.

See Fig. 1 and Table 1.

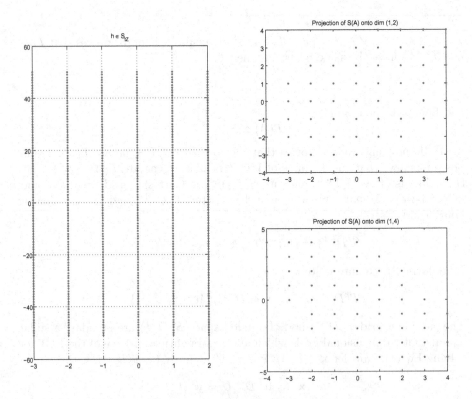

Fig. 1. The left frame displays the set $S(D^{-1})$. In the right frame we display two different 2-dim projections of the 4-dim index array $S(A)$. The 4 remaining 2-dim projections have a similar look.

Table 1. The relative error for the interpolating function when the grid is produced by the lattice in this section or by a regular, equidistant Cartesian grid using $N = 625$ gridpoints. The error is estimated by computing $\|f_k(\mathbf{x}) - If_k(\mathbf{x})\|_1/\|f_k(\mathbf{x})\|_1$ on a regular fine grid ($31 \times 31 \times 31 \times 31$ gridpoints). For f_3 we used $p = 3$ and for f_4, $\lambda = 7$.

Function	Lattice grid	Regular grid
$f_1(\mathbf{x}) = \prod_{\ell=1}^{s}(x_\ell - 1)^2 x_\ell^2$	0.031	0.038
$f_2(\mathbf{x}) = \prod_{\ell=1}^{s} e^{\sin(2\pi x_\ell) - 1}$	0.036	0.128
$f_3(\mathbf{x}) = \prod_{\ell=1}^{s}(2 + sign(x_\ell - \frac{1}{2})\sin(2\pi x_\ell)^p)$	0.250	0.367
$f_4(\mathbf{x}) = e^{-\lambda \prod_{\ell=1}^{s}(x_\ell - 1/2)^2}$	0.068	0.084

5   Numerical Examples

As an example, consider the lattice generated by A when

$$A^{-1} = \begin{pmatrix} 0 & -1 & -4 & -5 \\ 5 & 0 & -1 & -4 \\ 4 & 5 & 0 & -1 \\ 1 & 4 & 5 & 0 \end{pmatrix}$$

Its Smith Normal form, $\tilde{D} = \tilde{U}A^{-1}\tilde{V}$ is:

$$\begin{pmatrix} 1 & 0 & 0 & 0 \\ 0 & 1 & 0 & 0 \\ 0 & 0 & 6 & 0 \\ 0 & 0 & 0 & 102 \end{pmatrix} = \begin{pmatrix} 1 & -4 & 4 & 0 \\ 3 & 4 & -5 & 3 \\ -17 & 3 & 1 & -13 \\ -19 & -4 & 9 & -16 \end{pmatrix} \begin{pmatrix} 0 & -1 & -4 & -5 \\ 5 & 0 & -1 & -4 \\ 4 & 5 & 0 & -1 \\ 1 & 4 & 5 & 0 \end{pmatrix} \begin{pmatrix} -5 & 0 & -19 & 263 \\ -1 & 0 & -4 & 55 \\ 1 & -1 & 7 & -117 \\ 0 & 0 & 0 & 1 \end{pmatrix}.$$

This tells us that the lattice has rank 2, with invariants 6 and 102. Utilizing the reduced Smith Normal form, the rows of

$$D^{-1}U = \tilde{D}^{-1}(3:4, 3:4)\tilde{U}(3:4,:) = \begin{pmatrix} \frac{1}{6} & 0 \\ 0 & \frac{1}{102} \end{pmatrix} \begin{pmatrix} -17 & 3 & 1 & -13 \\ -19 & -4 & 9 & -16 \end{pmatrix},$$

are generators for $T(A)$. The columns of

$$A^{-1}VD^{-1} = \begin{pmatrix} 0 & -1 & -4 & -5 \\ 5 & 0 & -1 & -4 \\ 4 & 5 & 0 & -1 \\ 1 & 4 & 5 & 0 \end{pmatrix} \begin{pmatrix} -19 & 263 \\ -4 & 55 \\ 7 & -117 \\ 0 & 1 \end{pmatrix} \begin{pmatrix} \frac{1}{6} & 0 \\ 0 & \frac{1}{102} \end{pmatrix},$$

are generators for $S(A^T)$. To get the lattice points in $[0,1)^s$, needed for practical computation, the points obtained by (6) have to be taken modulo 1. Likewise, members of the non-aliasing set of Fourier coefficients obtained using (7) need to be shifted so that they represent the most significant Fourier modes, typically the lowest frequencies. As an illustration we have computed the interpolation on the above lattice grid for four 1-periodic functions and compared the error to the similar interpolation on a regular Cartesian grid. In all cases the lattice grid produce a more accurate interpolant. More exhaustive experimental results for these testfunctions on lattice grid versus regular grid in 2 and 3 dimension is given in [16], and they show a clear advantage for lattice grid. The advantage seems to increase with the dimension.

References

1. Bungartz, H., Griebel, M.: Sparse grids. Acta Numer. **13**, 1–123 (2004)
2. Fenrick, M.H.: Introduction to the Galois Correspondence. Birkhäuser, Boston (1992)
3. Kämmerer, L., Kunis, S., Potts, D.: Interpolation lattices for hyperbolic cross trigonometric polynomials. J. Complex. **28**(1), 76–92 (2012)
4. Kämmerer, L., Potts, D., Volkmer, T.: Approximation of multivariate functions by trigonometric polynomials based on rank-1 lattice sampling. J. Complex. **31**(4), 543–576 (2015)
5. Li, H., Sun, J., Xu, Y.: Discrete Fourier analysis, cubature and interpolation on a hexagon and a triangle SIAM. J. Numer. Anal. **46**, 1653–1681 (2008)
6. Li, H., Sun, J., Xu, Y.: Cubature formula and interpolation on the cubic domain. Numer. Math. Theory Method Appl. **2**, 119–152 (2009)
7. Li, H., Xu, Y.: Discrete Fourier analysis on a dodecahedron and a tetrahedron. Math. Comput. **78**, 999–1029 (2009)
8. Li, H., Xu, Y.: Discrete Fourier analysis on fundamental domain and simplex of A_d lattice in $d-$Variables. J. Fourier Anal. Appl. **16**, 383–433 (2010)
9. Li, H., Sun, J., Xu, Y.: Discrete Fourier analysis with lattices on planar domain. Numer. Algorithms **55**, 279–300 (2010)
10. Lyness, J.N.: An introduction to lattice rules and their generator matrices. IMA J. Numer. Anal. **9**, 405–419 (1989)
11. Lyness, J.N.: The canonical forms of a lattice rule. In: Brass, H., Hämmerlin, G. (eds.) Numerical Integration IV. ISNM International Series of Numerical Mathematics, vol. 112, pp. 225–240. Birkhäuser, Basel (1993). https://doi.org/10.1007/978-3-0348-6338-4_18
12. Lyness, J.N., Keast, P.: Application of the Smith normal form to the structure of lattice rules. SIAM J. Matrix Anal. Appl. **16**(1), 218–231 (1995)
13. Munthe-Kaas, H.Z.: On group Fourier analysis and symmetry preserving discretizations of PDEs. J. Phys. A **39**(19), 5563–5584 (2006)
14. Munthe-Kaas, H., Sørevik, T.: Multidimensional pseudo-spectral methods on lattice grids. Appl. Numer. Math. **62**, 155–165 (2012)
15. Smith, H.J.S.: On systems of linear indeterminate equations and congruences. Philos. Trans. Roy. Soc. A **151**, 293–326 (1861)
16. Sørevik, T., Nome, M.A.: Trigonometric interpolation on lattice grids. BIT Numer. Math. **56**(1), 341–356 (2016)
17. Sloan, I.H., Lyness, J.N.: The representation of lattice quadrature rules as multiple sums. Math. Comput. **52**(185), 81–94 (1989)
18. Zenger, C.: Sparse grids. Notes Numer. Fluid Mech. **31**, 241–251 (1991)

Monte Carlo and Quasi-Monte Carlo Methods

A Wigner Potential Decomposition in the Signed-Particle Monte Carlo Approach

Majid Benam$^{(\boxtimes)}$, Mihail Nedjalkov, and Siegfried Selberherr

Institute for Microelectronics, TU Wien, Vienna, Austria
benam@iue.tuwien.ac.at

Abstract. The description of the electron evolution, provided by the Wigner equation, involves a force-less Liouville operator, which is associated with particles moving over Newtonian trajectories, and a Wigner potential operator associated with generation of positive and negative particles. These concepts can be combined to develop stochastic algorithms for solving the Wigner equation, consolidated by the so-called signed particle approach. We investigate the option to split the Wigner potential into two parts and to approximate one of them by a classical force term. The purpose is two-fold: First, we search for ways to simplify the numerical complexity involved in the simulation of the Wigner equation. Second, such a term offers a way to a self-consistent coupling of the Wigner and the Poisson equations. The particles in the signed-particle approach experience a force through the classical component of the potential. A cellular automaton algorithm is used to update the discrete momentum of the accelerated particles, which is then utilized along with the Wigner-based generation/annihilation processes. The effect of the approximation on generic physical quantities such as current and density are investigated for different cut-off wavenumbers (wavelengths), and the results are promising for a self-consistent solution of the Wigner and Poisson equations.

Keywords: Wigner function · Potential splitting
Signed-particle approach

1 Introduction

The Wigner function is defined with the Fourier transform of the density matrix expressed in the mean and difference of coordinates in two dimensions:

$$f_{\mathrm{w}}(\mathbf{r}, \mathbf{k}, t) = \frac{1}{(2\pi)^2} \int_{-\infty}^{+\infty} d\mathbf{s} \, e^{-i\mathbf{k} \cdot \mathbf{s}} \rho\left(\mathbf{r} + \frac{\mathbf{s}}{2}, \mathbf{r} - \frac{\mathbf{s}}{2}, t\right).$$

Furthermore, the finite dimensions of the simulation domain allow the Wigner function to be calculated over finite dimensions and discretized $\mathbf{k}$ values. The physical domain in the simulations analyzed in this paper is a two-dimensional

© Springer Nature Switzerland AG 2019
G. Nikolov et al. (Eds.): NMA 2018, LNCS 11189, pp. 263–272, 2019.
https://doi.org/10.1007/978-3-030-10692-8_29

area of size $(L_x, L_y) = (20\,\text{nm}, 30\,\text{nm})$. We choose the center of the domain to be the origin, and therefore, the position and momentum vectors are discretized as:

$$\mathbf{r} \equiv (x\Delta r, y\Delta r) \qquad\qquad -\frac{M}{2} \le x < \frac{M}{2}, -\frac{N}{2} \le y < \frac{N}{2}$$

$$\mathbf{s} \equiv (m\Delta s, n\Delta s)$$

$$\mathbf{L} = (L_x, L_y) \equiv (M\Delta s, N\Delta s)$$

$$\mathbf{q}\Delta\mathbf{k} = (p, q).\left(\frac{\pi}{L_x}, \frac{\pi}{L_y}\right) \equiv \left(p\frac{\pi}{M\Delta s}, q\frac{\pi}{N\Delta s}\right) \qquad -\frac{M}{2} \le q < \frac{M}{2}, -\frac{N}{2} \le p < \frac{N}{2}$$

$$\mathbf{q'}\Delta\mathbf{k} = (p', q').\left(\frac{\pi}{L_x}, \frac{\pi}{L_y}\right) \equiv \left(p'\frac{\pi}{M\Delta s}, q'\frac{\pi}{N\Delta s}\right).$$

Δr and Δs represent the spatial spacing between nodes and are assumed to be equal. $\mathbf{s}$ and $\mathbf{q'}$ are used for performing summations over position and momentum variables. We use the short notation:

$$f(x, y, p, q) \equiv f(x\Delta r, y\Delta r, p\frac{\pi}{M\Delta s}, q\frac{\pi}{N\Delta s}).$$

The Wigner equation, which follows from the von Neumann equation for the density matrix [6], is written in the semi-discrete form as:

$$\left(\frac{\partial}{\partial t} + \frac{\hbar\mathbf{q}\Delta\mathbf{k}}{m^*}\nabla_\mathbf{r}\right) f_w(\mathbf{r}, \mathbf{q}, t) = \sum_\mathbf{q} V_W(\mathbf{r}, \mathbf{q} - \mathbf{q'}) f_w(\mathbf{r}, \mathbf{q'}, t). \tag{1}$$

The semi-discrete Wigner potential (WP), which is of central importance in the signed-particle approach, is defined as [4]:

$$V_W(\mathbf{r}, \mathbf{q}) \equiv \frac{1}{i\hbar\mathbf{L}} \int_{-\frac{L}{2}}^{\frac{L}{2}} ds e^{-i\mathbf{q}\Delta\mathbf{k}.\mathbf{s}} \left[V(\mathbf{r} + \frac{\mathbf{s}}{2}) - V(\mathbf{r} - \frac{\mathbf{s}}{2})\right]. \tag{2}$$

2   Wigner Potential Decomposition

Here, we focus on a full discretization, and thus use the fully discretized WP which must be computed at each node [1] becomes:

$$V_W(x, y, p, q) = \frac{1}{i\hbar MN} \sum_{m=-\frac{M}{2}}^{\frac{M}{2}-1} \sum_{n=-\frac{N}{2}}^{\frac{N}{2}-1} e^{-i(pm\frac{\pi}{M} + qn\frac{\pi}{N})}$$

$$\times \left[V\left(x + \frac{m}{2}, y + \frac{n}{2}\right) - V\left(x - \frac{m}{2}, y - \frac{n}{2}\right)\right]. \tag{3}$$

The two-dimensional discrete Fourier transform of a potential $V(x, y)$ in a region of size $M \times N$ is a function $\hat{V}(p, q)$. The pair is given by the equations [5]:

$$\hat{V}(p, q) = \sum_{x=-\frac{M}{2}}^{\frac{M}{2}-1} \sum_{y=-\frac{N}{2}}^{\frac{N}{2}-1} V(x, y) e^{-i2\pi(\frac{px}{M} + \frac{qy}{N})},$$

$$V(x,y) = \frac{1}{MN} \sum_{p=-\frac{M}{2}}^{\frac{M}{2}-1} \sum_{q=-\frac{N}{2}}^{\frac{N}{2}-1} \hat{V}(p,q) e^{i2\pi(\frac{px}{M}+\frac{qy}{N})}.$$

$\hat{V}(p,q)$ can be expressed in polar form, $\hat{V}(p,q) = A(p,q)e^{i\phi(q,p)}$. Since the two-dimensional discrete Fourier transform pair is periodic (k and l being integers),

$$\hat{V}(p,q) = \hat{V}(p+kM, q+lN), \quad V(x,y) = V(x+kM, y+lN),$$

we also assume the physical potential $V(x,y)$ to be periodic so that a shift equal to a multiple of the physical region lengths in the corresponding argument of the potential does not change the value of the potential. It holds:

$$\sum_{m=-\frac{M}{2}}^{\frac{M}{2}-1} \sum_{n=-\frac{N}{2}}^{\frac{N}{2}-1} e^{-i(\frac{\pi mp}{M}+\frac{\pi nq}{N})}\left[V\left(x+\frac{m}{2}, y+\frac{n}{2}\right)\right] = e^{i2(\frac{\pi xp}{M}+\frac{\pi yq}{N})}$$

$$\times \sum_{m'=x-\frac{M}{4}}^{x+\frac{M}{4}-1} \sum_{n'=y-\frac{N}{4}}^{y+\frac{N}{4}-1} e^{-i2(\frac{\pi m'p}{M}+\frac{\pi n'q}{N})}\left[V(m',n')\right] = e^{i2(\frac{\pi xp}{M}+\frac{\pi yq}{N})}\hat{V}(2p,2q),$$

$$\sum_{m=-\frac{M}{2}}^{\frac{M}{2}} \sum_{n=-\frac{N}{2}}^{\frac{N}{2}} e^{-i(\frac{\pi mp}{M}+\frac{\pi nq}{N})}\left[V\left(x-\frac{m}{2}, y-\frac{n}{2}\right)\right] = e^{-i2(\frac{\pi xp}{M}+\frac{\pi yq}{N})}$$

$$\times \sum_{m'=x-\frac{M}{4}}^{x+\frac{M}{4}-1} \sum_{n'=y-\frac{N}{4}}^{y+\frac{N}{4}-1} e^{i2(\frac{\pi m'p}{M}+\frac{\pi n'q}{N})}\left[V(m',n')\right] = \left[e^{i2(\frac{\pi xp}{M}+\frac{\pi yq}{N})}\hat{V}(2p,2q)\right]^{*}.$$

We have used the properties of the discrete Fourier transform, namely periodicity and scaling. Thus, the Wigner potential given in Eq. 3 becomes:

$$V_{\mathrm{w}}(x,y,p,q) = \frac{1}{i\hbar MN}\left\{e^{i2(\frac{\pi xp}{M}+\frac{\pi yq}{N})}\hat{V}(2p,2q) - \left[e^{i2(\frac{\pi xp}{M}+\frac{\pi yq}{N})}\hat{V}(2p,2q)\right]^{*}\right\}. \quad (4)$$

Using Euler's formula, Eq. 4 can be rewritten in a polar form as:

$$V_{\mathrm{w}}(x,y,p,q) = \frac{2}{\hbar MN}A(2p,2q)sin\left[\phi(2p,2q) + 2\frac{\pi xp}{M} + 2\frac{\pi yq}{N}\right]. \quad (5)$$

In the following, we show that treating the summation in the right-hand side of Eq. 1 in two separate regions results in the spectral decomposition of the potential profile into a slowly varying classical component and a rapidly varying quantum mechanical component [2]. For each direction we specify a cut-off wavenumber determining the sharpness of the corresponding low-pass filter which is discussed later in more details. The cut-off wavenumber is specified by a cut-off wavelength, $\lambda_{c_x} = \frac{2\pi}{q_{c_x}\Delta k_x}$ and $\lambda_{c_y} = \frac{2\pi}{q_{c_y}\Delta k_y}$. We assume q_{c_x} and q_{c_y} to be equal and use q_c for both directions. Applying the decomposition, the potential operator on the right-hand side of Eq. 1 can be rewritten as [3]:

$$Qf_w(x,y,p,q) = \sum_{p',q'} V_w(x,y,p',q')f_w(x,y,p-p',q-q') = \sum_{|q'|,|p'|\le\frac{q_c}{2}} + \sum_{|q'|,|p'|>\frac{q_c}{2}} = Q_{cl}f_w + Q_{qm}f_w$$

Q_{cl} and Q_{qm} represent the classical and potential parts of the potential operator, respectively. We recall that Lagrange's mean value theorem allows to express the increment of a continuous function on an interval through the value of the derivative at an intermediate point of the segment (for small $(m-n)\Delta x$)

$$f(m\Delta x) - f(n\Delta x) \simeq f'(k\Delta x)(m-n)\Delta x \simeq \Delta f(m-n), n \le k \le m.$$

Using the notations Δ_p and Δ_q, we calculate the classical potential operator as:

$$Q_{cl}f_w(x,y,p,q) = \sum_{|q'|,|p'|\le\frac{q_c}{2}} V_w(x,y,p',q')f_w(x,y,p-p',q-q')$$

$$\approx \sum_{|q'|,|p'|\le\frac{q_c}{2}} V_w(x,y,p',q')\left[f_w(x,y,p,q) - \frac{\pi p'}{M}\Delta_p f_w(x,y,p,q) - \frac{\pi q'}{N}\Delta_q f_w(x,y,p,q)\right].$$

In the second line, the summation over f_w vanishes as V_w is an odd function in both p and q. Using the polar form of the Wigner potential (Eq. 5), we obtain:

$$-\sum_{|q|,|p|\le\frac{q_c}{2}} \frac{\pi p}{M}V_w(x,y,p,q) = \frac{-2}{\hbar MN}\sum_{|q|,|p|\le\frac{q_c}{2}} \frac{\pi p}{M}A(2p,2q)\sin[\phi(2p,2q) + 2\frac{\pi xp}{M} + 2\frac{\pi yq}{N}]$$

$$= \frac{-1}{\hbar MN}\sum_{|q|,|p|\le q_c} \frac{\pi p}{M}A(p,q)\sin[\phi(p,q) + \frac{\pi xp}{M} + \frac{\pi yq}{N}]$$

$$= \Delta_x \frac{1}{\hbar MN}\sum_{|q|,|p|\le q_c} A(p,q)\cos[\phi(p,q) + \frac{\pi xp}{M} + \frac{\pi yq}{N}]$$

$$= \frac{1}{\hbar}\Delta_x\Re\left\{\frac{1}{MN}\sum_{|q|,|p|\le q_c} A(p,q)e^{i\phi(p,q)}e^{i\frac{\pi xp}{M}}e^{i\frac{\pi yq}{N}}\right\}$$

$$= \frac{1}{\hbar}\Delta_x\Re\left\{\frac{1}{MN}\sum_{|q|,|p|\le q_c} \hat{V}(p,q)e^{i(\frac{\pi xp}{M}+\frac{\pi yq}{N})}\right\} = \frac{1}{\hbar}\Delta_x V_{cl}(x,y).$$

With a similar approach, we can show that:

$$-\sum_{|q|\le\frac{q_c}{2}}\sum_{|p|\le\frac{q_c}{2}} \frac{\pi q}{N}V_w(x,y,p,q) = \frac{1}{\hbar}\Delta_y V_{cl}(x,y).$$

Here, we have introduced the classical potential component:

$$V_{cl}(x,y) = \frac{1}{MN}\sum_{|q|,|p|\le q_c} \hat{V}(p,q)e^{i(\frac{\pi xp}{M}+\frac{\pi yq}{N})}.$$

This function is real, as can be easily shown by substituting $\hat{V}(p,q)$:

$$V_{cl}(x,y) = \frac{1}{MN} \sum_{|q|,|p| \leq q_c} \sum_{m=-\frac{M}{2}}^{\frac{M}{2}-1} \sum_{n=-\frac{N}{2}}^{\frac{N}{2}-1} V(m,n) e^{i\frac{\pi(x-m)p}{M}} \cdot e^{i\frac{\pi(y-n)q}{N}}$$

$$= \sum_{m=-\frac{M}{2}}^{\frac{M}{2}-1} \sum_{n=-\frac{N}{2}}^{\frac{N}{2}-1} V(m,n) \cdot \frac{\sin\left[\frac{\pi q_c(x-m)}{M}\right]}{\pi(x-m)} \cdot \frac{\sin\left[\frac{\pi q_c(y-n)}{N}\right]}{\pi(y-n)} \cdot$$

In order to calculate V_{cl}, we have a convolution of real functions, the potential $V(x,y)$ and the *sinc* functions in both x and y directions acting as low-pass filters. The convolution involves an infinite summation for an ideal filter. However, we choose our low-pass filter to be bounded to the physical region. V_{cl} at each node in the spatial domain is then calculated as:

$$V_{cl}(x,y) = \sum_{m,n=-\infty}^{+\infty} V_{lp}(x-m, y-n) \cdot V(m,n) \simeq \frac{1}{\Omega(x,y)} \sum_{m=-\frac{M}{2}}^{\frac{M}{2}-1} \sum_{n=-\frac{N}{2}}^{\frac{N}{2}-1} \omega_{xymn} V(m,n).$$

The coefficients ω_{xymn} and $\Omega(x,y)$ are:

$$\omega_{xymn} = \frac{\sin\left[\frac{\pi q_c(x-m)}{M}\right]}{\pi(x-m)} \cdot \frac{\sin\left[\frac{\pi q_c(y-n)}{N}\right]}{\pi(y-n)}, \qquad \Omega(x,y) = \sum_{m=-\frac{M}{2}}^{\frac{M}{2}-1} \sum_{n=-\frac{N}{2}}^{\frac{N}{2}-1} \omega_{xymn}.$$

Therefore, through the introduction of q_c, the input potential $V(x,y)$ is split into a classical and a quantum-mechanical part.

$$V(x,y) = V_{cl}(x,y) + V_{qm}(x,y). \tag{6}$$

$V_{cl}(x,y)$ is the slowly varying part of the potential calculated above by filtering out the high frequency components. $V_{qm}(x,y)$ contains only the high-frequency components and represents the rapidly varying part of $V(x,y)$. It is easily calculated from Eq. 6 after knowing $V_{cl}(x,y)$. Using Eqs. 2 and 3, the fully discretized WP can then be computed at each node using our new approach:

$$V_{qm,W}(x,y,p,q) = \frac{1}{i\hbar MN} \sum_{m=-\frac{M}{2}}^{\frac{M}{2}-1} \sum_{n=-\frac{N}{2}}^{\frac{N}{2}-1} e^{-i\left(\frac{\pi mp}{M} + \frac{\pi qn}{N}\right)} \left[V\left(x+\frac{m}{2}, y+\frac{n}{2}\right) - V\left(x-\frac{m}{2}, y-\frac{n}{2}\right)\right].$$

The Wigner equation in the light of the spectral decomposition will be:

$$\left(\frac{\partial}{\partial t} + \frac{\hbar \mathbf{q} \Delta k}{m^*} \Delta_{\mathbf{r}} - \frac{1}{\hbar}\left[\Delta_{\mathbf{r}} V_{cl}(\mathbf{r})\right] \Delta_{\mathbf{q}}\right) f_w(\mathbf{r}, \mathbf{q}, t) = \sum_{\mathbf{q}} V_{qm,W}(\mathbf{r}, \mathbf{q} - \mathbf{q}') f_w(\mathbf{r}, \mathbf{q}', t).$$

As shown in the left-hand side of the modified Wigner equation, V_{cl} gives rise to a local force term which is calculated using the finite difference method. Furthermore, the new Wigner potential ($V_{qm,W}$) on the right-hand side of the equation is calculated from the non-local component of the potential (V_{qm}).

3   Calculating Local Force and Evolving the Particles

The finite difference method suggests to use the two classical potential values from two adjacent nodes to approximate the force:

$$F_x(x,y) = \frac{V_{cl}(x+1,y) - V_{cl}(x-1,y)}{2\Delta r}, \qquad F_y(x,y) = \frac{V_{cl}(x,y+1) - V_{cl}(x,y-1)}{2\Delta r}.$$

Here, we focus on the force due to the presence of a dopant in the center; as it is exerted on each particle, the value of the momentum in each direction is updated and lies somewhere between two momentum grid points in the corresponding direction. In the following, a probabilistic approach is discussed and the above-mentioned grid points in the x-direction are named A and B. When the particles change their momenta, they might jump to points on the momentum grid which are several Δk farther afield. To explain our algorithm, two variables are introduced. The first variable, k_{rnd}, is a random number between 0 and 1. The second one, k_{jump}, is the remaining fractional part of momentum after it is rounded to the smaller point on the momentum grid (A), and divided by Δk, therefore, k_{jump} is also always a real number between 0 and 1. Figure 1 illustrates these variables in more details.

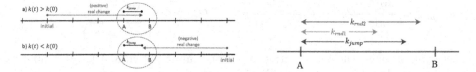

Fig. 1. *Left*: Two possible jumps on the momentum grid. *Right*: A schematic of a possible set of variables (Color figure online)

Without the probabilistic approach, for small momentum changes (compared to Δk), the new momentum is always rounded back to the same point on the momentum grid. In our approach, however, the decision whether to jump to the nearby point on the grid or not is based on a comparison. k_{rnd} is compared to k_{jump}; if k_{jump} is larger than k_{rnd} (the green case in Fig. 1) the momentum of the particle will jump to the nearby grid point (B). Otherwise (the red case in Fig. 1), it remains on the initial grid point A. Higher k_{jump} means higher probability of jumping to the nearby grid. The same approach is used in the y-direction.

4   Results

For comparison purposes, we introduced ($\frac{\lambda_c}{\Delta x}$) as a dimensionless input parameter in the ViennaWD simulator [7]. For $\frac{\lambda_c}{\Delta x} = 2$, V_{qm} vanishes as all the possible contributions of the adjacent nodes are canceled out due to the nature of the *sinc* function; however, as we increase $\frac{\lambda_c}{\Delta x}$, V_{cl} becomes smoother and V_{qm} gets closer to $V(x, y)$. Potential values at each node in the physical region contribute to V_{cl} (and hence V_{qm}) at all the other nodes through a weighted-average process. The contribution of each node depends on two factors: (1) How far is the node from the target node, (2) How big is q_c (i.e. how small is λ_c). The results for the potential decomposition in the case of a single charge in the center of a 20 nm × 30 nm region for three different values of λ_c are shown in Fig. 2:

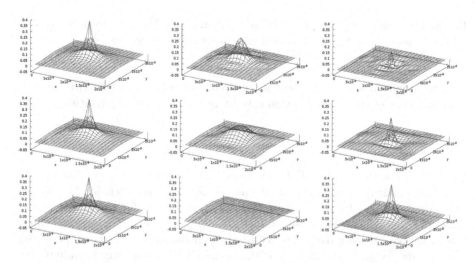

Fig. 2. The potential input (left) and its classical (middle) and quantum (right) components for a single charge in the center of a 20 nm × 30 nm region for $\frac{\lambda_c}{\Delta x} = 5$ (top row), $\frac{\lambda_c}{\Delta x} = 10$ (middle row), and $\frac{\lambda_c}{\Delta x} = 30$ (bottom row).

Increasing $\frac{\lambda_c}{\Delta x}$ results in a smoother V_{cl} as can be seen in Fig. 2. The right-most graphs show the rapidly varying quantum component of the potential (V_{qm}), which tends to $V(x, y)$ at higher values of λ_c. V_{qm} contains only the rapidly varying part of the potential and dictates the generation rate of particles (γ) at each node. Since the process of particle generation is exponentially related to γ, $N_{t_{n+1}} = N_{t_n} e^{2\gamma(t_{n+1} - t_n)}$, the increase in the number of particles becomes slower which results in fewer annihilation processes and hence improvements in the simulation time. It also reduces the undesirable effects of approximations inherent in the annihilation process [6]. In Fig. 3, the results for different λ_c values and the pure quantum case for $t = 95 fs$ are shown. The tendency towards pure quantum behaviour is noticed as we increase λ_c.

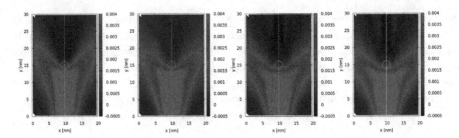

Fig. 3. The density of particles for different values of $\frac{\lambda_c}{\Delta x}$. From left to right: $\frac{\lambda_c}{\Delta x} = 2$ (classical case), $\frac{\lambda_c}{\Delta x} = 10$, $\frac{\lambda_c}{\Delta x} = 30$, and pure quantum (given for reference)

The density values at each point in the physical region are compared to the density values of the pure quantum case by introducing the error ratio:

$$Err_D(x_i, y_j) = \frac{D_{\lambda_c}(x_i, y_j) - D_{qm}(x_i, y_j)}{D_{qm}(x_i, y_j)}.$$

Figure 4 shows the results of the averaged value of this error for different λ_c values. Comparing Figs. 3 and 4, it can be noticed that even for low values of λ_c (high values of q_c), the error remains close to zero for the points in the physical region where the density values are significant. Therefore, we can claim that a spectral decomposition provides a promising step towards coupling the Poisson and Wigner equations. For the regions with lower density values, the improvement in Err_D is evident as we increase λ_c. If quantum effects such as tunneling are to be analyzed, higher values of λ_c are preferred and more reliable.

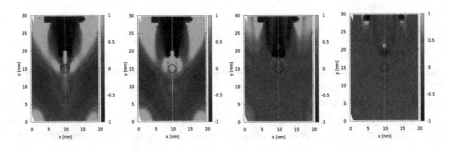

Fig. 4. The error ratio of particle density for different values of $\frac{\lambda_c}{\Delta x}$. From left to right: $\frac{\lambda_c}{\Delta x} = 2$ (classical case), $\frac{\lambda_c}{\Delta x} = 5$, $\frac{\lambda_c}{\Delta x} = 20$, and $\frac{\lambda_c}{\Delta x} = 30$.

As can be seen in Fig. 5, the current values come to a saturated value after around $50 fs$, and for different cut-off wavelengths, this saturated value lies within the 10% range of the quantum case.

It is important to note that the accuracy of the decomposition is not linearly related to the value of λ_c as depicted in Fig. 5.

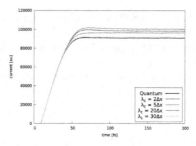

Fig. 5. Current curves for different values of λ_c and the pure quantum case (reference).

5 Discussion and Conclusion

The advantage of utilizing a potential splitting approximation is two-fold. On one hand, the statistics governing the generation of particles are modified so that the ensemble of particles experiences fewer annihilation processes. On the other hand, using this approach, the momentum values are updated according to V_{cl} and the local force term. Particles are accelerated in each time step and the value of the momentum, while remaining on the momentum grid, is no longer constant through the simulation time. Averaged physical quantities such as current and density show a decent similarity to the pure quantum case, especially for $\lambda_c \gg \Delta r$. Coupling the quantum character of carrier transport with the classical evolution of particles seems to be a promising step towards an efficient self-consistent coupling of the Wigner and Poisson equations.

Acknowledgements. This research has been supported by the Austrian Science Fund through the project FWF-P29406-N30.

References

1. Ellinghaus, P.: Two-Dimensional Wigner Monte Carlo Simulation for Time-Resolved Quantum Transport with Scattering. Dissertation, Institute for Microelectronics, TU Wien (2016)
2. Gehring, A., Kosina, H.: Wigner function-based simulation of quantum transport in scaled DG-MOSFETs using a Monte Carlo method. J. Comput. Electron. **4**(1–2), 67–70 (2005)
3. Nedjalkov, M., Querlioz, D., Dollfus, P., Kosina, H.: Wigner function approach. In: Vasileska, D., Goodnick, S. (eds.) Nano-Electronic Devices: Semiclassical and Quantum Transport Modeling, pp. 289–358. Springer, New York (2011). https://doi.org/10.1007/978-1-4419-8840-9_5
4. Querlioz, D., Dollfus, P.: The Wigner Monte Carlo Method for Nanoelectronic Devices. A Particle Description of Quantum Transport and Decoherence. Wiley, Hoboken (2010)
5. Gonzalez, R.C., Woods, R.E.: Digital Image Processing, 3rd edn, pp. 154–155. Prentice Hall, New York (2008)

6. Ellinghaus, P., Nedjalkov, M., Selberherr, S.: Optimized particle regeneration scheme for the Wigner Monte Carlo method. In: Dimov, I., Fidanova, S., Lirkov, I. (eds.) NMA 2014. LNCS, vol. 8962, pp. 27–33. Springer, Cham (2015). https://doi.org/10.1007/978-3-319-15585-2_3
7. ViennaWD - Wigner Ensemble Monte Carlo Simulator. http://www.iue.tuwien.ac.at/software/viennawd

Impact of the Trap Attributes
on the Gate Leakage Mechanisms
in a 2D MS-EMC Nanodevice Simulator

Cristina Medina-Bailon[1,2(✉)], Toufik Sadi[3], Carlos Sampedro[2],
Jose Luis Padilla[2], Luca Donetti[2], Vihar Georgiev[1], Francisco Gamiz[2],
and Asen Asenov[1]

[1] School of Engineering, University of Glasgow, Glasgow G12 8LT, Scotland, UK
Cristina.MedinaBailon@glasgow.ac.uk
[2] Nanoelectronics Research Group, Universidad de Granada,
18071 Granada, Spain
[3] Department of Neuroscience and Biomedical Engineering, Aalto University,
P.O. Box 12200, 00076 Aalto, Finland

Abstract. From a modeling point of view, the inclusion of adequate
physical phenomena is mandatory when analyzing the behavior of new
transistor architectures. In particular, the high electric field across the
ultra-thin insulator in aggressively scaled transistors leads to the possibility for the charge carriers in the channel to tunnel through the gate
oxide via various gate leakage mechanisms (GLMs). In this work, we
study the impact of trap number on gate leakage using the GLM model,
which is included in a Multi-Subband Ensemble Monte Carlo (MS-EMC)
simulator for Fully-Depleted Silicon-On-Insulator (FDSOI) field effect
transistors (FETs). The GLM code described herein considers both direct
and trap-assisted tunneling. This work shows that trap attributes and
dynamics can modify the device electrostatic characteristics and even
play a significant role in determining the extent of GLMs.

Keywords: Gate leakage mechanism · Direct tunneling
Trap assisted tunneling · MS-EMC · FDSOI

1 Introduction

Reducing the gate oxide thickness implies an increase in the field across the
oxide. The high electric field coupled with thin oxides leads to the possibility of
charge carriers traversing the barrier for transport set up by the dielectric layer,
resulting in tunneling processes from (to) substrate to (from) gate through the
gate oxide [1,2], and thus giving rise to a certain gate current. This effect is

The research leading to these results has received funding from the European Union's
Horizon 2020 research and innovation programme under grant agreement No 688101
SUPERAID7.

© Springer Nature Switzerland AG 2019
G. Nikolov et al. (Eds.): NMA 2018, LNCS 11189, pp. 273–280, 2019.
https://doi.org/10.1007/978-3-030-10692-8_30

known as the gate leakage mechanism (GLM) and includes both direct and trap-assisted tunneling.

The direct tunneling (DT) processes are always present, even if a dielectric film of perfect quality is assumed. In the case where a very thin oxide layer (less than 3–4 nm) is considered, electrons forming the inversion layer can tunnel to the gate through the energetically forbidden band gap of the dielectric material. Similarly, the trap assistant tunneling (TAT) processes are related to the existence of defect states which cause elastic or inelastic tunneling of electrons into and out of defects. In general, direct tunneling is the dominant phenomenon due to the small oxide thickness [3]. Nevertheless, trap attributes and dynamics can modify the device electrostatic properties.

Apart from the analysis of these tunneling mechanisms, the investigation of these processes on new technological nanodevices is mandatory. Currently, Fully-Depleted Silicon-On-Insulator (FDSOI) devices have been recognized as an alternative to bulk devices. However, the impact of GLM mainly depends on the electron confinement near the interface [3]. Accordingly, the number of electrons tunneling through the oxide is higher in the single gate FDSOI in contrast with other double gate devices, such as the vertical FinFET. Therefore, the study of this mechanism in FDSOI devices is of special interest.

The aim of this work is to perform a study of the impact of the trap attributes on the GLM and thus on the FDSOI performance. For this purpose, a detailed discussion of this transport mechanism is given together with the details of the stochastic simulation process in Sect. 2. The main findings are reported in Sect. 3 including a meticulous analysis of how the trap density modifies the GLM and the performance of FDSOI nano-transistors. Finally, conclusions are given in Sect. 4.

2   Methodology

The starting point of the simulation framework is a Multi-Subband Ensemble Monte Carlo (MS-EMC) code, which is based on the space-mode approach for quantum transport. The simulator solves the Schrödinger equation in the confinement direction and the Boltzmann Transport Equation (BTE) in the transport plane. The system is coupled by solving Poisson's equation in the 2D simulation domain. This tool has been widely used in different scenarios [4,5] including the study of other tunneling mechanisms such as source-to-drain tunneling (S/D tunneling) [6] or band-to-band tunneling (BTBT) [7]. The main advantage of our MS-EMC code is that the additional modules needed for taking into account the tunneling processes are included as separate transport mechanisms without increasing the computational time in comparison to purely quantum simulators. Apart from that, they can be activated or deactivated depending on the simulation scenario, giving us the possibility of independently studying GLM.

The noisy nature of the GLM, due to the random number of electrons affected by leakage, is included by implementing it as a stochastic mechanism evaluated for each particle at the end of the Monte Carlo cycle. However, it is necessary to define the input of both physical and simulation parameters before starting the Monte Carlo iterations.

Firstly, the number of traps is deterministically calculated according to the oxide dimensions and the trap density, which in turn depends on the material and the wafer orientation. The trap density is a particular input parameter in this approximation giving the possibility to the user to vary it in order to consider the fluctuating behavior of this quantity between samples. Secondly, the traps energy below the conduction band and its location along the oxide are chosen considering their random nature by reckoning a uniformly distributed random sequence of numbers. Their energy level is usually between 2.9 eV and 3.9 eV below the conduction band for the SiO_2 oxide [8,9]. Accordingly, the shift of the conduction band is calculated with the initial conditions and fixed during the whole simulation. Thirdly, it is indispensable to keep in mind that the MS-EMC code makes use of a 2D description, whereas an electron can be trapped only when it is located near a trap location in the oxide with 3D coordinates. Therefore, the dimension of the trap is defined as a cube where the assigned charge is estimated according to the trap density. This percentage n_{pery} will be compared to a random number in the MC iterations, so that it is possible to determine the probability of finding an electron located near the trap.

Then, when the traps are totally defined, the number of particles near the dielectric is required, given that the distance between their location and the interface modifies the tunnel probability. It is of note that the 2D MS-EMC code characterizes the semiclassical motion of the particles in the transport direction (x) even though its location in the confinement direction (z) is unknown. The simulated particles are distributed along the whole device and hence the percentage of the ones near the interface (n_{intf}) with respect to the total number of particles $(n(x, z))$ is estimated:

$$n_{intf} = \frac{\sum_{intf} \sum_x n(x, z)}{\sum_z \sum_x n(x, z)}, \tag{1}$$

where $intf$ represents the region near the interface in the z direction. In this study, $intf$ is taken as 10% of the T_{Si}.

The last step required before starting the Monte Carlo iterations is the calculation of the initial tunneling probabilities for each mechanism [9–11]. In general, the probability for the tunneling processes is calculated using the Wentzel Kramers Brillouin (WKB) approximation [12]. This transmission coefficient depends on the barrier thickness and height (which, in our case, is set up by the band gap of the dielectric material):

$$T_{WKB}(E) = \exp\left\{ -\frac{2}{\hbar} \int_a^b \sqrt{2m_z^*(E_{CB}(x, z) - E)}\, dx \right\}, \tag{2}$$

where a and b are the starting and ending points, E and m_z^* are the energy and the confinement effective mass of the electron, respectively, and $E_{CB}(x, z)$ corresponds to the energy of the conduction band at the point (x, z). Five tunneling processes have been implemented for the GLM in this simulation tool as illustrated in Fig. 1.

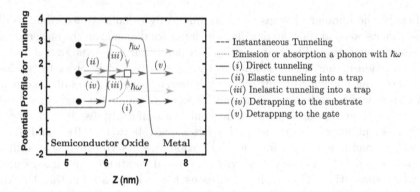

Fig. 1. Schematic band diagram of a MOS structure with metal gate and silicon substrate where the transport mechanisms implemented in the MS-EMC simulator are described: (i) direct tunneling, (ii) elastic tunneling and (iii) inelastic tunneling into a trap emitting or capturing a phonon with energy $\hbar\omega$, (iv) detrapping to the substrate, and (v) tunneling from the trap to the gate.

Direct tunneling probability is given directly by the WKB approximation. In general, when the ending point is the gate electrode, a Fermi-Dirac distribution of the electrons and available states at any given energy in the gate electrodes is assumed considering that, after tunneling, the electrons thermalize. For trap assisted tunneling, the probability depends on both the WKB approximation and some specific factors related to each mechanism. In the first place, the trap occupation must obey the Pauli exclusion principle so that no more than two particles as a maximum can be located in a trap. Secondly, if the tunneling is inelastic into a trap, it must emit or absorb a phonon. When the particle energy is higher (resp. lower) than the trap energy, a phonon is emitted (resp. absorbed). Finally, as a result of the doping level of the substrate and the large electric field at the oxide surface, the energy states within the semiconductor substrate are quantized. This leads to less occupied energy states from which electrons can tunnel and so this effect is forbidden if the energy trap state is lower than the first subband. Furthermore, the excess energy is transferred to a phonon via inelastic collisions. When a trapped electron tunnels again to the substrate, a new energy level must be chosen. As the carriers tend to be at the subband with lower kinetic energy, the approximation used in this mechanism calculates the subband in which the electron has lower kinetic energy. More details about how to calculate tunneling probabilities can be found in [9–11].

When all the initial parameters of the system are introduced, the Monte Carlo iterations begin and so the positions of each electron in the transport direction after a random flight time are calculated. There are two different scenarios regarding the GLM, as determined by the particle location:

– Particle in the channel: The first step is to determine if the particle is located near the substrate-dielectric interface using a uniformly distributed random number r_{ch1}. If $r_{ch1} > n_{intf}$, the particle continues with its normal motion, whereas if $r_{ch1} < n_{intf}$ the particle can undergo both DT and TAT or only DT. This choice is made again using another uniformly distributed random number r_{ch2}. If $r_{ch2} \leq n_{pery}$, the particle can undergo both DT and TAT, choosing the selected one from the comparison between the tunneling probabilities and another uniformly distributed random number. Otherwise, the particle undergoes DT through the insulator.

– Particle in a trap: In this scenario, the particle can experience three sub-scenarios: (i) going back to the substrate (only if the trap energy is higher than the lower subband energy), (ii) leaving the device going to the gate contact, or (iii) remaining in the trap. Due to its random nature, this choice is made again by comparing the tunneling probability with a uniformly distributed random number.

At this point, it is imperative to emphasize some global concepts. All probabilities must be recalculated when the conduction band changes, and when an electron is trapped or detrapped in each Monte Carlo iteration. Apart from that, the charge trapped is dynamically included in the 2D Poisson solution in order to preserve the self-consistency during the simulation time. Moreover, as the GLM has a very low frequency, the particles can only undergo this type of tunneling according to a certain period of occurrence and not after each integration step. Due to the negligible tunneling time through the thin oxide and the low frequency of these tunneling events, it is reasonable to assume that the electron goes directly from the starting point to the ending point at the same time step.

3 Results

3.1 Description of Simulated Devices and Processes

Device parameters, effective masses and orientation for the FDSOI structure herein analyzed are outlined in Fig. 2. The gate length ranges from $L_G = 7.5$ nm to $L_G = 20$ nm, whereas the rest of technological parameters have been fixed: the channel thickness T_{Si} is 3 nm, the gate oxide has an Equivalent Oxide Thickness $EOT = 1$ nm, and the gate work function is 4.385 eV. A Back-Plane with a $UTBOX = 10$ nm, a Back-Bias polarization (V_{BB}) of 0 V, and a Back-Plane work function of 5.17 eV have been chosen. The number of traps is estimated considering the typical trap density for a good quality gate oxide when the dielectric is SiO_2 and the wafer orientation is (100). In this particular work, the trap density (N_{Trap}) ranges from 10^{11} cm^{-2} and 10^{13} cm^{-2}.

3.2 Results and Discussion

An increase in the number of traps or their close location to the interface directly changes the tunneling probability from (to) the substrate to (from) the traps. Figure 3 shows the number of particles that can experience any GLM for different N_{Trap} values. Let us make several remarks. First, the number of particles that

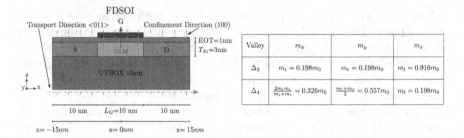

Fig. 2. (left) FDSOI structure analyzed in this work with $L_G = 10$ nm. 1D Schrödinger equation is solved for each grid point in the transport direction and BTE is solved by the MC method in the transport plane. (right) Effective masses in silicon for the device studied in this work: m_x is the transport mass, m_z is the confinement mass, m_0 is the electron free-mass, and the subindex in Δ represents the corresponding degeneracy factor, where Δ_2 is the most populated valley.

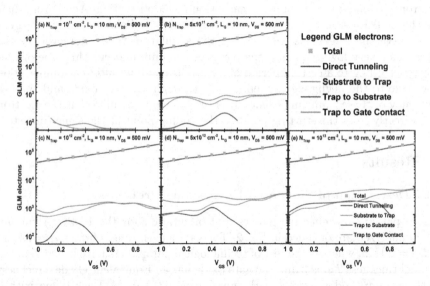

Fig. 3. Average number of electrons in arbitrary units affected by the total GLM, and by each individual mechanism as a function of V_{GS} in the 10 nm device, where $T_{Si} = 3$ nm and $V_{DS} = 500$ mV for different trap densities (N_{Trap}): 10^{11} cm^{-2} (a), 5×10^{11} cm^{-2} (b), 10^{12} cm^{-2} (c), 5×10^{12} cm^{-2} (d), and 10^{13} cm^{-2} (e).

suffers trap assisted tunneling is higher as the trap density increase. Second, the direct tunneling through the oxide is the dominant phenomenon due to the ultra-thin oxide even for the highest N_{Trap}. Third, the probability of a trapped electron to return to the substrate directly depends on the available energy states and so this type of TAT becomes forbidden as the gate bias increases.

In general, and as direct tunneling is the dominant mechanism, the particles that leave the device through the oxide reduce the drain current as depicted in Fig. 4a. This effect is almost negligible for this particular device because the total number of particles that undergoes any GLM is very reduced. However, the increase of the trap density can modulate the thermionic current as shown in the inset of Fig. 4a. The charge trapped in the oxide is dynamically included in the 2D Poisson solution in order to preserve self-consistency during the simulation time. It can be appreciated in Fig. 4b where the electron distribution is shown along the transport and confinement directions. Accordingly, an increase of the trapped charge reduces the subband levels (Fig. 4c) causing an enhanced thermionic current.

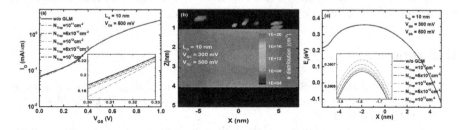

Fig. 4. (a) I_D vs. V_{GS} in the 10 nm FDSOI device at $V_{DS} = 500$ mV considering a simulation without GLM and others ones with GLM for different N_{it} values. (b) Electron distribution in cm^{-3} along the transport (X) and confinement (Z) directions in the same device as in (a) with $V_{GS} = 0.3$ V and $N_{Trap} = 10^{12}$ cm^{-2}. Recall that $X = 0$ nm corresponds to the center of the device. (c) Energy profiles of the lowest energy subband in the same device as in (a) at $V_{GS} = 0.3$ V, considering the case without GLM and others ones with GLM for different N_{Trap} values.

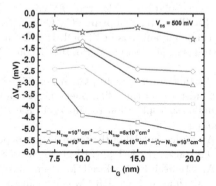

Fig. 5. Threshold voltage variation (ΔV_{th}) as a function of L_G calculated as the difference between simulations without and with GLM for different N_{Trap} values, for the FDSOI device at $V_{DS} = 500$ mV.

The impact of GLM on the threshold voltage variation (ΔV_{th}), as a function of the channel length, is shown in Fig. 5. It has been calculated as the difference between simulations without and with GLM for different N_{Trap} values. This mechanism is more important as the channel length increases because the area in which the particle can undergo GLM is higher and the number of traps increases.

4   Conclusions

This work presents the implementation of the gate leakage mechanism (GLM) including direct and trap assisted tunneling in a MS-EMC tool for the study of how the trap attributes can modify the device electrostatic properties in ultrascaled FDSOI devices. Our calculations show that direct tunneling is the dominant mechanism due to the ultra-thin oxide, resulting in the reduction of the drain current. However, the increase of the trap density slightly decreases the subband levels. Accordingly, this enhances thermionic current in comparison to the case where we only consider direct tunneling.

References

1. Roy, K., Mukhopadhyay, S., Mahmoodi-Meimand, H.: Leakage current mechanisms and leakage reduction techniques in deep-submicrometer CMOS circuits. Proc. IEEE **91**, 305–327 (2003)
2. Taur, Y., Ning, T.H.: Fundamentals of Modern VLSI Devices. Cambridge University Press, New York (2009)
3. Medina-Bailon, C., et al.: Assessment of gate leakage mechanism utilizing multi-subband ensemble Monte Carlo. In: 2017 Joint International EUROSOI and International Conference on Ultimate Integration on Silicon (EUROSOI-ULIS) (2017)
4. Sampedro, C., Gámiz, F., Godoy, A., Valín, R., García-Loureiro, A., Ruiz, F.G.: Multi-Subband Monte Carlo study of device orientation effects in ultra-short channel DGSOI. Solid State Electron. **54**(2), 131–136 (2010)
5. Sampedro, C., Gámiz, F., Godoy, A.: On the extension of ET-FDSOI roadmap for 22 nm node and beyond. Solid State Electron. **90**, 23–27 (2013)
6. Medina-Bailon, C., Padilla, J.L., Sampedro, C., Godoy, A., Donetti, L., Gamiz, F.: Source-to-drain tunneling analysis in FDSOI, DGSOI, and FinFET devices by means of multisubband ensemble Monte Carlo. IEEE Trans. Electron Devices (99), 1–7 (2018)
7. Medina-Bailon, C., Padilla, J., Sampedro, C., Alper, C., Gámiz, F., Ionescu, A.: Implementation of band-to-band tunneling phenomena in a multi-subband ensemble Monte Carlo simulator: application to silicon TFETs. IEEE Trans. Electron Dev. **64**(8), 3084–3091 (2017)
8. Vandelli, L., et al.: A physical model of the temperature dependence of the current through SiO 2 / HfO 2 stacks. IEEE Trans. Electron Dev. **58**(9), 2878–2887 (2011)
9. Sadi, T., Mehonic, A., Montesi, L., Buckwell, M., Kenyon, A., Asenov, A.: Investigation of resistance switching in SiOx RRAM cells using a 3D multi-scale kinetic Monte Carlo simulator. J. Phys. Condens. Matter **30**(8), 084005 (2018)
10. Jegert, G.C.: Modeling of leakage currents in high-κ dielectrics, Ph.D. dissertation, Technische Universität München, München, March 2012
11. Sadi, T., et al.: Advanced physical modeling of SiOx resistive random access memories. In: 2016 International Conference on Simulation of Semiconductor Processes and Devices (SISPAD), pp. 149–152 (2016)
12. Griffiths, D.J.: The WKB approximation. In: Introduction to Quantum Mechanics, chap. 8, pp. 274–297. Prentice Hall, New Jersey (1995)

Sensitivity Analysis of an Air Pollution Model by Using Quasi-Monte Carlo Algorithms for Multidimensional Numerical Integration

Tzvetan Ostromsky[1(✉)], Ivan Dimov[1], Venelin Todorov[1,2], and Zahari Zlatev[3]

[1] Department of Parallel Algorithms, Institute of Information
and Communication Technologies, Bulgarian Academy of Sciences (IICT-BAS),
Acad. G. Bonchev 25 A, 1113 Sofia, Bulgaria
{ceco,venelin}@parallel.bas.bg, ivdimov@bas.bg
[2] Department of Information Modelling, Institute of Mathematics and Informatics,
Bulgarian Academy of Sciences (IMI-BAS),
Acad. Georgi Bonchev Street, Block 8, 1113 Sofia, Bulgaria
vtodorov@math.bas.bg
[3] National Centre for Environment and Energy, University of Århus,
Frederiksborgvej 399, P.O. Box 358, 4000 Roskilde, Denmark
zz@dmu.dk

Abstract. Sensitivity analysis is a powerful tool for studying and improving the reliability of large and complicated mathematical models. Air pollution and meteorological models are in front places among the examples of such models, with a lot of natural uncertainties in their input data sets and parameters. We present here some results of our global sensitivity study of the Unified Danish Eulerian Model (UNI-DEM). One of the most attractive features of UNI-DEM is its advanced chemical scheme – the Condensed CBM IV, which consider in detail a large number of chemical species and numerous reactions between them.

Four efficient stochastic algorithms (Sobol QMC, Halton QMC, Fibonacci lattice rule and Latin hypercube sampling) have been used and compared by their accuracy in studying the sensitivity of ammonia and ozone concentration results with respect to the emission levels and some chemical reactions rates. The numerical experiments show that the stochastic algorithms under consideration are quite efficient for this purpose, especially for evaluating the contribution of small by value sensitivity indices.

1 Introduction

We discuss a systematic approach for sensitivity analysis studies in the area of air pollution modelling. The Unified Danish Eulerian Model (UNI-DEM) [15,16] is used in this particular study. Different parts of the large amount of output data, produced by the model, were used in various practical applications, where the

© Springer Nature Switzerland AG 2019
G. Nikolov et al. (Eds.): NMA 2018, LNCS 11189, pp. 281–289, 2019.
https://doi.org/10.1007/978-3-030-10692-8_31

reliability of this data should be properly estimated. Another reason to choose this model as a case study here is its sophisticated chemical scheme, where all relevant chemical processes in the atmosphere are accurately represented.

Four efficient stochastic algorithms (Sobol QMC, Halton QMC, Fibonacci lattice rule and Latin hypercube sampling) have been applied to sensitivity studies of concentration variations of air pollutants with respect to emission levels and some chemical reactions rates. More information on Sobol QMC algorithm can be found in [1]. For generating Sobol quasirandom sequences we use an adaption of INSOBL and GOSOBL routines, implemented respectively in ACM TOMS Algorithm 647 [7] and ACM TOMS Algorithm 659 [2]. The original code can only compute the "next" element of the sequence. The adapted code allows the user to specify the index of the desired element. The Halton sequence is completely described in [8,9]. Fibonacci lattice rule and Latin hypercube sampling are described in detail in our previous paper [6].

2   Description and Implementation of UNI-DEM and Its Sensitivity Analysis Version

UNI-DEM is a powerful large-scale air pollution model for calculating the concentrations of a large number of pollutants and other chemical species in the air, involved in chemical reactions with the pollutants. Among the most useful output results are the mean values of the pollutants' concentrations for certain time period (day, month, year). Other accumulative functions related to them as well as the peak values, are also calculated. These can be used in various application areas (environmental protection, agriculture, health care, etc.).

UNI-DEM is mathematically represented by the following system (1) of partial differential equations (PDE), in which the unknown concentrations c_s of a number of chemical species in the air (pollutants and other chemically active components) must be calculated. The main physical and chemical processes (advection, diffusion, chemical reactions, emissions and deposition) are represented in that system. It is computed in a large spatial domain (4800×4800 km.), which covers completely the European continent and the Mediterranean. Some typical background concentrations (which are varied both seasonally and diurnally) are used for boundary conditions. The large size of the computational domain and the fact that its west and north boundaries (from where the predominating winds blow) are above the ocean (where the concentrations of most pollutants are, in general, stable and much lower than over the continent) deminishes their effect on the results inside the domain. The I/O data arrays are structured by months, so the output concentrations at the end of an already calculated month are used as initial conditions for the next one. Initially, when there is no such data, calculations begin with a 5-day start-up period with some background initial concentrations and meteo data from the previous month in order to set up the inititial conditions.

$$
\begin{aligned}
\frac{\partial c_s}{\partial t} &= -\frac{\partial(uc_s)}{\partial x} - \frac{\partial(vc_s)}{\partial y} - \frac{\partial(wc_s)}{\partial z} \\
&+ \frac{\partial}{\partial x}\left(K_x \frac{\partial c_s}{\partial x}\right) + \frac{\partial}{\partial y}\left(K_y \frac{\partial c_s}{\partial y}\right) + \frac{\partial}{\partial z}\left(K_z \frac{\partial c_s}{\partial z}\right) \\
&+ E_s + Q_s(c_1, c_2, \ldots c_q) - (k_{1s} + k_{2s})c_s, \quad s = 1, 2, \ldots q \, .
\end{aligned}
\tag{1}
$$

c_s denote the concentrations of the chemical species; u, v, w are the wind components along the coordinate axes; K_x, K_y, K_z – the diffusion coefficients; E_s – the emissions; k_{1s}, k_{2s} – dry and wet deposition coefficients respectively; $Q_s(c_1, c_2, \ldots c_q)$ – non-linear functions, describing the chemical reactions between species under consideration.

The above PDE system is non-linear and stiff. Both non-linearity and stiffness are introduced mainly by the chemical scheme: the condensed CBM-IV (Carbon Bond Mechanism) [16]. It is quite detailed and accurate, but computationally expensive as well.

For the purpose of efficient numerical treatment, the system (1) is split according to the major physical and chemical processes and the following 3 submodels are formed: **Advection-diffusion**, **Chemistry & deposition** and **Vertical transport (vertical wind and convection)**.

Spatial and time discretization makes each of the submodels a tough computational task even for the most advanced supercomputer systems. Efficient parallelization has always been a crucial point in the computer implementation of UNI-DEM. The task became much more challenging with development of the sensitivity analysis version of the code – SA-DEM [11–13]. It consists of the following three parts:

- A modification of UNI-DEM with ability to modify certain parameters, subject to SA study. By now we have been interested in some chemical rate constants as well as in the input data for the anthropogenic emissions. A small number of input parameters is reserved for this purpose.
- A driver routine that automatically generates a set of tasks to produce the necessary results for a particular SA study. It allows to perform in parallel a large number of runs with common input data (reusing it), producing at once a whole set of values on a regular mesh (used later for calculating the sensitivity indices).
- An additional program for extracting the necessary mean monthly concentrations and computing the normalised ratios (to be analysed further on).

Significant improvements of the earlier versions of SA-DEM were made by introducing two additional levels of parallelism: top-level(MPI) and bottom-level(OpenMP). They allow us to use efficiently the computational power of the contemporary cluster supercomputers with multicore nodes. Other important improvement in the data management strategy reduced the number of I/O

Table 1. Time (T) and speed-up **Sp** of SA-DEM (MPI only) on the Spanish super-computer IBM MareNostrum III at BSC, Barcelona

#CPU	#Nodes	Advection		Chemistry		TOTAL		
		T [s]	**Sp**	T [s]	**Sp**	T [s]	**Sp**	E [%]
10	1	83460	**10**	77273	**10**	171707	**10**	100%
40	3	19448	**43**	16946	**46**	40471	**42**	106%
80	5	9874	**85**	9047	**85**	22261	**77**	96%
160	10	5250	**159**	4562	**169**	12875	**133**	83%
320	20	2895	**288**	2403	**322**	8233	**209**	65%
640	40	1522	**548**	1269	**609**	5387	**319**	50%
960	60	1215	**687**	822	**940**	4075	**421**	44%

operations and pipelined most of them with the computationally intensive stages, reducing significantly the CPU idle time in the parallel MPI processes.

In Table 1 we show some scalability results from experiments with SA-DEM on one of the largest supercomputers in Europe – IBM MareNostrum III (in BSC, Barcelona, Spain). It consists of 3028 nodes IBM dx360 M4 (16 core) with 32 GB RAM per node. It is seen from Table 1 that the chemical stage (the most computationally expensive) scales very well (shows almost linear speed-up in the whole range of experiments). Advection stage scales pretty well in most of the experiments, with understandable slow-down in the highly parallel experiments. It is caused by the significant boundary overlapping of the domain partitioning when approaching the inherent partitioning limitations. In general, SA-DEM performs quite efficiently and show relatively high scalability on such a large supercomputing system.

3   Sensitivity Studies with Respect to Emission Levels

In the huge output data stream of UNI-DEM are the mean monthly concentrations of more than 30 pollutants. We consider 2 of them: *ozone* (O_3) and *ammonia* (NH_3). In particular, we present some results of a sensitivity study of the mean monthly concentrations of ammonia in Milan.

In this section we present some results of our research on the sensitivity of UNI-DEM output (in particular, the ammonia mean monthly concentrations) with respect to the anthropogenic emissions input variation. The anthropogenic emissions input consists of 4 different components $\mathbf{E} = (\mathbf{E^A}, \mathbf{E^N}, \mathbf{E^S}, \mathbf{E^C})$ as follows:

$\mathbf{E^A}$ – ammonia (NH_3); $\mathbf{E^S}$ – sulphur dioxide (SO_2);
$\mathbf{E^N}$ – nitrogen oxides $(NO + NO_2)$; $\mathbf{E^C}$ – anthropogenic hydrocarbons.

The domain under consideration is the 4-dimensional hypercube $[0.5, 1]^4$. Poly-nomials of 2-nd degree have been used as an approximation tool [5]. The input data have been generated by the improved version of SA-DEM code, specialized for sensitivity studies (see the previous section).

Table 2. Relative error for the evaluation of $f_0 \approx 0.048$.

# samples n	Relative error			
	Sobol	Halton	FIBO	LHS
2^{10}	5.56e−04	3.15e−05	2.09e−04	5.37e−04
2^{12}	1.16e−04	1.14e−04	4.32e−05	2.27e−04
2^{14}	3.14e−05	1.27e−05	2.25e−05	6.28e−05
2^{16}	8.78e−06	8.20e−06	8.70e−06	7.74e−05
2^{18}	1.75e−06	2.40e−06	1.79e−06	3.80e−06
2^{20}	4.97e−07	1.03e−06	4.21e−07	7.16e−06

The results for relative errors for evaluation of the quantities f_0, total variances and first-order and total sensitivity indices using various stochastic approaches for numerical integration are presented in Tables 2, 3 and 4, respectively. The quantity f_0 is presented by 4-dimensional integral whereas the rest of quantities under consideration are presented by 2-dimensional integrals following the ideas of *correlated sampling* technique to compute sensitivity measures in a reliable way [10,14].

Homma and Saltelli discuss in [10] which of the two formulae below gives better estimation of $f_0^2 = \left(\int_{U^d} f(x) dx \right)^2$ in the expression for total variance and Sobol global sensitivity measures. The first formula is

$$f_0^2 \approx \frac{1}{n} \sum_{i=1}^{n} f(x_{i,1}, \ldots, x_{i,d}) \, f(x'_{i,1}, \ldots, x'_{i,d}) \tag{2}$$

where x and x' are two independent sample vectors, and the second one is

$$f_0^2 \approx \left\{ \frac{1}{n} \sum_{i=1}^{n} f(x_{i,1}, \ldots, x_{i,d}) \right\}^2 \tag{3}$$

In case of estimating sensitivity indices of a fixed order, the first formula (2) is better (as recommended in [10]).

The results in Table 2 show that the algorithms using generalized Fibonacci numbers and LHS simulate the behaviour of Sobol QMCA, but for higher dimensions their efficiency decrease. The particular case study confirms the conclusion that these algorithms are suitable and more efficient for smooth functions with comparatively low dimensions. From Tables 2 and 3 we can conclude that all stochastic approaches under consideration give reliable relative errors for sufficiently large number of samples. The most efficient in terms of computational

Table 3. Relative error for the evaluation of the total variance $\mathbf{D} \approx 0.0002$.

# samples n	Relative error			
	Sobol	Halton	FIBO	LHS
2^{10}	2.28e−03	1.40e−02	1.63e−01	1.74e−02
2^{12}	9.38e−04	7.81e−03	2.39e−02	1.04e−02
2^{14}	1.92e−04	1.77e−03	2.90e−03	1.04e−02
2^{16}	5.86e−05	5.96e−04	2.65e−04	3.65e−04
2^{18}	8.61e−06	1.48e−04	3.01e−04	1.21e−05
2^{20}	1.60e−06	4.77e−05	1.19e−04	5.96e−05

Table 4. Relative error for estimation of sensitivity indices of input parameters using various Monte Carlo and quasi-Monte Carlo approaches ($n \approx 65536$).

Sensit. index	Ref. value	Sobol	Halton	FIBO	LHS
S_1	9e−01	5.78e−06	2.95e−04	3.62e−04	9.79e−03
S_2	2e−04	1.52e−03	3.49e−02	1.74e−01	6.60e−01
S_3	1e−01	4.39e−05	2.30e−03	3.22e−03	8.65e−03
S_4	4e−05	2.87e−03	1.21e−01	4.87e−01	6.70e−01
S_1^{tot}	9e−01	5.19e−06	2.97e−04	4.61e−04	4.31e−04
S_2^{tot}	2e−04	1.36e−04	3.24e−02	3.45e−01	2.94e+01
S_3^{tot}	1e−01	4.65e−05	2.25e−03	1.96e−03	1.10e−02
S_4^{tot}	5e−05	1.57e−03	1.20e−01	5.06e−01	2.41e+02

complexity is the algorithm of Sobol, followed by Halton algorithm. The evaluated sensitivity measures, presented in the tables, are obtained either by multidimensional integrals (total variances) or by ratios of multidimensional integrals (Sobol global sensitivity indices). One can notice also from results in Table 4 that the order of relative error is different for different quantities of interest (see column *Reference value*) for the same sample size. It depends both on the integrand dimension and the magnitude of estimated quantity. The algorithms using generalized Fibonacci numbers and LHS are characterized with unreliable relative errors for small in value sensitivity measures.

4   Sensitivity Studies with Respect to Chemical Reactions Rates

Another part of our research was to study the sensitivity of the ozone concentration values in the air over Genova with respect to the rate variation of some chemical reactions of the condensed CBM-IV scheme [15], namely: ## 1, 3, 7, 22 (time-dependent) and $27, 28$ (time independent). The simplified chemical equations of those reactions are as follows:

[#1] $NO_2 + h\nu \Longrightarrow NO + O$;   [#22] $HO_2 + NO \Longrightarrow OH + NO_2$;
[#3] $O_3 + NO \Longrightarrow NO_2$;   [#27] $HO_2 + HO_2 \Longrightarrow H_2O_2$;
[#7] $NO_2 + O_3 \Longrightarrow NO_3$;   [#28] $OH + CO \Longrightarrow HO_2$.

The domain under consideration is the 6-dimensional hypercube $[0.6, 1.4]^6)$. Polynomials of second degree have been used for approximation again (see [4]).

Table 5. Relative error for the evaluation of $f_0 \approx 0.27$.

# samples n	Relative error			
	Sobol	Halton	FIBO	LHS
2^{10}	1.62e−04	1.60e−04	2.08e−03	3.73e−04
2^{12}	4.54e−05	5.55e−05	1.40e−04	2.41e−04
2^{14}	3.59e−06	2.70e−05	3.98e−04	7.53e−05
2^{16}	4.70e−06	1.60e−06	2.61e−04	2.02e−04
2^{18}	5.90e−07	1.02e−06	7.29e−06	2.82e−05
2^{20}	1.36e−07	5.56e−07	4.57e−07	1.04e−05

Table 6. Relative error for the evaluation of the total variance $\mathbf{D} \approx 0.0025$.

# samples n	Relative error			
	Sobol	Halton	FIBO	LHS
2^{10}	5.75e−03	4.86e−02	6.73e+00	1.91e−02
2^{12}	2.43e−03	1.25e−03	5.27e−01	9.99e−02
2^{14}	9.90e−05	1.65e−03	1.02e−01	1.62e−02
2^{16}	5.81e−05	4.34e−04	1.97e−03	3.56e−05
2^{18}	7.71e−06	3.79e−04	4.53e−03	7.78e−03
2^{20}	1.75e−06	3.34e−05	9.33e−03	2.78e−04

The relative errors for evaluation of the quantities f_0, total variances, first-order and total sensitivity indices using various stochastic approaches for numerical integration are presented in Tables 5, 6 and 7 respectively. The quantity f_0 is presented by 6-dimensional integral, whereas the rest of the quantities under consideration are presented by 2-dimensional integrals, following the ideas of *correlated sampling*.

From these tables we can see that Sobol QMCA gives better results than Halton QMCA and the difference is 1–2 orders. Quasi-MC lattice rule based on generalized Fibonacci numbers and Latin hypercube sampling produce better results for 6-dimensional integrals in comparison with 12-dimensional integrals. More results in favour of this conclusion can be found in [3].

Table 7. Relative error for estimation of sensitivity indices of input parameters using various Monte Carlo and quasi-Monte Carlo approaches ($n \approx 65536$).

Sensist. index	Ref. value	Sobol	Halton	FIBO	LHS
S_1	4e−01	1.83e−04	2.87e−03	3.82e−02	3.04e−02
S_2	3e−01	2.69e−05	3.76e−03	1.03e−02	7.35e−04
S_3	5e−02	1.08e−04	7.27e−03	5.48e−01	2.33e−02
S_4	3e−01	1.37e−04	2.19e−03	1.07e−02	2.47e−02
S_5	4e−07	2.69e−01	3.68e+01	3.40e+03	9.25e+02
S_6	2e−02	2.81e−03	1.30e−02	1.32e+00	3.81e−02
S_1^{tot}	4e−01	1.39e−04	2.79e−03	7.92e−02	2.03e−02
S_2^{tot}	3e−01	4.32e−05	3.26e−03	3.06e−02	1.45e−02
S_3^{tot}	5e−02	1.08e−04	6.43e−03	1.31e+00	1.55e−01
S_4^{tot}	3e−01	3.77e−04	2.11e−03	3.84e−01	1.11e−02
S_5^{tot}	2e−04	1.40e−03	1.38e−02	8.85e+01	1.45e+01
S_6^{tot}	2e−02	1.29e−05	1.04e−02	2.15e+00	9.75e−01
S_{12}	6e−03	6.03e−04	7.92e−03	3.21e+00	8.99e−02
S_{14}	5e−03	2.17e−03	9.12e−03	8.64e+00	2.74e−01
S_{15}	8e−06	9.33e+02	9.36e+02	9.19e+02	9.21e+02
S_{24}	3e−03	4.97e−04	1.83e−02	1.37e+01	7.10e−01
S_{45}	1e−05	1.48e−02	9.08e−01	4.25e+01	1.05e+01

Acknowledgements. The authors would like to thank Rayna Georgieva for her help. This work is supported by the Bulgarian Academy of Sciences through the Program for Career Development of Young Scientists, Grant DFNP-17-88 /28.07.2017; as well as by the Bulgarian NSF under Projects DN 12/4 -2017 and DN 12/5 -2017.

References

1. Antonov, I., Saleev, V.: An economic method of computing LP_τ-sequences. USSR Comput. Math. Phy. **19**, 252–256 (1979)
2. Bratley, P., Fox, B.: Algorithm 659: implementing Sobol's quasirandom sequence generator. ACM Trans. Math. Softw. **14**(1), 88–100 (1988)
3. Dimitrov, S., Dimov, I., Todorov, V.: Latin hypercube sampling and fibonacci based lattice method comparison for computation of multidimensional integrals. In: Dimov, I., Faragó, I., Vulkov, L. (eds.) NAA 2016. Lecture Notes in Computer Science, vol. 10187, pp. 296–303. Springer, Cham (2017). https://doi.org/10.1007/978-3-319-57099-0_32
4. Dimov, I., Georgieva, R., Ostromsky, Tz, Zlatev, Z.: Variance-based sensitivity analysis of the unified danish eulerian model according to variations of chemical rates. In: Dimov, I.T., Faragó, I., Vulkov, L. (eds.) NAA 2012. LNCS, vol. 8236, pp. 247–254. Springer, Heidelberg (2013). https://doi.org/10.1007/978-3-642-41515-9_26

5. Dimov, I.T., Georgieva, R., Ostromsky, T., Zlatev, Z.: Sensitivity studies of pollutant concentrations calculated by UNI-DEM with respect to the input emissions. Centr. Eur. J. Math. **11**(8), 1531–1545 (2013). https://doi.org/10.2478/s11533-013-0256-2. Numerical Methods for Large Scale Scientific Computing

6. Dimov, I.T., Georgieva, R., Todorov, V., Ostromsky, Tz.: Efficient stochastic approaches for sensitivity studies of an Eulerian large-scale air pollution model. In: AIP Conference Proceedings, vol. 1895, no. 1, p. 050004 (2017). https://doi.org/10.1063/1.5007376

7. Fox, B.: Algorithm 647: implementation and relative efficiency of quasirandom sequence generators. ACM Trans. Math. Softw. **12**(4), 362–376 (1986)

8. Halton, J.: On the efficiency of certain quasi-random sequences of points in evaluating multi-dimensional integrals. Numerische Mathematik **2**, 84–90 (1960)

9. Halton, J., Smith, G.B.: Algorithm 247: radical-inverse quasi-random point sequence. Commun. ACM **7**, 701–702 (1964)

10. Homma, T., Saltelli, A.: Importance measures in global sensitivity analysis of nonlinear models. Reliab. Eng. Syst. Saf. **52**, 1–17 (1996)

11. Ostromsky, T.z., Dimov, I.T., Georgieva, R., Zlatev, Z.: Air pollution modelling, sensitivity analysis and parallel implementation. Int. J. Environ. Pollut. **46**(1/2), 83–96 (2011)

12. Ostromsky, T.z., Dimov, I., Georgieva, R., Zlatev, Z.: Parallel computation of sensitivity analysis data for the Danish Eulerian model. In: Lirkov, I., Margenov, S., Waśniewski, J. (eds.) LSSC 2011. LNCS, vol. 7116, pp. 307–315. Springer, Heidelberg (2012). https://doi.org/10.1007/978-3-642-29843-1_35

13. Ostromsky, T.z., Dimov, I.T., Marinov, P., Georgieva, R., Zlatev, Z.: Advanced sensitivity analysis of the Danish Eulerian Model in parallel and grid environment. In: Proceedings of 3rd International Conference on AIP, AMiTaNS 2011, vol. 1404, pp. 225–232 (2011)

14. Sobol, I.M., Tarantola, S., Gatelli, D., Kucherenko, S., Mauntz, W.: Estimating the approximation error when fixing unessential factors in global sensitivity analysis. Reliab. Eng. Syst. Saf. **92**, 957–960 (2001)

15. Zlatev, Z.: Computer Treatment of Large Air Pollution Models. KLUWER Academic Publishers, Dorsrecht-Boston-London (1995)

16. Zlatev, Z., Dimov, I.T.: Computational and Numerical Challenges in Environmental Modelling. Elsevier, Amsterdam (2006)

Microscopic KMC Modeling
of Oxide RRAMs

Toufik Sadi[1,2(✉)] and Asen Asenov[2]

[1] Department of Neuroscience and Biomedical Engineering, Aalto University,
P.O. Box 12200, 00076 Aalto, Finland
toufik.sadi@aalto.fi

[2] School of Engineering, Electronic and Nanoscale Engineering,
University of Glasgow, Glasgow G12 8LT, Scotland, UK

Abstract. We investigate the microscopic behavior of oxide-based resistive random-access memory (RRAM) cells by using a unique three-dimensional (3D) physical simulator. RRAMs are attracting substantial attention and are considered as the next generation of non-volatile memory technologies. We study the operation of RRAM cells based on silica-rich silicon (SiO_x) and hafnia (HfO_x), by employing the stochastic kinetic Monte Carlo (KMC) approach for charge transport. The simulator self-consistently couples electron and ionic transport to the heat generation and diffusion phenomena and includes carefully the physics and random nature of vacancy generation and recombination, and trapping mechanisms. It models the dynamics of conductive filaments (CFs) in the 3D real space and captures correctly resistance switching regimes, including the CF formation (electroforming), set and reset processes. We describe the stochastic simulation process used for device analysis. We discuss the differences in the origin of switching between silica and hafnia based devices, and address the influence of the initial vacancy population on resistance switching in silicon rich silica RRAMs. We also emphasis the need for using 3D models and including thermal self-consistency to capture accurately the memristive nature of device switching.

Keywords: Kinetic Monte Carlo · RRAM · Nano-devices
Charge transport · Thermal effects

1 Introduction

In this work, we analyze the microscopic behavior of resistive random-access memory (RRAM) devices, based on silica-rich silicon (SiO_x) and transition metal oxides (TMOs) such as hafnia (HfO_x), using the stochastic kinetic Monte Carlo (KMC) approach, providing a deeper insight into RRAM physics and operation. RRAMs are viewed as the next generation of non-volatile memory devices, attracting considerable attention in recent years. Indeed, the ITRS report of 2013 [1] highlights the various incentives for developing RRAMs, including e.g. low cost and power dissipation, high endurance, and suitability for integration in

© Springer Nature Switzerland AG 2019
G. Nikolov et al. (Eds.): NMA 2018, LNCS 11189, pp. 290–297, 2019.
https://doi.org/10.1007/978-3-030-10692-8_32

three-dimensional (3D) crossbars. RRAMs are suitable for various applications, including e.g. novel processor architectures, neuromorphic computing or neural networks.

Due to their memristive behavior, RRAMs are best simulated using stochastic 'microscopic' methods such as KMC. Moreover, and to uncover the potential of RRAMs, it is imperative to use physical models instead of the widely used empirical approximations. Charge and thermal transport phenomena are explored using a thermally self-consistent 3D simulator, coupling a KMC description of ion and electron transport to device self-heating. The KMC solver employs the appropriate models describing vacancy generation and recombination, and trapping mechanisms in oxides. It allows the study of the dynamics of conductive filaments (CFs) in 3D, correctly reconstructing resistance switching features, including the electroforming, set and reset processes.

This work presents a comparison between two material systems: the widely used TMOs (currently the most attractive technology) and the promising yet less studied SiO_x (offering low-cost Si on-chip integration). In the background section (Sect. 2), we summarize progress in RRAM modeling, describe the originality of our simulation methodology, and give details about the studied structures. In Sect. 3, we discuss the details of the stochastic simulation process used for device analysis, and highlight the distinctive peculiarities of the numerical models employed in the simulator. In Sect. 4, we (i) highlight the differences in the origin of switching between SiO_x and HfO_x, (ii) address the role of initial vacancies in switching in SiO_x RRAMs, (iii) show the need for using 3D models to capture accurately switching, by comparison to two-dimensional (2D) models, and (iv) emphasize the need to account for coupled electro-thermal transport.

2 Background

During the operation of a RRAM device, resistance switching results from the creation and destruction of conductive filaments (CFs) [2–4]; CFs appear as oxygen ion-vacancy pairs are generated and ions are redistributed under the effect of the electric field and temperature. Most activities on RRAMs focus on transition metal oxide (TMO) devices, which are generally characterized by a high dielectric constant, which is a highly desirable feature towards high-density integration. Indeed, TMO RRAMs are nowadays considered as the most promising RRAM technology. However, they face numerous serious challenges, mainly Si microelectronics integration, hence our interest in SiO_x RRAMs, which can potentially result in cheap integration in Si CMOS chips. Furthermore, previous simulation work on RRAMs employed mostly phenomenological models, such as the resistor breaker network method [5,6], but also 2D approximations [7,8]. These models do not determine in a fully self-consistent manner the electric fields and do not consider accurately heat generation and conduction.

For an accurate understanding of the microscopic behavior of oxide-based RRAMs, we developed and refined a 3D KMC simulator. Our simulation framework has numerous unique capabilities distinguishing it from other established phenomenological models [5,7,8]. The framework uses a powerful combination of tools to describe electron-ion interactions and reconstruct in a realistic manner the CF formation and rupture in the 3D space. Additionally, it couples in a self-consistent fashion time-dependent electron and oxygen ion transport simulations, using the stochastic KMC approach, to the local electric field and temperature distributions in three-dimensions. Moreover, the dynamic nature of the ion-vacancy generation mechanisms in oxides is accounted for accurately, as described in Ref. [4]. Furthermore, the simulator accounts carefully for trapping dynamics and electron transport mechanisms in the oxide layer [9,10]. Last but not least, the simulation study is backed by experiments by various collaborators demonstrating switching in oxide RRAMs (see e.g. Refs. [11,12]).

We study the oxide-based structure shown in Fig. 1(a). The structure consists of an oxide (SiO_x or HfO_x) layer of a thickness $T \sim 10$ nm sandwiched between two titanium nitride (TiN) contacts. While experimental devices have areas as large as $100\,\mu m \times 100\,\mu m$ [11], we can limit the simulations to a much smaller (Si-rich) area (e.g. $L \times W = 20\,nm \times 20\,nm$), where a grain boundary may be present, as is the practice in the KMC modeling of RRAMs [3,4,7]. Coincidentally, experimental results suggest the existence of only one dominating CF per plate [11], justifying further the focus on a smaller contact area. Figure 1(b) summarizes the mechanisms governing electron transport in the oxide.

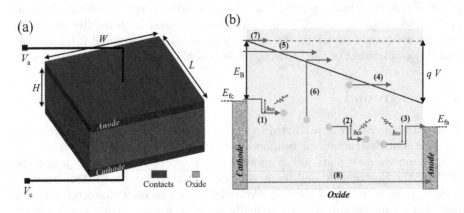

Fig. 1. (a) The studied RRAM structure, with an oxide (SiO_x or HfO_x) thickness $H = 10$ nm located between two contacts with an area $L \times W = 20\,nm \times 20\,nm$. (b) The processes governing electron transport, which include: (1) electrode-trap, (2) trap-trap, (3) trap-electrode, (4) electron to the conduction band (CB), and (5) Fowler-Nordheim tunneling mechanisms, as well as (6) Poole-Frenkel emission, (7) Schottky emission, and (8) direct electron tunneling.

3   Numerical Simulation Method

The device simulator uses a rigorous procedure self-consistently coupling the time-dependent electron and oxygen ion dynamics to the local electric field and temperature distribution in the oxide. Figure 2 is a flowchart illustrating the simulation framework. The simulator is calibrated using experimental results [11,12].

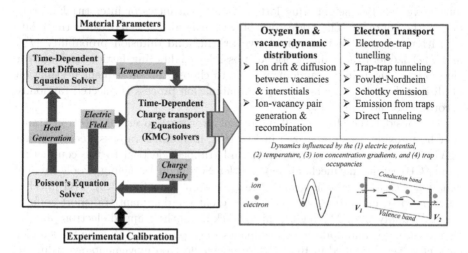

Fig. 2. The simulator, coupling the KMC description of electron and ion transport to the temperature and electric field distributions in the oxide, and accounting carefully for the vacancy generation process.

We use a stochastic KMC description to model the drifting and diffusion of ions between interstitial sites and oxygen vacancies, and ion-vacancy pair generation and recombination processes. The simulator effectively solves for the ionic transport equation given by:

$$\nabla \cdot [D\nabla n(\mathbf{r}, t) - vn(\mathbf{r}, t)] + R_G = \frac{\partial n(\mathbf{r}, t)}{\partial t}, \tag{1}$$

where n and v are the ion concentration and velocity, respectively, D is the diffusivity coefficient, and R_G is the net ion generation rate. $\mathbf{r}$ and t are the position and time, respectively. While the established vacancy generation model (through bond breakage) is used for hafnia [7,13], the reliable study of switching in SiO_x RRAM requires using the more refined generation model, as carefully detailed in Refs. [3,4]. As the field and temperature distributions are updated regularly, important physical quantities are re-calculated, including the attempt-to-escape rate for oxygen to jump over the barrier P_g (ion-vacancy generation rate), and the probability of ion-vacancy recombination P_r, as described in Ref. [3]. The generated ions move through lattice sites (interstitials or vacancies), either by drifting with an average velocity v (i.e. field and temperature-assisted ion hopping) and probability P_{dr}, or diffusing according to the diffusion constant D and probability P_{df}.

To apply the KMC algorithm for ion transport, the simulator first constructs the device geometry by assuming a 3D matrix of oxide molecules and (initial) oxygen vacancies at lattice sites, located between two contacts (where Dirichlet boundary conditions are applied). Ion-vacancy generation events are selected (in time) according to the probability P_g using random numbers. The generated ions move between neighboring lattice points (vacancies or interstitials), or to one of the contacts. The trajectory of an ion is also selected using the KMC algorithm, by constructing cumulative drifting and diffusion probability (P_{dr} and P_{df}) ladders (accounting for all possible neighboring lattice sites and/or electrodes), and using a random number to choose the subsequent destination of all ions. Equally, an ion-vacancy recombination process is selected randomly according to the probability P_r.

We consider the effect of all important processes governing electron transport in an oxide, as summarized in Fig. 1(b), including: trap-assisted tunneling (TAT) (i.e. electrode-trap, trap-trap, and trapped electron to the conduction band (CB) tunneling mechanisms), Fowler-Nordheim tunneling, Poole-Frenkel emission, Schottky emission, and direct electron tunneling [3,9]. TAT mechanisms need special attention, as they represent the dominant transport component. We apply the KMC algorithm to track down the trapped electron population in the oxide and update the occupancy of traps, instead of using the simpler current solver described in Ref. [7]. We model electron movements according to the tunneling rates given in Refs. [3,9]. For all electrons in the oxide, we generate a cumulative tunneling rate ladder including all possible destinations (vacancies or electrodes). The final electron destination is selected from the ladder using a random number. Electrons are injected from the electrodes (to occupy a vacancy or tunnel through) using the same cumulative ladder approach.

As the location of particles (electrons at traps and oxygen ions) are tracked in time and vacancies are generated, the electric potential and the lattice temperature distributions are calculated (in a self-consistent fashion) by solving Poisson's equation and the time-dependent heat diffusion equation (HDE), respectively. For this purpose, we use established finite-volume solvers developed by the authors [14]. The contributions of ions and electrons to the heat generation distribution (as needed to solve the HDE) are determined by the dot product of the local field and current density vectors.

4   Results and Discussion

To verify the reliability of the simulator, in modeling RRAMs based on various oxide materials, we show in Fig. 3(a) the time-to-failure (TF) variation with the applied electric field, as obtained using both our simulator and the procedure described in Ref. [13] at room temperature. The results are obtained by imposing a field on a 10 nm-thick metal-oxide-metal structure based on several pristine materials including SiO_2, HfO_2, Al_2O_3 and Ta_2O_5. We use the vacancy generation parameters (activation energy and field acceleration parameter) described in Ref. [13]. Clearly, the simulator reconstructs established TF data in the thin layers with reasonable accuracy. To show the corresponding device behavior, we show in Fig. 3(b) and (c) the $I-V$ characteristics during electroforming, for silica and hafnia, respectively, for various bias ramp rates. As expected, the forming biases are lower in hafnia, following the bond breakage behavior highlighted in Fig. 3(a). Similarly, forming occurs in general at lower biases, when the ramp rate is smaller.

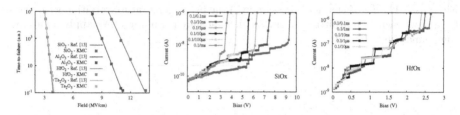

Fig. 3. (a) Time-to-failure variation with the applied field using our model and the method presented in [13], for SiO_2, HfO_2, Al_2O_3 and Ta_2O_5. The $I-V$ characteristics during electroforming for (b) silica and (c) hafnia.

Figure 4(a) shows the basic $I-V$ characteristics (for a bipolar mode) of a SiO_x RRAM device, using a bias ramping rate of $0.1\,V/\mu s$. Figure 4(b) shows the corresponding variation of the peak temperature in the oxide. The memristive behavior of the devices is captured, matching experimental trends [11]. We demonstrate two different regimes: the CF forming process and the subsequent set process. The peak temperatures follow the same trend as currents, reaching relatively high values ($\sim440\,K$) after the forming of the CF, when the device switches from a high-resistance state (HRS) to a low-resistance state (LRS). The results shown in Fig. 4(b) illustrate the need to correctly include thermal self-consistency (heating), to accurately reconstruct RRAM device characteristics.

As pointed out by the authors [3,4], the CFs are of a 3D nature, and hence using 2D models does not reflect the full physics of RRAMs. Figure 5 shows the variation of the forming bias with the initial (arbitrary) vacancy concentration, for the SiO_x structure. Figure 5 shows a strong dependence (for both 2D and 3D models) of the forming bias on initial concentrations, indicating that the CF creation occurs at lower biases as the initial concentration of vacancies is increased. As explained in Ref. [3], the activation energy for ion-vacancy generation E_a is lowered when the initial neutral oxygen vacancies are occupied by two electrons, facilitating the creation of a CF at lower biases in defect rich areas. The forming biases are highest in the pristine SiO_2 structure, enhancing the possibility of hard breakdown and hence irreversible ON/OFF transitions. Figure 5 illustrates how 2D models can lead to the erroneous prediction of the forming bias.

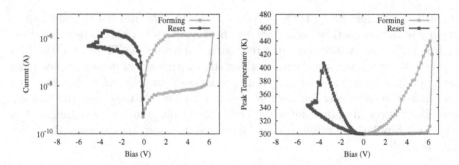

Fig. 4. (a) The $I - V$ characteristics and (b) the peak temperature variation with bias, for a device with an arbitrary initial oxygen vacancy population. We apply a current compliance limit 10^{-6} A.

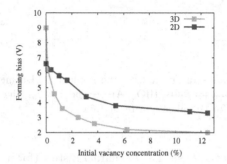

Fig. 5. The variation of the forming bias with the initial vacancy concentrations. The results are from both 2D and 3D models.

5   Conclusion

We used an electrothermal KMC simulator to investigate the switching behavior of oxide RRAM devices. We analyzed the differences in the origin of switching between SiO_x and HfO_x structures, and discussed the influence of the initial vacancy population on resistance switching in SiO_x RRAMs. We demonstrated the need for using 3D models and accounting for device self-heating to capture correctly device operation. The simulator is very well-suited for the analysis and optimization of new RRAM designs, with a focus on performance, variability and reliability.

Acknowledgment. The authors thank the Engineering and Physical Sciences Research Council (EPSRC−UK) for funding under grant agreement EP/K016776/1.

References

1. The ITRS Report 2013. http://www.itrs2.net/2013-itrs.html. Accessed 5 June 2018
2. Buckwell, M., Montesi, L., Hudziak, S., Mehonic, A., Kenyon, A.J.: Conductance tomography of conductive filaments in intrinsic silicon-rich silica RRAM. Nanoscale 7(43), 18030–18035 (2015)
3. Sadi, T., Mehonic, A., Montesi, L., Buckwell, M., Kenyon, A., Asenov, A.: Investigation of resistance switching in SiO$_x$ RRAM cells using a 3D multi-scale kinetic Monte Carlo simulator. J. Phys. Condens. Matter 30(8), 084005 (2018)
4. Sadi, T., et al.: Advanced physical modeling of SiO$_x$ resistive random access memories. In: International Conference on Simulation of Semiconductor Processes and Devices (SISPAD 2016), 6–8 September 2016, Nuremberg, Germany, pp. 149–152 (2016)
5. Chae, S.C., et al.: Random circuit breaker network model for unipolar resistance switching. Adv. Mater. 20, 1154–1159 (2008)
6. Brivio, S., Spiga, S.: Stochastic circuit breaker network model for bipolar resistance switching memories. J. Comput. Electron. 16(4), 1154–1166 (2017)
7. Yu, S., Guan, X., Wong, H.-S.P.: On the stochastic nature of resistive switching in metal oxide RRAM: physical modeling, Monte Carlo simulation, and experimental characterization. In: 2011 IEEE International Electron Devices Meeting (IEDM), Washington, DC, USA, 5–7 December 2011, p. 17.3.1 (2011)
8. Kim, S., et al.: Physical electro-thermal model of resistive switching in bi-layered resistance-change memory. Sci. Rep. 3, 1680 (2013)
9. Jegert, G.C.: Modeling of leakage currents in high-k dielectrics. PhD Dissertation, Technical University Munich, Germany (2011). http://www.iaea.org/inis/collection/NCLCollectionStore/_Public/44/011/44011419.pdf. Accessed 5 June 2018
10. Vandelli, L., Padovani, A., Larcher, L., Southwick, R.G., Knowlton, W.B., Bersuker, G.: A physical model of the temperature dependence of the current through SiO2/HfO2 stacks. IEEE Trans. Electron. Devices 58, 2878–2887 (2011)
11. Mehonic, A., et al.: Resistive switching in silicon sub-oxide films. J. Appl. Phys. 111, 074507 (2012)
12. Mehonic, A., et al.: Structural changes and conductance thresholds in metal-free intrinsic SiO$_x$ resistive random access memory. J. Appl. Phys. 117, 124505 (2015)
13. McPherson, J., Kim, J.-Y., Shanware, A., Mogul, H.: Thermochemical description of dielectric breakdown in high dielectric constant materials. Appl. Phys. Lett. 82, 2121–2123 (2003)
14. Sadi, T., Thobel, J.-L., Dessenne, F.: Self-consistent electrothermal Monte Carlo simulation of single InAs nanowire channel metal-insulator field-effect transistors. J. Appl. Phys. 108, 084506 (2010)

Contributed Talks

Numerical Simulation of the Stiff System of Equations Within the Spintronic Model

Pavlina Kh. Atanasova[1], Stefani A. Panayotova[1(✉)], Elena V. Zemlyanaya[2,3],
Yury M. Shukrinov[2,3], and Ilhom R. Rahmonov[2,4]

[1] University of Plovdiv Paisii Hilendarski, 24 Tzar Asen, 4000 Plovdiv, Bulgaria
stefani.panaiotova93@gmail.com
[2] Joint Institute of Nuclear Research, Dubna, Moscow Region 141980, Russia
[3] Dubna State University, Dubna 141980, Russia
[4] Umarov Physical Technical Institute, TAS, Dushanbe 734063, Tajikistan

Abstract. We consider a stiff system of ordinary differential equations within a spintronic model of the superconductor-ferromagnetic-/superconductor Josephson junction (SFS JJ). For some values of parameters, the explicit algorithms failed for numerical solution of this system and special numerical approaches like the implicit two-stage Gauss-Legendre method are required. In our study, we use both explicit and implicit numerical schemes which have been implemented in the respective interactive software on the basis of Wolfram Mathematica technique. In this software, we employ the 4-step explicit Runge-Kutta algorithm and the two-stage Gauss–Legendre method of the 4th accuracy order (also known as the implicit Runge-Kutta scheme), combined with the fixed point method. We analyze the effectiveness of two numerical approaches and demonstrate an advantage of implicit method over the explicit scheme. Results of numerical simulation of superconducting processes in the SFS JJ depending on parameters are presented.

Keywords: Stiff system · Implicit method · Spintronic model

1 Introduction

Superconducting properties of spintronic systems (Josephson junctions with magnetic moment) are of great interest due to a perspective of various applications in nanoelectronics and quantum computing [1,2]. Spintronic models are described by multiparameter systems of nonlinear differential equations. They cannot be solved analytically in most cases. Thus the only way of investigation of such models is a numerical study that can be a hard process. This makes important a development of appropriate numerical approach in respective user software with effective visualization of the results. Beside, in some cases the systems under study can be stiff and require a special methods for numerical solution.

In this paper we consider an implicit method of numerical solution of the stiff Cauchy problem within the spintronic model. Effectiveness of this

© Springer Nature Switzerland AG 2019
G. Nikolov et al. (Eds.): NMA 2018, LNCS 11189, pp. 301–308, 2019.
https://doi.org/10.1007/978-3-030-10692-8_33

approach in comparison with standard explicit numerical scheme is demonstrated. This method is incorporated into the *Wolfram Mathematica* user software that was developed for numerical study of Josephson junctions with magnetic momenta [5].

2   Mathematical Model

Geometry of the system of the superconductor-ferromagnetic-superconductor Josephson junction (SFS JJ) under consideration is presented in Fig. 1. The ferromagnetic easy-axis is directed along the z-axis, which is also the direction n of the gradient of the spin-orbit potential. The magnetization component m_y is coupled with Josephson current I_s, which is along the x-axis. The magnetic moment $\mathbf{M}$ is investigated by its three spatial components along the axises.

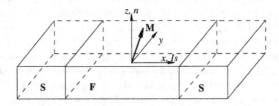

Fig. 1. Geometry of the considered SFS JJ system. Here S is superconductor and F – ferromagnetic.

The dynamics of the magnetic moment and phase difference in the SFS JJ is described by a system of first order ordinary differential equations [3]. The system in dimensionless form is as follows:

$$\frac{dm_x}{dt} = -\frac{\omega_F}{1 + M\alpha^2}\{(m_y H_z - m_z H_y) + \alpha[m_x(m_x H_x + m_y H_y + m_z H_z) - H_x]\}$$

$$\frac{dm_y}{dt} = -\frac{\omega_F}{1 + M\alpha^2}\{(m_z H_x - m_x H_z) + \alpha[m_y(m_x H_x + m_y H_y + m_z H_z) - H_y]\}$$

$$\frac{dm_z}{dt} = -\frac{\omega_F}{1 + M\alpha^2}\{(m_x H_y - m_y H_x) + \alpha[m_z(m_x H_x + m_y H_y + m_z H_z) - H_z]\}$$

$$\frac{d\varphi}{dt} = \frac{1}{w}(I_{pulse}(t) - \sin(\varphi - rm_y)),$$

$$I_{pulse}(t) = \begin{cases} A_s, & t \in [t_0 - 1/2\Delta t, t_0 + 1/2\Delta t]; \\ 0, & \text{otherwise}, \end{cases} \quad (1)$$

where $t \geq 0$ is the time, the components of the effective magnetic field $H_x(t)$, $H_y(t)$, $H_z(t)$ are determined as follows: $H_x(t) = 0$, $H_y(t) = Gr\sin(\varphi(t) - rm_y(t))$, $H_z(t) = m_z(t)$.

The physical parameters of the system (1) are following: ω_F – a feromagnetic resonance frequency, α – a damping parameter, G – parameter of the phase difference coupling, r – the spin-orbit coupling parameter, $w = V_F/(I_cR) = \omega_F/\omega_R$, $V_F = \hbar\omega_F/(2e)$, I_c – critical current, R – resistance of JJ, $\omega_R = 2eI_cR/\hbar$ – characteristic frequency. The parameters of amplitude A_s, the middle time of influence t_0 and the time interval of influence Δt characterize the electric current pulse. The unknown functions are the magnetic moment components $m_x(t), m_y(t), m_z(t)$ and the phase difference $\varphi(t)$. Initial conditions for these time-dependent functions are:

$$m_x(0) = 0, \quad m_y(0) = 0, \quad m_z(0) = 1, \quad \varphi(0) = 0. \tag{2}$$

The superconducting current $I_s(t)$ is calculated via the function $\varphi(t)$ by the following formula $I_s(t) = I_c \sin\left(\varphi(t) - rm_y(t)\right), t \geq 0$.

3 Numerical Approach

Let us start to describe the numerical method with the general formulation of the Cauchy problem [4]:

$$\bar{y}' = \bar{f}(x, \bar{y}), \quad \bar{y}(0) = \bar{y}_0, \quad x \geq 0. \tag{3}$$

In the problem (1), (2), we have $\bar{y} = (m_x, m_y, m_z, \varphi)$, $x = t$, and $\bar{f}$ corresponds to the right hand side of the system (1).

In many cases, explicit Runge-Kutta algorithm can be successfully used for numerical solution of system (1), (2) (see, e.g. [3,5]). However, for some values of parameters (for example, $G = 500\pi$, $r = 0.1; \alpha = 0.1; t_0 = 25; \Delta t = 6; A_s = 1.1; \omega_F = 1$) it becomes stiff and explicit numerical scheme requires a significant decreasing of the time-stepsize to avoid an overflowing. It is demonstrated in Fig. 2 ($h = 0.1$) and Fig. 3 ($h = 0.001$).

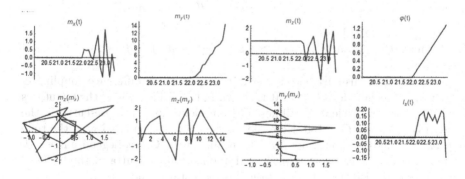

Fig. 2. Results of the simulations for the case $G = 500\pi$, $r = 0.1; \alpha = 0.1; t_0 = 25;$ $\Delta t = 6; A_s = 1.1; \omega_F = 1, h = 0.1$.

The computations with large step $h = 0.1$ reach overflowing for only 15 steps after pulsing the current at $t = 22$. This can be seen at the graphics in first line of Fig. 2 (for the functions $m_x(t), m_y(t), m_z(t)$ and $\varphi(t)$). Here the computation stops at $t = 23.5$. Also the values of $m_y(t)$ exponentially grow. One can see that they reach 14 at the end of the computations. This could not be possible because $m_y \in [-1, 1]$. Chaotic movements are also good seen (see, for example, the parametric curve $m_z(m_x)$).

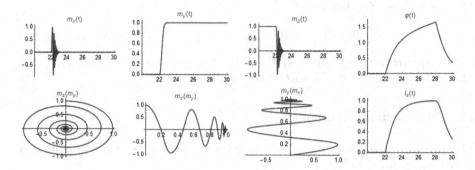

Fig. 3. Results of the simulations for the case $G = 500\pi$, $r = 0.1$; $\alpha = 0.1$; $t_0 = 25$; $\Delta t = 6$; $A_s = 1.1$; $\omega_F = 1$, $h = 0.001$.

It can be seen that the problem can be solved within the explicit method only in case of sufficiently small step ($h = 0.001$ in Fig. 3) that takes a lot of computer time. So, special numerical approach is required in case when our system becomes stiff. For our purposes we use the following definition of the stiffness [4]:

Definition 1: *A system is called stiff when for every x and $\bar{y}$ in certain area (for all solutions of Cauchy problem (3)) the eigenvalues $\lambda_j, j = \overline{1, n}$ of the n-dimension Jacobi matrix $A = \partial \bar{f}/\partial \bar{y}$ satisfy the conditions:*

1. $Re(\lambda_j) < 0, \quad j = \overline{1, n}$,
2. $C = \max\limits_j |Re(\lambda_j)| / \min\limits_k |Re(\lambda_k)| >> 1$ – *coefficient of stiffness.*

System (1), (2) usually becomes stiff when the phase difference coupling G is growing, especially for $G \geq 50$. It is shown in the Table 1 where the quantities $Re(\lambda_j)$ and the coefficient of stiffness C from Definition 1 are presented in the cases of $G = 50$ and $G = 150$. Here the solution $\bar{y}$ was calculated with the time stepsize $h = 0.001$ and accuracy 10^{-8} by means of the Gauss-Legendre method.

In our study we use both explicit and implicit Runge-Kutta methods. They are represented with the following formulas for s-stages and N nodes:

$$\bar{y}_{n+1} = \bar{y}_n + h \sum_{i=1}^{s} b_i \bar{k}_i, \qquad n = \overline{0, N-1} \tag{4}$$

$$\bar{k}_i = \bar{f}(t_n + c_i h_i, \bar{y}_n + h \sum_{j=1}^{s} a_{ij} \bar{k}_j) \qquad i = \overline{1, s}. \tag{5}$$

Table 1. Values of $Re(\lambda_j), j = 1, 2, 3, 4$ and C from Definition 1 in the case of $r = 0.1; \alpha = 0.3; t_0 = 5; \Delta t = 10; A_s = 2.1; \omega_F = 1$, for two values of G.

t	$G = 50$					$G = 150$				
	$Re(\lambda_1)$	$Re(\lambda_2)$	$Re(\lambda_3)$	$Re(\lambda_4)$	C	$Re(\lambda_1)$	$Re(\lambda_2)$	$Re(\lambda_3)$	$Re(\lambda_4)$	C
2	-0.93	-0.93	-0.60	-0.001	612.82	-3.09	-3.09	-0.62	-5.01E-6	617
5	-0.93	-0.93	-0.61	-0.009	95.77	-3.09	-3.09	-0.62	-1.59E-7	1.94E7
10	-0.91	-0.91	-0.61	-0.007	128.11	-3.09	-3.09	-0.62	-4.72E-9	6.54E8

The method is explicit in case of $s \leq i$ in (5), otherwise we have implicit scheme. The coefficients c_i, b_i, a_{ij} when $i, j = 1, 2, ..., s$ are given in the form of Butcher Tableau (see Table 2(a)).

Table 2. Butcher Tableau. The coefficients c_i, b_i, a_{ij} $(i, j = 1, 2, ..., s)$ of the scheme (4), (5).

(a) General				(b) Runge-Kutta 4					(c) Gauss-Legendre			
c_1	a_{11}	...	a_{1s}	0	0	0	0	0	$\frac{1}{2} - \frac{1}{6}\sqrt{3}$	$\frac{1}{4}$	$\frac{1}{4} - \frac{1}{6}\sqrt{3}$	
c_2	a_{21}	...	a_{2s}	1/2	1/2	0	0	0	$\frac{1}{2} + \frac{1}{6}\sqrt{3}$	$\frac{1}{4} + \frac{1}{6}\sqrt{3}$	$\frac{1}{4}$	
...	...	...	...	1/2	0	1/2	0	0				
c_s	a_{s1}	...	a_{ss}	1/2	0	0	1	0				
	b_1	...	b_s		1/6	2/6	2/6	1/6		1/2	1/2	

The case of the 4-step explicit Runge-Kutta(RK4) is given in Table 2(b). The implicit two-stage Runge-Kutta scheme (also known as the Gauss-Legendre(GL2) method and Hammer-Hollingsworth method [4][p.137]) is shown in Table 2(c). In order to find $\bar{k}_i, i = \overline{1, s}$ in the implicit scheme, we have to solve the nonlinear system via the fixed point method:

$$\bar{k}_i^{(m+1)} = \bar{f}(x + c_i h, \bar{y}_n + h \sum_{j=1}^{s} a_{ij} \bar{k}_j^{(m)}), \quad i = 1, ..., s, m = 0, 1, 2.... \quad (6)$$

Table 3. Difference between the stability for RK4 and GL2 methods for the case $G = 500, r = 0.2; \alpha = 0.1; t_0 = 5; \Delta t = 10; A_s = 1; \omega_F = 1$ and four different values of h.

	$h = 0.1$		$h = 0.02$		$h = 0.01$		$h = 0.001$	
	RK4	GL2	RK4	GL2	RK4	GL2	RK4	GL2
n	2	7	2500	2500	5000	5000	50000	50000
error	$\approx 10^{-2}$	$\approx 10^{-6}$	$\approx 10^{-3}$	$\approx 10^{-11}$	$\approx 10^{-5}$	$\approx 10^{-12}$	$\approx 10^{-9}$	$\approx 10^{-78}$

Advantage of Gauss-Legendre method is that with only 2 stage we receive accuracy of order $O(h^4)$. The stability of the numerical scheme is also better than that of the Runge-Kutta method. Results of the reached computed values while solving the system (1), (2) and the accuracy are given in Table 3. Here n is the number of the last calculated point. Taking into account that the magnetic moment is normalized (the length $\|M\| = \sqrt{m_x^2 + m_y^2 + m_z^2}$ should be always equal to 1), error represents the difference $|\|M\| - 1|$ for the last computed values. Iterations (6) stop when the previously set accuracy of h^4 is received ($\max\limits_{i=\overline{1,s}} \|\bar{k}_i^{(m+1)} - \bar{k}_i^{(m)}\| \le h^4$.) The last found $\bar{k}_i^{(m)}, i = \overline{1,s}$ are used in formula (4) for executing the next values for $\bar{y}_{n+1}$.

The privilege of implicit method over the explicit one is very good demonstrated comparing the errors at the end of the computations. The order of accuracy for GL2 method is approximately 3 times higher than that using the standard RK4 method. When the step is small ($h = 0.001$) the received accuracy is even much more higher ($\approx 10^{-78}$). For large values of the step (for example $h = 0.1$ in Table 3) GL2 gives the opportunity of computing the solutions for more points than using RK4.

Implementation of the Gausse-Legendre implicit method sufficiently extends capabilities of the *Wolfram Mathematica* user software which was developed for numerical study of Josephson junctions with magnetic momenta on the basis of system (1), (2) [5].

4   Numerical Results

Below we present results of phase and magnetization dynamics simulations and demonstrate that developed numerical scheme works properly at rather large step in time: $h = 0.01$ for $G = 50\pi$ and $h = 0.005$ for $G = 500\pi$.

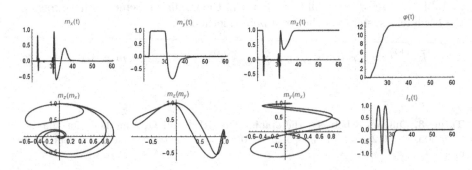

Fig. 4. Results of the simulations for the case $G = 50\pi$, $r = 0.1$; $\alpha = 0.4$; $t_0 = 25$; $\Delta t = 6$; $A_s = 2.1$; $\omega_F = 1$, $h = 0.01$.

In Fig. 4 we show time dependence of magnetic moment components $m_x(t)$, $m_y(t)$, $m_z(t)$ and phase difference $\varphi(t)$ together with trajectories in planes

$m_z - m_x$, $m_z - m_y$, $m_x - m_y$ and time dependence of superconducting current $I_s(t)$. Magnetic moments and phase difference start at their initial states (2) but after pulse of current magnetic moment components $m_x(t), m_y(t), m_z(t)$ and phase difference $\varphi(t)$ are getting perturbated. They can be stabilized in their initial state (see Fig. 4) or they come to another stable state as shown in Figs. 5, 6 and 7. As we see in Fig. 4, after current pulse the values of m_x and m_y are getting equal to zero, and $m_z = 1$.

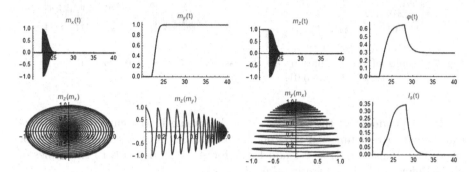

Fig. 5. Results of the simulations for the case $G = 500\pi$, $r = 0.3; \alpha = 0.02; t_0 = 25; \Delta t = 6; A_s = 0.35; \omega_F = 1$, $h = 0.005$.

Figure 5 shows how x-component returns to zero state ($m_x = 0$), but y and z components of the magnetic moment change their state to $m_y = 1$ and $m_z = 0$, respectively. It demonstrates the well known Kapitza pendulum feature [3].

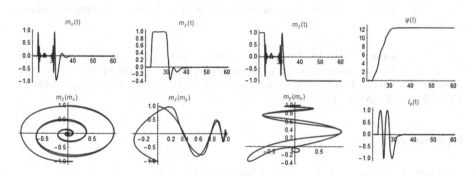

Fig. 6. Results of the simulations for the case $G = 50\pi$, $r = 0.1; \alpha = 0.2; t_0 = 25; \Delta t = 6; A_s = 2.1; \omega_F = 1$, $h = 0.01$.

The effect shown in Fig. 6 where only z-component turns from $m_z = 1$ to $m_z = -1$ is known as the magnetic moment reversal effect. This effect can be used to create a memory's cell demonstrating a bit of information and used in the quantum computing models [3].

The state of the magnetic moment is very sensitive of the parameters values. For example, changing only the value of parameter α from 0.4 (Fig. 4) to $\alpha = 0.2$ (Fig. 6) causes the significant difference between returning in the initial state and the magnetic reversal.

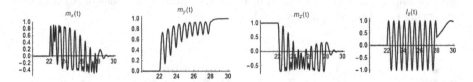

Fig. 7. Results of the simulations for the case $G = 50\pi$, $r = 0.1$; $\alpha = 0.3$; $t_0 = 25$; $\Delta t = 6$; $A_s = 10$; $\omega_F = 1$, $h = 0.01$.

The dependence of the number of local extreme in the curve of the super-conducting current function $I_s(t)$ on the value of A_s is clearly seen in Fig. 6 for $A_s = 0.35$, Fig. 5 for $A_s = 2.1$ and Fig. 7 for $A_s = 10$ depends to the values of phase difference in the corresponding figures [3].

5   Conclusions

By detailed investigation of the phase and magnetization dynamics we have demonstrated that the developed numerical scheme works properly at rather large step in time. The developed numerical scheme allowed to find all stable states of magnetization of the investigated system: orientation along the easy axes, manifestation of Kapitza pendulum features and complete magnetization reversal.

Acknowledgement. The work is supported by project FP17-FMI-008, Bulgaria, by the JINR–Bulgaria cooperation program, by the grant of AYSS of JINR with the project 18-302-08 and by the Russian Foundation for Basic Research (projects 18-52-45011_Ind, 17-01-00661, 18-02-00318).

References

1. Linder, J., Robinson, J.W.A.: Nat. Phys. **11**, 307 (2015)
2. Waintal, X., Brouwer, P.W.: Phys. Rev. B **65**, 054407 (2002)
3. Shukrinov, Yu.M., Rahmonov, I.R., Sengupta, K., Buzdin, A.: Appl. Phys. Lett. **110**, 182407 (2017)
4. Hall, G., Watt, J.M.: Modern Numerical Methods for Ordinary Differential Equations. Clarendon Press, Oxford (1976)
5. Atanasova, P., Panayotova, S., Shukrinov, Y., Rahmonov, I., Zemlyanaya, E.: EPJ Web of Conf, vol. 173, p. 05002 (2018)

Parallel SFF-SANS Study of Structure of Polydispersed Vesicular Systems

Maxim Bashashin[1,2]([✉]) [ID], Elena Zemlyanaya[1,2], and Mikhail Kiselev[1,2]

[1] Joint Institute for Nuclear Research, Joliot-Curie Str. 6, 141980 Dubna, Russia
bashashinmv@jinr.ru
[2] Dubna State University, University Str. 19, 141980 Dubna, Russia

Abstract. One of the important trends of modern nanobiophysics is the development of the drug delivery systems on the basis unilamellar vesicles (ULVs) of phospholipids. The small angle scattering of neutrons (SANS) and of X-rays (SAXS) are well known tools for investigation of the structure of the nanosystems like ULVs. In our study, analysis of SANS/SAXS experimental data is based on the separated form factors method (SFF). Effectiveness of parallel implementation of the SFF approach on the basis of MPI-version of the local minimization procedure is investigated; the results of SFF-SANS analysis of structure of the phospholipid ULVs are presented.

Keywords: Phospholipid · Small angle scattering
Minimization procedure · Parallel algorithm

1 Introduction

Phospholipids play important role in biological systems as a part of the cell membranes. In liquid media, phospholipid molecules can form unilammelar or multilammelar vesicles (liposomes): a nanosize breather-like objects with the shell (bilayer) of special structure. This property of phospholipids has a wide range of applications in biochemistry, pharmacology, biomedicine. One of important applications of unilammelar vesicles (ULVs) is the transport of medical drugs incorporated into ULVs [1,2]. Thus, the information about structure and properties of phospholipid ULVs is of great practical interest.

The small angle scattering of neutrons (SANS) is the well known experimental tool for study of structure of nanosize objects like ULVs. An efficient method of analysis of SANS data is the separated form factors method (SFF) [3]. In this approach, basic parameters of polydispersed vesicular systems are determined by minimization of a discrepancy between experimental data and calculated characteristics of SANS.

The work is performed under the grant of Russian Science Foundation (project No 14-12-00516).

G. Nikolov et al. (Eds.): NMA 2018, LNCS 11189, pp. 309–317, 2019.
https://doi.org/10.1007/978-3-030-10692-8_34

Depending on complexity of the problem, global or local minimization procedure should be used. The global minimization is based on the asynchronous differential evolution algorithm (ADE) [4]; it is employed in case of the complex multi-parameter model of bilayer (see, e.g. [5]). For the local minimization, the iterative procedure [6,7] can be successfully used (see, e.g. [3,8–11]). This approach provides a quick convergence to the local minimum but in massive calculations, the minimization process can take a significant computer time. So, for parallel optimization, we employ the parallel version of the same algorithm implemented in the code PFUMILI [13].

The effectiveness of parallel implementation of the SFF-PFUMILI fitting procedure was analysed in [12]. In this contribution, we extend this analysis considering both SANS and SAXS spectra with different number of experimental points. Also, the SANS-estimations of basic parameters of two polydispersed ULV systems are presented: the system of dipalmitoylphosphatidylcholine (DPPC) vesicles and the population of ULVs of the phospholipid transport nanosystem (PTNS). The PTNS drug delivering system with extremely small ULV size has been obtained on the basis of the soybean phosphatidylcholine in the Orekhovich Research Institute of Biomedical Chemistry (Moscow, Russia). Incorporation of drugs into PTNS ULVs is found to increase sufficiently the therapeutic effectiveness [14].

The calculations have been made at the cluster HybriLIT which is a part of the heterogeneous platform of the Multipurpose information and computing complex (MICC) of the Laboratory of Information Technologies of JINR. We used the CPU-node of this cluster consisting of two CPUs Intel Xeon E5-2695v2 (2.4 GHz, 12 cores).

2   Separated Form Factors Method

The SFF method for analysis of SANS data is determined by the following formulas (details are given in [3,5,15]). The expression of the intensity $I(q)$ has the following form:

$$I^{\text{theor}}(q) = I_m(q) + \frac{1}{2}\Delta^2 \frac{d^2 I_m(q)}{dq^2} + I_B, \tag{1}$$

$$I_m(q) = \left[\int_{R_{\min}}^{R_{\max}} I_{\text{SFF}}(q,R)G\left(R,\bar{R}\right) dR\right] \times \left[\int_{R_{\min}}^{R_{\max}} G\left(R,\bar{R}\right) dR\right]^{-1}, \tag{2}$$

q is the scattering vector, I_B characterizes the incoherent background, Δ^2 is the second momentum of spectrometer resolution function. I_m is the ULV system SANS intensity with average radius $\bar{R}$. Polydispersity σ is accounted for by the Schulz distribution $G(R,\bar{R})$ with coefficient m:

$$G\left(R,\bar{R}\right) = \frac{R^m}{m!}\left(\frac{m+1}{\bar{R}}\right)^{m+1} \exp\left[-\frac{(m+1)R}{\bar{R}}\right], \quad \sigma = \frac{1}{\sqrt{m+1}}. \tag{3}$$

I_{SFF} is the intensity of monodispersed system of ULVs:

$$I_{\mathrm{SFF}}(q) = n\, I_o\, F_s(q)\, F_b(q)\, S(q), \tag{4}$$

where I_o is the intensity of the incident beam, n is a number of vesicles in cm^3, $S(q) \sim 1$ is a structural factor, F_s is the form factor of the spherical surface with radius R:

$$F_s(q) = \left(4\pi R^2 \frac{\sin(qR)}{qR}\right)^2, \tag{5}$$

and F_b is the form factor of the symmetric lipid bilayer of thickness d:

$$F_b(q) = \left(\int\limits_{-d/2}^{+d/2} \rho_c(x)\cos(qx)\,dx\right)^2. \tag{6}$$

Here $\rho_c(x)$ is the contrast between the scattering length density of the lipid bilayer and the density of the solvent. Structure of the density function depends on the temperature and chemical properties of the phospholipid molecules. In our analysis, we use the hydrophilic-hydrophobic model (HH) of the density distribution across bilayer (see Fig. 1(a)) and the "step" density shown in Fig. 1(b). The HH model with "smooth" bound between hydrophilic and central hydrophobic region of bilayer is appropriate in case of the liquid phase of phospholipid [3] while the "step" density is used in the gel phase case.

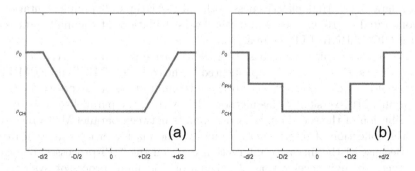

Fig. 1. Models of the scattering length density across bilayer: d is the bilayer thickness, D is the thickness of hydrophobic region, ρ_{CH} is the hydrocarbon chains density, ρ_{PH} is the polar head group density, ρ_0 is the solvent density. (a) HH-density, (b) "step" density.

3   Parallel Implementation

Parameters of ULVs are adjusted by means of the minimization of the least square discrepancy between theoretical and experimental values of SANS (SAXS) intensity at the ULV system:

$$\chi^2 = \frac{1}{N-k} \sum_{j=1}^{N} \left(\frac{I^{\text{theor}}(q_j) - I^{\text{exper}}(q_j)}{\delta^{\text{exper}}(q_j)} \right)^2, \qquad (7)$$

where $I^{\text{theor}}(q_j)$ and $I^{\text{exper}}(q_j)$ are, respectively, the theoretical and the experimental small angle intensity values at the points q_j of the scattering vector, N is the number of experimental points, k is the number of parameters to be fitted, δ are the experimental errors. Parameters to be fitted are following: average radius $\bar{R}$, coefficient of polydispersity m, thickness of bilayer d, incoherent background IB, number of vesicles per volume unit n, and parameters of the density function. In the HH-model, only the thickness of hydrophobic region D is to be fitted. In case of the "step"-density, we have one more varied parameter, the polar head group density ρ_{PH}.

As mentioned in Sect. 1, we employ both global and local minimization for the ULVs SFF-parameters adjustment in the investigations. The *global* minimization is based on the asynchronous differential evolution algorithm (ADE). The ADE minimizer with adaptive correlation matrix automatically adapts to the landscape of the optimized objective function, see [4] and references therein for details. This approach was successfully employed for the SFF analysis of SAXS data [5,16]. High effectiveness of the SFF-ADE parallel implementation is demonstrated in [16] by means of methodical calculations at the multi-processor cluster CICC (JINR, LIT, Dubna).

The *local* minimization is based on the iterative procedure described in [6]. In many cases, this approach implemented in the code DFUMIL[1] and FUMILIM [7], provides a quick convergence to the minimum of the χ^2 expression (7). In the parallel MPI-version of this method, PFUMILI[2], the parallelism is based on a distribution of the set of experimental points between parallel MPI-processes, with the collection of intermediate results to the master process at each iteration. Because of the intensive interaction between the MPI-processes, one cannot expect so high acceleration of calculations on multi-processor systems as observed in the ADE-minimization case. Nevertheless, one expects a noticeable decrease of computer time in case the number of parallel processes is sufficiently smaller the number of experimental points. In our SFF-SANS analysis problem, for each experimental point q_i at each iteration of the fitting process, the theoretical values of $I^{\text{theor}}(q)$ are calculated according to the formulas in Sect. 2 that requires a significant computational load. Therefore, one can expect the acceleration of computations when working in parallel mode, even with a relatively

---

[1] https://wwwinfo.jinr.ru/programs/jinrlib/d510.
[2] https://wwwinfo.jinr.ru/programs/jinrlib/pfumili.

small number of experimental points. Obviously, the parallel implementation efficiency should increase as the number of experimental points is growing.

In order to test the effectiveness of the PFUMILI-based parallel implementation of the SFF-parameters fitting procedure, we have made methodical calculations for three sets of experimental points. Two of them are the same as in [12]: the SANS experimental data on a polydispersed ULVs population of dimirostoilphosphatidilholin (DMPC) in D_2O. The first spectrum with 60 experimental points was measured at the YuMO spectrometer (Dubna), the second one (227 experimental points) was obtained at the SANS-I PSI facility (Paul Scherrer Institute, Villingen, Switzerland). Both measurements have been made at the temperature $30\,^\circ C$. In calculations, we used the HH-model of the scattering length density across bilayer, with $\rho_{CH} = -0.36 \cdot 10^{10} cm^{-2}$, $\rho_0 = 6.4 \cdot 10^{10} cm^{-2}$. Both sets of experimental data in comparison with the respective theoretical curves are shown in Fig. 2(a) (reproduced from [12]). The adjusted parameters of average radius, polydispersity and bilayer thickness are given in Table 1. The PFUMILI fitting results are in agreement with the results of [3, 8, 10]. This confirms correctness of the parallel implementation of the SFF approach.

The third spectrum of 1115 experimental points of the small angle X-ray scattering (SAXS) was collected at the Kurchatov Synchrotron Radiation Source of the NRC "Kurchatov Institute" (Moscow, Russia) from the polydispersed population of PTNS ULVs in water solvent with 20% maltose concentration. This measurement was made at temperature $20\,^\circ C$. The calculations are performed with the "step"-model of the scattering length density across bilayer, with $\rho_{CH} = 7.95 \cdot 10^{10} cm^{-2}$, $\rho_0 = 10.7 \cdot 10^{10} cm^{-2}$. The experimental and theoretical curves are shown in Fig. 2(b). The results of fitting (see Table 1) are close to the ADE-estimations in [17].

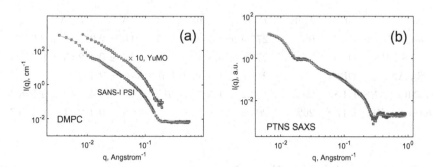

Fig. 2. (a) Theoretical and experimental SANS spectra of the polydispersed DMPC ULV system in D_2O. SANS data have been obtained at spectrometers YuMO (Dubna) and SANS-I PSI (Villingen). (b) Theoretical and experimental SAXS spectra of polydispersed population of PTNS ULVs in the 20% maltose concentration solvent.

Figure 3(a) presents the speedup of calculations versus the number of parallel MPI-processes in cases of 60, 227 and 1115 experimental points. The maximal speedup is about 6.6 times for the 60-point YuMO spectrum, about 8.6 time for the 227-points SANS-I spectrum, and about 13 times for the 1115-points SAXS-PTNS spectrum. Fig. 3(b) demonstrates the graph of effectiveness calculated as the speedup divided by the number of MPI-processes. Analysis of both graphs determines a preferable number of parallel MPI-processes depending on the number of experimental points (8, 16 and 24 MPI-processes in case of 60, 227 and 1115 experimental points, respectively).

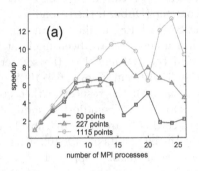 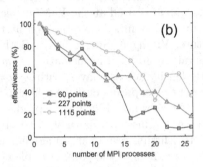

Fig. 3. (a) Speedup of PFUMILI-SFF calculations versus the number of MPI-processes in case of analysis of the 60-point YuMO spectrum, the 227-points SANS-I spectrum, and 1115-points SAXS spectrum (reproduced from [12]). (b) Effectiveness of parallel implementation versus the number of MPI-processes.

Table 1. Parameters of the average radius $\bar{R}$, the polydispersity coefficient m, the bilayer thickness d, and the hydrophobic region thickness D for the vesicular systems of DMPC, DPPC, PTNS.

Setup	ULV	Maltose	$\bar{R}$, Å	m	d, Å	D, Å	χ^2
YuMO SANS	DMPC	0%	277.0 ± 5.0	9.8 ± 0.7	49.0 ± 2.0	20.0 ± 3.0	1.1
SANS-I PSI	DMPC	0%	275.6 ± 0.5	12.6 ± 0.3	47.8 ± 0.2	20.5 ± 0.4	1.7
YS SANS	DPPC	0%	261.2 ± 1.6	14.1 ± 0.5	55.0 ± 2.4	20.0	3.5
YuMO SANS	PTNS	20%	192.2 ± 1.5	7.7 ± 0.7	38.9 ± 3.4	18.0 ± 1.9	8.2
KISI SAXS	PTNS	20%	205 ± 3	18 ± 1	44.9 ± 0.4	27.1 ± 0.2	0.7

4   Analysis of ULVs Structure

Within the SFF model, the structure of the polydispersed population of the DPPC ULVs in heavy water has been analysed on the basis of the SANS data collected at the "Yellow submarine" (YS) small angle spectrometer, Budapest, Hungary (temperature 20 °C). Fig. 4(a) demonstrates that the SFF theoretical curve well reproduce the experimental data. As in [11], the calculation was performed with the "step"-model of the scattering length density, $\rho_{CH} = -0.36 \cdot 10^{10} \mathrm{cm}^{-2}$, $\rho_0 = 6.4 \cdot 10^{10} \mathrm{cm}^{-2}$, parameter D was fixed $D = 20\text{Å}$ from the data of the X-ray small angle diffraction. The values of adjusted parameters (given in Table 1) are in reasonable agreement with the estimations in [11] on the basis of analysis of the YuMO DPPC spectrum (a discrepancy between values of parameters $\bar{R}$, m, d, D fitted to the YS data and to the YuMO data is about 10%).

The results the SFF analysis of the PTNS ULVs structure on the basis of the YuMO SANS data collected at temperature 20 °C are presented in Fig. 4(b). Previously, the structure of PTNS ULVs was analysed in [16,18,19] on the basis of SAXS data collected at the Kurchatov Synchrotron Radiation Source of the NRC "Kurchatov Institute" (Moscow, Russia). Here we present the SANS estimations of basic parameters of PTNS ULVs in the heavy water solvent of the 20% maltose concentration on the basis of the "step"-model of bilayer. In this calculation, $\rho_0 = 5.33 \cdot 10^{10} \text{cm}^{-2}$, $\rho_{CH} = -0.36 \cdot 10^{10} \text{cm}^{-2}$. Figure 4(b) demonstrates a reasonable agreement of the SFF theoretical curve with experimental data. The results are close to the ADE-analysis of the same spectrum in [20] and to the PFUMILI-calculations within the HH bilayer model [12]. Values of basic parameters of PTNS system are given in Table 1. It is seen from the Table 1 that the estimations of average radius of the PTNS ULVs on the basis of SANS and SAXS data are close to each other. Discrepancy between SANS and SAXS values of d is about 15%. Note that the SAXS- and SANS-estimations of m and D are not in good agreement that requires the further experimental and numerical study.

It is also seen from the Table 1 that the average radius $\bar{R}$ of PTNS ULVs is significantly smaller than the radius of "classical" DMPC and DPPC ULVs.

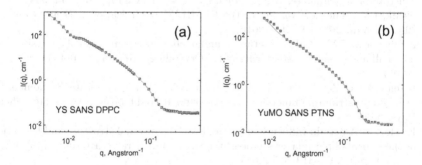

Fig. 4. Theoretical and experimental SANS spectra: (a) for the polydispersed DPPC ULV system in D_2O and (b) for the polydispersed population of PTNS ULVs in the 20% maltose concentration heave water solvent.

5   Summary

- We have found that the PFUMILI-based parallel implementation of the SFF-parameters fitting procedure can provide the 6–13 times speedup in comparison with the serial calculations.
- The SFF analysis of structure of DPPC ULVs on the basis of the YS SANS data confirmed the results of analysis of the YuMO SANS spectrum in [11].

– Estimations of basic parameters of the PTNS ULVs within the SFF analysis of the YuMO SANS data have been made. The calculation confirms the small size of PTNS vesicle in comparison with the "classical" ULVs of DMPC and DPPC.

References

1. Das, S., Chaudhury, A.: Recent advances in lipid nanoparticle formulations with solid matrix for oral drug delivery. AAPS PharmSci. Tech. **12**, 62–76 (2011)
2. Martins, S., Sarmento, B., Ferreira, D.C., Souto, E.B.: Lipid-based colloidal carriers for peptide and protein delivery - liposomes versus lipid nanoparticles. Int. J. Nanomed. **2**, 595–607 (2007)
3. Kiselev, M.A., Zemlyanaya, E.V., Aswal, V.K., Neubert, R.: What can we learn about the lipid vesicle structure from the small angle neutron scattering experiment? Eur. Biophys. J. **35**, 477–493 (2006)
4. Zhabitskaya, E.I., Zhabitsky, M.V.: Asynchronous differential evolution with adaptive correlation matrix. In: Proceedings of the 15th Annual Conference on Genetic and Evolutionary Computation, NY, USA, pp. 455–462 (2013). ISBN 978-1-4503-1963-8
5. Kiselev, M.A., Zemlyanaya, E.V., Zhabitskaya, E.I., Aksenov, V.L.: Investigation into the structure of unilamellar dimyristoylphosphocholine vesicles in aqueous sucrose solutions by small-angle neutron and X-ray scattering. Crystallogr. Rep. **60**, 143–147 (2015)
6. Kurbatov, V.S., Silin, I.N.: New method for minimizing regular functions with constraints on parameter region. Nucl. Instr. Meth. A **345**, 346–350 (1994)
7. Sitnik, I.M.: The new version of the FUMILIM minimization package. Comput. Phys. Comm. **209**, 199–204 (2016)
8. Zemlyanaya, E.V., et al.: Structure of unilamellar vesicles:numerical analysis based on small-angle neutron scattering data. Crystallogr. Rep. **51**(Suppl. 1), S22–S26 (2006)
9. Kiselev, M.A., Zemlyanaya, E.V., Ryabova, N.Y., Hauss, T., Dante, S., Lombardo, D.: Water distribution function across the curved lipid bilayer: SANS study. Chem. Phys. **345**, 185–190 (2008)
10. Kiselev, M.A., et al.: Influence of ceramide on the internal structure and hydration of the phospholipid bilayer studied by neutron and X-ray scattering. Appl. Phys. A Mater. Sci. **116**, 319–325 (2014)
11. Kiselev, M.A., Zemlyanaya, E.V.: Dimethyl Sulfoxide-Induced dehydration of the intermembrane space of dipalmitoylphosphatidylcholine multilamellar vesicles: neutron and synchrotron diffraction study. Crystallogr. Rep. **62**(5), 763–767 (2017)
12. Bashashin, M., Zemlyanaya, E., Zhabitskaya, E., Kiselev, M., Sapozhnikova, T.: Determination of the vesicular systems parameters: parallel implementation and analysis of the PTNS vesicle structure. Eur. Phys. J. Web Conf. **173**, 05003 (2018)
13. Sapozhnikov A. P.: Parallel version of Fumili program. JINR LIT Scientific report 2008–2009. JINR, Publishing Department, Dubna, pp. 96–98 (2009)
14. Archakov, A.I. et al.: Based on botanical phospholipids nanosystem for activation of biologically active compounds, and method of its manufacture (versions). Patent RU 2391966 1, Russian Federation
15. Kiselev, M.A., et al.: Influence of trehalose on the structure of unilamellar DMPC vesicles. Colloids Surf. A **256** 1–7 (2005)

16. Zhabitskaya, E., Zemlyanaya, E., Kiselev, M., Gruzinov, A.: The parallel asynchronous differential evolution method as a tool to analyze synchrotronous scattering experimental data from vesicular systems. In: EPJ Web of Conference , vol. 108, p. 02047 (2016)
17. Kiselev, M.A., Zemlyanaya, E.V., Gruzinov, A.Yu., Zhabitskaya, E.I., Ipatova, O.M., Aksenov, V.L.: Analysis of Vesicular Structure of Nanoparticles in the Phospholipid Based Drug Delivery System using SAXS data. JINR Preprint P3–2017-32, Dubna (2017). Accepted to Crystallography Reports
18. Kiselev, M.A., et al.: Application of small-angle X-ray scattering to the characterization and quantification of the drug transport nanosystem based on thesoybean phosphatidylcholine. J. Pharm. Biomed. Anal. 114, 288–291 (2015)
19. Zemlyanaya, E.V., et al.: SFF analysis of the small angle scattering data for investigation of a vesicle systems structure. J. Phys. Conf. Ser. 724, 012056 (2016)
20. Zemlyanaya, E.V., et al.: The small-angle neutron scattering data analysis of the phospholipid transport nanosystem structure. J. Phys. Conf. Ser. 1023, 012017 (2018)

Molecular Dynamic Modeling
of Long-Range Effect in Metals Exposed
to Nanoclusters

Balt Batgerel[3], Igor Puzynin[1], Taisia Puzynina[1], Ivan Hristov[2],
Radoslava Hristova[2], Zafar Tukhliev[1(✉)], and Zarif Sharipov[1]

[1] Joint Institute for Nuclear Research, 141980 Dubna, Russia
`{zafar,zarif}@jinr.ru`
[2] Sofia University "St. Kliment Ohridski", Sofia, Bulgaria
[3] Mongolian State University of Science and Technology, Ulaanbaatar, Mongolia

Abstract. Molecular dynamic simulation of the long-range effect in a
metal target, irradiated by nanoclusters, is carried out. Calculations have
shown that under simultaneous irradiation by several nanoclusters, having different areas of interaction with the target surface, fusion of high
temperature moving regions occurs in depth. The temperature in the
fusion region rises sharply, exceeding the melting temperature of the
target. As a result, structural changes in the crystal lattice at a target
depth, exceeding the penetration depth of the nanoclusters, can occur.

Keywords: Molecular dynamics · Long-range effect · Nanoclusters

1   Introduction

In materials irradiated by particles, often structural changes occur at a depth of
the irradiated target, that exceeds the penetration depth of the particles [1–6].
This phenomenon, called long-range effect, has been studies experimentally and
theoretically by several research groups in Russia for more than twenty years.
There is a large number of experimental works in this area, using various types
of materials and irradiation sources. However, a unified theoretical model for
describing the long-range effect has not been developed yet [3,4].

In the works [1–6], a special attention is given to research of ion implantation into metals, which led to the initial study of the long-range effects [1,2].
A theoretical model of this phenomenon was discussed in [3]. This model is
based on acoustic wave propagation, characterized by dependance on the thermal processes and on the structural changes, that occur in the metal, due to
the accelerated ions implantation. In our works [7,8] the wave dynamics of the
heat transfer inside a copper target irradiated by copper nanoclusters of energies
10–50 eV/atom, was simulated using the molecular dynamics method [9]. The
effect of a moving region with a high temperature (or a shock wave), described
by a nonlinear heat equation [10] and found by numerical simulations in 1D–
3D models, was confirmed in [7,8]. Therefore, it is of great interest to study

© Springer Nature Switzerland AG 2019
G. Nikolov et al. (Eds.): NMA 2018, LNCS 11189, pp. 318–325, 2019.
https://doi.org/10.1007/978-3-030-10692-8_35

the long-range interaction phenomenon by numerical simulations based on the molecular dynamics method.

We are not aware of investigations of the long-range effect, using the molecular dynamic modeling. Lack of such studies is partly due to impossibility of the direct application of this method, since for irradiation beams, starting with ion energy 1 keV, the fraction of the inelastic interactions of the ions, incident with the target, increases. Accounting for inelastic interactions requires the introduction of an electronic subsystem for the target (the heat equation of the electron gas of the target), and, accordingly, the modeling problem becomes more complicated. At present, such an approach (introduction of the electron subsystem) has been developed and applied upon irradiation with high-energy heavy ions (10–1000 MeV) [11], when the elastic interactions of the incident ion with the target can be neglected (their fraction is 5–0.1%, respectively). At irradiation energies 1-1000 keV/ion, elastic and inelastic interactions of the incident ions with the target are substantial. As the simulation shows, the irradiation of metal targets by nanoclusters [7,8] is similar to the irradiation by other sources: high-energy heavy ions and pulsed beams. In both cases, irradiation with nanoclusters and irradiation with beams of high-energy heavy ions, a large amount of energy per unit volume is released in the local target region. Thus, ion beam irradiation can be treated as nanocluster irradiation within the framework of energy release in a small volume, which makes it possible to use the methods of molecular dynamics [9].

This work differs from the previous works [7,8] in that the metal target is irradiated not by one nanocluster, but simultaneously by two and four nanoclusters, whose interaction areas with the target surface are separated from each other at a certain distance (4–10 nm). That gives the possibility to intensify the long-range effect and to show its dependence on the irradiations intensity.

2   Molecular Dynamics Method

At present, for application of the molecular dynamics methods, there are ready-made software packages (LAMMPS, DL.POLY, NAMD, etc.). In this work, the simulation was carried out using the LAMMPS package [12], installed on the HybriLit cluster [13]. Multi-particle systems are considered in the package, where all particles (atoms or molecules) are material points. The behavior of an individual particle is described by the classical Newton motion equations:

$$m_i \frac{d^2 \mathbf{r}_i}{dt^2} = \mathbf{f}_i. \tag{1}$$

Here i is the order number of the particle ($1 \leq i \leq N$), N is the total number of particles, m_i is the mass of the particle, $\mathbf{f}_i$ is the resultant of all forces, acting on the particle:

$$\mathbf{f}_i = -\frac{\partial U(\mathbf{r_1}, \dots, \mathbf{r_N})}{\partial \mathbf{r_i}} + \mathbf{f}_i^{ex},$$

where U is the interaction potential between the particles, $\mathbf{f}_i^{ex}$ is the force of the external fields (neglected further).

Under the assumption that the force of interaction between any pair of particles depends only on the distance between them, the interaction potential U takes the form:

$$U(\mathbf{r_1},\ldots,\mathbf{r_N}) = \sum_{i=1}^{N-1} \sum_{j=i+1}^{N} V(r_{ij}), \quad r_{ij} = ||\mathbf{r_i} - \mathbf{r_j}||, \tag{2}$$

$V(r_{ij})$ is the potential of the pair interaction between particles i and j.

To integrate the equations of particle motion the Verlet method is used [14]. The discretization of the classical equations of motion (1) is performed as follows:

$$\mathbf{f}_i = -\frac{\partial}{\partial \mathbf{r_i}} \sum_{j=1,j\neq i}^{N} V(r_{ij}).$$

Then the new coordinates of the particles are calculated, from which the resultant forces are determined:

$$\mathbf{r}_i(t + \Delta t) = \mathbf{r}_i(t) + \mathbf{v}_i(t)\Delta t + \frac{\mathbf{a}_i(t)}{2}\Delta t^2. \tag{3}$$

Here, $\mathbf{a}$ is the acceleration, $\mathbf{a}(t + \Delta t) = \mathbf{f}(t + \Delta t)/m$. Further the velocities of the particles are determined:

$$\mathbf{v}(t + \Delta t) = \mathbf{v}(t) + \frac{\mathbf{a}(t + \Delta t) + \mathbf{a}(t)}{2}\Delta t. \tag{4}$$

For the numerical solution of the system (1) by means of the difference scheme (3)–(4) it is necessary to set initial conditions. A general approach to statement of initial conditions is given in [14]. In the present work, the simulation of the long-range effect in the copper target irradiated by copper nanoclusters is carried out. For this purpose, the LAMMPS package has been modified. By using the methodology from [15], additional programs for nanocluster formation were constructed and incorporated in the package. The EAM (Embedded atom model) potential for copper, from the LAMMPS package [12] was used as the interatomic potential (2). Initial conditions were used from the LAMMPS package according to the selected EAM potential [12].

By using the velocities (4), the instant temperature T is calculated:

$$T = \frac{1}{dNk}\sum_{i=1}^{N} m_i \mathbf{v}_i^2,$$

where k is the Boltzmann constant, d is the space dimension, here $d = 3$.

After averaging over time interval Δt, we obtain the temperature of the system:

$$\bar{T} = \frac{1}{3Nk\Delta t} \int_{t_0}^{t_0+\Delta t} \sum_{i=1}^{N} m_i \mathbf{v}_i^2 dt.$$

3   Methodology of Modeling and Obtained Results

The simulation of the interaction of nanoclusters with a metal target was carried out as follows. Within the framework of the molecular dynamic approach, we solve the problem of irradiating a metal target by several nanoclusters with different energies. In the first stage, the simulation was carried out for region of interaction of irradiation by two nanoclusters. In the simulation, the distance between clusters was varied to obtain the desired effect. In the second stage, the simulation was carried out for simultaneous irradiation by four nanoclusters. In the previous works of the authors, at the study of thermal processes in the target material from copper irradiated with copper nanoclusters, the effect of the wave process of thermal conductivity (a moving region with a high temperature) was obtained [7]. The wave transfer of heat propagates diagonally, and the fusion of moving regions with a high temperature in the depth of the target takes place. The computational domain is a parallelepiped with sides of 22 × 18 × 11 nm and a number of particles is 373,000 in a copper target when it is irradiated by two nanoclusters of copper (Fig. 1.a). If the irradiation by four nanoclusters is simulated, the computational domain is a parallelepiped with sides 22 × 22 × 11 nm with a particle number 447,000 (Fig. 1.b). The EAM copper potential built into the LAMMPS package was used as the interatomic potential [12]. The simulation was carried out for nanoclusters with a particle number of 141 and irradiation energy 10 eV/atom to 50 eV/atom.

The results of simulation, visualized by means of the OVITO program [16], are presented on Figs. 2, 3, 4 and 5.

Figures 2 and 3 show the results obtained when the irradiation is by two nanoclusters of energy 50 eV/atom. The figures demonstrate the dynamics of the structural changes at the surface (Fig. 2) and at a depth of 5 nm (Fig. 3) of the irradiated target. It can be seen on Fig. 3.d, that the main changes in the target in the depth take place at the center, where the regions with a high temperature merge.

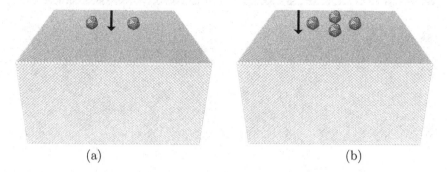

(a)   (b)

Fig. 1. Scheme of target irradiation with two (a) and four (b) nanoclusters. The arrows show the direction of irradiation.

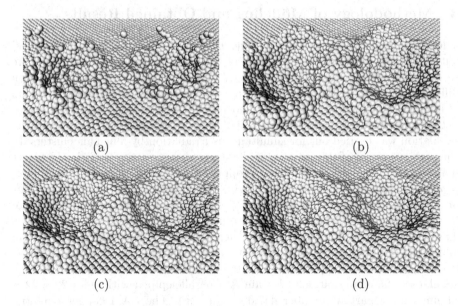

Fig. 2. Structural changes of the target and formation of craters on the surface of the target when irradiated with two nanoclusters of energy 50 eV/atom at instants of time 1 ps (a), 4 ps (b), 7 ps (c), and 10 ps (d). The distance between clusters is 6 nm.

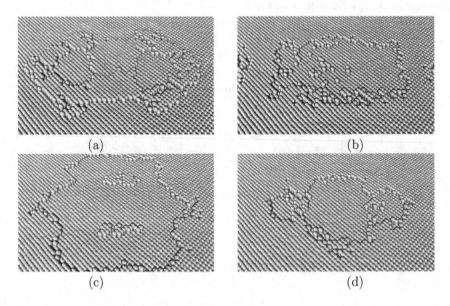

Fig. 3. Concentration of the structural changes in the target at a depth of 5 nm upon irradiation with two nanoclusters of energy 50 eV/atom at instants of time 1 ps (a), 4 ps (b), 7 ps (c), and 10 ps (d).

Figures 4 and 5 show the results obtained after irradiation of the target by four nanoclusters of energy 50 eV/atom. These figures also demonstrate the dynamics of structural changes on the surface (Fig. 4) and in a depth of 5 nm (Fig. 5) of the irradiated target. Our calculations show, that a single nanocluster of energy 50 eV/atom penetrates the target to a depth of 2–2.5 nm. The results show how structural changes occur in the depth of the target due to wave heat transfer. In addition, the target undergoes strong elastic interactions with the nanoclusters and some of the heat transfer is related to the thermoelastic effect.

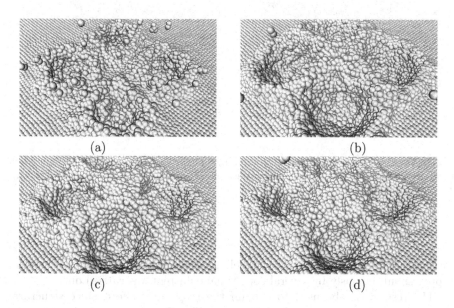

Fig. 4. Structural changes of the target and formation of craters on the surface of the target when irradiated with four nanoclusters of energy 50 eV/atom at instants of time 1 ps (a), 4 ps (b), 7 ps (c), and 10 ps (d).

The temperature in the fusion region rises sharply, exceeding the melting temperature of the target, that leads to structural changes in the target. It is numerically determined that this region is located in a depth exceeding the depth of penetration of the cluster. This result can be used to explain the long-range effect. The concentration of the structural changes in the case of four nanoclusters is higher than that in the case of two nanoclusters. Thus, we can expect more significant structural changes in the case of four nanoclusters.

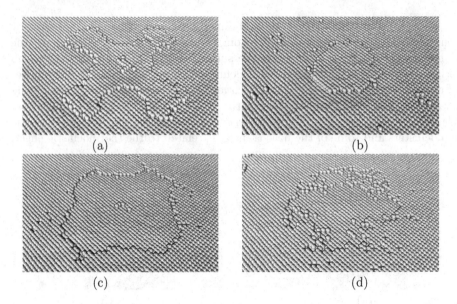

Fig. 5. Concentration of the structural changes in the target at a depth of 5 nm when irradiated with four nanoclusters of energy 50 eV/atom at instants of time 1 ps (a), 4 ps (b), 7 ps (c), and 10 ps (d).

4   Conclusions

In the work, molecular dynamic modeling of the irradiation of a target from copper simultaneously by several copper nanoclusters was carried out.

The results can be used to explain the long-range effect, since structural changes can occur at a target depth exceeding the penetration depth of the nanoclusters.

Acknowledgement. The work was financially supported by RFBR grant No. 17-01-00661-a and by a grant of the Plenipotentiary Representative of the Republic of Bulgaria at the JINR.

References

1. Tetelbaum, D.I.: Long-range effect at low-dose ion and electron irradiation of metals. Nucl. Instrum. Methods Phys. Res. B **127/128**, 153–156 (1997)
2. Sharkeev, Yu.P., et al.: The long-range effect in metals during ion implantation. Metals **1**, 109–115 (1998). (Russian)
3. Tetelbaum, D.I., Bayankin, V.Ya.: Long-range effect. Nature **4**, 9–17 (2005). (Russian)
4. Aparina, N.P., Gusev, M.I., et al.: Some aspects of the long-range effect. Probl. Atomic Sci. Eng. (PASE) Ser. Thermonucl. Fusion **3**, 18–27 (2007). (Russian)
5. Stepanov, V.A., Khmelevskaya, V.S.: Radiation-induced plastic deformation and the long-range effect. J. Tech. Phys. **81**(9), 52–56 (2011). (Russian)

6. Ovchinnikov, V.V.: Radiation-dynamic effects: the possibilities for the formation of unique structural states and properties of condensed matter. Adv. Phys. Sci. **39**, 991–1001 (2008). (Russian)
7. Batgerel, B., Dimova, S., Puzynin, I., et al.: Modeling thermal effects in metals irradiated by copper nanoclusters. In: EPJ Web Conference, vol. 173, p. 06001 (2018)
8. Batgerel, B., Didyk, A.Yu., Puzynin, I.V., et al.: Simulation of the interaction of nanoclusters with metal films. J. Surf. Investig. **9**, 1026–1030 (2015)
9. Holmurodov, H.T., Puzynin, I.V., et al.: Methods of molecular dynamics for modeling physical and biological processes. PEPAN **2**(34), 472–515 (2010). (Russian)
10. Samarskii, A.A., Sobol, I.M.: Examples of numerical calculation of temperature waves. J. Comput. Math. Math. Phys. **3**(4), 702–719 (1963). (Russian)
11. Batgarel, B., Puzynin, I.V., Puzynina, T.P., et al.: Development of the continuously-atomic approach for modeling the processes of interaction of heavy high-energy ions with metals. J. Surf. Investig. **7**, 103–107 (2018). (Russian)
12. Plimpton, S.: Fast parallel algorithms for short-range molecular dynamics. J. Comp. Phys. **117**, 1–19 (1995)
13. Hybrilit Homepage. http://hybrilit.jinr.ru/. Accessed 2016
14. Gould, H., Tobochnik, J.: Computer Modeling in Physics, 1st edn. Mir, Moscow (1990). (Russian)
15. Batgerel, B., Nikonov, E.G., Puzynin, I.V.: Simulation of interaction of colliding nanoclusters beam with solid surface. Vestnik RUDN J. Math. Inf. Sci. Phys. **1**, 47–51 (2014)
16. Stukowski, A.: Visualization and analysis of atomistic simulation data with OVITO-the Open Visualization Tool. Model. Simul. Mater. Sci. Eng. **18**, 015012 (2010)

Classification Studies in Various Areas of Science

Agnieszka Bielińska[1], Mikołaj Majkowicz[2], Dorota Bielińska-Wąż[3($\boxtimes$)], and Piotr Wąż[4]

[1] Department of Quality of Life Research, Medical University of Gdańsk, Gdańsk, Poland
[2] Department of Public Health, Pomeranian University in Słupsk, Słupsk, Poland
[3] Department of Radiological Informatics and Statistics, Medical University of Gdańsk, Gdańsk, Poland
djwaz@gumed.edu.pl
[4] Department of Nuclear Medicine, Medical University of Gdańsk, Gdańsk, Poland

Abstract. Classifications studies are performed for various kinds of objects. Several 2-dimensional classification diagrams have been created. It is shown that this kind of mathematical models can be useful in different areas of science.

Keywords: Mathematical modeling · Classification Distribution moments

1 Introduction

Classification is one of the ways to see the analogs and the properties of considered objects. Initially, it was based on the observations only. Recently it became an inspiration for the creation of various mathematical models. The problem of classification is strictly related to the notion of similarity. The objects can be classified in many different ways depending on the similarity criteria. In particular, this kind of mathematical models has been developed in chemistry, where the theory of molecular similarity, with various applications, has been created [1–10]. The basis of the molecular classifications are numerical characteristics called descriptors. In particular, we have proposed moments of the intensity distributions as molecular descriptors [11]. This kind of descriptors we have also applied for the classifications of the biological sequences [12], of the solutions in chaotic systems [13], and of the stellar spectra [14]. In the present work, we continue our studies related to the classification. The classified objects are of different nature. In this work we consider molecules and groups of individuals for which we have already performed some pilot studies [15,16].

2 Methods

As far as the molecules are concerned, the criterion of the classification (the descriptor) considered in the present work is the moment of the intensity

© Springer Nature Switzerland AG 2019
G. Nikolov et al. (Eds.): NMA 2018, LNCS 11189, pp. 326–333, 2019.
https://doi.org/10.1007/978-3-030-10692-8_36

distribution. The n-th moment of a discrete intensity distribution I_ν is defined as:

$$M_n = \mathcal{N} \sum_i I_{\nu_i} \nu_i^n, \quad n = 0, 1, 2, \ldots \tag{1}$$

I_{ν_i} is the intensity of the i-th line, ν_i is the corresponding frequency and

$$\mathcal{N} = \left(\sum_i I_{\nu_i} \right)^{-1} \tag{2}$$

is the normalization constant. We also consider the moments which are normalized to the mean value equal to zero ($M_1' = 0$) (*centered moments*):

$$M_n' = \mathcal{N} \sum_i I_{\nu_i} (\nu_i - M_1)^n. \tag{3}$$

The moments for which, additionally, the variance is equal to 1 ($M_1'' = 0$, $M_2'' = 1$) are defined as

$$M_n'' = \mathcal{N} \sum_i I_{\nu_i} \left[\frac{(\nu_i - M_1)}{\sqrt{M_2 - (M_1)^2}} \right]^n. \tag{4}$$

Similar descriptors (the distribution moments) we have also introduced for the characterization of the biological sequences within the *2D-Dynamic Representation of DNA/RNA Sequences* [17].

Such a choice of the descriptors leads to the classifications of two groups of the DNA sequences: histone H4 coding sequences of plants and of vertebrates. Since by using the standard sequence alignment methods, for example BLAST [18], CLUSTAL [19] one obtains large similarities between these two groups of sequences, it is difficult to find descriptors which distinguish between them. Therefore, obtaining the classification in this case, is a good test of the quality of the descriptors used in this work.

In the 2D-dynamic method, the nucleotides in the sequence are represented by unit vectors: adenine $= (-1, 0)$, guanine $= (1, 0)$, cytosine $= (0, 1)$, and thymine/uracil $= (0, -1)$. The method is based on walks in a 2D space. We start the walk at $(0, 0)$ point and then continue moving along consecutive vectors representing nucleotides in the sequence. After each single step of the walk a point-mass with m $= 1$ is put at the end of the unit vector corresponding to this step. If the ends of the vectors meet several times at the same point then the mass of this point is equal to the sum of the single masses. As a consequence, the sequence is represented by a set of point-masses in the 2D space (2D-dynamic graph). Non-standard treatment of the biological sequences considered as rigid bodies, analogously as in the classical dynamics, has been also generalized to 3 dimensions [20].

In the case of groups of individuals the criteria of the classification are answers to questionnaire questions. The classification diagrams (similarity maps) have been obtained using the *correspondence analysis* [21–23]. The questions have been taken from the World Health Organization Quality of Life-BREF

(WHOQOL-BREF) questionnaire [24, 25]. They concern the *Overall Quality of Life and General Health*, and the quality of life in four domains: *Physical Health, Psychological, Social Relationships*, and *Environment*. To each question the respondents could choose between 5 answers. We consider 3-point scale of answers: *negative, neutral,* and *positive*. Two answers that correspond to the least and the most favorable quality of life we combined into one *negative*, and *positive*, respectively. The same methodology can be applied to the 5-point scale. The studies were performed from February to May, 2017. We considered 449 individuals, citizens of Bydgoszcz - a city in Poland: 160 employees (100 females and 60 males) and 289 retirees (186 females and 103 males). The group of retirees has been split to two subgroups: 106 students of the University of the Third Age *retirees2* (79 females and 27 males) and 183 non-students of the University of the Third Age *retirees1* (107 females and 76 males).

3   Results and Discussion

A typical way of presentations of the results related to the answers to particular questions in the questionnaires is showing the number of answers given by subgroups of individuals. This information can be shown in the form of the spine plots. Two examples for the *Physical Health* domain are shown in Fig. 1 (males - left panel; females - right panel). One can compare the numbers of answers (normalized to 1) for particular subgroups. The χ^2 test performed by us means that the results are statistically significant: $p - value = 7.33 * 10^{-6}$ (Fig. 1, left panel) and $p - value < 2.2 * 10^{-16}$ (Fig. 1, right panel).

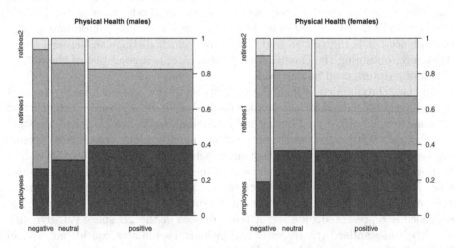

Fig. 1. Normalized numbers of answers for the groups of individuals.

In order to show graphically a global description of the system we introduce similarity maps obtained using a computational tool known as the correspondence analysis. In this method, one creates 2-dimensional maps in which the objects under consideration cluster in a specific way. The structure and the distribution of the *clusters* carry information about the objects.

Figures 2 and 3 show the maps for different groups of individuals.

For example, in the left panel of Fig. 2, the first dimension of our correspondence analysis explains 98.5% of the variance in the data and the second 1.5% (the numbers in the brackets shown in the labels of the axes of the map).

The left panels describe the males and the right ones – the females. Considering gender in the *Physical Health* domain, the differences between females and males are small. The structures of the maps are similar for males and females (Fig. 2, top panels).

This means that gender is not an important factor determining the quality of life in the *Physical Health* domain. In the *Psychological* domain gender plays an important role. The structures of the two maps are different for males and females (Fig. 2, bottom panels). For females we observe a cluster: *positive—neutral—employees—retirees2*, and for males two clusters: *employees—positive* and *retirees1—neutral*. Analogously, in the domain *Social Relationships*: the structures of the two maps for males and females are different. The cluster appears only for females: *retirees2—positive—employees* (Fig. 3, top panels). Gender is an important factor of determining the quality of life in the *Social Relationships* domain. In the domain *Environment* the differences are not large. For males we observe a cluster: *neutral—positive—retirees1—employees*. For females *positive* is excluded from this cluster (Fig. 3, bottom panels). Gender has an influence on the quality of life in this domain but this influence is weaker than in the *Psychological* and *Social Relationships* domains.

Figure 4 shows 2-dimensional maps for the chemical compounds. We consider two groups of molecules, nitriles and amides. Their infrared spectra have been calculated using *the Density Functional Theory* (DFT) method [11]. This invaluable tool in organic structure determination and verification involves the class of electromagnetic (EM) radiation with frequencies between 4000 and 400 cm^{-1} (wavenumbers). The category of EM radiation is termed infrared (IR) radiation, and its application to organic chemistry known as IR spectroscopy. DFT is a physical theory in which the description of a many-electron system is performed using the information contained in the density function only. As a consequence, the complexity of the description is considerably reduced. For more details see [26].

The main idea of our approach has been taken from the *Statistical Theory of Spectra* [27]. In the Statistical Theory of Spectra a spectrum is treated as a statistical ensemble and is characterized by the distribution moments. This results in a very substantial reduction of data. The maps have been obtained using high-order moments defined in Eq. 4. As we see, Statistical Theory of Spectra leads to the classification of the chemical compounds. The two groups of molecules (nitriles and amides) are located in different parts of the maps. This proves the efficiency and adequacy of the proposed approach. The descriptors characterizing different groups of molecules are separated by lines and denoted by different symbols in Fig. 4.

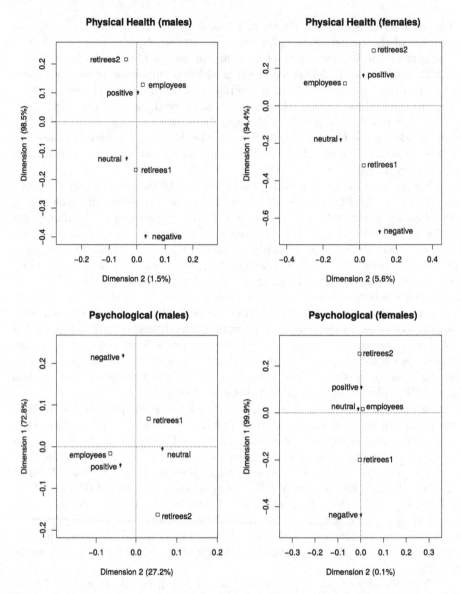

Fig. 2. Correspondence analysis maps: *Physical Health* - top panels, *Psychological* - bottom panels.

We notice that the patterns in the two maps are similar. This means that higher-order moments do not introduce any new essential information. The problem of correlations in spectral statistics we discussed in [28].

Summarizing, the classification maps are a source of information in different areas of studies. One can find new aspects of similarities between the considered objects using such an approach.

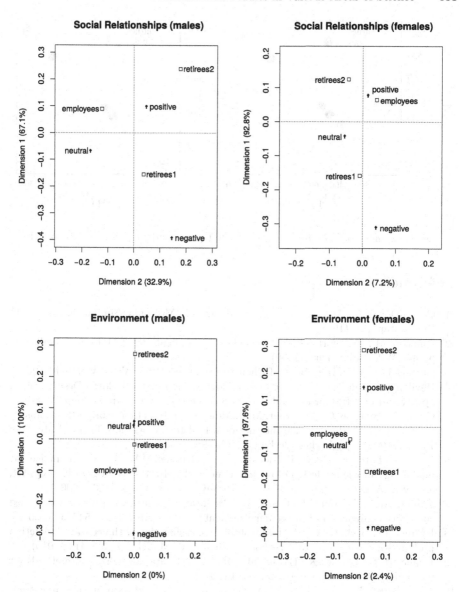

Fig. 3. Correspondence analysis maps: *Social Relationships* - top panels, *Environment* - bottom panels.

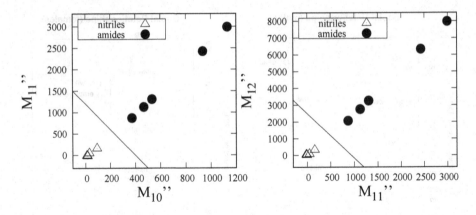

Fig. 4. Moment-based maps: $M_{10}'' - M_{11}''$ left panel and $M_{11}'' - M_{12}''$ right panel.

References

1. Carbo-Dorca, R., Mezey, P.G. (eds.): Advances in Molecular Similarity, vol. 2. JAI Press, Stamford (1998)
2. Johnson, M., Maggiora, G.M.: Concepts and Applications of Molecular Similarity, p. 393. Wiley, New York (1990)
3. Gasteiger, J. (ed.): Handbook of Chemoinformatics. Wiley-VCH, Weinheim (2003)
4. Devillers, J., Balaban, A.T. (eds.): Topological Indices and Related Descriptors in QSAR and QSPR. Gordon and Breach Science Publishers, Amsterdam (1999)
5. Clark, T., Byler, K.G., de Groot, M.J.: Molecular interactions - bringing chemistry to life. In: Proceedings of the International Beilstein Workshop, Bozen, Italy, May 15–19, 2006, Logos Verlag, Berlin, pp. 129–146 (2008)
6. Clark, T., Ford, M., Essex, J., Richards, W.G., Ritchie, D.W.: QSAR and molecular modelling in rational design of bioactive molecules. In: Proceedings of EuroQSAR, ed. by E. Aki, I. Yalcin, CADDDS in Turkey, Ankara, pp. 536–537 (2006)
7. Jakobi, A.-J., Mauser, H., Clark, T.: ParaFrag - an approach for surface-based similarity comparison of molecular fragments. J. Mol. Model. **14**, 547–558 (2008)
8. Kramer, C., Beck, B., Clark, T.: Insolubility classification with accurate prediction probabilities using a MetaClassifier. J. Chem. Inf. Model. **50**, 404–414 (2010)
9. Huang, H.-C., Qin, L.-X.: Empirical evaluation of data normalization methods for molecular classification. PerrJ **6**, e4584 (2018)
10. Jayalakshmi, R., Natarajan, R., Vivekanandan, M.: Extension of molecular similarity analysis approach to classification of DNA sequences using DNA descriptors. SAR and QSAR Environ. Res. **22**, 21–34 (2011)
11. Bielińska-Wąż, D., Nowak, W., Pepłowski, Ł., Wąż, P., Basak, S.C., Natarajan, R.: Statistical spectroscopy as a tool for the study of molecular spectroscopy. J. Math. Chem. **43**, 1560–1572 (2008)
12. Bielińska-Wąż, D., Wąż, P.: 2D-dynamic representation of DNA sequences as a graphical tool in bioinformatics. In: Todorov, M.D. (ed.) 8th International Conference on Promoting the Application of Mathematics in Technical and Natural Sciences (AMiTaNS 2016), June 22–27, 2016, Albena, Bulgaria, AIP Conference Proceedings, vol. 1773, Art. No. 060004 (2016)

13. Wąż, P., Bielińska-Wąż, D.: Asymmetry coefficients as indicators of chaos. Acta Phys. Pol. A **116**, 987–991 (2009)
14. Wąż, P., Bielińska-Wąż, D., Strobel, A., Pleskacz, A.: Statistical indicators of astrophysical parameters. Acta Astron. **60**, 283–293 (2010)
15. Bielińska, A., Majkowicz, M., Wąż, P., Bielińska-Wąż, D.: Overall quality of life and general health - changes related to the retirement threshold. In: Murata, Y. et al. (eds.) eTELEMED 2018 The Tenth International Conference on eHealth, Telemedicine, and Social Medicine, March 25–29, 2018, Rome, Italy, pp. 1–5. XPS IARIA Press, Rome (2018)
16. Bielińska, A., Majkowicz, M., Wąż, P., Bielińska-Wąż, D.: Influence of the education level on health of elderly people. In: Murata, Y., et al. (eds.) eTELEMED 2018 The Tenth International Conference on eHealth, Telemedicine, and Social Medicine, March 25–29, 2018, Rome, Italy, pp. 6–11. XPS IARIA Press, Rome (2018)
17. Bielińska-Wąż, D., Nowak, W., Wąż, P., Nandy, A., Clark, T.: Distribution moments of 2D-graphs as descriptors of DNA sequences. Chem. Phys. Lett. **443**, 408–413 (2007)
18. Altschul, S.F., Gish, W., Miller, W., Myers, E.W., Lipman, D.J.: Basic local alignment search tool. J. Mol. Biol. **215**, 403–410 (1990)
19. Chenna, R., et al.: Multiple sequence alignment with the clustal series of programs. Nucleic Acids Res. **31**, 3497–3500 (2003)
20. Wąż, P., Bielińska-Wąż, D.: Non-standard similarity/dissimilarity analysis of DNA sequences. Genomics **104**, 464–471 (2014)
21. Hirschfeld, H.O.: A connection between correlation and contingency. Proc. Cambridge Philos. Soc. **31**, 520–524 (1935)
22. Benzécri, J.-P.: L'Analyse des Données. L'Analyse des Correspondances, Paris, France, Dunod, vol. II (1973)
23. Greenacre, M.: Correspondence Analysis in Practice, 2nd edn. Chapman & Hall/CRC, London (2007)
24. http://www.who.int/substance_abuse/research_tools/whoqolbref/en/
25. Wołowicka, L.: Quality of Life in Medical Sciences, pp. 231–281. Wydawnictwo Uczelniane AM w Poznaniu, Poznań (2001)
26. Parr, R., Yang, W.: Density-Functional Theory of Atoms and Molecules. Oxford University Press, Oxford (1989)
27. Bielińska-Wąż, D.: Statistical method of generating envelopes of the electronic hot bands. Int. J. Quant. Chem. **115**, 1726–1732 (2015)
28. Bielińska-Wąż, D., Wąż, P.: Correlations in spectral statistics. J. Math. Chem. **43**, 1287–1300 (2008)

Mathematical Modeling: Interdisciplinary Similarity Studies

Agnieszka Bielińska[1], Mikołaj Majkowicz[2], Piotr Wąż[3（✉）],
and Dorota Bielińska-Wąż[4]

[1] Department of Quality of Life Research, Medical University of Gdańsk,
Gdańsk, Poland
[2] Department of Public Health, Pomeranian University in Słupsk, Słupsk, Poland
[3] Department of Nuclear Medicine, Medical University of Gdańsk, Gdańsk, Poland
phwaz@gumed.edu.pl
[4] Department of Radiological Informatics and Statistics,
Medical University of Gdańsk, Gdańsk, Poland

Abstract. Similarity maps are proposed as mathematical models that
can be used in different areas of science. Using this kind of a graphical
representation one can reveal the properties that determine similarity or
dissimilarity of the considered objects. Several new similarity maps have
been created.

Keywords: Mathematical modeling · Similarity studies · Descriptors

1 Introduction

Different kinds of problems in many areas of science can be treated as similarity studies. Such kind of mathematical modeling may reveal new properties of the investigated systems. The objects in different similarity maps (classification diagrams) may be of entirely different characters but the objects located close to each other on the maps are similar in some way.

One of the objects considered here are the biological (DNA/RNA, protein) sequences. Recently, we have created a new mathematical model called by us *3D-Dynamic Representation of DNA/RNA Sequences* [1–3]. We have proved high accuracy of this approach which offers both numerical and graphical tool for similarity/dissimilarity analysis of the biological sequences. Other objects analyzed by us are various kinds of spectra [4–6]. As a consequence, a new method of similarity analysis of the stellar spectra, more accurate than the commonly used approach, has been formulated. We have also performed pilot studies in which the objects under considerations are groups of people with similarity studies based on their answers to specific questions [7,8].

In the present work we continue these studies. In the subsequent sections we describe the methods which are the basis for the creation of new similarity maps. The objects in the maps are either biological sequences and their structures or groups of individuals and their questionnaire answers.

© Springer Nature Switzerland AG 2019
G. Nikolov et al. (Eds.): NMA 2018, LNCS 11189, pp. 334–341, 2019.
https://doi.org/10.1007/978-3-030-10692-8_37

2   Methods

One of the computational methods we used for obtaining the similarity maps is *3D-Dynamic Representation of DNA/RNA Sequences*. This approach belongs to nonstandard (alignment-free) groups of methods aiming at similarity/dissimilarity analysis of the biological sequences [9–13]. Within this approach the sequence is represented by the 3D-dynamic graph composed of point masses in a 3D space. The DNA sequence is a sequence composed of four letters corresponding to four nucleotides: A - adenine, C - cytosine, G - guanine, T - thymine. In this method we represent the nucleotides by the basis vectors: $A = (-1, 0, 1)$, $G = (1, 0, 1)$, $C = (0, 1, 1)$, and $T = (0, -1, 1)$. The process of the creation of the 3D-dynamic graph is based on shifts in a 3D-space. We start the shifts from the origin of the coordinate system. After the first shift representing the first base in the sequence a point-mass is located at the end of the vector representing this shift. For example, if the first base is A, then the vector representing the first shift is $(-1, 0, 1)$ and the coordinates of the first point-mass are $(-1, 0, 1)$. The next shift, defined by the next base in the chain, originates from the end-point of the previous shift. At its end the next point-mass is located. The procedure continues until the end of the chain. An example is shown in Fig. 1. The shape and the location in the space of the graph is specific for the sequence described by this graph.

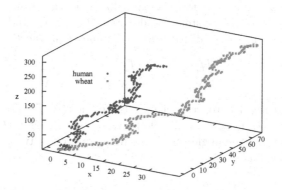

Fig. 1. 3D-dynamic graphs representing human and wheat histone H4 coding sequences.

For the numerical characterization of the sequences different kinds of characteristics called descriptors have been designed [1–3]:

- Coordinates of the centers of mass of the 3D-dynamic graphs (μ_x, μ_y, μ_z),
- Normalized principal moments of inertia of the graphs (r_1, r_2, r_3),
- Cosines of the angles between the planes C_{ij}, $i, j = 1, 2, 3$.

The coordinates of the center of mass of the 3D-dynamic graph, in the XYZ coordinate system are defined as

$$\mu_x = \frac{\sum_i m_i x_i}{\sum_i m_i}, \quad \mu_y = \frac{\sum_i m_i y_i}{\sum_i m_i}, \quad \mu_z = \frac{\sum_i m_i z_i}{\sum_i m_i}, \tag{1}$$

where (x_i, y_i, z_i) are the coordinates of the mass m_i. The tensor of the moment of inertia is defined by the matrix

$$\hat{I} = \begin{pmatrix} I_{xx} & I_{xy} & I_{xz} \\ I_{yx} & I_{yy} & I_{yz} \\ I_{zx} & I_{zy} & I_{zz} \end{pmatrix} \tag{2}$$

with

$$I_{aa} = \sum_i^N m_i \left[(b_i')^2 + (c_i')^2 \right], \quad I_{ab} = I_{ba} = -\sum_i^N m_i a_i' b_i', \tag{3}$$

where $\{a, b, c\} = \{x, y, z\}$, $a \neq b \neq c$. (x_i', y_i', z_i') are the coordinates of m_i in the Cartesian coordinate system with the origin selected at the center of mass, and N is the length of the sequence.

The *principal moments of inertia* are defined as the eigenvalues of $\hat{I}$

$$\hat{I}\omega_k = I_k \omega_k, \quad k = 1, 2, 3. \tag{4}$$

The eigenvectors ω_1, ω_2, ω_3 form a basis for a new coordinate system. The corresponding axes of this system are denoted Ω_1, Ω_2, Ω_3 and they are called *the principal axes*. The *normalized principal moments of inertia* are defined as

$$r_1 = \sqrt{\frac{I_1}{N}}, \quad r_2 = \sqrt{\frac{I_2}{N}}, \quad r_3 = \sqrt{\frac{I_3}{N}}. \tag{5}$$

Let M_1, M_2 and M_3 denote, respectively, the planes (X, Y), (X, Z) and (Y, Z). Analogously, Q_1, Q_2, Q_3 denote the planes (Ω_1, Ω_2), (Ω_1, Ω_3), (Ω_2, Ω_3), respectively. The cosines of the angles between the planes of the two systems of coordinates are defined as

$$C_{ij} \equiv \cos(M_i, Q_j), \quad i, j = 1, 2, 3. \tag{6}$$

Another computational method used for obtaining the similarity maps is the *correspondence analysis* [14–16]. The objects in these similarity maps are particular subgroups of individuals and their answers to some specific questions. The questions are related to the quality of life in different domains. The subgroups of individuals are the *retirees* and the *employees*. By such a choice of individuals one can study changes of the quality of life at the retirement threshold. The individuals were evaluated using the Polish version of the World Health Organization Quality of Life-BREF (WHOQOL-BREF) questionnaire [17,18]. The questionnaire is composed of 26 questions. Two questions are related to the Overall Quality of Life and the General Health. The remaining 24 questions concern four domains: *Physical Health (Domain 1)*; *Psychological (Domain 2)*; *Social*

Relationships (Domain 3); Environment (Domain 4). One could choose one of five answers to each question. Two answers that correspond to the least favorable quality of life we combined into one *negative* and two answers that correspond to the most favorable quality of life we denote as *positive*. As a consequence we consider 3-point scale of answers: *negative, neutral,* and *positive*. The group of *retirees* has been divided to two subgroups: the one of not attending lectures at the University of the Third Age (*retirees1*) and the one attending the University (*retirees2*).

3   Results and Discussion

Using the methods described in the previous section we have already obtained several similarity maps [1,3,7,8]. In the present work we continue the studies and show several new maps for the biological sequences and for the groups of individuals and their answers to the questions. We can see that the points representing specific objects cluster in a specific way.

In Figs. 2 and 3 histone H4 coding sequences available in the GenBank database for different species are used. These protein coding sequences are very similar for plants and for vertebrates and it is difficult to find a property which distinguishes between these two groups of sequences. Therefore these sequences may be used as a test for new methods. Our aim is to design descriptors which would be able to separate them. As we can see, the descriptors defined in Eq. (6) and used for the creation of the maps lead to a separation of the two subgroups of the sequences (the lines in the Figures separate the two subgroups). This proves the efficiency of the method. Analogous similarity maps for histone H4 coding sequences have also been obtained using the 2D-dynamic method [19].

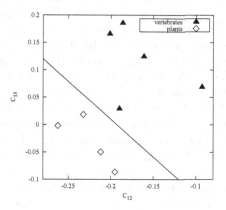

Fig. 2. Map obtained using the 3D-dynamic method: $C_{12} - C_{13}$.

As a consequence, good descriptors which lead to the classification of histone H4 coding sequences of plants and of vertebrates are, among others: the distribution moments, the moments of inertia, angles between the x axis and the principal axis of inertia of the graphs, or the cosines of the angles between the planes used in this work.

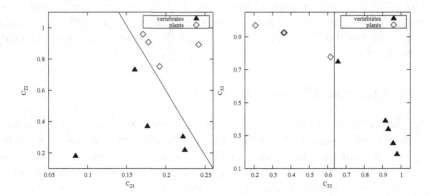

Fig. 3. Maps obtained using the 3D-dynamic method: $C_{21} - C_{22}$ (left panel), $C_{32} - C_{33}$ (right panel).

As far as the subgroups of individuals are concerned, the standard way of presenting the results is showing the percentage of answers to a given question by each subgroup. The numbers of answers (normalized to 1) to the questions by different subgroups are shown in the spine plots in Fig. 4. As we can see, the total number of positive answers is the largest, and the number of the negative answers is the smallest both for *married* and for *others*. We have performed the χ^2 test and the results are statistically significant: $p - value < 2.2 * 10^{-16}$ (Fig. 4, left panel) and $p - value = 1.05 * 10^{-11}$ (Fig. 4, right panel).

The histograms show only the number of answers, separately for each object. In the correspondence analysis particular objects are treated as a single system. For obtaining the maps the information about all objects is taken into account simultaneously. Therefore, by studying the structure of the maps we obtain the global description of the system.

In Figs. 5 and 6 two subgroups of individuals are represented by the correspondence analysis maps: *married* (left panels of the Figures) and *others* (right panels of the Figures). Comparing the structures of the Figures we can see some aspects of the differences between the estimations of the quality of life by different subgroups of the individuals. For example, the structures of the top left and the top right panels of Fig. 5 are similar. This means that the subgroups *married* and *others* estimate their quality of life in a similar way in the Physical Health domain (the marital status does not influence the quality of life in this domain). However, this factor is important in other domains (the structures of the Figures are different).

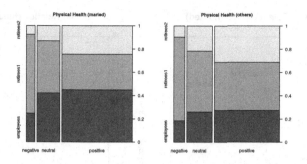

Fig. 4. Spine plots for the subgroups of individuals.

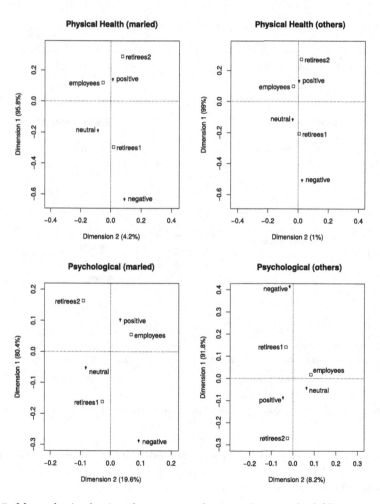

Fig. 5. Maps obtained using the correspondence analysis method (*Domains 1, 2*).

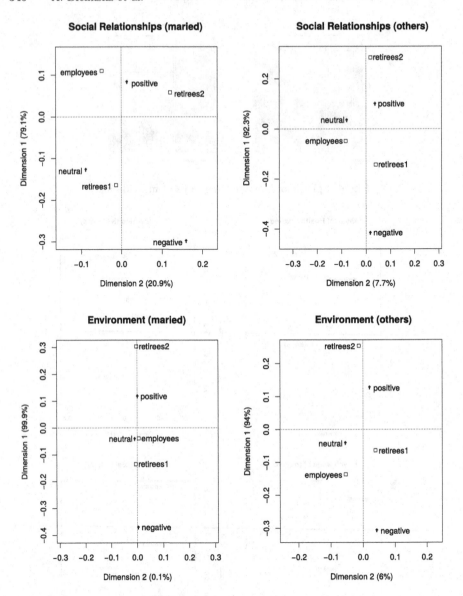

Fig. 6. Maps obtained using the correspondence analysis method (*Domains 3, 4*).

Summarizing, as we can see, the presented graphical representation of the results is helpful in the comparative studies performed on many diverse sets of objects in different areas of science.

References

1. Wąż, P., Bielińska-Wąż, D.: 3D-dynamic representation of DNA sequences. J. Mol. Model. **20**, 2141 (2014)
2. Wąż, P., Bielińska-Wąż, D.: Non-standard similarity/dissimilarity analysis of DNA sequences. Genomics **104**, 464–471 (2014)
3. Wąż, P., Bielińska-Wąż, D.: 3D-dynamic graphs as a classification tool of DNA sequences. In: Todorov, M.D. (ed.) 8th International Conference on Promoting the Application of Mathematics in Technical and Natural Sciences (AMiTaNS 2016), 22–27 June 2016, Albena, Bulgaria, AIP Conference Proceedings, vol. 1773, Art. no. 060007 (2016)
4. Wąż, P., Bielińska-Wąż, D., Pleskacz, A., Strobel, A.: Identification of stellar spectra using methods of statistical spectroscopy. Acta Phys. Pol. B **39**, 1993–2001 (2008)
5. Wąż, P., Bielińska-Wąż, D., Strobel, A., Pleskacz, A.: Statistical indicators of astrophysical parameters. Acta Astron. **60**, 283–293 (2010)
6. Bielińska-Wąż, D., Wąż, P., Basak, S.C.: Similarity studies using statistical and genetical methods. J. Math. Chem. **42**, 1003–1013 (2007)
7. Bielińska, A., Majkowicz, M., Wąż, P., Bielińska-Wąż, D.: Overall quality of life and general health - changes related to the retirement threshold. In: Murata, Y., et al. (ed.) eTELEMED 2018 The Tenth International Conference on eHealth, Telemedicine, and Social Medicine, 25–29 March 2018, Rome, Italy, pp. 1–5. XPS IARIA Press, Rome (2018)
8. Bielińska, A., Majkowicz, M., Wąż, P., Bielińska-Wąż, D.: Influence of the education level on health of elderly people. In: Murata, Y., et al. (ed.) eTELEMED 2018 The Tenth International Conference on eHealth, Telemedicine, and Social Medicine, 25–29 March 2018, Rome, Italy, pp. 6–11. XPS IARIA Press, Rome (2018)
9. Randić, M., Plavšić, D.: Novel spectral representation of RNA secondary structure without loss of information. Chem. Phys. Lett. **476**, 277–280 (2009)
10. Zhang, Z., et al.: ColorSquare: a colorful square visualization of DNA sequences. MATCH Commun. Math. Comput. Chem. **68**, 621–637 (2012)
11. Liao, B., Xiang, Q., Cai, L., Cao, Z.: A new graphical coding of DNA sequence and its similarity calculation. Phys. A **392**, 4663–4667 (2013)
12. Ping, P.Y., Zhu, X.Y., Wang, L.: Similarities/dissimilarities analysis of protein sequences based on PCA-FFT. J. Biol. Syst. **25**, 29–45 (2017)
13. Yang, L.P., Zhang, W.L.: A multiresolution graphical representation for similarity relationship and multiresolution clustering for biological sequences. J. Comput. Biol. **24**, 299–310 (2017)
14. Hirschfeld, H.O.: A connection between correlation and contingency. Proc. Camb. Philos. Soc. **31**, 520–524 (1935)
15. Benzécri, J.-P.: L'Analyse des Données. L'Analyse des Correspondances, vol. II, Paris, France, Dunod (1973)
16. Greenacre, M.: Correspondence Analysis in Practice, 2nd edn. Chapman & Hall/CRC, London (2007)
17. http://www.who.int/substance_abuse/research_tools/whoqolbref/en/
18. Wołowicka, L.: Quality of Life in Medical Sciences, pp. 231–281. Wydawnictwo Uczelniane AM w Poznaniu, Poznań (2001)
19. Wąż, P., Bielińska-Wąż, D., Nandy, A.: Descriptors of 2D-dynamic graphs as a classification tool of DNA sequences. J. Math. Chem. **52**, 132–140 (2014)

Model-Based Stabilization of a Fermentation Process Using Output Feedback with Discrete Time Delay

Milen K. Borisov[1(✉)], Neli S. Dimitrova[1], and Mikhail I. Krastanov[1,2]

[1] Institute of Mathematics and Informatics, Bulgarian Academy of Sciences,
Acad. G. Bonchev Str. Bl. 8, 1113 Sofia, Bulgaria
{milen_kb,nelid}@math.bas.bg
[2] Faculty of Mathematics and Informatics, Sofia University,
"St. Kl. Ohridski" 5 James Bourchier Blvd., 1164 Sofia, Bulgaria
krastanov@fmi.uni-sofia.bg

Abstract. The present study is devoted to the stabilization of a bioreactor model, describing an anaerobic fermentation process for biological degradation of organic wastes with methane production. The stabilization is realized by means of a feedback control law related to the model output and involving a discrete time delay. We determine a nontrivial equilibrium point of the closed-loop system and investigate its asymptotic stability as well as the appearance of bifurcations with respect to the delay parameter. We establish the existence of an attracting and invariant region around the equilibrium such that all trajectories enter this region in finite time for some values of the delay and remain there. An iterative numerical extremum seeking algorithm is applied to the closed-loop system aimed to maximize the methane flow rate in real time. Simulation results are presented to illustrate the theoretical studies.

1 Introduction

Delayed mathematical models of bioprocesses have been extensively studied in recent years in order to explain the appearance of different phenomena in the real process, cf. [8,9] and the references therein. In the same time, feedback control of bioreactor models provides many advantages in operating a plant and is used to increase its efficiency. In the present paper we combine the above mentioned approaches in studying a bioprocess mathematical model, namely we use a feedback control related to the (on-line measurable) process output for the dynamics stabilization, and introduce a discrete time delay. The time delay is involved in the feedback, because there is always a delay between the output measurements and the system's response (cf. [4]).

We consider one of the bench-mark mathematical models of the continuous methane fermentation, the so-called "single biomass/single substrate" model.

© Springer Nature Switzerland AG 2019
G. Nikolov et al. (Eds.): NMA 2018, LNCS 11189, pp. 342–350, 2019.
https://doi.org/10.1007/978-3-030-10692-8_38

It is described by two nonlinear ordinary differential equations

$$\frac{ds}{dt} = -k_1\mu(s)x + u(s_{in} - s)$$

$$\frac{dx}{dt} = (\mu(s) - \alpha u)x$$

(1)

and one algebraic equation for the gaseous output

$$Q(s, x) = k_2\mu(s)x.$$

(2)

The state variables $x = x(t)$ and $s = s(t)$ represent biomass concentration [mg/dm^3] and substrate concentration [mg/dm^3] respectively, s_{in} is influent substrate concentration [mg/dm^3], u is dilution rate [day^{-1}], k_1 is yield coefficient, k_2 is coefficient [(dm^3)^2/mg], and Q is methane gas flow rate [dm^3/day]. The parameter $\alpha \in (0, 1)$ accounts for the biomass retention. The model function $\mu(s)$ presents the specific growth rate of the biomass.

The paper is organized as follows. The next Sect. 2 contains the assumptions imposed on the model. Section 3 is devoted to stability and bifurcation analysis of the equilibrium points with respect to the delay parameter. In Sect. 4 we proof the existence of an attracting and invariant set in the phase plane, such that all trajectories enter it in finite time and remain there for sufficiently small values of the delay. The last Sect. 5 demonstrates the applicability of the model-based extremum seeking algorithm using a numerical example.

2   Assumptions on the Model

The theoretical studies of the model (1) are carried out under several assumptions presented below.

Assumption 1. The function μ is defined for $s \in [0, +\infty)$, $\mu(0) = 0$, $\mu(s) > 0$ for $s > 0$, and $\mu(s)$ is continuously differentiable for all $s \geq 0$.

Assumption 2. Lower bounds s_{in}^- and k_2^- for the values of s_{in} and k_2 respectively, and an upper bound k_1^+ for the value of k_1 are known.

Denote $\beta^- = \dfrac{k_1^+}{k_2^- s_{in}^-}$ and consider the feedback control law

$$\kappa(s(t), x(t)) = \beta\, k_2\, \mu(s(t))\, x(t) \quad \text{with} \quad \beta \in (\beta^-, +\infty).$$

(3)

Obviously, $\kappa(\cdot) = \beta Q(\cdot)$ holds true. Replacing in the model (1) the dilution rate u by the feedback $\kappa(s(t-\tau), x(t-\tau))$, where $\tau > 0$ is a discrete delay, we obtain

$$\frac{ds}{dt} = -k_1\mu(s(t))x(t) + \kappa(s(t-\tau), x(t-\tau))(s_{in} - s(t))$$

(4)

$$\frac{dx}{dt} = \mu(s(t))x(t) - \alpha\kappa(s(t-\tau), x(t-\tau))x(t).$$

(5)

Choose some $\beta \in (\beta^-, +\infty)$ and define

$$\bar{s} = s_{in} - \frac{k_1}{k_2 \beta}, \quad \bar{x} = \frac{1}{\alpha \beta k_2}, \quad \bar{p}_\beta = (\bar{s}, \bar{x}). \tag{6}$$

It is straightforward to see that $\bar{p}_\beta$ is an equilibrium point for (4)–(5), and $\bar{s}$ belongs to the interval $(0, s_{in})$.

Assumption 3. There exist points s^- and s^+ such that $s^- < \bar{s} < s^+ < s_{in}$ and

(i) the function μ is strictly increasing on the interval $(s^-, s^+]$;
(ii) $\mu(s) < \mu(s^-) < \mu(s_{in})$ for each $s \in [0, s^-)$;
(iii) there exists $\varepsilon > 0$ such that $\mu(s^+) < \mu(s)$ for each $s \in (s^+, s^+ + \varepsilon)$.

Denote

$$u^- = \mu(s^-)/\alpha, \quad u^+ = \mu(s^+)/\alpha. \tag{7}$$

Assumption 3 implies that $u^- < u^+$.

Assumption 4. Each point from the interval $[u^-, u^+]$ is an admissible value for the control function u.

Denote further $\bar{u} = \kappa(\bar{s}, \bar{x}) = \mu(\bar{s})/\alpha$; obviously $\bar{u} \in (u^-, u^+)$ holds true.

3   Stability and Bifurcations of the Equilibrium Point

We shall investigate the local asymptotic stability of the equilibrium point $\bar{p}_\beta$ from (6) with respect to the parameters of the system (4)–(5).

The characteristic equation of system (4)–(5) evaluated at the equilibrium point $\bar{p}_\beta$ has the form (cf. [5,7])

$$\lambda^2 + a\lambda + b + (c\lambda + d)e^{-\lambda \tau} = 0, \tag{8}$$

where λ is a complex number, and

$$a = a(\beta) = k_1 \bar{x} \mu'(\bar{s}) + \frac{1}{\alpha}\mu(\bar{s}), \quad b = b(\beta) = k_1 \bar{x} \mu(\bar{s})\mu'(\bar{s}),$$

$$c = c(\beta) = \mu(\bar{s}) - k_1 \bar{x}\mu'(\bar{s}), \quad d = d(\beta) = \mu(\bar{s})\left(\frac{1}{\alpha}\mu(\bar{s}) - k_1\bar{x}\mu'(\bar{s})\right).$$

Theorem 1. *Let Assumptions 1, 2 and 3 be satisfied. (i) If $b \geq d$ then the equilibrium point $\bar{p}_\beta$ is locally asymptotically stable for any value of the delay $\tau \geq 0$. (ii) If $b < d$ then there exists $\tau_0 > 0$ such that the equilibrium point $\bar{p}_\beta$ is locally asymptotically stable for all values τ such that $0 < \tau < \tau_0$; the equilibrium is locally unstable if $\tau \geq \tau_0$, and a Hopf bifurcation occurs for $\tau = \tau_0$.*

Proof. First we shall show that if $\tau = 0$ then the characteristic equation does not possess roots λ with nonnegative real part. For $\tau = 0$ Eq. (8) takes the form

$$\lambda^2 + (a + c)\lambda + b + d = 0. \tag{9}$$

Assumption 3(i) implies $a > 0$ and $b > 0$. Since $a + c = \left(1 + \frac{1}{\alpha}\right)\mu(\bar{s}) > 0$ and $b + d = \frac{1}{\alpha}\mu^2(\bar{s}) > 0$ it follows that the roots of the quadratic Eq. (9) have negative real parts.

Let $\tau > 0$. We are looking for purely imaginary roots $\lambda = \pm i\omega$ of (8) with $\omega > 0$. We obtain consecutively:

$$-\omega^2 + ai\omega + b + (ci\omega + d)e^{-i\omega\tau} = 0,$$
$$-\omega^2 + ai\omega + b + (ci\omega + d)(\cos(\omega\tau) - i\sin(\omega\tau)) = 0.$$

Separating the real and the imaginary parts of the last equation leads to

$$-\omega^2 + b = -c\,\omega\sin(\omega\tau) - d\cos(\omega\tau)$$
$$a\,\omega = -c\,\omega\cos(\omega\tau) + d\sin(\omega\tau). \tag{10}$$

Squaring both sides of the Eq. (10) and adding them together implies

$$\omega^4 - (c^2 - a^2 + 2b)\omega^2 + b^2 - d^2 = 0.$$

With $v := \omega^2$ we obtain the quadratic equation

$$v^2 - (c^2 - a^2 + 2b)v + b^2 - d^2 = 0. \tag{11}$$

It is straightforward to see that the discriminant $\Delta = (c^2 - a^2)(c^2 - a^2 + 4b) + 4d^2$ of (11) is strongly positive, i. e. the quadratic Eq. (11) possesses two real roots v_1 and v_2, say $v_1 < v_2$, satisfying the relations $v_1 + v_2 = c^2 - a^2 + 2b < 0$, $v_1 v_2 = b^2 - d^2 = (b - d)(b + d)$.

Case 1: $0 < b = d$. In this case $v_1 = c^2 - a^2 + 2b < 0$ and $v_2 = 0$, thus the characteristic Eq. (8) does not possess purely imaginary roots for any $\tau > 0$. However, $\lambda = 0$ is not a root of (8) since $b + d = 2b > 0$ holds. Hence, there is no stability switch of the equilibrium point $\bar{p}_\beta$ for any $\tau > 0$.

Case 2: $b > d$. Now the two real roots v_1 and v_2 are strongly negative, so the characteristic Eq. (8) does not have purely imaginary roots for any $\tau > 0$. The equilibrium point $\bar{p}_\beta$ is locally asymptotically stable for any $\tau > 0$.

Case 3: $0 < b < d$. Equation (11) has one negative and one positive root; the positive root is $v_2 = \frac{1}{2}\left(c^2 - a^2 + 2b + \sqrt{\Delta}\right)$, i. e. the characteristic Eq. (8) possesses a purely imaginary root when τ takes certain values. Denoting $\omega_+ = \sqrt{v_2}$, these values of τ can be determined from system (10):

$$\sin(\omega_+\tau) = \frac{\omega_+(c\omega_+ + ad - bc)}{c^2\omega_+^2 + d^2}, \quad \cos(\omega_+\tau) = \frac{(d - ac)\omega_+^2 - bd}{c^2\omega_+^2 + d^2}. \tag{12}$$

If $c < 0$ then $ad - bc > 0$; if $c > 0$ then $ad - bc > d(a - c) > 0$ holds. Therefore, we have $\sin(\omega_+\tau) > 0$. Denote $\theta = \omega_+\tau$, $0 < \theta < \pi$. If $\cos(\theta) > 0$, then we take $0 < \theta < \pi/2$, otherwise we take $\pi/2 < \theta < \pi$. Hence,

$$\theta = \operatorname{arccot}\frac{(d - ac)\omega_+^2 - bd}{\omega_+(c\omega_+^2 + ad - bc)}.$$

Denote $\tau_0 = \dfrac{\theta}{\omega_+} > 0$. We shall see whether a Hopf bifurcation occurs at $\tau = \tau_0$.

To check the transversality condition for a Hopf bifurcation (cf. [5,7]), we need to determine the sign of the derivative of $\mathrm{Re}\,\lambda(\tau)$ at the point where $\lambda(\tau)$ is purely imaginary. Differentiating implicitly (8) with respect to τ we obtain

$$\frac{d\lambda}{d\tau} = \frac{(c\lambda + d)\lambda e^{-\lambda\tau}}{2\lambda + a + (c - (c\lambda + d)\tau)e^{-\lambda\tau}}.$$

For convenience, we shall study the sign of $\left(\dfrac{d\lambda}{d\tau}\right)^{-1}$. We have

$$\left(\frac{d\lambda}{d\tau}\right)^{-1} = \frac{2\lambda + a + (c - (c\lambda + d)\tau)e^{-\lambda\tau}}{(c\lambda + d)\lambda e^{-\lambda\tau}} = \frac{(2\lambda + a)e^{\lambda\tau} + c}{\lambda(c\lambda + d)} - \frac{\tau}{\lambda}$$

$$= -\frac{2\lambda + a}{\lambda(\lambda^2 + a\lambda + b)} + \frac{c}{\lambda(c\lambda + d)} - \frac{\tau}{\lambda},$$

and further

$$\mathrm{sign}\left(\frac{d\,(\mathrm{Re}\,\lambda)}{d\tau}\right)_{\lambda = i\omega_+} = \mathrm{sign}\left\{\mathrm{Re}\left(\frac{d\lambda}{d\tau}\right)^{-1}\right\}_{\lambda = i\omega_+}$$

$$= \mathrm{sign}\left\{\mathrm{Re}\left(-\frac{2\lambda + a}{\lambda(\lambda^2 + a\lambda + b)}\right) + \mathrm{Re}\left(\frac{c}{\lambda(c\lambda + d)}\right)\right\}_{\lambda = i\omega_+}$$

$$= \mathrm{sign}\left\{\mathrm{Re}\left(-\frac{(2i\omega_+ + a)(-a\omega_+^2 - i\omega_+(b - \omega_+^2))}{\omega_+^2(a^2\omega_+^2 + (b - \omega_+^2)^2)}\right) + \mathrm{Re}\left(\frac{c\omega_+(-c\omega_+ - id)}{\omega_+^2(c^2\omega_+^2 + d^2)}\right)\right\}$$

$$= \mathrm{sign}\left\{\frac{a^2 - 2(b - \omega_+^2)}{a^2\omega_+^2 + (b - \omega_+^2)^2} - \frac{c^2}{c^2\omega_+^2 + d^2}\right\} = \mathrm{sign}\left\{2\omega_+^2 - (c^2 - a^2 + 2b)\right\} = +1.$$

The last result means that all roots that cross the imaginary axis at $i\omega_+$, cross this axis from left to right as τ increases. The proof is completed.

4 Asymptotic Stabilization of the Model Solutions

In practice, the dilution rate u is proportional to the speed of the input mechanism which feeds the bioreactor with substrate. Thus u is always lower- and upper-bounded [3]. Let u^- and u^+ be determined according to (7).

Define the set

$$\Omega = \{\zeta = (s, x) : s > 0, \; x > 0\}.$$

Let $\tau > 0$ and $\zeta^0 = (s^0, x^0) \in \Omega$ be an arbitrary point such that $s(t) = s^0 > 0$, $x(t) = x^0 > 0$ for each $t \in [-\tau, 0]$. Consider the following closed-loop system Σ

$$\dot{s}(t) = -k_1\mu(s(t))x(t) + \chi(t)(s_{in} - s(t)) \tag{13}$$

$$\dot{x}(t) = (\mu(s(t)) - \alpha\chi(t))x(t), \tag{14}$$

where $\chi(t)$ is defined in the following way:

$$\chi(t) = \begin{cases} u^-, & \text{if } \kappa(s(t-\tau), x(t-\tau)) \leq u^-, \\ \kappa(s(t-\tau), x(t-\tau)), & \text{if } u^- \leq \kappa(s(t-\tau), x(t-\tau)) \leq u^+, \\ u^+, & \text{if } \kappa(s(t-\tau), x(t-\tau)) \geq u^+. \end{cases} \quad (15)$$

Obviously, $\bar{p}_\beta = (\bar{s}, \bar{x})$ is an equilibrium point of Σ, i. e. of (13)–(14). Denote by $\varphi(\cdot, \zeta^0) = (s(\cdot), x(\cdot))$ the solution of Σ starting from ζ^0. Important properties of $\varphi(\cdot, \zeta^0)$ are given in the next Lemma 1. Similar assertions can be found e.g. in [3,8,9] for various bioreactor models.

Lemma 1. *For each point $\zeta^0 = (s^0, x^0) \in \Omega$ the solution $\varphi(t, \zeta^0) = (s(t), x(t))$ of Σ is defined for each $t > 0$, and*

(i) for each $\varepsilon_1 > 0$ there exists $T_1 > 0$ such that for each $t > T_1$ the inequalities $s_{in} - \varepsilon_1 < s(t) + k_1 x(t) < s_{in}/\alpha + \varepsilon_1$ hold true.

(ii) there exist $\varepsilon_2 > 0$ and $T_2 > 0$ such that for each $t > T_2$ the estimates $s(t) < s_{in}$ and $x(t) \geq \varepsilon_2/k_1 =: x_{\min} > 0$ hold true.

For the proof of the next theorem we need the following lemma.

Barbălat's Lemma (cf. [2]). *If $f : (0, \infty) \to R$ is Riemann integrable and uniformly continuous, then $\lim_{t\to\infty} f(t) = 0$.*

Theorem 2. *Let Assumptions 1, 2, 3 and 4 be fulfilled. Then there exists $\bar{\tau} > 0$ such that for each $\tau \in (0, \bar{\tau})$ and for each point $\zeta^0 = (s^0, x^0) \in \Omega$ the solution $\varphi(t, \zeta^0)$ of Σ has the following property: there exists $T > 0$ such that for each $t > T$,*

$$\varphi(t, \zeta^0) \in \Omega_{s^-, s^+} := \{(s, x) : s \in [s^-, s^+], x > 0\}.$$

Proof. Let us fix an arbitrary $\tau > 0$ and assume that $s(t) \leq s^-$ for each $t \geq 0$. Then Assumption 3(ii) and (7) imply that $\mu(s(t)) \leq \mu(s^-) = \alpha u^-$. The definition of $\chi(\cdot)$ implies $\chi(t) \geq u^-$ for each $t \geq 0$. Then $\mu(s(t)) - \alpha\chi(t) \leq \mu(s(t)) - \alpha u^- \leq 0$ for each $t \geq 0$, and hence $\dot{x}(t) = (\mu(s(t)) - \alpha\chi(t))x(t) \leq 0$. Thus the function $x(\cdot)$ is non increasing and there exists $\tilde{x} = \lim_{t\to\infty} x(t)$. According to Barbălat's Lemma, we obtain that $\dot{x}(t) \to 0$ as $t \to +\infty$. Since $x(t) \geq x_{\min} > 0$ (see Lemma 1(ii)), Eq. (14) implies that $(\mu(s(t)) - \alpha u^-) + \alpha(u^- - \chi(t)) \to 0$ as $t \to +\infty$. The last relation leads to

$$\mu(s(t)) \to \alpha u^- \text{ and } \chi(t) \to u^- \text{ as } t \to +\infty.$$

Applying again Assumption 3(ii) we obtain that $s(t) \to s^-$ as $t \to +\infty$. It follows from Barbălat's Lemma that $\dot{s}(t) = \chi(t)(s_{in} - s(t)) - k_1\mu(s^-)x(t) \to 0$ as $t \to \infty$, and hence

$$u^-(s_{in} - s^-) - k_1\mu(s^-)\tilde{x} = 0, \text{ i. e. } u^-(s_{in} - s^-) - \alpha k_1 u^- \tilde{x} = 0.$$

Therefore, $s_{in} = s^- + \alpha k_1 \tilde{x}$. We also have $\chi(t) \to u^-$ as $t \to +\infty$. This is possible iff for each $\varepsilon > 0$ there exists $T_\varepsilon > 0$ such that $\kappa(s(t-\tau), x(t-\tau)) < u^- + \varepsilon$ for each $t > T_\varepsilon$ for which $\kappa(s(t-\tau), x(t-\tau)) > u^-$ (if $\kappa(s(t-\tau), x(t-\tau)) \leq u^-$, then $\chi(t) = u^-$). Then for each $t > T_\varepsilon$ we have

$$u^- + \varepsilon > \kappa(s(t-\tau), x(t-\tau)) = \beta k_2 \mu(s(t-\tau)) x(t-\tau) = \frac{\mu(s(t-\tau)) x(t-\tau)}{\alpha \bar{x}}.$$

Taking a limit in the latter inequality we obtain $\dfrac{\mu(s^-)\tilde{x}}{\alpha\bar{x}} \leq u^-$, i. e. $\dfrac{\alpha u^- \tilde{x}}{\alpha\bar{x}} \leq u^-$

or $\tilde{x} \leq \bar{x}$. But this is impossible because $\tilde{x} = \dfrac{s_{in} - s^-}{\alpha k_1} > \dfrac{s_{in} - \bar{s}}{\alpha k_1} = \bar{x}$.

Assuming that $s(t) \geq s^+$ for each $t \geq 0$, we obtain a contradiction in a similar way.

If $\tau > 0$ is sufficiently small, one can easily apply the Lyapunov functions approach to prove that $(s(t), x(t))$ tends to $(\bar{s}, \bar{x})$ as $t \to \infty$. This completes the proof of Theorem 2.

Remark 1. It follows from Theorem 2 that the feedback (15) ensures attractivity and invariance of the region Ω_{s^-,s^+} for some values of the time delay $\tau \geq 0$. If s^- and s^+ are sufficiently close to each other and $\bar{p}_\beta = (\bar{s}, \bar{x})$ is locally asymptotically stable, then the trajectories remain close to $\bar{p}_\beta$ because $\bar{s} \in (s^-, s^+)$ holds true. The existence of Ω_{s^-,s^+} is important for the practical applications (cf. [3]) and we shall exploit it in the next section.

5   Numerical Extremum Seeking

Consider the Haldane model function for the specific growth rate (cf. [6])

$$\mu(s) = \frac{m_1 s}{k_s + s + s^2/k_i},$$

where m_1 is the maximum specific growth rate of the microorganisms [1/day], k_s and k_i are the saturation and inhibition constants respectively. We use the following values for the model coefficients (cf. [6]):

$$k_1 = 3, \; s_{in} = 2, \; m_1 = 0.35, \; k_s = 0.7, \; k_i = 0.6, \; \alpha = 0.5, \; k_2 = 5.6.$$

With $k_1^+ = 3.1$, $s_{in}^- = 1.95$ and $k_2^- = 5.59$ we obtain $\beta^- \approx 0.2844$.

The function $\mu(s)$ achieves its maximum at the point $s_{\mu_{max}} = \sqrt{k_s k_i} \approx 0.6481$ and $\mu(s)$ is strongly increasing for $s \in (0, s_{\mu_{max}})$. Solving the equation $\bar{s} = s_{\mu_{max}}$ with respect to β implies $\beta = \beta_{\mu_{max}} \approx 0.3963$. Since $\bar{s}$ is an increasing function of β, it suffice to consider $\beta \in (\beta^-, \beta_{\mu_{max}})$ in order to have Assumption 3 satisfied.

Consider Eq. (2) describing the process output, and evaluate the function Q on the set of all equilibrium points $\bar{p}_\beta$, parameterized with respect to β. The so obtained function $Q(\beta)$ is called input-output static characteristic of the model. It is straightforward to see that $Q(\beta)$ is strongly unimodal, i. e. there exists a unique point $\beta_{max} \approx 0.3411 < \beta_{\mu_{max}}$, where $Q(\beta)$ takes a maximum, $Q_{max} = Q(\beta_{max})$, the function strongly increases in the interval (β^-, β_{max}) and strongly decreases in $(\beta_{\max}, \beta_{\mu_{\max}})$. Denote by $p_{\beta_{\max}} = (s_{\max}, x_{\max})$ the steady state where $Q_{\max}$ is achieved. We have $s_{\max} \approx 0.4294$, $x_{\max} \approx 1.047$, and $Q_{\max} = Q(p_{\beta_{\max}}) \approx 0.6134$. Our goal is to stabilize in real time the system (13)–(14) towards this (unknown) equilibrium point $p_{\beta_{\max}}$ and therefore to the maximum methane flow rate. This is realized by applying a numerical model-based extremum seeking algorithm (ESA). The ESA is described in details in [1] for the same model (1) with another feedback and without delay. Now ESA is adopted to Σ and implemented in the programme language *Python*.

In the simulation process we consider $\beta \in (0.29, 0.39)$ and take $s^- = 0.1527$, $u^- = 0.1199$, $s^+ = 0.6264 < s_{\mu_{\max}}$, $u^+ = 0.2214$; obviously $s_{\max} \in (s^-, s^+)$. We choose $\beta = 0.37$. According to Theorem 1(ii) a stability switch of $\bar{p}_\beta$ may occur at $\tau_0 = 121$ [days], and $\bar{p}_\beta$ is locally asymptotically stable if $\tau < \tau_0$. The delay $\tau_0 = 121$ [days] is however rather large and not feasible in actual practice. The numerical results from ESA are visualized in Fig. 1 for $\tau = 4$ and $\tau = 7$.

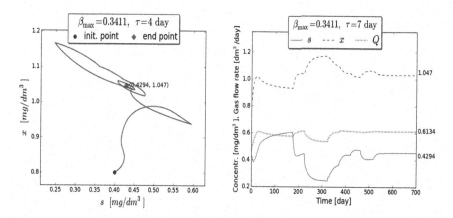

Fig. 1. A trajectory in the (s, x) phase plane for $\tau = 4$ (left); time evolution of $s(t)$, $x(t)$, $Q(t)$ towards $s_{\max}$, $x_{\max}$, $Q_{\max}$ respectively for $\tau = 7$ (right)

Acknowledgements. The work of the first and the second author has been partially supported by the Bulgarian Academy of Sciences, the Program for Support of Young Scientists and Scholars, grant No. DFNP-17-25/25.07.2017. The work of the third author has been partially supported by the Sofia University "St. Kl. Ohridski" under contract No. 80-10-133/25.04.2018.

References

1. Dimitrova, N., Krastanov, M.: Nonlinear adaptive control of a model of an uncertain fermentation process. Int. J. Robust Nonlinear Control **20**, 1001–1009 (2010)
2. Gopalsamy, K.: Stability and Oscillations in Delay Differential Equations of Population Dynamics. Kluwer Academic Publishers, Dordrect (1992)
3. Grognard, F., Bernard, O.: Stability analysis of a wastewater treatment plant with saturated control. Water Sci. Technol. **53**(1), 149–157 (2006)

4. Robledo, G.: Feedback stabilization for a chemostat with delayed output. Preprint INRIA Sophia Antipolis RR No 5844 (2004)
5. Ruan, S.: On nonlinear dynamics of predator-prey models with discrete delay. Math. Model. Nat. Phenom. **4**(2), 140–188 (2009)
6. Simeonov, I.S.: Mathematical modeling and parameters estimation of anaerobic fermentation process. Bioprocess Eng. **21**, 377–381 (1999)
7. Smith, H.: An Introduction to Delay Differential Equations with Applications to the Life Sciences. Springer, New York (2011)
8. Wolkowicz, G.S.K., Xia, H.: Global asymptotic behaviour of a chemostat model with discrete delays. SIAM J. Appl. Math. **57**, 1281–1310 (1997)
9. Xia, H., Wolkowicz, G.S.K., Wang, L.: Transient oscillations induced by delayed growth response in the chemostat. J. Math. Biol. **50**, 489–530 (2005)

Some Convergence Results of a Multidimensional Finite Volume Scheme for a Semilinear Parabolic Equation with a Time Delay

Abdallah Bradji[1]([⊠]) and Tarek Ghoudi[2]

[1] Department of Mathematics, Faculty of Sciences,
University of Annaba, Annaba, Algeria
abdallah.bradji@gmail.com, abdallah.bradji@etu.univ-amu.fr
[2] LAGA (Laboratoire d'Analyse, Géométrie, et Applications),
University of Paris 13, Paris, France
tarekghoudi@gmail.com, ghoudi@math.univ-paris13.fr
https://www.i2m.univ-amu.fr/~bradji/

Abstract. Delay differential equations occur in many applications such as ecology and biology. They have long played important roles in the literature of theoretical population dynamics, and they have been continuing to serve as useful models.

There is a huge literature on the approximation of ODDEs (Ordinary Delay Differential Equations) whereas a few contributions, w.r.t. ODDEs, dealt with DPDEs (Delay Partial Differential Equations). Some of these works which dealt with the numerical approximation of DPDEs consider only the one dimensional case.

In this contribution we construct a linearized implicit scheme, in which the space discretization is performed using a general class of nonconforming finite volume meshes, to approximate a semilinear parabolic equation with a time delay. We prove the existence and uniqueness of the discrete solution. We derive a discrete *a priori estimate* which allows to derive error estimates in discrete seminorms of $L^\infty(H_0^1)$ and $W^{1,2}(L^2)$.

Keywords: Delay equation · SUSHI scheme · Discrete gradient

1 Description of the Model, Aim of This Contribution, and Motivation

We consider the following semilinear parabolic equation with a delay in time:

$$u_t(\boldsymbol{x}, t) - \Delta u(\boldsymbol{x}, t) = f(\boldsymbol{x}, t, u(\boldsymbol{x}, t), u(\boldsymbol{x}, t - \tau)), \ (\boldsymbol{x}, t) \in \Omega \times (0, T), \quad (1)$$

where Ω is an open polygonal bounded subset in $\mathbb{R}^d$, f is a given function, and $T > 0$ and $\tau > 0$ are given.

Initial condition is given by, for a given function u^0 defined on Ω

$$u(\boldsymbol{x}, t) = u^0(\boldsymbol{x}, t), \quad \boldsymbol{x} \in \Omega, \quad -\tau \le t \le 0, \quad (2)$$

© Springer Nature Switzerland AG 2019
G. Nikolov et al. (Eds.): NMA 2018, LNCS 11189, pp. 351–359, 2019.
https://doi.org/10.1007/978-3-030-10692-8_39

Homogeneous Dirichlet boundary conditions are given by

$$u(\boldsymbol{x}, t) = 0, \quad (\boldsymbol{x}, t) \in \partial\Omega \times (0, T). \tag{3}$$

The following assumption on the function f is needed in our analysis:

Assumption 1 (Assumption on f) *We assume that the function $f(\boldsymbol{x}, t, s, r)$ is Lipschitz continuous with respect to (s, r) with constant κ, i.e.*

$$|f(\boldsymbol{x}, t, s, r) - f(\boldsymbol{x}, t, s', r')| \leq \kappa\left(|s - s'| + |r - r'|\right), \quad \forall(\boldsymbol{x}, t, s, r), (\boldsymbol{x}, t, s', r') \in \Omega \times \mathbb{R}^3.$$

Note that the Assumption 1 is satisfied when for instance $f \in \mathscr{C}^1(\Omega \times \mathbb{R}^3)$ and

$$\sup_{\Omega \times \mathbb{R}^3} \left|\frac{\partial f}{\partial s}\right| + \left|\frac{\partial f}{\partial r}\right| \leq \kappa.$$

One dimensional case of Eq. (1) (i.e. $d = 1$) is considered in [7] where a compact multisplitting finite difference scheme is analyzed. Ordinary Differential version for the Eq. (1), i.e. without a consideration for the spacial domain Ω, is considered for instance in [6] where discontinuous Galerkin methods are applied.

We first establish a linear numerical scheme in which the space discretization is performed using the SUSHI method [4] whereas the time stepping is Euler implicit. We then provide a convergence analysis for this scheme. The SUSHI developed in [4] refers to the "Scheme Using Stabilization and Hybrid Interfaces" uses general nonconforming meshes in which the control volumes can only be assumed to be polyhedral (the boundary of each control volume is a finite union of subsets of hyperplanes). In addition to this, the formulation of SUSHI involves a consistent and stable Discrete Gradient.

Delay differential equations occur in several applications such as ecology, biology, medicine, see [1,5,7] for more details. However, the numerical methods which are carried out with Partial (or Ordinary) Differential Equations are not enough to deal with Delay Partial Differential Equations. Indeed, the implementation of schemes and some desirable accuracy and stability results which are known for Partial Differential Equations can be destroyed when applying these methods to Delay Partial Differential Equations, cf. [1, Pages 9–19].

In addition, numerical methods for the delay equations are well developed for the case of Ordinary Differential Equations but the subject of numerical analysis for Delay Partial Differential Equations has not attracted the attention it merits yet, see [1,7].

To the best of our knowledge, we are not aware of the existing of any previous work which deals with Finite Volume methods for Delay Partial Differential equations.

2   Space and Time Discretizations and Some Preliminaries

Definition 1 (Space discretization, cf. [4]). *Let Ω be a polyhedral open bounded subset of $\mathbb{R}^d$, where $d \in \mathbb{N} \setminus \{0\}$, and $\partial\Omega = \overline{\Omega} \setminus \Omega$ its boundary. A discretization of Ω, denoted by $\mathscr{D}$, is defined as the triplet $\mathscr{D} = (\mathscr{M}, \mathscr{E}, \mathscr{P})$, where:*

1. *$\mathscr{M}$ is a finite family of non empty connected open disjoint subsets of Ω (the "control volumes") such that $\overline{\Omega} = \cup_{K \in \mathscr{M}} \overline{K}$. For any $K \in \mathscr{M}$, let $\partial K = \overline{K} \setminus K$ be the boundary of K; let $\mathrm{m}(K) > 0$ denote the measure of K and h_K denote the diameter of K.*
2. *$\mathscr{E}$ is a finite family of disjoint subsets of $\overline{\Omega}$ (the "edges" of the mesh), such that, for all $\sigma \in \mathscr{E}$, σ is a non empty open subset of a hyperplane of $\mathbb{R}^d$, whose $(d-1)$–dimensional measure is strictly positive. We also assume that, for all $K \in \mathscr{M}$, there exists a subset $\mathscr{E}_K$ of $\mathscr{E}$ such that $\partial K = \cup_{\sigma \in \mathscr{E}_K} \overline{\sigma}$. For any $\sigma \in \mathscr{E}$, we denote by $\mathscr{M}_\sigma = \{K, \sigma \in \mathscr{E}_K\}$. We then assume that, for any $\sigma \in \mathscr{E}$, either $\mathscr{M}_\sigma$ has exactly one element and then $\sigma \subset \partial\Omega$ (the set of these interfaces, called boundary interfaces, denoted by $\mathscr{E}_{\mathrm{ext}}$) or $\mathscr{M}_\sigma$ has exactly two elements (the set of these interfaces, called interior interfaces, denoted by $\mathscr{E}_{\mathrm{int}}$). For all $\sigma \in \mathscr{E}$, we denote by $\boldsymbol{x}_\sigma$ the barycentre of σ. For all $K \in \mathscr{M}$ and $\sigma \in \mathscr{E}$, we denote by $\boldsymbol{n}_{K,\sigma}$ the unit vector normal to σ outward to K.*
3. *$\mathscr{P}$ is a family of points of Ω indexed by $\mathscr{M}$, denoted by $\mathscr{P} = (\boldsymbol{x}_K)_{K \in \mathscr{M}}$, such that for all $K \in \mathscr{M}$, $\boldsymbol{x}_K \in K$ and K is assumed to be $\boldsymbol{x}_K$–star-shaped, which means that for all $\boldsymbol{x} \in K$, the property $[\boldsymbol{x}_K, \boldsymbol{x}] \subset K$ holds. Denoting by $d_{K,\sigma}$ the Euclidean distance between $\boldsymbol{x}_K$ and the hyperplane including σ, one assumes that $d_{K,\sigma} > 0$. We then denote by $\mathscr{D}_{K,\sigma}$ the cone with vertex $\boldsymbol{x}_K$ and basis σ.*

The time discretization is performed with a constrained time step-size k such that $\dfrac{\tau}{k} \in \mathbb{N}$. We set then $k = \dfrac{\tau}{M}$, where $M \in \mathbb{N} \setminus \{0\}$. Denote by N the integer part of $\dfrac{T}{k}$, i.e. $N = \left[\dfrac{T}{k}\right]$. We shall denote $t_n = nk$, for $n \in [\![-M, N]\!]$. As particular cases $t_{-M} = -\tau$, $t_0 = 0$, and $t_N \leq T$. One of the advantages of this time discretization is that the point $t = 0$ is a mesh point which is suitable since we have Eq. (1) defined for $t \in (0, T)$ and initial condition (2) defined for $t \in (-\tau, 0)$. We denote by ∂^1 the discrete first time derivative given by

$$\partial^1 v^{j+1} = \frac{v^{j+1} - v^j}{k}. \tag{4}$$

Throughout this paper, the letter C stands for a positive constant independent of the parameters of the space and time discretizations.

We define the discrete space $\mathscr{X}_{\mathscr{D},0}$ as the set of all $v = \left((v_K)_{K \in \mathscr{M}}, (v_\sigma)_{\sigma \in \mathscr{E}}\right)$, where $v_K, v_\sigma \in \mathbb{R}$ and $v_\sigma = 0$ for all $\sigma \in \mathscr{E}_{\mathrm{ext}}$. Let $H_{\mathscr{M}}(\Omega) \subset L^2(\Omega)$ be the space of functions which are constant on each control volume K of the mesh $\mathscr{M}$. For all

$v \in \mathscr{X}_\mathscr{D}$, we denote by $\Pi_\mathscr{M} v \in H_\mathscr{M}(\Omega)$ the function defined by $\Pi_\mathscr{M} v(\boldsymbol{x}) = v_K$, for a.e. $\boldsymbol{x} \in K$, for all $K \in \mathscr{M}$.

In order to analyze the convergence, we need to consider the size of the discretization $\mathscr{D}$ defined by $h_\mathscr{D} = \sup\{\text{diam}(K), K \in \mathscr{M}\}$ and the regularity of the mesh given by

$$\theta_\mathscr{D} = \max\left(\max_{\sigma \in \mathscr{E}_{\text{int}}, K, L \in \mathscr{M}} \frac{d_{K,\sigma}}{d_{L,\sigma}}, \max_{K \in \mathscr{M}, \sigma \in \mathscr{E}_K} \frac{h_K}{d_{K,\sigma}}\right). \tag{5}$$

The scheme we want to consider is based on the use of the discrete gradient given in [4]. For $u \in \mathscr{X}_\mathscr{D}$, we define, for all $K \in \mathscr{M}$

$$\nabla_\mathscr{D} u(\boldsymbol{x}) = \nabla_K u + \left(\frac{\sqrt{d}}{d_{K,\sigma}}\left(u_\sigma - u_K - \nabla_K u \cdot (\boldsymbol{x}_\sigma - \boldsymbol{x}_K)\right)\right) \mathbf{n}_{K,\sigma}, \quad \text{a.e. } \boldsymbol{x} \in \mathscr{D}_{K,\sigma}, \tag{6}$$

where $\nabla_K u = \dfrac{1}{\text{m}(K)} \displaystyle\sum_{\sigma \in \mathscr{E}_K} \text{m}(\sigma)\left(u_\sigma - u_K\right) \mathbf{n}_{K,\sigma}$.

Using the discrete gradient $\nabla_\mathscr{D}$, we are able to define the following bilinear form defined on $\mathscr{X}_\mathscr{D} \times \mathscr{X}_\mathscr{D}$ by

$$\langle u, v \rangle_F = \int_\Omega \nabla_\mathscr{D} u(\boldsymbol{x}) \cdot \nabla_\mathscr{D} v(\boldsymbol{x}) d\boldsymbol{x}, \quad \forall (u, v) \in \mathscr{X}_\mathscr{D} \times \mathscr{X}_\mathscr{D}. \tag{7}$$

3   Formulation of a New Finite Volume Scheme for a Delay Equation

We now set a formulation of an implicit linear finite volume scheme for problem (1)–(3). The unknowns of this scheme are the set $\{u_\mathscr{D}^n; n \in [\![-M, N]\!]\}$ which are expected to approximate the set of the unknowns

$$\{u(t_n); n \in [\![-M, N]\!]\}.$$

1. **Approximation of initial condition (2).** The discretization of initial condition (2) can be performed as: for any $n \in [\![-M, 0]\!]$

$$\langle u_\mathscr{D}^n, v \rangle_F = -\left(\Delta u^0(t_n), \Pi_\mathscr{M} v\right)_{L^2(\Omega)}, \quad \forall v \in \mathscr{X}_{\mathscr{D},0}. \tag{8}$$

2. **Approximation of (1) and (3).** For any $n \in [\![0, N-1]\!]$, find $u_\mathscr{D}^n \in \mathscr{X}_{\mathscr{D},0}$ such that, for all $v \in \mathscr{X}_{\mathscr{D},0}$

$$\left(\partial^1 \Pi_\mathscr{M} u_\mathscr{D}^{n+1}, \Pi_\mathscr{M} v\right)_{L^2(\Omega)} + \langle u_\mathscr{D}^{n+1}, v \rangle_F$$
$$= \left(f(t_{n+1}, \Pi_\mathscr{M} u_\mathscr{D}^n, \Pi_\mathscr{M} u_\mathscr{D}^{n+1-M}), \Pi_\mathscr{M} v\right)_{L^2(\Omega)}, \tag{9}$$

where $f(t_{n+1}, \Pi_\mathscr{M} u_\mathscr{D}^n, \Pi_\mathscr{M} u_\mathscr{D}^{n+1-M})$ denotes the function

$$\boldsymbol{x} \mapsto f(\boldsymbol{x}, t_{n+1}, \Pi_\mathscr{M} u_\mathscr{D}^n(\boldsymbol{x}), \Pi_\mathscr{M} u_\mathscr{D}^{n+1-M}).$$

4   Convergence Order of Scheme (8)–(9)

The main result of this note is the following theorem, that is the existence, uniqueness, and convergence order of the finite volume scheme (8)–(9).

Theorem 1. *(Error estimates for the finite volume scheme (8)–(9)) Let Ω be a polyhedral open bounded subset of $\mathbb{R}^d$, where $d \in \mathbb{N} \setminus \{0\}$, and $\partial\Omega = \overline{\Omega} \setminus \Omega$ its boundary. Assume that the solution of (1)–(3) satisfies $u \in \mathscr{C}^2([0,T]; \mathscr{C}^2(\overline{\Omega}))$. Let $k = \dfrac{\tau}{M}$, where $M \in \mathbb{N} \setminus \{0\}$. Denote by N the integer part of $\dfrac{T}{k}$, i.e. $N = \left[\dfrac{T}{k}\right]$. We shall denote $t_n = nk$, for $n \in [\![-M, N]\!]$. As particular cases $t_{-M} = -\tau$ and $t_0 = 0$. Let $\mathscr{D} = (\mathscr{M}, \mathscr{E}, \mathscr{P})$ be a discretization in the sense of Definition 1. Assume that $\theta_{\mathscr{D}}$ (given by (5)) satisfies $\theta \geq \theta_{\mathscr{D}}$. Let $\nabla_{\mathscr{D}}$ be the discrete gradient given by (6) and denote by $\langle \cdot, \cdot \rangle_F$ the bilinear form defined by (7).*
Then there exists a unique solution $(u_{\mathscr{D}}^n)_{n=-M}^N \in \mathscr{X}_{\mathscr{D},0}^{M+N+1}$ for the discrete problem (8)–(9). If we assume in addition that Assumption 1 is satisfied, then the following error estimates hold:

- $L^\infty(H_0^1)$–estimate. For all $n \in [\![-M, N]\!]$

$$\|\nabla_{\mathscr{D}} u_{\mathscr{D}}^n - \nabla u(t_n)\|_{L^2(\Omega)} \leq C(k + h_{\mathscr{D}})\|u\|_{\mathscr{C}^2([0,T]; \mathscr{C}^2(\overline{\Omega}))}. \tag{10}$$

- $W^{1,2}(L^2)$–estimate.

$$\left(\sum_{n=-M+1}^N k\,\|u_t(t_n) - \Pi_{\mathscr{M}}\,\partial^1 u_{\mathscr{D}}^n\|_{L^2(\Omega)}^2\right)^{\frac{1}{2}} \leq C(k + h_{\mathscr{D}})\|u\|_{\mathscr{C}^2([0,T]; \mathscr{C}^2(\overline{\Omega}))}. \tag{11}$$

To prove Theorem 1, we need to use the following discrete *a priori* estimate:

Lemma 1 (A priori estimate for the discrete problem). *Under the same hypotheses of Theorem 1, assume that there exists $(\eta^n)_{n=0}^N \in (\mathscr{X}_{\mathscr{D},0})^{N+1}$ such that for all $n \in [\![0, N-1]\!]$*

$$\left(\partial^1 \Pi_{\mathscr{M}} \eta_{\mathscr{D}}^{n+1}, \Pi_{\mathscr{M}} v\right)_{L^2(\Omega)} + \langle \eta_{\mathscr{D}}^{n+1}, v \rangle_F = \left(\mathscr{S}^{n+1}, \Pi_{\mathscr{M}} v\right)_{L^2(\Omega)}, \tag{12}$$

where $\mathscr{S}^{n+1} \in L^2(\Omega)$, for all $n \in [\![0, N-1]\!]$.
Then, the following estimate holds, for all $J \in [\![1, N]\!]$

$$\sum_{n=0}^{J-1} k\|\partial^1 \Pi_{\mathscr{M}} \eta_{\mathscr{D}}^{n+1}\|_{L^2(\Omega)}^2 + \|\nabla_{\mathscr{D}} \eta_{\mathscr{D}}^J\|_{L^2(\Omega)}^2 \leq \sum_{n=0}^{J-1} k\left(\mathscr{S}^{n+1}\right)^2 + \|\eta_{\mathscr{D}}^0\|_{L^2(\Omega)}^2. \tag{13}$$

Proof. To derive an estimate for the solution $\eta_{\mathscr{D}}^{n+1}$ of (12), we take $v = \partial^1 \eta_{\mathscr{D}}^{n+1}$ in (12) to get, for all $n \in [\![0, N-1]\!]$

$$\|\partial^1 \Pi_{\mathscr{M}} \eta_{\mathscr{D}}^{n+1}\|_{L^2(\Omega)}^2 + \langle \eta_{\mathscr{D}}^{n+1}, \partial^1 \eta_{\mathscr{D}}^{n+1} \rangle_F = \left(\mathscr{S}^{n+1}, \partial^1 \Pi_{\mathscr{M}} \eta_{\mathscr{D}}^{n+1}\right)_{L^2(\Omega)}. \tag{14}$$

The following rule will be useful:

$$\langle \eta_{\mathscr{D}}^{n+1}, \partial^1 \eta_{\mathscr{D}}^{n+1} \rangle_F = \frac{1}{2k} \langle \eta_{\mathscr{D}}^{n+1} - \eta_{\mathscr{D}}^n, \eta_{\mathscr{D}}^{n+1} - \eta_{\mathscr{D}}^n \rangle_F$$

$$+ \frac{1}{2k} \left(\langle \eta_{\mathscr{D}}^{n+1}, \eta_{\mathscr{D}}^{n+1} \rangle_F - \langle \eta_{\mathscr{D}}^n, \eta_{\mathscr{D}}^n \rangle_F \right). \tag{15}$$

Gathering now (14) and (15) yields

$$2k \| \partial^1 \Pi_{\mathscr{M}} \eta_{\mathscr{D}}^{n+1} \|_{L^2(\Omega)}^2 + \langle \eta_{\mathscr{D}}^{n+1}, \eta_{\mathscr{D}}^{n+1} \rangle_F - \langle \eta_{\mathscr{D}}^n, \eta_{\mathscr{D}}^n \rangle_F$$

$$\leq 2k \left(\mathscr{S}^{n+1}, \partial^1 \Pi_{\mathscr{M}} \eta_{\mathscr{D}}^{n+1} \right)_{L^2(\Omega)}. \tag{16}$$

Summing the previous inequality over $n \in [\![0, J-1]\!]$ where $J \in [\![1, N]\!]$ yields

$$2 \sum_{n=0}^{J-1} k \| \partial^1 \Pi_{\mathscr{M}} \eta_{\mathscr{D}}^{n+1} \|_{L^2(\Omega)}^2 + \| \nabla_{\mathscr{D}} \eta_{\mathscr{D}}^J \|_{L^2(\Omega)}^2$$

$$\leq 2 \sum_{n=0}^{J-1} k \left(\mathscr{S}^{n+1}, \partial^1 \Pi_{\mathscr{M}} \eta_{\mathscr{D}}^{n+1} \right)_{L^2(\Omega)} + \| \eta_{\mathscr{D}}^0 \|_{L^2(\Omega)}^2. \tag{17}$$

Using the Cauchy Schwarz inequality together with the Young inequality implies that

$$\left(\mathscr{S}^{n+1}, \partial^1 \Pi_{\mathscr{M}} \eta_{\mathscr{D}}^{n+1} \right)_{L^2(\Omega)} \leq \| \mathscr{S}^{n+1} \|_{L^2(\Omega)} \| \partial^1 \Pi_{\mathscr{M}} \eta_{\mathscr{D}}^{n+1} \|_{L^2(\Omega)}$$

$$\leq \frac{\| \mathscr{S}^{n+1} \|_{L^2(\Omega)}^2}{2} + \frac{\| \partial^1 \Pi_{\mathscr{M}} \eta_{\mathscr{D}}^{n+1} \|_{L^2(\Omega)}^2}{2}. \tag{18}$$

Using this inequality, inequality (17) yields the desired estimate (13).   ∎

SKETCH OF PROOF OF THEOREM 1.

1. Existence and uniqueness for scheme (8)–(9). The existence and uniqueness of $(u_{\mathscr{D}}^n)_{n \in [\![-M, 0]\!]}$ for (8) is straightforward (see [4]). To prove the existence and uniqueness of the solution $(u_{\mathscr{D}}^n)_{n \in [\![1, N]\!]}$ for the linear scheme (9) with (8), we set (as usual to prove the uniqueness for linear systems) $f(t_{n+1}, \Pi_{\mathscr{M}} u_{\mathscr{D}}^n, \Pi_{\mathscr{M}} u_{\mathscr{D}}^{n+1-M}) = 0$ and $u_{\mathscr{D}}^n = 0$. Taking $v = u_{\mathscr{D}}^{n+1}$ in (9) yields $\| \nabla_{\mathscr{D}} u_{\mathscr{D}}^{n+1} \|_{L^2(\Omega)} = 0$. This implies that, since $u_{\mathscr{D}}^{n+1} \in \mathscr{X}_{\mathscr{D},0}$, $u_{\mathscr{D}}^{n+1} = 0$. This yields the uniqueness of the solution $u_{\mathscr{D}}^{n+1}$ for (9) for given $u_{\mathscr{D}}^n$ and $u_{\mathscr{D}}^{n+1-M}$. The existence of $u_{\mathscr{D}}^{n+1}$ follows the uniqueness, since (9) is a finite dimensional linear system with respect to the unknowns $\{ (u_K^{n+1}, u_\sigma^{n+1}); K \in \mathscr{M}, \sigma \in \mathscr{E}_{\text{int}} \}$ (with as many unknowns as many equations). This implies, successively on n, the existence and uniqueness of $u_{\mathscr{D}}^n$ for all $n \in [\![0, N]\!]$.

2. Proof of estimates (10)–(11). To prove (10)–(11), we compare (8)–(9) with the following auxiliary scheme: For any $n \in [\![-M, N]\!]$, find $\bar{u}_{\mathscr{D}}^n \in \mathscr{X}_{\mathscr{D},0}$ such that

$$\langle \bar{u}_{\mathscr{D}}^n, v \rangle_F = (-\Delta u(t_n), \Pi_{\mathscr{M}} v)_{L^2(\Omega)}, \quad \forall v \in \mathscr{X}_{\mathscr{D},0}. \tag{19}$$

2.1. Comparison between the solution (19) and the solution of problem (1)–(3). The following convergence results hold, see [3,4]:

- Discrete $L^\infty(L^2)$–error estimate. For all $n \in [\![-M, N]\!]$

$$\|u(t_n) - \Pi_{\mathscr{M}} \bar{u}_{\mathscr{D}}^n\|_{L^2(\Omega)} \leq C h_{\mathscr{D}} \|u\|_{\mathscr{C}([0,T]; \mathscr{C}^2(\overline{\Omega}))}. \tag{20}$$

- $\mathscr{W}^{1,\infty}(L^2)$–error estimate. For all $n \in [\![-M + 1, N]\!]$

$$\|u_t(t_n) - \partial^1 \Pi_{\mathscr{M}} \bar{u}_{\mathscr{D}}^n\|_{L^2(\Omega)} \leq C(h_{\mathscr{D}} + k) \|u\|_{\mathscr{C}^2([0,T]; \mathscr{C}^2(\overline{\Omega}))}. \tag{21}$$

- Error estimate in the gradient approximation. For all $n \in [\![-M, N]\!]$

$$\|\nabla u(t_n) - \nabla_{\mathscr{D}} \bar{u}_{\mathscr{D}}^n\|_{(L^2(\Omega))^d} \leq C h_{\mathscr{D}} \|u\|_{\mathscr{C}([0,T]; \mathscr{C}^2(\overline{\Omega}))}. \tag{22}$$

2.2. Comparison between the solution of (8)–(9) and the auxiliary scheme (19). Let us define the *auxiliary* error

$$\eta_{\mathscr{D}}^n = u_{\mathscr{D}}^n - \bar{u}_{\mathscr{D}}^n \in \mathscr{X}_{\mathscr{D},0}. \tag{23}$$

Comparing (19) with (8) and using the fact that $u(t_n) = u^0(t_n)$ for all $n \in [\![-M, 0]\!]$ (subject of (2)) imply that, for all $n \in [\![-M, 0]\!]$

$$\eta_{\mathscr{D}}^n = 0. \tag{24}$$

Writing scheme (19) in the level $n+1$ and subtracting the result from (9) to get, for all $v \in \mathscr{X}_{\mathscr{D},0}$

$$\left(\partial^1 \Pi_{\mathscr{M}} u_{\mathscr{D}}^{n+1}, \Pi_{\mathscr{M}} v\right)_{L^2(\Omega)} + \langle \eta_{\mathscr{D}}^{n+1}, v \rangle_F$$
$$= \left(f(t_{n+1}, \Pi_{\mathscr{M}} u_{\mathscr{D}}^n, \Pi_{\mathscr{M}} u_{\mathscr{D}}^{n+1-M}) + \Delta u(t_{n+1}), \Pi_{\mathscr{M}} v\right)_{L^2(\Omega)}. \tag{25}$$

Subtracting $\left(\partial^1 \Pi_{\mathscr{M}} \bar{u}_{\mathscr{D}}^{n+1}, \Pi_{\mathscr{M}} v\right)_{L^2(\Omega)}$ from both sides of the previous equation and replacing $\Delta u(\boldsymbol{x}, t_{n+1})$ by $u_t(\boldsymbol{x}, t_{n+1}) - f(\boldsymbol{x}, t_{n+1}, u(\boldsymbol{x}, t_{n+1}), u(\boldsymbol{x}, t_{n+1} - \tau))$ (which stems from (1)), we get, for all $v \in \mathscr{X}_{\mathscr{D},0}$

$$\left(\partial^1 \Pi_{\mathscr{M}} \eta_{\mathscr{D}}^{n+1}, \Pi_{\mathscr{M}} v\right)_{L^2(\Omega)} + \langle \eta_{\mathscr{D}}^{n+1}, v \rangle_F = \left(\mathscr{S}^{n+1}, \Pi_{\mathscr{M}} v\right)_{L^2(\Omega)}, \tag{26}$$

where

$$\mathscr{S}^{n+1}(\boldsymbol{x}) = \mathbb{T}^1(\boldsymbol{x}) + \mathbb{T}^2(\boldsymbol{x}), \tag{27}$$

with

$$\mathbb{T}^1(\boldsymbol{x}) = u_t(\boldsymbol{x}, t_{n+1}) - \partial^1 \Pi_{\mathscr{M}} \bar{u}_{\mathscr{D}}^{n+1} \tag{28}$$

and (recall that $-\tau = t_{-M}$)

$$\mathbb{T}^2(\boldsymbol{x}) = f(\boldsymbol{x}, t_{n+1}, \Pi_{\mathscr{M}} u_{\mathscr{D}}^n, \Pi_{\mathscr{M}} u_{\mathscr{D}}^{n+1-M}) - f(\boldsymbol{x}, t_{n+1}, u(\boldsymbol{x}, t_{n+1}), u(\boldsymbol{x}, t_{n+1} - \tau))$$
$$= f(\boldsymbol{x}, t_{n+1}, \Pi_{\mathscr{M}} u_{\mathscr{D}}^n(\boldsymbol{x}), \Pi_{\mathscr{M}} u_{\mathscr{D}}^{n+1-M}(\boldsymbol{x})) - f(\boldsymbol{x}, t_{n+1}, u(\boldsymbol{x}, t_{n+1}), u(\boldsymbol{x}, t_{n+1-M})). \tag{29}$$

Using (21) yields

$$\|\mathbb{T}^1\|_{L^2(\Omega)} \leq C(h_{\mathscr{D}} + k)\|u\|_{\mathscr{C}^2([0,T];\,\mathscr{C}^2(\overline{\Omega}))}. \tag{30}$$

Using Assumption 1 together with (20) implies that

$$\|\mathbb{T}^2\|_{L^2(\Omega)} \leq C \left(\|\Pi_{\mathscr{M}}\, u_{\mathscr{D}}^n - u(t_{n+1})\|_{L^2(\Omega)} + \|\Pi_{\mathscr{M}}\, u_{\mathscr{D}}^{n+1-M} - u(t_{n+1-M})\|_{L^2(\Omega)} \right)$$
$$\leq C \left(\|\Pi_{\mathscr{M}}\, \eta_{\mathscr{D}}^n\|_{L^2(\Omega)} + \|\Pi_{\mathscr{M}}\, \eta_{\mathscr{D}}^{n+1-M}\|_{L^2(\Omega)} + (h_{\mathscr{D}} + k)\|u\|_{\mathscr{C}^1([0,T];\,\mathscr{C}^1(\overline{\Omega}))} \right). \tag{31}$$

Since $(\eta^n)_{n=0}^N \in (\mathscr{X}_{\mathscr{D},0})^{N+1}$ is satisfying (26), hence it satisfies the hypothesis (12) of Lemma 1. Applying now (13) of Lemma 1 and using (24) and (30)–(31) to get, for all $J \in [\![1, N]\!]$

$$\sum_{n=0}^{J-1} k\|\partial^1 \Pi_{\mathscr{M}}\, \eta_{\mathscr{D}}^{n+1}\|_{L^2(\Omega)}^2 + \|\nabla_{\mathscr{D}}\eta_{\mathscr{D}}^J\|_{L^2(\Omega)}^2 \leq \sum_{n=0}^{J-1} k\,(\mathscr{S}^{n+1})^2$$
$$\leq \sum_{n=0}^{J-1} k\|\Pi_{\mathscr{M}}\, \eta_{\mathscr{D}}^n\|_{L^2(\Omega)}^2 + \sum_{n=0}^{J-1} k\|\Pi_{\mathscr{M}}\, \eta_{\mathscr{D}}^{n+1-M}\|_{L^2(\Omega)}^2 + C(h_{\mathscr{D}} + k)^2\|u\|_{\mathscr{C}^2([0,T];\,\mathscr{C}^2(\overline{\Omega}))}^2$$
$$\leq 2\sum_{n=0}^{J-1} k\|\Pi_{\mathscr{M}}\, \eta_{\mathscr{D}}^n\|_{L^2(\Omega)}^2 + C(h_{\mathscr{D}} + k)^2\|u\|_{\mathscr{C}^2([0,T];\,\mathscr{C}^2(\overline{\Omega}))}^2.$$

This with the Poincaré inequality [4, Lemma 5.4, Page 1038] imply that

$$\sum_{n=0}^{J-1} k\|\partial^1 \Pi_{\mathscr{M}}\, \eta_{\mathscr{D}}^{n+1}\|_{L^2(\Omega)}^2 + \|\nabla_{\mathscr{D}}\eta_{\mathscr{D}}^J\|_{L^2(\Omega)}^2$$
$$\leq C\sum_{n=0}^{J-1} k\|\nabla_{\mathscr{D}}\eta_{\mathscr{D}}^n\|_{L^2(\Omega)}^2 + C(h_{\mathscr{D}} + k)^2\|u\|_{\mathscr{C}^2([0,T];\,\mathscr{C}^2(\overline{\Omega}))}^2. \tag{32}$$

This implies that

$$\|\nabla_{\mathscr{D}}\eta_{\mathscr{D}}^J\|_{L^2(\Omega)}^2 \leq C\sum_{n=0}^{J-1} k\|\nabla_{\mathscr{D}}\eta_{\mathscr{D}}^n\|_{L^2(\Omega)}^2 + C(h_{\mathscr{D}} + k)^2\|u\|_{\mathscr{C}^2([0,T];\,\mathscr{C}^2(\overline{\Omega}))}^2 \tag{33}$$

and

$$\sum_{n=0}^{N} k\|\partial^1 \Pi_{\mathscr{M}}\, \eta_{\mathscr{D}}^n\|_{L^2(\Omega)}^2 \leq C\sum_{n=0}^{N-1} k\|\nabla_{\mathscr{D}}\eta_{\mathscr{D}}^n\|_{L^2(\Omega)}^2 + C(h_{\mathscr{D}} + k)^2\|u\|_{\mathscr{C}^2([0,T];\,\mathscr{C}^2(\overline{\Omega}))}^2. \tag{34}$$

Using a discrete form of Gronwall's Lemma together with the fact that $\sum_{n=0}^{N-1} k \leq T$, (33) implies that, for all $J \in [\![1, N]\!]$

$$\|\nabla_{\mathscr{D}}\eta_{\mathscr{D}}^J\|_{L^2(\Omega)}^2 \leq C(h_{\mathscr{D}} + k)^2\|u\|_{\mathscr{C}^2([0,T];\,\mathscr{C}^2(\overline{\Omega}))}^2. \tag{35}$$

Using this estimate, (34) implies that

$$\sum_{n=0}^{N} k \|\partial^1 \Pi_{\mathscr{M}} \eta_{\mathscr{D}}^n\|_{L^2(\Omega)}^2 \leq C(h_{\mathscr{D}} + k)^2 \|u\|_{\mathscr{C}^2([0,T];\,\mathscr{C}^2(\overline{\Omega}))}^2. \tag{36}$$

Using now the triangle inequality, (21)–(22), and (35)–(36) yield the desired estimates (10)–(11) of Theorem 1.   ∎

Remark 1. (A possible extension) The present results can be extended to the following general semilinear parabolic equation with delays which occur not only in the exact solution but also in its gradient:

$$u_t(\boldsymbol{x},t) - \Delta u(\boldsymbol{x},t) = f\left(\boldsymbol{x},t,u(\boldsymbol{x},t),u(\boldsymbol{x},t-\tau),\nabla u(\boldsymbol{x},t),\nabla u(\boldsymbol{x},t-\tau)\right).$$

5 Conclusion and Perspectives

We considered the convergence of an implicit finite volume scheme, in any space dimension, for a simple semi-linear delay parabolic equation. The order is proved to be one (both in time and space). One of the main tasks we will work on is the use of Crank Nicolson method in order to improve the order in time. Another interesting path to be followed is to consider the semi-linear time fractional diffusion equation with time delay, i.e. the time derivative u_t in (1) is replaced by a fractional derivative $\partial_t^\alpha u$ with $0 < \alpha < 1$.

References

1. Bellen, A., Zennaro, M.: Numerical Methods for Delay Differential Equations. Numerical Mathematics and Scientific Computation. Oxford University Press, Oxford (2003)
2. Bradji, A.: An analysis for the convergence order of gradient schemes for semilinear parabolic equations. Comput. Math. Appl. **72**(5), 1287–1304 (2016)
3. Bradji, A., Fuhrmann, J.: Some abstract error estimates of a finite volume scheme for a nonstationary heat equation on general nonconforming multidimensional spatial meshes. Appl. Math. **58**(1), 1–38 (2013)
4. Eymard, R., Gallouët, T., Herbin, R.: Discretization of heterogeneous and anisotropic diffusion problems on general nonconforming meshes. IMA J. Numer. Anal. **30**(4), 1009–1043 (2010)
5. Kuang, Y.: Delay Differential Equations: with Applications in Population Dynamics. Mathematics in Science and Engineering, vol. 191. Academic Press, Boston (1993)
6. Li, D., Zhang, C.: L^∞-error estimates of discontinuous Galerkin methods for delay differential equations. Appl. Numer. Math. **82**, 1–10 (2014)
7. Zhang, Q., Zhang, C.: A new linearized compact multisplitting scheme for the nonlinear convection-reaction-diffusion equations with delay. Commun. Nonlinear Sci. Numer. Simul. **18**(12), 3278–3288 (2013)

STARDEX and ETCCDI Climate Indices Based on E-OBS and CARPATCLIM
Part One: General Description

Hristo Chervenkov[1]($\boxtimes$) (iD), Kiril Slavov[1], and Vladimir Ivanov[2]

[1] National Institute of Meteorology and Hydrology, Bulgarian Academy of Sciences,
66, Tsarigradsko Shose blvd, 1784 Sofia, Bulgaria
{hristo.tchervenkov,kiril.slavov}@meteo.bg
[2] National Institute in Geophysics, Geodesy and Geography,
Bulgarian Academy of Sciences, 1113 Acad. G. Bonchev str., bl. 3, Sofia, Bulgaria
vivanov@geophys.bas.bg
http://www.meteo.bg
http://www.niggg.bas.bg

Abstract. The paper presents a suite of 26 datasets of climate indices on monthly, seasonal and annual basis, as well as linear trend and statistical significance estimation for the considered time windows. They are calculated with the standard software of the STARDEX and ETCCDI international projects correspondingly, with data from the ECA&D E-OBS and CARPATCLIM gridded databases. The database of climate indices presented in this paper, named ClimData, is intended to serve as a convenient, barrier free and versatile tool for research. The present article, which is part one of more common study, is dedicated on the description of the motivation for the creation, the content, structure and the access point of ClimData.

Keywords: Climate indices · E-OBS · CARPATCLIM · STARDEX ETCCDI · ClimData database

1 Introduction

The oncoming climate changes are the biggest challenge that the mankind faces. For decades, most analyses of long-term global climate change using observational temperature and precipitation data are focused on changes in mean values. However, immediate damages to humans and their properties are not obviously caused by gradual changes in these variables but mainly by so-called extreme climate events. The extreme weather phenomena are discussed in all reports of the Intergovernmental Panel on Climate Change (see, for example, [17]). There are various methods to investigate extreme events, but the computation and analysis of climate indices (CIs) derived from daily data is probably the most widely used non-parametric approach. In order to detect changes in climate extremes, it is important to develop a set of indices that are statistically robust, cover a wide

© Springer Nature Switzerland AG 2019
G. Nikolov et al. (Eds.): NMA 2018, LNCS 11189, pp. 360–367, 2019.
https://doi.org/10.1007/978-3-030-10692-8_40

range of climate conditions, and have a high signal-to-noise ratio. CIs include absolute-thresholds indices, percentile-based indices, and indices based on the duration of an event. They are used in several projects on climate change with focus on different spatial scales, from planetary to continental, regional, national or local scale, as prevailing indicators of changes of the extreme events. Thus, the number of publications on this topic is very large and the main results will be summarized and only the major conclusions will be presented very concisely. Earlier studies as these from Groisman et al. [7] and Frich et al. [6] are hampered by the data scarcity over big territories. The comprehensive global study of Alexander et al. [2] shows widespread significant changes in temperature extremes for the period 1951–2003, especially those related to daily minimum temperatures. Changes in daily maximum temperature are less marked, implying that many regions around the world has become less cold rather than hotter. Precipitation changes have been much less coherent than temperature changes, but annual precipitation has shown a widespread significant increase. Several recent studies, as the most cited above, have concentrated on the analysis of CIs based on observational data from weather stations (e.g., [6,16,21] and many others). Thus, for example, Moberg et al. [16] computes a set of CIs from records for stations in Europe west of 60 °E as well as linear trends over the period 1901–2000. The study shows a warming for all temperature indices but, large regional differences in temperature trend patterns. It is revealed also that the winter areal averaged precipitation totals have increased significantly and absence of overall long-term trend in summer precipitation ones. Other group of studies are dedicated primarily on the assessment of the present climate [1] and/or in future climate projections [18,22] using CIs, derived from climate models (CMs) output.

As far as many of these studies (e.g. [18]) use partially pre-existing datasets of CIs, the availability of such databases could facilitate any future work, which relies more or less on CIs-based analysis of the present climate. The objectives of the present study are, first, to construct and present to the expert community for barrier free use a comprehensive suite of climate indices datasets (called ClimData), computed from reliable input data from one side and well elaborated and internationally accepted methodology from other. The second objective is to demonstrate some of the possibilities of ClimData for CIs-based analysis for selected key indicators. Hence the importance of assessing trends in climate extremes is often emphasized (e.g. [16]), estimations of the magnitude of the trend as well as its statistical significance, are accepted as 'natural' supplement to the CIs-time series. Thus, such information for all indices on seasonal and annual basis, is also included in ClimData. Although, as it will be commented in Sect. 3, similar sets are already available, partially from the input data vendors, their completeness are not full and/or are based on outdated data. In contrast, our intend is to provide consolidated database, based on the most recent source.

The gridded time series of the necessary parameters from the CARPATCLIM and E-OBS projects are used as input and the procedures from the STARDEX and ETCCDI initiatives are applied for computation of the CIs. The paper is organized as follows. The both CIs suites, as well as the motivation for their selection, are briefly described in Sect. 2, followed by a discussion of the input data for their calculation in Sect. 3. Previous climatological applications of the same sources as well as some issues are also presented. The performed calculations, validation and results as form of complete list of all 26 datasets in ClimData, are placed in Sect. 4. The access to the complete ClimData is commented in Sect. 5.

2   Climate Indices Selection

The relative big number of CIs and the popularity of the CIs-based analysis among the expert community impose standardization of the definitions world-wide. The importance of the of exact formulation of an internationally accepted set of indices of climate extremes related to precipitation and temperature, obtained from daily data, was recognized during the last decades. The use of approved indices allows comparison of analyses conducted in any part of the world and seamless merging of index data to produce a global picture as well. Many attempts were made some in frames of international collaborative projects – such are the European Commission funded CIRCE (Climate change and impact research: the Mediterranean environment, https://www.cmcc.it/projects/circe-climate-change-and-impact-research-the-mediterranean-environment) and STARDEX (STAtistical and Regional dynamical Downscaling of EXtremes for European regions, http://www.cru.uea.ac.uk/projects/stardex). STARDEX is focused on relatively moderate extremes rather than the most extreme events. The project uses in total 57 CIs calulated on annual (noted further as Y-basis) and seasonal basis (S-basis). The STARDEX core subset consists of 10 indices. Additionally is computed the slope of the linear trend by means of Least Squares Estimation (LSE) and, second, is analyzed the statistical significance of trends with the Mann-Kendall (MK) test for each CI. The MK test [10,15] is a non-parametric and rank-based procedure, especially suitable for non-normally distributed data, data containing outliers and nonlinear trends. Consequently, this test is widely used in the geosciences as standard tool for trend significance estimation [3,4]. The two tailed test is applied in this study. Based on [6], the Commission for Climatology (CCl)/CLIVAR/JCOMM Expert Team on Climate Change Detection and Indices (ETCCDI) (previously known as the Expert Team on Climate Change, Detection, Monitoring and Indices (ETCCDMI), http://www.clivar.org/organization/etccdi/etccdi.php) defined a suite of indices that have subsequently become known as the ETCCDI-indices. These indices were chosen to sample a wide variety of climates. However, the definitions and usefulness of some of these indices, although meant to be globally valid, became a subject of discussion. As a result, definitions of some of them as well as their calculations were reconsidered. The ETCCDI-indices are obtained on Y-basis and monthly basis (M-basis) and the threshold-based ones that have to be calculated relative to a base period are calculated according to the bootstrap method [23].

The main features of the both CIs-suites are summarized in Table 1.

Table 1. Main features of the STARDEX and ETCCDI CIs-suites. See the cited above web-pages for details.

Feature	STARDEX	ETCCDI
Total number	57	29
on Y-basis	57	29
on S-basis	54	0
on M-basis	0	13
Temperature based	24	18
Precipitation based	33	11
Input variables	tn, tx, td, prec.	tn, tx, prec.
Bootstrap correction	No	Yes
Trend & MK test	Yes	No

Complete lists of all CIs, together with the definitions, could be found on the web-pages of the corresponding projects cited above.

A key moment of the both initiatives, STARDEX and ETCCDI, is that they supply standard software, in form of open source, for computation of the corresponding CIs-suite. This possibility ensures effective and reliable calculation of the considered indicators from arbitrary input dataset and this will be demonstrated in Sect. 4.

3 Input Data

Daily maximum, minimum and mean temperatures (tx, tn, td) as well as the daily precipitation sum (prec.) are core climatic parameters particularly involved in determining climate change impacts on society and ecosystems. We use two data sets in this study: both of them are based on surface measurements, are in form of gridded database and, last but not least, they are free available. Although some criticism exists, calculating of CIs from gridded data is not unusual [2,14]. First of the datasets, used in this study, is CARPATCLIM (http://www.carpatclim-eu.org/), which is a high-resolution homogeneous gridded database covering 1961–2010 for the Carpathian region (44°N-50°N and 17°E-27°E) with 0.1° horizontal resolution, containing all the major surface meteorological variables [19]. The commonly used methods and software were the method MASH (Multiple Analysis of Series for Homogenization) for homogenization, quality control, completion of the observed daily data series; and the method MISH (Meteorological Interpolation based on Surface Homogenized Data Basis; [20]) for gridding of homogenized daily data series. Besides the common software, the harmonization of the results across country borders was promoted also by near border data exchange. The second data set is the well known and widely used in the meteorological community E-OBS (https://www.ecad.eu/download/ensembles/download.php) of the European Climate Assessment & Dataset (ECA&D) project [8]. Unlike the CARPATCLIM, E-OBS is updated periodically and version 16.0, spanning from 1950 til the end of 2016, for domain, covering whole Europe (30.125 °N - 71.875 °N and 11.875 °W - 59.875 °E) with 0.25° horizontal resolution, is selected. The E-OBS production procedure includes two step spatial interpolation of station observations (thin-plate spline interpolation of monthly means/totals; kriging of daily anomalies) after the quality control. Although intended primarily for validation of regional CMs, the E-OBS is synoptic scale dataset. In contrast, the CARPATCLIM is a typical mesoscale application. Due to this reason, along the embedded in the both projects different treatment chains of raw data selection, quality control and processing, is reasonable to expect differences in the output estimates. E-OBS may be affected by some potentially important limitations, such as heterogeneities (both spatial and temporal) and large absolute and relative differences over regions where dense station networks exist [9] or undercatching in mountain areas [13]. When compared to a high-density network over the Czech Republic, Kyselý and Plavcová [11] found that E-OBSv2.0 shows large biases, especially for tn at the tail of the probability distribution function. Despite these issues, CARPATCLIM and diverse versions of E-OBS are widely used in climatology, partially applying CIs-based analysis. The web-portals of the both projects offers CIs in form of digital maps or interactive figures. Many of the key indicators as well as atmospheric drought indicators are available also in CARPATCLIM; ECA&D provides 31 climate indices for the E-OBSv11.0 using, however, 1981–2010 as normal period instead of 1961–1990. Dosio [5] uses E-OBSv10.0 as reference in work, where the outputs of an ensemble of regional CMs from the Coordinated Regional-climate Downscaling Experiment over Europe (EURO-CORDEX, http://www.cordex.org/) have been bias adjusted. A number of ETCCDI-CIs have been calculated for the present and projected future climate. Using CARPATCLIM, Lakatos et al. [12] calculates several temperature and precipitation CIs. The obtained trends together with the confidence are also demonstrated. In order to identify changes in the annual temperature extremes, the MK test has been applied in [3] to several thermal indices, computed from CARPATCLIM. Our approach is similar to these studies. Intending to assess the influence of the input data, the same CIs, calculated from E-OBSv16.0 and CARPATCLIM, are demonstrated simultaneously in the second part of this article.

4   Calculations, Validation and Results

The implementation of standard, own for the each of the corresponding projects, software for calculation of the CIs, ensures the reliability of the computational results. STARDEX provides FORTRAN 90 source code and ETCCDI-written in R script RClimDex as well as FORTRAN 90 source code FClimDex. The FClimDex is console application and thus is better than RClimDex situated for embedding in bigger projects. Such method is applied in [18], but in this case, we have not altered even the input/output interface. The pre-build FClimDex as well as the STARDEX code are invoked as external procedures and the CIs and additional quantities are calculated for each CARPATCLIM and E-OBS gridcell individually. The World Meteorological Organization standard period 1961–1990 is set as reference. Hence FClimDex does not perform calculation of CIs of S-basis, as well as linear trend and significance measure by means of the MK test, these quantities are obtained *a posteriori* the CIs with purposely written by the authors procedures. The output datasets, composed in such way, are summarized in Table 2.

Table 2. List of the ClimData output datasets. These, calculated *a posteriori*, are marked with asterisk

CIs set Input Data	STARDEX	ETCCDI
CARPATCLIM		CIs - M-basis
	CIs - S-basis	*CIs - S-basis
	CIs - Y-basis	CIs - Y-basis
	LT estimation - S-basis	*LT estimation - S-basis
	LT estimation - Y-basis	*LT estimation - Y-basis
	MK test - S-basis	*MK test - S-basis
	MK test - Y-basis	*MK test - Y-basis
E-OBSv16.0		CIs - M-basis
	CIs - S-basis	*CIs - S-basis
	CIs - Y-basis	CIs - Y-basis
	LT estimation - S-basis	*LT estimation - S-basis
	LT estimation - Y-basis	*LT estimation - Y-basis
	MK test - S-basis	*MK test - S-basis
	MK test - Y-basis	*MK test - Y-basis

Seasonal indices are calculated as average from the monthly indices, where for December-January-February (DJF) the December of the previous year is used. As in STARDEX, the probability value of the MK test from our computations is stored, which is most important for the null hypothesis testing. The validation of the outcomes is performed in different ways. All the calculations are checked periodically for randomly selected gridcells for correct output of the CIs. Although the applied procedures uses the STARDEX-subroutines for the MK test, the STARDEX dataset with MK test results is used as benchmark: Our program computes the MK test probability value independently with the STARDEX CIs-time series, delivering the same output as from the STARDEX procedure. Finally, the spatial distributions of some key parameters (tropical nights, number of frost/ice days, growing season length) are visually compared with same figures from independent sources (e.g. [3]).

5   Results Dissemination

The barrier free access to all datasets of ClimData is key point of the initiative and is in accordance with the tendencies in the modern geophysical sciences. The VI–SEEM project (https://vi-seem.eu/), which is focused on the scientific communities of Life Sciences, Climatology and Digital Cultural Heritage, is convenient single point access to ClimData and thus it is made available at https://repo.vi-seem.eu/handle/21.15102/VISEEM-343.

References

1. Alexander, L.V., Arblaster, J.M.: Assessing trends in observed and modelled climate extremes over Australia in relation to future projections. Int. J. Climatol. **29**(3), 417–435 (2009). https://doi.org/10.1002/joc.1730
2. Alexander, L.V., et al.: Global observed changes in daily climate extremes of temperature and precipitation. J. Geophys. Res. **111**(D5) (2006). https://doi.org/10.1029/2005jd006290
3. Birsan, M.V., Dumitrescu, A., Micu, D.M., Cheval, S.: Changes in annual temperature extremes in the Carpathians since AD 1961. Nat. Hazards **74**(3), 1899–1910 (2014). https://doi.org/10.1007/s11069-014-1290-5
4. Chervenkov, H., Tsonevsky, I., Slavov, K.: Drought events assessment and trend estimation-results from the analysis of long-term time series of the standardized precipitation index. C. R. Acad. Bulg. Sci. **69**(8), 983–994 (2016)
5. Dosio, A.: Projections of climate change indices of temperature and precipitation from an ensemble of bias-adjusted high-resolution EURO-CORDEX regional climate models. J. Geophys. Res.: Atmos. **121**(10), 5488–5511 (2016). https://doi.org/10.1002/2015jd024411
6. Frich, P., et al.: Observed coherent changes in climatic extremes during the second half of the twentieth century. Clim. Res. **19**(3), 193–212 (2002)
7. Groisman, P.Y., et al.: Changes in the probability of heavy precipitation: Important indicators of climatic change. Clim. Change **42**(1), 243–283 (1999). https://doi.org/10.1023/a:1005432803188
8. Haylock, M.R., et al.: A European daily high-resolution gridded data set of surface temperature and precipitation for 1950–2006. J. Geophys. Res.: Atmos. **113**(D20), D20119 (2008). https://doi.org/10.1029/2008JD010201
9. Hofstra, N., New, M., McSweeney, C.: The influence of interpolation and station network density on the distributions and trends of climate variables in gridded daily data. Clim. Dyn. **35**(5), 841–858 (2009). https://doi.org/10.1007/s00382-009-0698-1
10. Kendall, M.G.: A new measure of rank correlation. Biometrika **30**(1–2), 81–93 (1938). https://doi.org/10.1093/biomet/30.1-2.81
11. Kyselý, J., Plavcová, E.: A critical remark on the applicability of E-OBS European gridded temperature data set for validating control climate simulations. J. Geophys. Res. **115**(D23) (2010). https://doi.org/10.1029/2010jd014123
12. Lakatos, M., Szentimrey, T., Bihari, Z., Szalai, S.: Investigation of climate extremes in the Carpathian region on harmonized data. In: International Scientific Conference on Environmental Changes and Adaptation Strategies, September 2013

13. Lenderink, G.: Exploring metrics of extreme daily precipitation in a large ensemble of regional climate model simulations. Clim. Res. **44**(2–3), 151–166 (2010). https://doi.org/10.3354/cr00946

14. Malcheva, K., Chervenkov, H., Marinova, T.: Winter severity assessment on the basis of measured and reanalysis data. In: 16th International Multidisciplinary Scientific GeoConference SGEM 2016, pp. 719–726. SGEM (2016)

15. Mann, H.B.: Nonparametric tests against trend. Econometrica **13**(3), 245–259 (1945)

16. Moberg, A., et al.: Indices for daily temperature and precipitation extremes in Europe analyzed for the period 1901–2000. J. Geophys. Res. **111**(D22), 1–25 (2006). https://doi.org/10.1029/2006jd007103

17. Pachauri, R.K., Meyer, L.A.: Climate Change 2014: Synthesis Report. Contribution of Working Groups I, II and III to the 5th Assessment Report of the Intergovernmental Panel on Climate Change. Tech. rep., IPCC (2014)

18. Sillmann, J., Röckner, E.: Indices for extreme events in projections of anthropogenic climate change. Clim. Change **86**(1–2), 83–104 (2007). https://doi.org/10.1007/s10584-007-9308-6

19. Szalai, S., et al.: Climate of the Greater Carpathian Region. Final Technical Report. Tech. rep., JRC (2013)

20. Szentimrey, T., Bihari, Z., Lakatos, M., Szalai, S.: Mathematical, methodological questions concerning the spatial interpolation of climate elements. In: Proceedings from the Second Conference on Spatial Interpolation in Climatology and Meteorology (2009)

21. Tank, A.M.G.K., Können, G.P.: Trends in indices of daily temperature and precipitation extremes in Europe, 1946–99. J. Clim. **16**(22), 3665–3680 (2003). 10.1175/1520-0442(2003)016<3665:tiiodt>2.0.co;2

22. Tebaldi, C., Hayhoe, K., Arblaster, J.M., Meehl, G.A.: Going to the extremes, an intercomparison of model-simulated historical and future changes in extreme events. Clim. Change **79**(3–4), 185–211 (2006). https://doi.org/10.1007/s10584-006-9051-4

23. Zhang, X., Hegerl, G., Zwiers, F.W., Kenyon, J.: Avoiding in homogeneity in percentile-based indices of temperature extremes. J. Clim. **18**(11), 1641–1651 (2005). https://doi.org/10.1175/jcli3366.1

STARDEX and ETCCDI Climate Indices Based on E-OBS and CARPATCLIM
Part Two: ClimData in Use

Hristo Chervenkov$^{(\boxtimes)}$ (iD) and Kiril Slavov

National Institute of Meteorology and Hydrology, Bulgarian Academy of Sciences,
66, Tsarigradsko Shose blvd, 1784 Sofia, Bulgaria
{hristo.tchervenkov,kiril.slavov}@meteo.bg
http://www.meteo.bg

Abstract. In part one of the present article are described the motivation for the creation, the content, structure and the access point to the free available database of climate indices ClimData. Part two is dedicated on the possibilities of climate indices-based analysis with ClimData. They are demonstrated with some thermal absolute-thresholds and percentile-based indices. Most significant outcome of the study, beside the creation of ClimData itself, is the clearly expressed warming signal in the considered climate indices. This outcome agrees generally with the prevailing number recent studies. This signal is spatially dominating over Europe, almost everywhere statistically significant and coherent in the STARDEX and ETCCDI variants of the same indices. The latter suggest that relatively small modifications of the index-definitions or selection of the input data are not sufficient enough to change the general picture of revealed changes.

Keywords: Climate indices · E-OBS · CARPATCLIM · STARDEX ·
ETCCDI · ClimData database

1 Examples for ClimData in Use

Detailed description of the database of climate indices ClimData is performed in [3] and ClimData is available at https://repo.vi-seem.eu/handle/21.15102/VISEEM-343. The sets of climate indices (CIs) were introduced and discussed in [3]. However, the number of these indices is rather big. Therefore, it is necessary to reduce their number in each of these sets by choosing only the most representative of them. The problem is obviously not straightforward. STARDEX outlines a subset of 10 ('top10') indices, 6 precipitation and 4 temperature-based. Sillmann and Röckner [7] uses the minimum tn (TNn) and maximum tx (TXx) as well as the number of tropical nights (TR) from the ETCCDI suite, Birsan et al. [2] - the number of frost/ice/summer days, tropical nights, as well as cold/warm spell duration index and growing season length as thermal representatives. The tendency of the change of ice and summer days is investigated also in [4]. Regardless of their practical importance, however, the absolute-thresholds as well as the

© Springer Nature Switzerland AG 2019
G. Nikolov et al. (Eds.): NMA 2018, LNCS 11189, pp. 368–374, 2019.
https://doi.org/10.1007/978-3-030-10692-8_41

indices based on the duration of an event suffers from common issue: They are not transferable across the range of climatic regimes experienced across Europe. Thus, for example, the tropical night index is substantial only in low-elevation areas (below 800 m) and not over big areas in N and NE Europe. Similarly, the number of ice/frost days index have little meaning along the Mediterranean cost. To avoid this problem, we will demonstrate the usefulness of ClimData with four percentile-based indices: the relative share of days in year with tn below/above the $10^{th}/90^{th}$ percentile, called often cold/warm nights threshold and the relative share of days in year with tx below/above the $10^{th}/90^{th}$ percentile, known also as cold/warm days threshold. The cold nights as well as the warm days threshold, are in the STARDEX top10 subset.

1.1 Selected Cases

The aim of the following demonstration is both to assess the biases in the spatial distribution of the considered CIs, caused from the different definitions in the STARDEX and ETCCDI, and to examine the influence of the meteorological input dataset. According the Climate Indicator Bulletins (http://cib.knmi.nl/mediawiki/index.php/CIBs), 1976 is one of the coldest and 1989 - one of the warmest in pan-European context years and thus they are selected for the demonstration, as shown on Fig. 1.

Many conclusions could be drawn from these examples, but the most important of them are obvious. First and foremost, there are not significant discrepancies between the STARDEX and ETCCDI suite in the general pattern of the spatial distribution of the concerned CIs, obtained from E-OBS, nor from CARPAT-CLIM. The biases between the corresponding fields are dominantly between −5% and 5% and seems not systematic. The difference between the STARDEX and ETCCDI suite, concerning percentile-based CIs, is in the procedures in the calculation of the threshold in the base period. According [9], the problem occurs because it is affected by sampling error. The authors of this study states, that the proposed by them bootstrap procedure, which is consequently embedded in the FclimDex, effectively removes the inhomogeneity. Second, the values of the CIs, obtained with the E-OBS data for the CARPATCLIM domain, shown as frame on Fig. 1, are consistent with the corresponding distributions on Fig. 2. Intending to asses this consistency for other CIs, we present on Fig. 3 the fields of the number of frost and ice days as well as the growing season length for both years. They are valuable agrometeorological indicators relevant for cultivated plants phenology and active growth of crops. Hence the procedure, both in STARDEX and ETCCDI, for computation of the absolute-thresholds CIs is simple counting the events below/above the corresponding threshold, the fields of the frost/ice days from the same inputs are identical. The small differences in the growing season length could be rooted in the fact, that in the ETCCDI-suite the average from the tn and tx is used as a proxy for the td, instead of td in STARDEX, where it is independent input parameter. The differences for the same indicators, obtained from the both sources, either E-OBSv16.0 or CARPATCLIM, are inherited from the input data. They are more prominent near the Carpathian

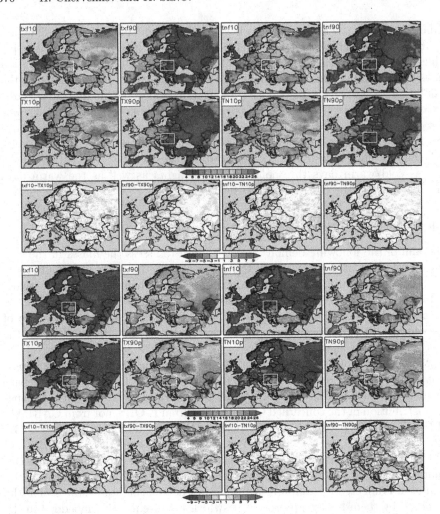

Fig. 1. Cold/warm days and cold/warm nights threshold (unit: %) for 1976 (lower group) and 1989 (upper group). The CIs from STARDEX are on the first, these of ETCCDI - on the second and the biases - on the third rows correspondingly. The original CIs-notation are preserved. The CARPATCLIM domain is marked with frame.

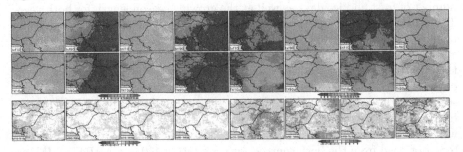

Fig. 2. Same as Fig. 1, but for CARPATCLIM

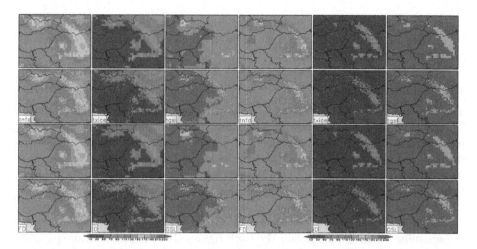

Fig. 3. Number of frost, ice days and growing season length for 1976 (left group) and 1989 (right group). The CIs from E-OBS/STARDEX, CARPATCLIM/STARDEX, E-OBS/ETCCDI and CARPATCLIM/ETCCDI are on the first second, third and fourth row correspondingly.

ridge, where the impact of the interpolation applied, altitude of concrete pixel and the influence of the orientation of slopes on climatology becomes visible.

The overall consistency between the STARDEX and ETCCDI CIs-suites for the considered indicators could be confirmed not only in space, but also in time. The spatially averaged time series of the cold/warm days as well as these of the cold/warm nights, obtained from E-OBS for the CARPATCLIM-domain and these from CARPATCLIM, are presented together on Fig. 4. Hence the

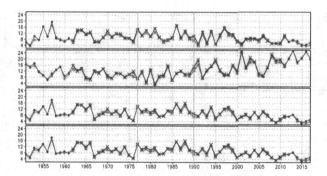

Fig. 4. Time series (unit: %) of the cold days (first row), warm days (second row), cold nights (thirth row) and warm nights (fourth row) thresholds. The E-OBS based values are marked with dots and these of CARPATCLIM - with crosses. The STARDEX-CIs are shown in red and these of ETCCDI - in blue. The green lines denotes the end of the years 1976 and 1989. (Color figure online)

corresponding chronograms are almost overlapping for the whole time-span of the E-OBS and CARPATCLIM, it could be concluded that systematic differences are absent.

1.2 Trend Analysis

The importance of assessing trends in weather extremes is often emphasized. The principal reason is that extreme weather conditions related to temperature, precipitation, storms or other aspects of climate, can cause loss of life, severe damage and large economic and societal losses (see [5] and references therein). As mentioned in [3], ClimData contains estimation linear trend by means of LSE and estimation of the statistical significance of trends with the MK test for all CIs-datasets on annual and seasonal basis according Table 2 in [3]. Figures 5 and 6 are examples for the considered in SubSect. 1.1 CIs. First and foremost, the warming trend is obvious on the both figures. It is almost overall spatially dominating and positive for the warm day/nights threshold and negative for the cold day/nights threshold. Generally, the trend for the nights, both cold and warm, is stronger than for the days. No remarkable differences between the STARDEX and ETCCDI indices could be outdrawn in this case as a whole. These conclusions are strengthened by the fact, that the trends are almost overall significant at the 5% level. From the few exceptions from the general picture for continental Europe, most interesting is the warm nights trend anomaly over Romania. The analysis shows that this is caused by very high values of the minimum temperature in couple adjacent gridcells in the early 1950s. Although presented in narrow time window, these outliers are strong enough to change the sign of the trend. In CARPATCLIM, which do not covers this period and uses other raw data and methods, this defect, as shown on Fig. 6, is completely absent. This problem, which is also apparent on the ECA&D maps, could be treated as E-OBS issue.

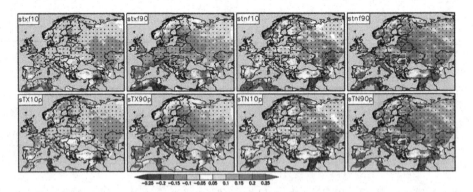

Fig. 5. LSE of the trend slope (unit: %/year) for the cold/warm days and cold/warm nights threshold from STARDEX (first row) and ETCCDI (second row) CIs calculated from E-OBS. The gridcells with significance at the 5% level are marked with dots.

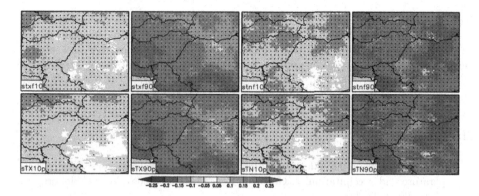

Fig. 6. Same as Fig. 5, but for CARPATCLIM

2 Conclusion

The CARPATCLIM and ECA&D projects contributes to the availability of a viable sets of spatially and temporally representative data to prepare relevant climate change studies in the corresponding domains. These datasets are reliable sources for input data for calculation of climate indices, which can be organized as continuous, both in space and time, digital maps.

Most significant outcome of the presented study, beside the creation of ClimData itself, is the clearly expressed warming signal in the considered climate indices. It is spatially dominating over Europe, almost everywhere statistically significant and coherent in the STARDEX and ETCCDI variants of the same indices. The last outcome suggest that small, but methodologically significant from point of view of the statistics (see [10] again) modifications in the CIs-definitions, could not alter the overall picture.

Although most widely used, the LSE of the slope of the linear trend is nonrobust technique and have to be applied with caution in presence of even few outliers, which is the common case in the climatology. This fact is also stated in [1], which recommends the use the outlier-resistant Theil-Sen method [6,8] as alternative. This method is used also in other studies, as the cited in [3] CIRCE project for example.

ClimData is intended to be barrier free and convenient versatile for broad range of experts - meteorologists, climatologists, hydrologist, which scientific research includes Cis-based analysis. Thus far, the ClimData datasets, together with their descriptors, are in GrADS binary form, but could be supplied also for the corresponding user in standard (netCDF) format upon request. They are available via the repository service of the VI–SEEM project (https://vi-seem. eu/) at https://repo.vi-seem.eu/handle/21.15102/VISEEM-343.

Acknowledgment. We thank to the anonymous reviewers for their comments and suggestions which led to an overall improvement of the original manuscript. The project for creation of ClimData and this study is entirely based on free available data and software. The authors would like to express their deep gratitude to the organizations and institutes (CARPATCLIM, ECA&D, STARDEX, ETCCDI, Unidata and others), which provides free of charge software and data. Without their innovative data services and tools this project would be not possible. This work was partially supported by the European Commission under H2020 project VI–SEEM (Contract No. 675121) and by the Bulgarian National Science Fund (grant DN-14/3/13.12.2017).

References

1. Alexander, L.V., et al.: Global observed changes in daily climate extremes of temperature and precipitation. J. Geophys. Res. **111**(D5) (2006). https://doi.org/10.1029/2005jd006290
2. Birsan, M.V., Dumitrescu, A., Micu, D.M., Cheval, S.: Changes in annual temperature extremes in the carpathians since AD 1961. Nat. Hazards **74**(3), 1899–1910 (2014). https://doi.org/10.1007/s11069-014-1290-5
3. Chervenkov, H., Slavov, K., Ivanov, V.: STARDEX and ETCCDI Climate Indices Based on E-OBS and CARPATCLIM-part one: general description. In: Numerical Methods and Applications (in press)
4. Lakatos, M., Szentimrey, T., Bihari, Z., Szalai, S.: Investigation of climate extremes in the Carpathian region on harmonized data. In: International Scientific Conference on Environmental Changes and Adaptation Strategies (2013)
5. Moberg, A., et al.: Indices for daily temperature and precipitation extremes in Europe analyzed for the period 1901–2000. J. Geophys. Res. **111**(D22), 1–25 (2006). https://doi.org/10.1029/2006jd007103
6. Sen, P.K.: Estimates of the regression coefficient based on kendall's tau. J. Am. Stat. Assoc. **63**(324), 1379–1389 (1968). https://doi.org/10.1080/01621459.1968.10480934
7. Sillmann, J., Röckner, E.: Indices for extreme events in projections of anthropogenic climate change. Clim. Change **86**(1–2), 83–104 (2007). https://doi.org/10.1007/s10584-007-9308-6
8. Theil, H.: A rank-invariant method of linear and polynomial regression analysis. Advanced Studies in Theoretical and Applied Econometrics, pp. 345–381. Springer, Netherlands (1992). https://doi.org/10.1007/978-94-011-2546-8-20
9. Zhang, X., Hegerl, G., Zwiers, F.W., Kenyon, J.: Avoiding in homogeneity in percentile-based indices of temperature extremes. J. Clim. **18**(11), 1641–1651 (2005). https://doi.org/10.1175/jcli3366.1
10. Zhang, X., et al.: Indices for monitoring changes in extremes based on daily temperature and precipitation data. Wiley Interdisc. Rev. Clim. Change **2**(6), 851–870 (2011). https://doi.org/10.1002/wcc.147

Global Sensitivity Analysis for a Chronic Myelogenous Leukemia Model

Gabriel Dimitriu[✉]

Department of Medical Informatics and Biostatistics, University of Medicine
and Pharmacy Grigore T. Popa, 700115 Iaşi, Romania
gabriel.dimitriu@umfiasi.ro

Abstract. The goal of this paper is to carry out a global sensitivity analysis applied to a mathematical model for chronic myelogenous leukemia (CML) dynamics with T cell interaction. The interaction mechanism between naïve T cells, effector T cells, and CML cancer cells in the body is modeled by a system of ordinary differential equations which defines rates of variation for the three cell populations. We explain how to globally analyse the sensitivity of this complex system by means of two graphical objects: the sensitivity heat map and the parameter sensitivity spectrum.

1 Introduction

Sensitivity analysis methods have a long history and it has been widely applied in different fields such as environmental modeling study [2], economical modeling for decision making [16], parameter estimation and control [3–5,12], chemical kinetics [19], and biological modeling analysis, with metabolic networks, signaling pathways and genetic circuits [14]. Local sensitivity analyses in biological models were performed in [7] (applied for a SEIT - Susceptible, Exposed, Infectious, and Treated - epidemic model), [8] (applied for an activated T cell model describing virus dynamics), and [6] (applied to a mathematical model for chronic myelogenous leukemia (CML) dynamics with T cell interaction).

In comparison with the local sensitivity analysis performed in [6], the aim of this study is to carry out a global sensitivity analysis [18] applied to the same model (proposed by Moore and Li in [15]), with the help of two graphical objects: the sensitivity heat map and the parameter sensitivity spectrum. The model defines the response of the human immune system to CML in a hypothetical patient. The immune system is known to play an important role in the dynamics of CML. That is why great efforts have been made to design new targeted and effective immunotherapies, possibly combined with controlled chemotherapies, capable to generate an active systemic immune response leading to elimination of residual and malignant cells [9,13].

© Springer Nature Switzerland AG 2019
G. Nikolov et al. (Eds.): NMA 2018, LNCS 11189, pp. 375–382, 2019.
https://doi.org/10.1007/978-3-030-10692-8_42

2 Application to a Chronic Myelogenous Leukemia Model with T Cell Interaction

2.1 Defining the Model

The model proposed in [15] consists of a system of three nonlinear ordinary differential equations which define the rates of change for naïve and effector T cell populations, as well as the CML cancer cell population. It is based in the blood circulation system. We assume the effector T cells come randomly into contact with CML cancer cells in the blood.

We introduce the following three populations of cells existed in the circulating blood system, measured as concentrations of cells per μl: T_n – naïve T cells, T_e – effector T cells specific to CML, and C – chronic myelogenous leukemia (CML) cancer cells. The system of ODEs is presented below (more details about the model are given in [15]):

$$\frac{dT_n}{dt} = s_n - d_n T_n - k_n T_n \left(\frac{C}{C + \eta} \right), \tag{1}$$

$$\frac{dT_e}{dt} = \alpha_n k_n T_n \left(\frac{C}{C + \eta} \right) + \alpha_e T_e \left(\frac{C}{C + \eta} \right) - d_e T_e - \gamma_e C T_e, \tag{2}$$

$$\frac{dC}{dt} = r_c C \ln \left(\frac{C_{max}}{C} \right) - d_c C - \gamma_c C T_e. \tag{3}$$

The parameter values are taken from available experimental data and estimates and the model is analyzed for sensitivity to changes in the parameters (see Table 1, p. 516 in [15] for more details). The Eqs. (1)–(3) define the rates of change with respect to time variable of the three cell populations. The system (1)–(3) also contains 12 parameters, which are all positive constants (Table 1). The saturation effect of CML cells is generated in this model by the presence of the Michaelis-Menten term $C/(C + \eta)$ in the equations (1) and (2). The growth of CML cells is modeled in the last Eq. (3) by using a term defining a Gompertz distribution. The Gompertz law is considered the best choice for leukemic cancer

Table 1. Description of parameters of the CML model

Parameter	Biological meaning	Parameter	Biological meaning
s_n	T_n source term	α_n	T_n proliferation rate
d_n	T_n death rate	α_e	T_e recruitment rate
d_e	T_e death rate	C_{max}	maximum C
d_c	C death rate	r_c	C growth rate
k_n	T_n differentiation rate	γ_e	T_e loss rate (due to C)
η	Michaelis-Menten constant	γ_c	C loss rate (due to T_e)

behavior, and generally for cancerous tumors proliferation (see, e.g., [22]). Afenya and Calderón [1] proved that the shape of the Gompertz curve gives a much better fit for clinical data in leukemic cancer data than other well-known curves (e.g., logistic, exponential, or polynomial).

2.2 Numerical Simulations

We evaluate the sensitivity of the 12 parameters in the system (1)–(3) using the toolbox PeTTSy (Perturbation Theory Toolbox for Systems, see [10]). A matrix M is generated by stacking the variable time series, so that the j-th column represents the derivative of each variable time series with respect to the j-th selected parameter. A singular value decomposition is then performed on this matrix, such that $M = U\Sigma V^T$ (V^T denotes the transpose of the matrix V).

Matrix U contains the principal components (PC) of M, it will have one column for each selected parameter and a number of rows equal to the product of the number of selected time points and selected variables. The matrix Σ contains the diagonal entries σ_j (known as the singular values of M) equal to the number of selected parameters. We denote by W, the inverse of V – a square matrix with dimension equal to the number of selected parameter.

The sensitivity heat maps (see Figs. 1, 3 and 5) show the sensitivity of the variables to the principal components, versus time. These sensitivities are defined as the product of the j-th singular value and the j-th PC of M, $\sigma_j U_j$.

The parameter sensitivity spectrum, or strength values, are calculated by multiplying the j-th row of W by the j-th singular value, $\sigma_j W_j$. This represents the effect of perturbing each parameter on the j-th principal component. Strengths are very useful when you have identified a particular principal component to which the time series in very sensitive (see Figs. 2, 4 and 6).

The model (1)–(3) was solved numerically by choosing the solver ode15s in Matlab for stiff systems. The initial values for the three populations (measured in cells/μl) have been set as [15]

$$(T_n(0), T_e(0), C(0)) = (1510, 20, 10000). \tag{4}$$

These initial estimates of $T_n(0)$, $T_e(0)$, $C(0)$ have been set in accordance with numerous clinical databases. In what follows, the components of the vector state (T_n, T_e, C) are represented in Figs. 1, 3 and 5 by (y_1, y_2, y_3). We focus our sensitivity analysis on three scenarios.

Scenario #1: Here we consider the situation when CML are kept under control, in the sense that CML cell counts continuously decreases in time. The stability condition [15] regarding the number of equilibrium solutions

$$\frac{d_c}{r_c} < \ln\left(\frac{\gamma_e C_{max}}{d_n}\right) \tag{5}$$

is not fulfilled. In this case, the system (1)–(3) has only one equilibrium, the trivial (healthy) state $P_1 = (1, 0, 0)$, and this equilibrium is asymptotically stable.

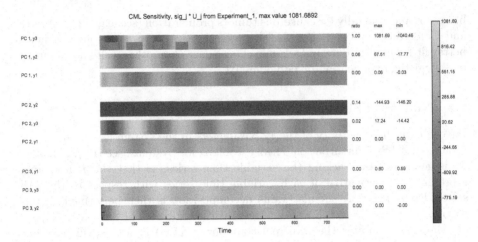

Fig. 1. The sensitivity heat map of the variables with respect to time corresponding to the first three principal components. Taking into account the first principal component, the variable y_3 (C) presents the highest sensitivity, followed by y_2 (T_e), and then y_1 (T_n). We also remark that only variable y_3 (CML cells) exceed 75% of the global maximum (regions highlighted by four small rectangles). – Scenario #1.

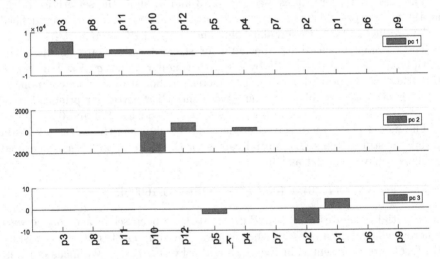

Fig. 2. The parameter sensitivity spectrum illustrates the effect of each parameter with respect to first three principal components. Focusing only on the first principal component, the parameter p_3 (d_e) is the most sensitive among all the model parameters, followed by p_8 (α_e) and p_{11} (γ_e) – Scenario #1.

The sensitivities are evaluated around the nominal parameter value (see [15]):

$$\tilde{p}_0 = [p_1\ p_2\ p_3\ p_4\ p_5\ p_6\ p_7\ p_8\ p_9\ p_{10}\ p_{11}\ p_{12}]^T$$
$$= [s_n\ d_n\ d_e\ d_c\ k_n\ \eta\ \alpha_n\ \alpha_e\ C_{max}\ r_c\ \gamma_e\ \gamma_c]^T$$
$$= [0.37; 0.23; 0.3; 0.028; 0.0062; 720; 0.14; 0.98; 2.3 \times 10^5; 0.0057; 0.057; 0.0034]^T.$$

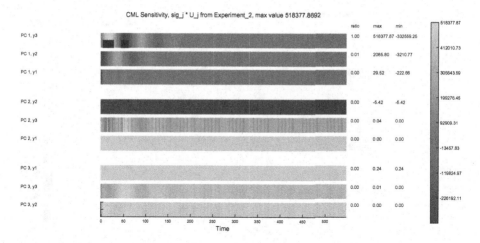

Fig. 3. The sensitivity heat map of the variables with respect to time corresponding to the first three principal components. – Scenario #2.

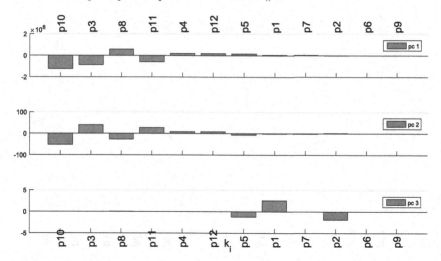

Fig. 4. The most sensitive parameter is p_{10} (r_c) taking into account only the first principal component – Scenario #2.

Scenario #2: In this experiment the parameter sensitivities are computed around the nominal parameter value:

$$\tilde{p}_0 = [s_n \; d_n \; d_e \; d_c \; k_n \; \eta \; \alpha_n \; \alpha_e \; C_{max} \; r_c \; \gamma_e \; \gamma_c]^T$$
$$= [0.29; 0.35; 0.40; 0.012; 0.066; 140; 0.39; 0.65; 1.6 \times 10^5; 0.011; 0.079; 0.058]^T.$$

It corresponds to the situation when CML cells increase over the time period.

Scenario #3: The parameter values used in this simulation lead to a remission and a relapse or rebound of CML. Initially, CML cell counts decrease, but

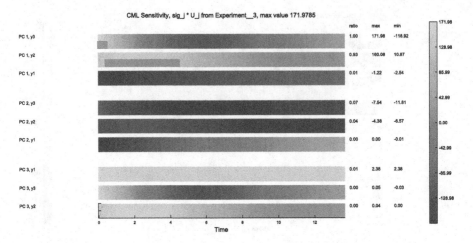

Fig. 5. The sensitivities of the variables y_3 (C) and y_2 (T_e) exceed for a longer period of time during treatment 75% of the global maximum – Scenario #3.

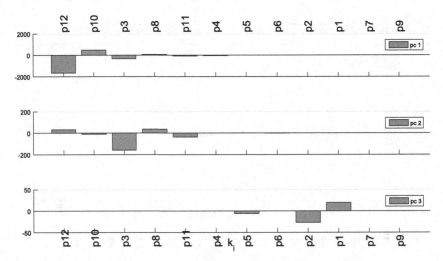

Fig. 6. The parameter sensitivity spectrum indicates p_{12} (γ_c) as the most sensitive parameter – Scenario #3.

later rebound. The parameter sensitivities were computed around the nominal parameter value:

$$\tilde{p}_0 = [s_n \ d_n \ d_e \ d_c \ k_n \ \eta \ \alpha_n \ \alpha_e \ C_{max} \ r_c \ \gamma_e \ \gamma_c]^T$$
$$= [0.071; 0.05; 0.12; 0.68; 0.0063; 43; 0.56; 0.53; 1.9 \times 10^5; 0.23; 0.0077; 0.047]^T.$$

The condition (5) is satisfied in the scenarios #2 and #3, and it was proved in [15] that the state system (1)–(3) exhibits two asymptotically stable equilibria, $P_1 = (1, 0, 0)$ and $P_2 = (T_n, T_e, C)$.

We notice that in all three scenarios the sensitivity heat maps indicate the same decreasing ordering of importance of the variables: y_3 (C), y_2 (T_e), and y_1 (T_n) with respect to their sensitivities on the first principal component.

3   Conclusions

In this work, we carried out a global sensitivity analysis with respect to parameters for a chronic myelogenous leukemia model with T cell interaction. Three scenarios have been taken into account in this sensitivity analysis.

In the first scenario, corresponding to the situation when the disease is kept under control, the parameter d_e (T_e death rate) was found the most sensitive among all the model parameters, followed by α_e (T_e recruitment rate) and γ_e (T_e loss rate due to C).

The sensitivity analysis applied to the second scenario (the case when CML cells increase over time period) revealed that the most sensitive parameter is r_c (the growth rate of CML) followed by d_e (T_e death rate) and α_e (T_e recruitment rate). These findings are in agreement with the observation in [15] that r_c, and the natural death rate, d_c, are the most significant parameters in the control of CML model. However, our sensitivity analysis placed the parameter d_c on the fifth position on a decreasing ordering of the sensitivity magnitudes (see Fig. 4, parameter p_4 corresponds to d_c). This result matches clinical observations by which increasing d_c may not be sufficient in all cases in controlling CML.

The third scenario simulated a remission of CML. The parameter γ_c (C loss rate due to T_e) was found the most sensitive parameter, followed by r_c (the growth rate of C cells) and d_e (the death rate of the effector T cells).

Other global sensitivity analyses (e.g., multi-parametric sensitivity analysis, Morris sensitivity analysis, weighted average of local sensitivities and Sobol sensitivity analysis) are also recommended to explore the impact of large model input variations [17,18,20,21]. At the same time, the extension to models including stochastic terms, nonlinear incidence rates of infection and distributed delays [11,23] could find new and plausible strategies for clinical relevance and laboratory research on treatment protocols with CML.

References

1. Afenya, E., Calderón, C.: Diverse ideas on the growth kinetics of disseminated cancer cells. Bull. Math. Biol. **62**, 527–542 (2000)
2. Daescu, D.N., Navon, I.M.: Sensitivity analysis in nonlinear variational data assimilation: theoretical aspects and applications. In: Farago, I., Zlatev, Z. (eds.) Advanced Numerical Methods for Complex Environmental Models: Needs and Availability, pp. 1–16. Bentham Science Publishers (2013)
3. Dimitriu, G.: Determination of the optimal inputs for an identification problem using sensitivity analysis. In: Proceedings of the Second International Symposium on Sensitivity Analysis of Model Output (SAMO), Venice, 19–22 April 1998, pp. 99–102 (1998)

4. Dimitriu, G.: Numerical approximation of the optimal inputs for an identification problem. Intern. J. Comput. Math. **70**, 197–209 (1998)
5. Dimitriu, G.: Convergence rate for a convection parameter identified using Tikhonov regularization. In: Vulkov, L., Yalamov, P., Waśniewski, J. (eds.) NAA 2000. LNCS, vol. 1988, pp. 246–252. Springer, Heidelberg (2001). https://doi.org/10.1007/3-540-45262-1_30
6. Dimitriu, G.: Identifiability and Sensitivity Analyses for a Chronic Myelogenous Meukemia Model with T Cell Interaction. In: Proceedings of the IEEE International Conference ICSTCC, Sinaia, 19–21 October 2017, pp. 704–710 (2017)
7. Dimitriu, G., Boiculese, V.L.: Sensitivity study for a SEIT epidemic model. In: Proceedings of the 5th IEEE International Conference on e-Health and Bioengineering, Iaşi, 19–21 November 2015
8. Dimitriu, G., Moscalu, M., Boiculese, V.L.: A local sensitivity study for an activated T-cell model. In: Proceedings of the International Conference on e-Health and Bioengineering, Sinaia, 22–24 June 2017
9. Dimitriu, G., Lorenzi, T., Ştefănescu, R.: Evolutionary dynamics of cancer cell populations under immune selection pressure and optimal control of chemotherapy. Math. Model. Nat. Phenom. **9**(04), 88–104 (2014)
10. Domijan, M., Brown, P., Shulgin, B., Rand, D.: Using PeTTSy, Perturbation Theory Toolbox for Systems. Warwick Systems Biology Centre
11. Georgescu, P., Hsieh, Y.-H.: Global stability for a virus dynamics model with nonlinear incidence of infection and removal. SIAM J. Appl. Math. **67**(2), 337–353 (2006)
12. Kalaba, R., Spingarn, K.: Control, Identification, and Input Optimization. Plenum Press, New York (1982)
13. Kirschner, D., Panetta, J.C.: Modeling immunotherapy of the tumor-immune interaction. J. Math. Biol. **37**, 235–252 (1998)
14. Li, Z.: Sensitivity analysis approaches applied to systems biology models. IET Syst. Biol. **5**(6), 336–346 (2011)
15. Moore, H., Li, N.K.: A mathematical model for chronic myelogenous leukemia (CML) and T cell interaction. J. Theor. Biol. **227**, 513–523 (2004)
16. Pannell, D.J.: Sensitivity analysis of normative economic models: theoretical framework and practical strategies. Agric. Econ. **16**(2), 139–152 (1997)
17. Robert, C.P., Casella, G.: Monte Carlo Statistical Methods. Springer, New York (2004). https://doi.org/10.1007/978-1-4757-4145-2
18. Saltelli, A., Campolongo, F., Cariboni, J., et al.: Global Sensitivity Analysis: The Primer. Wiley-Interscience, Chichester (2008)
19. Saltelli, A., Ratto, M., Tarantola, S., Campolongo, F.: Sensitivity analysis for chemical models. Chem. Rev. **105**(7), 2811–2827 (2005)
20. Sellier, J.M., Dimov, I.: A sensitivity study of the Wigner Monte Carlo method. J. Comput. Appl. Math. **277**, 87–93 (2015)
21. Sobol, I.M.: Global sensitivity indices for nonlinear mathematical models and their Monte Carlo estimates. Math. Comput. Simul. **55**(1–3), 271–280 (2001)
22. Steel, G.G.: Growth of Kinetic Tumors. Oxford University Press, Oxford (1977)
23. Zhang, H., Xia, J., Georgescu, P.: Stability analyses of deterministic and stochastic SEIRI epidemic models with nonlinear incidence rates and distributed delay. Nonlinear Anal. Model. Control **22**(1), 64–83 (2017)

Numerical Analysis of a Pollution and Environment Interaction Model

Ivan Dimov[1], Juri Kandilarov[2(✉)], Venelin Todorov[1,3], and Lubin Vulkov[2]

[1] Institute of Information and Communication Technologies, BAS, Sofia, Bulgaria
ivdimov@bas.bg
[2] Department of Mathematics, University of Ruse, Ruse, Bulgaria
{ukandilarov,lvalkov}@uni-ruse.bg
[3] Institute of Mathematics and Informatics, BAS, Sofia, Bulgaria
venelintodorov@fmi.uni-sofia.bg

Abstract. Finite difference and finite element approximations for solving numerically the systems of partial differential equations, by which comprehensive models for studying complex environmental problems are studied, are proposed and discussed in this paper. First, we establish a minimum principle for the differential problem and then nonnegativity of the semidiscrete solutions. Algorithms of explicit-implicit and fully explicit schemes are realized for solution of the discrete systems. Numerical experiments are provided to illustrate the efficiency of the algorithms.

Keywords: Environment interaction model · Parabolic system
Finite difference scheme · FEM · Immersed interface method

1 Introduction

Modern air pollution models can be used to simulate the evolution of the concentrations of contaminants in the atmosphere and living environment, see e.g. [1–3,7,9,11]. Our objective is the numerical investigation of the interaction between a pollutant and the living environment. It is known that the animate nature can absorb pollutant up to certain limits (threshold value) and after this amount this capability is loosed. Various experiments show that the dependence between the emitted quantity of a pollutant and remaining quantity can be described by a certain function. If some quantity of the pollutant is emitted regularly, one obtain the iterative process for a sequence of functions describing the dependence between the emitted and the remaining quantities of the pollutant. It is proved in [1,2] that this sequence converges. Using this discrete functional model, the authors construct a system of two parabolic equations of Lotka-Volterra types and derived two- and three-dimensional models.

The patterns exhibited by the conceptual model lay the foundation for introducing a dynamic model based on differential equations. The authors of [1,2] consider the distributed model of the interaction between the pollutant and environment on the plane. Let x_k, y_k be coordinates of the pollutant's sources

© Springer Nature Switzerland AG 2019
G. Nikolov et al. (Eds.): NMA 2018, LNCS 11189, pp. 383–391, 2019.
https://doi.org/10.1007/978-3-030-10692-8_43

on the plane, let Q_k be the intensities of these sources, $k = 1, 2, \ldots, m$. Taking as a basis the differential model of air pollution [7,11] and using the equation of biomass concentration, they obtain the following initial value problem for the system of two semi-linear parabolic equations

$$\frac{\partial w}{\partial t} = k_x \frac{\partial^2 w}{\partial x^2} + k_y \frac{\partial^2 w}{\partial y^2} + \nu_x \frac{\partial w}{\partial x} + \nu_y \frac{\partial w}{\partial y} + f(w, v) + \sum_{i=1}^{m} Q_i \delta(x - x_i, y - y_i),$$

$$\frac{\partial v}{\partial t} = d_x \frac{\partial^2 v}{\partial x^2} + d_y \frac{\partial^2 v}{\partial y^2} + g(w, v) \tag{1}$$

$$w(x, y, 0) = \varphi(x, y), \quad v(x, y, 0) = \psi(x, y),$$

where, following the modeling in [1,2], the nonlinear terms have the form

$$f(w, v) = -g_0 w - \frac{wv}{\lambda + w} + u^0(x, t), \quad g(w, v) = u_0 v - dv^2 - wv. \tag{2}$$

Here w is the concentration of the substance (the pollution level) and v is the concentration of the biological mass, k_x, k_y are diffusion coefficients, ν_x, ν_y are spreading velocities of the pollutant and d_x, d_y are diffusion coefficients, while $\varphi(x, y), \psi(x, y)$ are some initial distributions of the pollutant and the animate nature on the plane, respectively. If the pollutant sources cover some set on the plane, then in the first equation of (1) the sum becomes the product of the values of intensity Q by the delta-function of this set. For instance, if the pollution is a highway, then the term with the delta-function has the form

$$Q\delta(y - \chi(x)) \quad \text{on the curve} \quad y = \chi(x), \tag{3}$$

which is determined by the location of the highway on the plane and $Q = Q(t) > 0$. For application of atmosphere dispersion models in the estimations of vehicle emissions from cars driving along busy highways, which can be approximated as continuous curves, see e.g. in [2,3].

In this paper, we focus our attention on the application of parabolic systems to modeling in the interaction between pollution and environment. For spatial discretization we have implemented two variants of the Galerkin FEM scheme and also simple difference schemes in order to investigate the possible benefits of the *immersed interface* FEM approach. In Sect. 2, we introduce the models we will investigate throughout the remainder of the paper. In Sect. 3, we discuss the basic properties of the differential solutions concentrating on the *positivity property*. Section 4 is devoted to the space FDM and FEM semidiscretizations with positivity conservation. Numerical results are analyzed in Sect. 5. Finally, the last section presents some conclusions.

2   One-Dimensional Problem

In this paper we present our methods on the 1D reaction-diffusion-convection system

$$\frac{\partial w}{\partial t} = a\frac{\partial^2 w}{\partial x^2} + b\frac{\partial w}{\partial x} + f(w,v) + Q(t)\delta(x - x_0), \tag{4}$$

$$\frac{\partial v}{\partial t} = c\frac{\partial^2 v}{\partial x^2} + g(w,v), \quad x > 0, \quad t > 0. \tag{5}$$

Here $a, b, c, d, f_0, g_0, \lambda$ and u^0 in (2) we take as positive constants and also we suppose that the process doesn't depend of the space variable y. We assume that the particle concentration is negligible at the distances far enough from the sources, so that we can impose the far-field boundary condition [7,11]

$$w(x,t) \to 0 \text{ as } x \to \infty. \tag{6}$$

At $x = 0$ we impose for w a mixed (Robin) boundary condition

$$\left(a\frac{\partial w}{\partial x} + bw\right)\bigg|_{x=0} = w_e(t), \quad 0 < t < T. \tag{7}$$

For biomass concentration we take the boundary conditions

$$c\frac{\partial v}{\partial x}\bigg|_{x=0} = v_l(t), \quad v(X,t) = v_r(t), \quad X \text{ large}, \quad 0 < t < T. \tag{8}$$

Multiplying both sides of (4) and (5) by a functions $\varphi, \psi \in H^1(\Omega)$, $\Omega = (0, X)$, after integration by parts and taking boundary conditions ((6), (7) and (8)) into account, we obtain the following variational formulation:

$$\int_0^X \frac{\partial w}{\partial t}\varphi(x)dx = -\int_0^X \left(a\frac{\partial w}{\partial x} + bw\right)\varphi'(x)dx - w_l(t)\varphi(0)$$

$$+ \int_0^X f(w,v)\varphi(x)dx + Q\varphi(x_0), \ \forall \ \varphi \in H^1(\Omega), \ \varphi(X) = 0, \tag{9}$$

$$\int_0^X \frac{\partial v}{\partial t}\psi(x)dx = -\int_0^X c\frac{\partial v}{\partial x}\psi'(x)dx + v_l(t)\psi(0)$$

$$+ \int_0^X g(w,v)\psi(x)dx, \ \forall \ \psi \in H^1(\Omega), \ \psi(X) = 0. \tag{10}$$

Existence and uniqueness of solutions in appropriate functional spaces can be treated on the base of the theory in [8], but the maximum (minimum) principle requires additional consideration.

Theorem 1 *(Minimum principle). Let $u^0 \geq 0$, $Q(t) \geq 0$ and w, v be any pair of solutions of system (4), (5) with initial conditions, such that*

$$w(x,0) = w_0(x) \geq 0, \quad v(x,0) = v_0(x) \geq 0.$$

Then, $w \geq 0$, $v \geq 0$ for all $x \in \Omega$, $t \in (0,T)$.

Proof *(Outline).* It uses linearization of the functions $f(w,v)$, $g(w,v)$ around $(0,0)$, see e.g. [8], and the *positivity property* of f and g as in Lemma 1 below.

3   Space Discretizations and Positivity

In this section we derive and analyze FDM and FEM space approximations of problem (4)–(8).

3.1   Second-Order Finite Difference Discretization

We can now discretize (4), (5) by FDM in order to obtain a system of ordinary differential equation (ODEs). The spatial domain $\Omega = (0, X)$ is divided into N equal intervals of length h and we consider points in $\overline{\Omega} = [0, 1] : x_i = (i - 1)h, \quad i = 1, \ldots, N + 1$. Using an artificial grid node $x_{-1} = -h$ we derive the following second-order FDM scheme:

$$\dot{w}_1 = -\left(\frac{2b}{h} + \frac{2a}{h^2} + \frac{b^2}{a} + g_0\right) w_1 + \frac{2a}{h^2} w_2 - \frac{w_1 v_l}{\lambda + w_1} + \frac{2}{h} w_l(t),$$

$$\dot{w}_i = \left(+\frac{a}{h^2} - \frac{b}{2h}\right) w_{i-1} - \left(\frac{2a}{h^2} + g_0\right) w_i + \left(\frac{a}{h^2} - \frac{b}{2h}\right) w_{i+1} - \frac{w_i v_i}{\lambda + w_i} + \frac{1}{h} Q(t) \delta_{i i_0},$$

$$\text{for } i = 2, \ldots, N, \quad \text{and } w_{N+1} = w_r(t),$$

$$\dot{v}_1 = -\left(\frac{2c}{h^2} + u_0\right) v_1 + \frac{2c}{h^2} v_2 - d v_1^2 - w_1 v_1 + \frac{2}{h} v_l(t), \tag{11}$$

$$\dot{v}_i = \frac{c}{h^2} v_{i-1} + \left(-\frac{2c}{h^2} + u_0\right) v_i + \frac{c}{h^2} v_{i+1} - d v_i^2 - w_i v_i,$$

$$\text{for } i = 2, \ldots, N, \quad \text{and } v_{N+1} = v_r(t).$$

3.2   Galerkin FEM Discretizations

We denote $\Omega_i = (x_i, x_{i+1}), \quad i = 1, \ldots, N$ and introduce the standard linear functions

$$S_1 = \begin{cases} (x_2 - x)/h, & x \in \Omega_1, \\ 0, & x \in \Omega \backslash \Omega_1, \end{cases}$$

$$S_{N+1} = \begin{cases} (x - x_N)/h, & x \in \Omega_N, \\ 0, & x \in \Omega \backslash \Omega_N, \end{cases} \quad S_i = \begin{cases} (x - x_{i-1})/h, & x \in \Omega_{i-1}, \\ (x_{i+1} - x)/h, & x \in \Omega_i, \\ 0, & x \in \Omega \backslash (\Omega_{i-1} \cup \Omega_i). \end{cases}$$

We also introduce the *modified basis functions* of the **immersed interface method** (IIM) [6], concerning the interface x_0 and $x_0 \in [x_{i_0}, x_{i_0+1})$:

$$S_{i_0} = \begin{cases} A_1 x + B_1, & x_0 \in [x_{i_0}, x_0], \\ C_1 x + D_1, & x \in [x_0, x_{i_0+1}], \end{cases} \quad S_{i_0+1} = \begin{cases} A_2 x + B_2, & x \in [x_{i_0}, x_0], \\ C_2 x + D_2, & x \in [x_0, x_{i_0+1}]. \end{cases}$$

where the constants $A_1, A_2, B_1, B_2, C_1, C_2, D_1, D_2$ are found from the relations

$$[S_i] = 0, \quad [a S_i'] = Q, \quad i = i_0, i_0+1,$$

$$S_{i_0}(x_{i_0}) = 1, \quad S_{i_0}(x_{i_0+1}) = 0, \quad S_{i_0+1}(x_{i_0}) = 0, \quad S_{i_0+1}(x_{i_0+1}) = 1.$$

We carry out the following approximations of w and v

$$w(x,t) \approx w_h(x,t) = \sum_{i=1}^{N} w_i(t)S_i(x), \quad v(x,t) \approx v_h(x,t) = \sum_{i=1}^{N} v_i(t)S_i(x).$$

Substituting these approximations in Eqs. (9), (10) and taking successively $\varphi(x) = S_j(x)$ and $\psi(x) = S_j(x)$, $j = 1, \ldots, N$ and performing a special treatment of the nonlinear terms of $f(w,v)$ and $g(w,w)$ we obtain:

$$\frac{1}{3}\dot{w}_1 + \frac{1}{6}\dot{w}_2 = \left(-\frac{a}{h^2} + \frac{b}{2h}\right)w_1 + \left(\frac{a}{h^2} + \frac{b}{2h}\right)w_2 + \frac{\lambda v_1}{2(\lambda + w_1)}$$
$$-g_0\left(\frac{1}{3}w_1 + \frac{1}{6}w_2\right) - \left(\frac{1}{3}v_1 + \frac{1}{6}v_2\right) + w_l(t),$$

$$\frac{1}{6}\dot{w}_{i-1} + \frac{2}{3}\dot{w}_i + \frac{1}{6}\dot{w}_{i+1} = \left(\frac{a}{h^2} - \frac{b}{2h}\right)w_{i-1} - \frac{2a}{h^2}w_i + \left(\frac{a}{h^2} + \frac{b}{2h}\right)w_{i+1}$$
$$-\frac{g_0}{6}(w_{i-1} + 4w_i + w_{i+1}) - \frac{1}{6}(v_{i-1} + 4v_i + v_{i+1}) + \frac{\lambda v_i}{2(\lambda + w_i)} + Q(t)S_i(x_0),$$

$$\text{for } i = 2, \ldots, N, \quad w_{N+1} = w_r(t);$$

$$\frac{1}{3}\dot{v}_1 + \frac{1}{6}\dot{v}_2 = \frac{c}{h^2}(v_1 - v_2) + u_0\left(\frac{1}{3}v_1 + \frac{1}{6}v_2\right) \qquad (12)$$

$$-d\left(\frac{1}{4}v_1^2 + \frac{1}{6}v_1 v_2 + \frac{1}{12}v_2^2\right) - \frac{1}{12}(3v_1 w_1 + w_1 v_2 + w_2 v_1 + w_2 v_2),$$

$$\frac{1}{6}v_{i-1} + \frac{2}{3}v_i + \frac{1}{6}v_{i+1} = c\frac{v_{i-1} - 2v_i + v_{i+1}}{h^2} + \frac{1}{6}u_0(v_{i-1} + 4v_i + v_{i+1})$$
$$- \frac{d}{12}\left(v_{i-1}^2 + 2v_{i-1}v_i + 6v_i^2 + 2v_i v_{i+1} + v_{i+1}^2\right)$$

$$-\frac{1}{12}(w_{i-1}v_{i-1} + w_{i-1}v_i + w_i v_{i-1} + 6w_i v_i + w_i v_{i+1} + w_{i+1}v_i + w_{i+1}v_{i+1}),$$

$$\text{for } i = 1, \ldots, N, \quad \text{and } v_{N+1} = v_r(t).$$

Following (9), we treated the fractional nonlinear term as follows:

$$\int_0^X \frac{wv}{\lambda + w}S_i(x)dx \approx \sum_{l=1}^{N} v_i(t)\int_0^X S_i(x)S_l(x)dx$$
$$- \lambda \sum_{j=1}^{N} v_i(t)\int_0^X \frac{S_j(x)S_i(x)}{\lambda + \sum_{k=1}^{N} w_k(t)S_k(x)}dx$$

For the computation of the integrals we used the compact support of $S_i(x)$ and the integrals of the second sum are approximated by the trapezoidal rule.

3.3   Positivity

Following the physical motivation we will discuss the *non-negativity* of the semidiscrete solutions. Consider the initial value problem (IVP)

$$\frac{\partial P}{\partial \tau} = \mathcal{F}(\tau, P), \quad \tau \geq 0, \quad P(0) = P_0, \quad P_0 \in R^\mu, \tag{13}$$

where $\mathcal{F} : R^+ \times R^\mu \to R^\mu$ is continuous and (13) has unique solution for all P_0.

In the following we will write $v \geq 0$ for a vector $v \in R^\mu$ if all the components are non-negative. The ODE system (13) **will be called positive** (short for "non-negative preserving") if $P(t) \geq 0$ holds for all $t \geq 0$ whenever $P_0 \geq 0$. An often used criteria is given by the following lemma:

Lemma 1 *(Theorem 7.1 in [4]). The IVP (13) is positive if and only if*

$$P_i = 0, \quad P_j \geq 0 \text{ for all } j \neq i \Longrightarrow \mathcal{F}_i(\tau, P) \geq 0, \quad i = 1, \dots, M$$

holds for all τ.

As a preparation for our analysis we express the semidiscrete problems in (11) and (12) in matrix form

$$M^w \frac{dW}{dt} = S^w W + F_w(W, V), \quad M^v \frac{dV}{dt} = S^v V + F_v(W, V), \tag{14}$$

where M^w, M^v and S^w, S^v are *mass* and *stiffness* matrices. We also use the lumped mass method, which results from replacing the mass matrices $M^w(m_{ij}^w)$ and $M^v(m_{ij}^v)$ by diagonal matrices D^w, D^v with diagonal elements $d_{ii}^w = \sum_{j=1}^N m_{ij}^w$ and respectively $d_{ii}^v = \sum_{j=1}^N m_{ij}^v$, see e.g. [10]. We now give sufficient conditions for positivity of the discrete solutions of problems (11) and (12).

Theorem 2. *Let the conditions of Theorem 1 are fulfilled. If $0.5\,\text{h} < a/b$ then the ODE systems (11) and ODE mass lumped system and (14) are positive.*

4   Numerical Results

We consider the problem (4)–(8) with the following values of the parameters: $a = 0.4$, $b = 4$, $c = 0.001$, $d = 1$, $\lambda = 1$, $u^0 = 0$, $u_0 = 0.5$, $g_0 = 1$ and $x_0 = 1/3$, see e.g. [2]. We take the initial conditions of the form: $v(x, 0) = sin(\pi x)$ and

$$w(x,0) = \begin{cases} 1 + sin(\pi x)/sin(\pi/3) + Q(1/a - 2/3)x, & 0 \leq x \leq x_0, \\ 1 + sin(\pi(1-x))/sin(2\pi/3) + Q(1/(3a) - 2x/3), & x_0 \leq x \leq 1. \end{cases}$$

In Fig. 1 we present the numerical solutions obtained by the implicit-explicit (IMEX) scheme in the case of Dirichlet boundary conditions $w(0, t) = 1$, $w(1, t) = 1 + Q(1/(3a) - 2/3)$, $v(0, t) = v(1, t) = 0$ and $Q = 17$. If we take

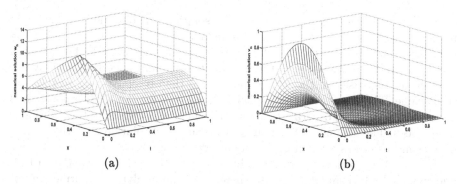

Fig. 1. Numerical solution with mesh parameters $N = 32$, $M = 40$, and constant Dirichlet boundary conditions and IMEX FDS: (a) for w_h; (b) for v_h.

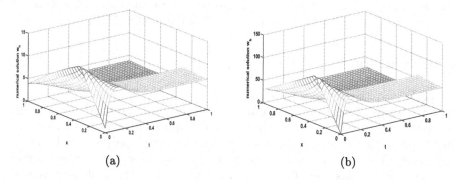

Fig. 2. Numerical solution with mesh parameters $N = 32$, $M = 40$, and mixed boundary conditions: (a) for w_h, $Q = 17$ and implicit FDS (b) for w_h, $Q = 170$ and FEM

Table 1. The errors in maximum norm for the numerical example

M	N	IMEX			Standard FEM			Modified FEM		
		w_h	Diff.	Order	w_h	Diff.	Order	w_h	Diff.	Order
32	40	6.6757			6.4142	–	–	6.4663		
64	80	6.9912	0.3155		6.7112	0.2970	–	6.7119	0.2456	
128	160	6.8436	0.1476	1.0959	6.5601	0.1511	0.9749	6.6232	0.0886	1.47
256	320	6.9198	0.0762	0.9538	6.6348	0.0747	1.0163	6.6519	0.0286	1.63
512	640	6.9634	0.0436	0.8054	6.5972	0.0376	0.9904	6.6431	0.0087	1.71

bigger values of the local source term $Q = 170$, to preserve the positivity of the numerical solution we must decrease the space mesh parameter h. The position of the source term is chosen so that it is not a node of the stencil. Then instead the term $1/hQ\delta_{ii_0}$ in (11) using the ideas from the IIM we add respectively for the $i = i_0$ and $i = i_0 + 1$ the terms $\frac{(x_{i_0+1} - x_0)}{h^2}(1 + bh)\frac{Q}{a}\delta_{i,i_0}$, $\frac{(x_0 - x_{i_0})}{h^2}(1 - bh)\frac{Q}{a}\delta_{i,i_0+1}$.

In Fig. 2(a) we present the numerical solution w_h obtained by the fully implicit scheme in the case of mixed boundary condition $w_l(0) = 0$, $w(1, t) = 1 + Q(1/(3a) - 2/3)$, $v_l(t) = v_r(t) = 0$, $Q = 17$ and in Fig. 2(b) - for $Q = 170$ and FEM. In both cases due to the mixed left boundary condition initially the solution w_h increases on the left boundary and then it slowly goes to the constant value. The results for the implicit scheme and for the FEM are obtained using Matlab function **ode15s**. In all experiments the positivity of the solution is clearly seen. The common behaviour for the computed solution v_h is the rapidly decreasing to zero as a result of the pollutant source term.

In Table 1 we present the results, obtained by IMEX scheme, standard FEM, and modified FEM for $Q = 17$. As there is not an analytical solution we use the double mesh principle. We control the value of the numerical solution w_h at final time $t = 1$ and at $x = 10/32 = 0.3125$ - the mesh point, closed to the interface $x_0 = 1/3$ and common for all meshes. The results show first order of the standard schemes and near second order for the modified FEM.

5   Conclusion

We performed the rigorous numerical analysis of the semi-discrete in space approximations of the FDS and FEM schemes for a pollution and environment interaction model. The semidiscretizations are coupled with explicit-implicit and fully implicit schemes and they preserve the non-negativity property of the differential problem solution as numerical experiments also show. Detail experimental and theoretical analysis will be very interesting as well as the implementation of the *immersed interface method* [6] if the pollutant is a highway, see (3).

Acknowledgements. The first, third and fourth authors are supported by the Bulgarian National Fund of Science under Projects DN 12/5-2017 and DN 12/4-2017, and the second author - by the Bilateral Project DNTS/Russia 02/12 from 2018.

References

1. Bratus, A., Mescherin, A., Novozhilov, A.: Mathematical models of interaction between pollutant and environment. In: Proceedings of the International Conference on "Control of Oscillations and Chaos'2000", vol. 3, St. Petersburg (2000)
2. Bratus, A., Mescherin, A., Novozhilov, A.: Mathematical models of interaction between pollutant and environment, Vest. MGU, Vych. Mat. Kybern. vol. 1, pp. 23–28 (2001)
3. Ganev, K.G., Syrakov, D.E., Zlatev, Z.: Effective indices for emissions from road transport. In: Lirkov, I., Margenov, S., Waśniewski, J. (eds.) LSSC 2007. LNCS, vol. 4818, pp. 401–409. Springer, Heidelberg (2008). https://doi.org/10.1007/978-3-540-78827-0_45
4. Hundstorfer, W., Vrewer, J.G.: Numerical Solution of Time-Dependent Advection-Diffuison-Reaction Equations, vol. 33. Springer, Heidelberg (2003). https://doi.org/10.1007/978-3-662-09017-6

5. Kandilarov, J.D., Vulkov, L.G.: The immersed interface method for two-dimansional heat-diffusion equation with singular own sources. Appl. Numer. Math. **57**, 486–497 (2007)
6. Li, Z., Ito, K.: The Immersed Interface Method - Numerical Solutions of PDEs Involving Interfaces and Irregular Domains. SIAM, Philadelphia (2006)
7. Marchuk, G.I.: Mathematical Modeling in Environmental Problems. Science, Moscow (1977). (in Russian)
8. Pao, C.V.: Nonlinaer Parabolic and Elliptic Equations. Plenum Press, New York (1992)
9. Penenko, A.V.: Consistent numerical scheme for solving nonlinear inverse source problems with gradient-type algorithms and Newton-Kantorovich methods. Numer. Anal. Appl. **11**(1), 73–88 (2018)
10. Thomée, V.: On positivity preservation in some finite element methods for the heat equation. In: Dimov, I., Fidanova, S., Lirkov, I. (eds.) NMA 2014. LNCS, vol. 8962, pp. 13–24. Springer, Cham (2015). https://doi.org/10.1007/978-3-319-15585-2_2
11. Zlatev, Z., Dimov, I.: Computational and Numerical Challenges in Environmental Modeling. Elsevier, Amsterdam (2006)

Computer Simulation of Heat Exchange Through 3D Fabric Layer

Aušra Gadeikytė$^{(\boxtimes)}$ (iD) and Rimantas Barauskas

Department of Applied Informatics, Kaunas University of Technology, Studentu Str. 50-407, 51368 Kaunas, Lithuania
{ausra.gadeikyte,rimantas.barauskas}@ktu.lt

Abstract. In this study the mathematical model of coupled problem of fluid flow and heat transfer through 3D polypropylene textile layer at micro-scale is presented. The purpose of the study is to investigate the influence of the fabric structural pattern on the heat resistance between human skin and the outer environment. The heat transfer mechanism includes the heat conduction of the solid fabric structure, as well as the convective heat transfer by means of the air and water vapor flow through the structure. The finite element model was created by using COMSOL Multiphysics software. The simulation results are analyzed to verify heat transfer properties, which make influence on the wearing comfort of the clothes. Simultaneously, the heat conduction, air and water vapor permeability and other important parameters used for the fabric properties characterization at macro-scale can be obtained on the base of the micro-scale analysis results.

Keywords: 3D fabric layer · Heat exchange · Finite element
COMSOL Multiphysics

1 Introduction

The aim of this work is to investigate the influence of 3D polypropylene textile layer on the heat loss between human skin and the fabric. Polypropylene is widely used in products such as plastic parts, carpeting, reusable products, paper, laboratory equipment, technology, thermoplastic fiber reinforced composites, etc. [5]. Polypropylene does not absorb moisture, has high heat resistance and is mechanically flexible. One of the important challenges for clothing designers is to ensure the wearing comfort by providing necessary thermal and moisture concentration balance at the human skin surface referred to as the microclimate layer [2].

In the recent decades various aspects of heat exchange have been investigated. Heat transfer by conduction and radiation has been well understood and documented. The heat transfer by convection through porous media layer is not straightforward as there is no exact solution to this heat transfer problem. Most of the solutions are approximate or based on empirically obtained very

© Springer Nature Switzerland AG 2019
G. Nikolov et al. (Eds.): NMA 2018, LNCS 11189, pp. 392–399, 2019.
https://doi.org/10.1007/978-3-030-10692-8_44

simplified models [4,11]. The heat resistance of a fabric can be determined by applying the standard hot plate chamber experiment. The heat power necessary for maintaining the prescribed steady temperature value of the hot plate covered by the fabric sample is measured during the experiment [3]. Bhattacharjee et al. [4] used computational fluid dynamics (CFD) software for obtaining the equivalent heat transfer coefficient of the fabric under natural and forced heat convection by simulation. Alptekin et al. [1] numerically simulated 1D coupled transient heat transfer inside multi-layer firefighter protective clothing and skin layer. Dennis et al. [6] numerically simulated the time relationships of head and neck cooling with local tissue properties. They presented the 3D head-cooling helmet model. The surface convection boundary conditions were used in the regions where the helmet would contact the head. The tissue perfusion has been neglected in the model, and the heat transfer coefficient as h $= 25\,\mathrm{W/m^2\,°C}$ has been derived from real data. Verleye et al. [10] investigated the fluid flow through the woven structure, where the flow through air pores and around the yarns was considered. Generally, the flow through textiles is governed by the Navier-Stokes equations in the fluid domain and by the Brinkman equations in the porous domain. Venkataraman et al. [9] modelled the insulation behavior of nonwoven fabrics without and with aerogel. The heat transfer mechanism was simulated by using ANSYS and COMSOL Multiphysics finite element software. However, more precise theoretical investigations of heat exchange through 3D polypropylene fabric layer are not straightforward due to coupled heat and mass transfer and other physical processes, complex internal structure, unknown material properties which should be obtained from micro-scale analysis [2].

In this paper the finite element model of heat exchange through the 3D polypropylene textile layer at micro-scale was investigated including heat conduction through solid structures of the textile, heat convection due to the flow of air and water vapor. Numerical simulations were used to evaluate the heat flux and temperature distribution in the fibrous insulating material by using COMSOL Multiphysics software.

2 Methodology

2.1 Governing Equations

The air flow was modelled by Navier-Stokes equations in the free flow spaces of the textile structure and by Brinkman equations in the porous parts of the structure. The Navier-Stokes momentum equation reads as

$$0 = \nabla \cdot [-p\mathbf{I} + \mu(\nabla\mathbf{u} + (\nabla\mathbf{u})^T)],$$

and is solved together with the continuity equation as

$$\rho\nabla \cdot \mathbf{u} = 0, \tag{1}$$

where ∇ is the gradient operator, $\mathbf{u}$ is the fluid flow velocity, ρ is the fluid mass density, p is the fluid pressure, μ is the fluid dynamic viscosity, $\mathbf{I}$ is the identity matrix.

The Brinkman equation for steady-state in the porous media flow reads as

$$0 = \nabla \cdot [-p\mathbf{I} + \frac{\mu}{\epsilon_p}(\nabla\mathbf{u} + (\nabla\mathbf{u})^T] - (\mu\kappa^{-1} + \frac{Q_{br}}{\epsilon_p^2})\mathbf{u},$$

$$\rho\nabla \cdot \mathbf{u} = Q_{br}, \tag{2}$$

where ϵ_p- constant porosity, κ- permeability. For zero mass source ($Q_{br} = 0$) the equation of continuity, Eq. 2 reduces to Eq. 1.

For a homogeneous fluid region, the governing heat transfer equation is expressed as

$$\rho C_p \mathbf{u} \cdot \nabla T + \nabla \cdot \mathbf{q} = Q,$$

$$\mathbf{q} = -k\nabla T,$$

where C_p- specific heat capacity, Q- overall heat transfer, k- thermal conductivity of fluid. In porous media

$$\mathbf{q} = -k_{eff}\nabla T$$

where $k_{eff} = \theta_p k_p + (1 - \theta_p)k$ is effective thermal conductivity of the fluid-solid mixture, k_p is the thermal conductivity of the porous medium and θ_p is the volume fraction.

2.2   Finite Element Model

We developed three finite element models to analyze heat exchange mechanism in 3D textile layer at micro-scale. All models include two porous media domains which are 1.4 mm × 1.4 mm × 1 mm in length (x direction), width (y direction) and height (z direction) (see Fig. 1). In model 1 air domain is 1.4 mm × 1.4 mm × 6.3 mm in length, width, and height. In model 2 and model 3 the air domain is 1.4 mm × 1.4 mm × 5.3 mm in length, width, and height. The initial material properties of air and polypropylene are given in Table 1. The boundary conditions of heat transfer in porous media and the boundary conditions of free and porous media flow used in the three models are shown in Fig. 1. Model 1 simulates the heat transfer in porous media driven by the change of temperature from T = 37 °C at the human skin to the ambient temperature T = 20 °C at the opposite surface of the model. Model 2 presents the situation when convective heat flux from the human skin is given as a boundary condition. In Model 3 the heat source boundary condition at the human skin was employed.

Table 1. The characteristics of air and polypropylene at T = 20 °C [7,8]

Material	Density kg · m^{-3}	Thermal conductivity W/(m · K)	Dynamic viscosity kg/(m · s)	Specific heat J/(kg · K)
Air	1.2047	0.0256	$1.8205 \cdot 10^{-5}$	1006.1
Polypropylene	900	0.22	–	1700

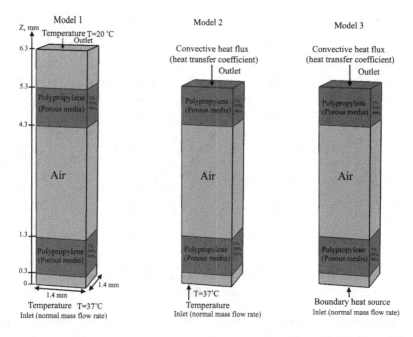

Fig. 1. Model 1, Model 2, Model 3 geometry and boundary conditions

3   Numerical Results

All three Models were constructed using the laminar flow mode combined with heat transfer in porous media flow mode. Model 1 presents the heat exchange process at constant temperature of skin (heating plate). Figure 2 shows the temperature distribution along Oz when porosity is 1%, 50% and 99% correspondingly. Figure 3 presents the dependencies of the temperature distribution along Oz where the heat transfer coefficient was varied from h = 20 W/m²K to h = 25 W/m²K at porosity 99%.

Model 1 and Model 2 are comparable when h = 25 W/m²K (without normal mass flow rate m = 0 kg/s) and h = 22 W/m²K (with normal mass flow rate m = 1.5092 · 10⁻⁸ kg/s). In our opinion, the assumed value h = 25(W/m²K) of the surface convection coefficient could be regarded as realistic basing on previous works of several researchers [6]. At porosity value 50% the same heat transfer coefficient values (h = 25 W/m²K and h = 22 W/m²K) were obtained (Fig. 4).

In model 3 we applied the surface convection coefficient value 22 W/m²K at normal mass flow rate m = 1.5092 · 10⁻⁸ kg/s. Figure 5 demonstrates temperature distribution and z position dependencies on general source Q. Model 1 and Model 3 are comparable when general source Q varies from 41 W/m² to 42 W/m².

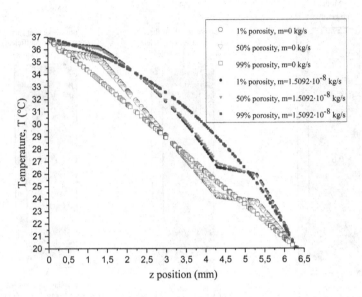

Fig. 2. Temperature distribution along Oz in Model 1, normal mass flow rate $m = 0\,kg/s$, $m = 1.5092 \cdot 10^{-8}$ kg/s

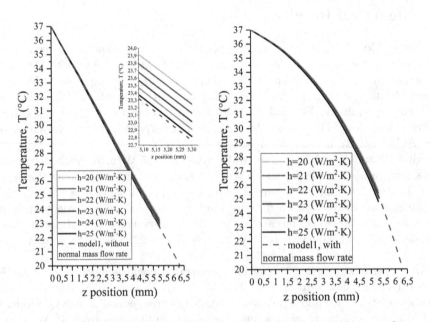

Fig. 3. Temperature distribution along Oz in Model 2, normal mass flow rate $m = 0\,kg/s$, $m = 1.5092 \cdot 10^{-8}$ kg/s, porosity 99%

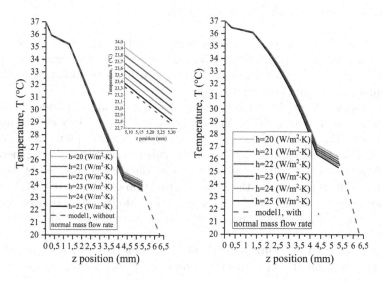

Fig. 4. Temperature distribution along Oz in Model 2, normal mass flow rate m $= 0\,\mathrm{kg/s}$, m $= 1.5092 \cdot 10^{-8}$ kg/s, porosity 50%

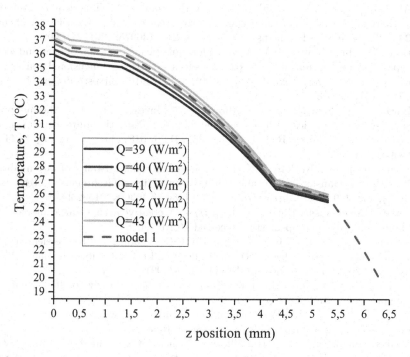

Fig. 5. Temperature distribution along Oz in Model 3, normal mass flow rate m $= 1.5092 \cdot 10^{-8}$ kg/s, porosity 50%, heat transfer coefficient h $= 22(\mathrm{W/m^2 K})$

4   Conclusions and Future Work

Computer simulations of heat exchange were carried out through 3D polypropylene textile layer by COMSOL Multiphysics software. The simulation results showed that Model 1, Model 2 and Model 3 can provide similar results when using appropriate boundary conditions.

The comparison of the results among the three models provided a certain background for obtaining the reasonable types of the boundary conditions and the values of the heat transfer coefficients. Skin (heating plate) temperature $T = 37\,^{\circ}\text{C}$ (Model 1) approximately corresponds to the surface heat generation power density as 41–42 W/m^2 (Model 3). From the results provided by Model 2 we obtained that the value of the surface convection coefficient h = 22 $\text{W/m}^2\text{K}$ could be regarded as reasonable. As a structure, Model 3 is the most realistic model which presents heat exchange between human skin and 3D textile layer at micro-scale.

Further research is to investigate convective heat transfer and the effect of polypropylene geometry structure (pore sizes) in Model 3.

References

1. Alptekin, E., Ezan, M.A., Gül, B.M., Kurt, H., Ezan, A.C.: Numerical investigation of thermal regulation inside firefighter protective clothing. Tekstil ve Mühendis **24**(106), 94–100 (2017). https://doi.org/10.7216/1300759920172410606
2. Barauskas, R., Abraitiene, A.: A model for numerical simulation of heat and water vapor exchange in multilayer textile packages with three-dimensional spacer fabric ventilation layer. Text. Res. J. **0**(00), 1–21 (2010). https://doi.org/10.1177/0040517510392468
3. Barauskas, R., Sankauskaite, A., Abraitiene, A.: Investigation of the thermal properties of spacer fabrics with bio-ceramic additives using the finite element model and experiment. Text. Res. J. **88**(3), 293–311 (2018). https://doi.org/10.1177/0040517516677228
4. Bhattacharjee, D., Kothari, V.K.: Prediction of thermal resistance of woven fabrics. Part II: heat transfer in natural and forced convective environments. J. Text. Inst. **99**(5), 433–449 (2008). https://doi.org/10.1080/00405000701582596
5. Creative Mechanisms Homepage. https://www.creativemechanisms.com/blog/all-about-polypropylene-pp-plastic. Accessed 30 Apr 2018
6. Dennis, B.H., Eberhart, R.C., Dulikravich, G.S., Radons, W.: Finite-element simulation of cooling of realistic 3-D human head and neck. J. Biomech. Eng. **125**(6), 832–840 (2003). https://doi.org/10.1115/1.1634991
7. Global Supplier for Materials Homepage. http://www.goodfellow.com/E/Polypropylene.html. Accessed 1 June 2018
8. University of Waterloo Homepage. http://www.mhtl.uwaterloo.ca/old/onlinetools/airprop/airprop.html. Accessed 1 June 2018
9. Venkataraman, M., Mishra, R., Militky, J., Behera, B.K.: Modelling and simulation of heat transfer by convection in aerogel treated nonwovens. J. Text. Inst. **108**(8), 1442–1453 (2017). https://doi.org/10.1080/00405000.2016.1255124

10. Verleye, B., Klitz, M., Crose, R., Roose, D., Lomov, S., Verpoest, I.: Computation of permeability of textile reinforcements. In: Proceedings, Scientific Computation IMACS, Paris, 11–15 July 2005
11. Zhu, G., Kremenakova, D., Wang, Y., Militky, J., Mishra, R., Wiener, J.: 3D Numerical simulation of laminar flow and conjugate heat transfer through fabric. AUTEX Res. J. **17**(1), 53–60 (2017). https://doi.org/10.1515/aut-2015-0052

Statistically Significant Comparative Performance Testing of Julia and Fortran Languages in Case of Runge–Kutta Methods

Migran N. Gevorkyan[1] , Anna V. Korolkova[1] , Dmitry S. Kulyabov[1,2]([⊠]) ,
and Konstantin P. Lovetskiy[1]

[1] Department of Applied Probability and Informatics,
Peoples' Friendship University of Russia (RUDN University),
6 Miklukho-Maklaya St, Moscow 117198, Russian Federation
{gevorkyan_mn,korolkova_av,kulyabov_ds,lovetskiy_kp}@rudn.university
[2] Laboratory of Information Technologies, Joint Institute for Nuclear Research,
6 Joliot-Curie St., Dubna, Moscow region 141980, Russian Federation

Abstract. In this paper we compare the performance of classical Runge–Kutta methods implemented in Fortran and Julia languages. We use the technique described in technical report by Tomas Kalibera and Richard E. Jones from University of Kent. This technique allows to solve the following problems. 1. The determination of the number of runs required by the program to pass the warm-up stage (e.g. JIT-compilation, memory buffers filling). 2. The determination of the optimal number of levels of the experiment and the number of repetitions at each level for robust testing. 3. The construction of the confidence interval for the resulting average run time. For the numerical experiment we implement 6-th order classical Runge–Kutta methods in both languages in the most similar way. We also study unvectorized versions of our functions. For Julia we tested not only built-in vectorization capabilities, but also external library. For processing the results of measurements Python 3 with Matplotlib, NumPy and SciPy (stats module) were used. We carried out experiments for variety of ODE dimensions (from 2 to 64) and different types of processors. Our work may be interesting not only for the results of comparison of the new Julia language with Fortran, but also for the robust testing method demonstration.

Keywords: Runge–Kutta scheme · Julia language
Fortran language · Performance

The publication has been prepared with the support of the "RUDN University Program 5-100" and funded by Russian Foundation for Basic Research (RFBR) according to the research project No 16-07-00556.

1   Introduction

In this paper, we compare the possibilities of vectorization of operations with arrays in Fortran and Julia [6,7] languages and the application of vectorization for implementation of classical Runge–Kutta schemes [9,11,12]. Average time measurements are made by the method described in papers [1,15]. This technique allows to obtain the statistically significant estimator of average time and determine the optimal number of measurements. In [1,15] the authors described in detail all the steps of their methodology for obtaining statistically significant estimators of the average time of programs execution. They also provided some numerical examples. In the first part of our paper, we will describe how all necessary statistical computations may be implemented by using Python [5,16] with NumPy [4], SciPy [14] and Matplotlib [3]. Our source code is open and available at https://bitbucket.org/mngev/fortran-vs-julia. The second part of the article describes the possibilities of arrays vectorization in Fortran and Julia languages. The programs that we used to compare these two languages are presented. We calculate statistically significant estimators of programs average time execution and the confidence interval for each measurement.

In the third part of the article, we compare the performance of two implementation of classical Runge-Kutta method — in Fortran and Julia languages. SIMD Vectorization is the most natural way to improve the performance of such methods, so the language (strictly speaking the compiler) that implements the best vectorization of actions with small arrays (vectors) will show better performance.

2   The Method of Program Execution Time Measuring

When measuring the execution time of the program the repetition of trials is generally accepted, as well as skipping a certain number of runs for so-called "warm-up". However, the number of trials and the number of runs for warm-up is often determined heuristically and the measurement results may be statistically insignificant.

In our work, we use the [1,15] methodology, which we briefly will describe in this part of paper, focusing on the implementation with Python and stack of scientific libraries.

2.1   The Number of the Experiment's Levels

In our case, the experiment will be understood as execution time measurement for a function, subprogram, program or software complex. Before starting an experiment, one should determine the maximum number of *experiment levels*.

For example, if we want to estimate the execution time of some function in a compiled programming language, we can distinguish three levels of the experiment. The first level is r_1 function calls inside the program, the second level is r_2 executions of program and the third is r_3 completions of the program source file.

In general, we assume that there is $n+1$ level of repetition of the experiment. At each level i, r_i tests are performed, where $i = 1, \ldots, n+1$. Number of tests at the highest level is r_{n+1}. The described technique is divided into three stages.

2.2   The Results of the Time Measurements

At the first stage, a preliminary experiment with a heuristic choice of the number of levels $n+1$ and the number of tests r_i at each i-th level is carried out. Each test gives a value of the measurement time $X_{j_{n+1}j_n \ldots j_2 j_1}$, where $j_i = 1, \ldots, r_i$. For example, if we measure the time spent on the third call of a function when the program is run for the second time using the fourth compiled executable file, it will be X_{423}. To store the results of measurements we used a multidimensional NumPy array. Every element of this array may be considered independent random number. At each level, a one-dimensional sample of the obtained time measurements is considered.

2.3   First Step: Visual Analysis

After the measurements are completed, one should proceed to a visual evaluation of the data, which will give an opportunity to estimate the number of pre-runs required for the system warm-up. We denote the number of pre-runs for each level by $c_1, c_2, \ldots, c_{r_i}$. For each level i one should draw three plots: run-sequence plot, lag-plot and auto-correlation function plot (ACF-plot).

To plot the autocorrelation graph, the use of the `acorr` function will give an incorrect result, since it uses a different algorithm for calculating autocorrelation in signal processing. So the calculation of autocorrelation must be implemented independently. The result of its work can be displayed on the chart using the `vlines` function.

2.4   Second Step: Biased and Unbiased Estimators

S_i^2 is biased estimator for variance at level i. Despite the cumbersome mathematical formulas, the computation of S_i^2 in terms of NumPy arrays is a trivial task. For example, let us measure time with three levels of experiment. As a result of measurements we obtain a three-dimensional array X. To calculate S_i^2, use `mean` function for 2 and 3 dimensions, and then `var` function to find the unbiased sample variance.

3   Vectorization of Action with Arrays in Fortran and Julia

3.1   SIMD Instructions

Most modern processors support SIMD (single instruction stream/multiple data stream) instructions. This means that a single processor core can apply the same operation to multiple numbers at the same time. To do this, the data should be written to special *vector* registers. For x86 architecture the most widespread technologies are: MMX (MultiMedia eXtension), SEE (Streaming SIMD Extensions) and AVX (Advanced Vector Extensions) of different revisions. For the measurements we used processors with AVX support, which gives 16 registers with a total volume of 256 bits.

3.2   Support for SIMD Instructions in Fortran and Julia

To measure the efficiency of SIMD capabilities utilization, let's consider several ways to add one array to another in Fortran and Julia languages. Let's begin from Fortran and set three dynamic arrays and then fill them with random numbers. You may also use the `concurrent` statement added to the Fortran 2008 [8,10,13] standard to explicitly point out the loop independence.

There are three options for similar actions in Julia. Two of them are comparable with Fortran. The third option allows using @simd and @inbounds macros to explicitly specify the usage of SIMD instructions.

3.3   Benchmarking

For Fortran program tests we used GNU Fortran [2] compiler version 7.3.0. To get a file with a report about loops vectorization attempts, it is necessary to add the option `-fopt-info-vec-all=file.log` to compiler. We have implemented pre-mentioned examples as separate pure functions. The Fortran source code is located in the `fortran` directory in the `src` folder. Each function takes numeric arrays as arguments, performs their addition and returns the result as an array. To measure the time required to call functions, we used the intrinsic subroutine `cpu_time`.

Summation was performed over arrays with floating point numbers of single precision (32 bits). The length of the array was calculated from the total volume of vector AVX registers (256 bits). The division by 32 bits gives us exact 8. The time required for 10^4 function calls was measured. Each measurement was repeated 100 times, the program was executed 20 times and compilation was performed also 20 times. In general, it turned out $100 \cdot 20 \cdot 20$ measurements.

Note also that optimization flags -O1, -O2 and -O3 were used during compilation.

Table 1. Results for Fortran time measurements. Confident interval was calculated for $\alpha = 0.01$

Overall mean			Conf. interval sizes			Optimum		
Function	01	02	03	01	02	03	01	02	03
Iter.	0.000436	0.000405	0.000400	0.000009	0.000010	0.000008	34	41	33
Vect.	0.000246	0.000374	0.000379	0.000011	0.000009	0.000009	27	29	36

The obtained results are summarized in the Table 1. In table one can find the overall mean time of execution for each of two functions, the confidence interval and the optimal number of repetitions for all stages, calculated by the method described above. For clarity, the overall mean time is shown as bar charts in the Fig. 1.

The bar chart clearly shows that when adding arrays without loops and without optimization (flag -01), the compiler automatically applied vectorization. At higher optimization levels, the compiler apply vectorization also to loops. It is noteworthy that the performance of the function with non-index addition excels slightly from more aggressive optimization (-02 and -03 flags).

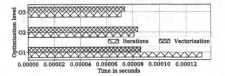   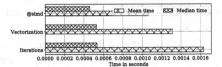

Fig. 1. Overall mean from Table 1   **Fig. 2.** Overall mean from Table 2

Similar measurements were carried out for programs in Julia. Due to the use of JIT compilation, the experiment levels were reduced from three to two: 20 program launches with 100 measurements at each launch. Also, there are no levels of optimization.

The results of measurements are summarized in the Table 2, and the overall mean time is clearly shown in the form of bar chart 2.

Table 2. Results for Julia time measurements. Confident interval was calculated for $\alpha = 0.01$

	Iterations	Vectorization	@simd
Overall mean	0.00163201	0.00131059	0.00105772
Conf. interval sizes	0.0011087924109	0.00087504387	0.00071035994
Optimal	11	11	11

The bar chart shows that the average time of Julia program is almost two orders of magnitude lower than Fortran. However, the median time is less than the average Fortran program execution time by an order of magnitude. If we consider the run-sequence plot, it becomes clear that a some values of measurements differ greatly from the average time. It is these anomalous measurements that lead to an increase in the average time.

As mentioned in the official documentation of the Julia language, the use of non-index operations in the Julia language does not provide any significant increase in productivity (see median time), as at the moment is only a syntactic sugar. However, the use of the @simd macro allows to get slightly better loop performance.

4   Performance Measurements for Runge-Kutta Schemes

Let us first introduce the numerical scheme of the explicit Runge-Kutta method for Cauchy problem [9, 12]. Let's consider two smooth functions: $\mathbf{x}(t)\colon [t_0, T] \to \mathbb{R}^N$ and $\mathbf{f}(t, \mathbf{x}(t))\colon \mathbb{R} \times \mathbb{R}^N \to \mathbb{R}^N$, where $[t_0, T] \in \mathbb{R}$. The initial value is $\mathbf{x}_0 = \mathbf{x}(t_0)$. Then the Cauchy problem for a system of N ordinary differential equations is formulated as follows:

$$\begin{cases} \dfrac{\mathrm{d}\mathbf{x}(t)}{\mathrm{d}t} = \mathbf{f}(t, \mathbf{x}), \\ \mathbf{x}(t_0) = \mathbf{x}_0. \end{cases} \tag{1}$$

The explicit s-stage Runge–Kutta method for the Cauchy problem (1) with constant step is given by the following formulas:

$$\mathbf{k}^1 = \mathbf{f}(t_m, \mathbf{x}_m),$$
$$\mathbf{k}^2 = \mathbf{f}(t_m + c^2 h, \mathbf{x}_m + h a_1^2 \mathbf{k}^1),$$
$$\dots$$
$$\mathbf{k}^s = \mathbf{f}\big(t_m + c^s h, \mathbf{x}_m + h(a_1^s \mathbf{k}^1 + a_2^s \mathbf{k}^2 + \dots + a_{s-1}^s \mathbf{k}^{s-1})\big),$$
$$\mathbf{x}_{m+1} = \mathbf{x}_m + h(b_1 \mathbf{k}^1 + b_2 \mathbf{k}^2 + \dots + b_{s-1} \mathbf{k}^{s-1} + b_s \mathbf{k}^s)$$

For this scheme it is difficult to build a parallel algorithm based on processes, since each $\mathbf{k}^i, i = 1, \dots, s$ depends explicitly on all previous ones and cannot be calculated independently. However, vectorization with SIMD instructions can give a performance benefit, since expression $a_1^i \mathbf{k}^1 + a_2^i \mathbf{k}^2 + \dots + a_{s-1}^i \mathbf{k}^{i-1}, i = 1, \dots, s$ can be efficiently vectorized at each stage.

We implemented an explicit numerical scheme with order $p = 6$ and stage $s = 7$ in Fortran and Julia languages in the most similar way (the function is called RKp6). In Fortran, we set the coefficients of the method as separate variables with the **parameter** attribute, and in Julia we used the **const** statement. This will allow Fortran and Julia compiles to optimize the program, as the values of the coefficients will be known at the compilation stage and can not be changed during the program run time.

For testing we used simple linear oscillator equation:

$$\frac{\mathrm{d}x_1(t)}{\mathrm{d}t} = -x_2, \quad \frac{\mathrm{d}x_2(t)}{\mathrm{d}t} = x_1,$$

on the time interval $[0, 10]$ with initial value $\mathbf{x} = (1, 1/2)^T$ and with the method step $h = 10^{-4}$. Was performed 1000 calls to the function RKp6, with 10 runs of programs (for Julia and for Fortran). We didn't repeat the compilations for Fortran, because the estimator S_3^2 showed that this level of experiment does not have a statistically significant effect on the result.

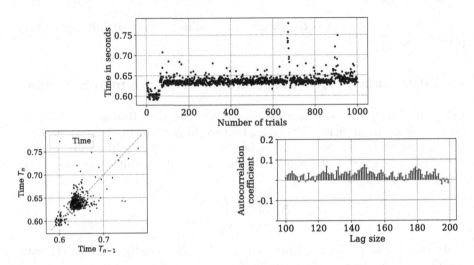

Fig. 3. The run-sequence plot, lag-plot and ACF-plot for Julia Runge-Kutta program measurements.

The Fig. 3 shows three plots required for visual estimation of measurements. The first run-sequence plot shows that approximately the first 100 measurements yielded results that were significantly different from the next ones, so we had to discard them. Moreover, the autocorrelation coefficient for the first 100 measurements was also much higher than the allowed values $[-0.1, 0.1]$. According to lag-plot, the measured values are distributed randomly and do not form any regular structures.

Fortran program was compiled with different optimization flags. The vectorization report showed that the compiler found beneficial to vectorized the addition of $\mathbf{k}^i$ inside the functions. As in our previous measurements, the Julia program was much slower (Fig. 4).

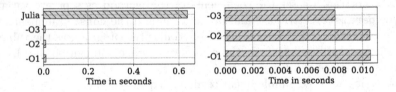

Fig. 4. Fortran and Julia Runge–Kutta performance comparison

5 Conclusion

We have measured the performance of Fortran and Julia languages in tasks that benefit from the use of vector instructions of modern processors. In addition, we briefly described the methodology [1,15], which we used during for our measurements.

Despite the fact that Julia showed worse results than Fortran program, it is necessary to make a number of notes in favor of Julia.

- Julia Language is still under heavy development and major changes may be expected in the syntax and the individual aspects of performance.
- We have performed measurements only for a specific array vectorization task, so for other tasks the performance difference may not be so significant.
- We have not considered the external library for static arrays for Julia, which could potentially bring performance boost.
- Julia is a dynamic language, so in most cases it will win against Fortran in development speed.

The source code of the examples and additional performance measurements can be found by link https://bitbucket.org/mngev/fortran-vs-julia.

References

1. Rigorous Benchmarking in Reasonable Time. ACM, New York, June 2013
2. Gnu fortran (2018). https://gcc.gnu.org/fortran/
3. Matplotlib home site (2018). https://matplotlib.org/
4. Numpy home site (2018). http://www.numpy.org//
5. Python home site (2018). https://www.python.org/
6. Bezanson, J., Edelman, A., Karpinski, S., Shah, V.B.: Julia: A fresh approach to numerical computing. SIAM Rev. **59**(1), 65–98 (2017). https://doi.org/10.1137/141000671
7. Bezanson, J., Karpinski, S., Shah, V.B., Edelman, A.: Julia: A Fast Dynamic Language for Technical Computing, September 2012
8. Brainerd, W.S.: Guide to Fortran 2008 Programming. Springer, London (2015). https://doi.org/10.1007/978-1-4471-6759-4
9. Butcher, J.: Numerical Methods for Ordinary Differential Equations, 2nd edn. Wiley, New Zealand (2003)
10. Chapman, S.J.: Fortran for Scientists and Engineers. McGraw-Hill Education, New York (2018)
11. Gevorkyan, M.N., Velieva, T.R., Korolkova, A.V., Kulyabov, D.S., Sevastyanov, L.A.: Stochastic Runge–Kutta software package for stochastic differential equations. In: Zamojski, W., Mazurkiewicz, J., Sugier, J., Walkowiak, T., Kacprzyk, J. (eds.) Dependability Engineering and Complex Systems. AISC, vol. 470, pp. 169–179. Springer, Cham (2016). https://doi.org/10.1007/978-3-319-39639-2_15
12. Hairer, E., Nørsett, S.P., Wanner, G.: Solving Ordinary Differential Equations I, 2nd edn. Springer, Heidelberg (2008). https://doi.org/10.1007/978-3-540-78862-1
13. Hanson, R.J., Hopkins, T.: Numerical Computing With Modern Fortran. SIAM, Philadelphia (2013)
14. Jones, E., Oliphant, T., Peterson, P., et al.: SciPy: Open source scientific tools for Python (2001). http://www.scipy.org/
15. Kalibera, T., Jones, R.E.: Quantifying Performance Changes with Effect Size Confidence Intervals. Technical report 4–12, University of Kent, June 2012
16. Rossum, G.: Python reference manual. Technical report, Amsterdam, The Netherlands (1995)

A Numerical Model for Random Fibre Networks

Mark Houghton$^{(\boxtimes)}$ ⓘ, David Head ⓘ, and Mark Walkley ⓘ

University Of Leeds, Leeds, UK
mm12m2h@leeds.ac.uk

Abstract. Modelling a random fibre network representative of a real world material leads to a large sparse linear matrix system with a high condition number. Current off-lattice networks are not a realistic model for the mechanical properties of the large volume of random fibres seen in actual materials. In this paper, we present the numerical methods employed within our two-dimensional and three-dimensional models that improve the computational time limitations seen in existing off-lattice models. Specifically, we give a performance comparison of two-dimensional random fibre networks solved iteratively with different choices of preconditioner, followed by some initial results of our three-dimensional model.

Keywords: Fibre network · Iterative · Preconditioning

1 Introduction

Many real world materials can be represented at the microscopic level by a random network of fibres, including paper and felt, non-woven fabrics, tissue scaffolds, and the cytoskeletons of eukaryotic cells. A good understanding of the mechanical behaviour of these networks is key to understanding physical properties at the macroscopic level and for the development of new materials.

2 Modelling a Random Fibre Network

In many applications, especially biological, it is appropriate to model individual fibres as semiflexible polymers [5]. With this assumption in place, the extensible *Wormlike Chain* model has been shown to adequately describe the elastic behaviour of individual semiflexible fibres [6]. Discretising this we can define fibre stretching, compression and bending, but we choose to neglect thermal contributions for simplicity. To expand this theory from single fibres to whole networks we adopt the Mikado model [3,4,7], which allows us to express the energy of the

---

M. Houghton—Supported by an EPSRC (UK) DTP studentship.

G. Nikolov et al. (Eds.): NMA 2018, LNCS 11189, pp. 408–415, 2019.
https://doi.org/10.1007/978-3-030-10692-8_46

system as the total sum of stretching energy added to the total sum of bending energy of the network,

$$E = \frac{\mu}{2} \sum_{ij} \frac{\delta \ell_{ij}^2}{\ell_{ij}} + \frac{1}{2} \sum_{\langle ijk \rangle} \frac{\kappa_{ijk} \theta_{ijk}^2}{\bar{\ell}_{ijk}}, \tag{1}$$

where the left sum considers all segments, ij, of every fibre of the network, and the right sum considers all of the consecutive segments, ij, jk, along each fibre, for stretching constant μ, bending modulus κ_{ijk}, segment length ℓ_{ij}, average length of two consecutive segments $\bar{\ell}_{ijk}$, and angular deflection θ_{ijk}. To better understand (1), it is useful to first consider individual fibres.

2.1   Fundamentals

Defining fibres as slender elastic bodies with uniform circular cross-sections, we can consider a two-dimensional plane or three-dimensional cuboid wherein a predetermined number of fibres are generated with random position and orientation. After cross-linking these fibres (see Sect. 2.5), a mechanical structure remains in which nodes are identified as freely rotating points, and categorised as one of three types. Boundary nodes occur as points fixed at the aperiodic boundaries of the domain, dangling nodes are attributed to the end points of fibres not on the boundary, and internal nodes are the points associated with cross-links. For adjacent nodes φ, ψ the tangent vector along the segment is

$$\hat{\mathbf{t}}_{\varphi\psi} = \frac{\mathbf{s}_\psi - \mathbf{s}_\varphi}{\ell_{\varphi\psi}},$$

where $\mathbf{s}_\varphi$ is the position vector at φ and $\ell_{\varphi\psi} = |\mathbf{s}_\psi - \mathbf{s}_\varphi|$ is the segment length. We denote $\mathbf{u}_\varphi$ as the displacement of an individual node after a load or perturbation has been applied to the network, and refer to individual components as $u_{\varphi_x}, u_{\varphi_y}, u_{\varphi_z}$.

2.2   Local Stretching Behaviour

Consider a segment $\varphi\psi$ comprising of an adjacent node pair φ, ψ respectively, treated as a simple Hookean spring. The stretching energy for this segment is

$$E_{\varphi\psi}^{\text{stretch}} = \frac{k_{\varphi\psi}}{2} [(\mathbf{u}_\psi - \mathbf{u}_\varphi) \cdot \hat{\mathbf{t}}_{\varphi\psi}]^2$$

with a stretching constant $k_{\varphi\psi} = \frac{\mu}{\ell_{\varphi\psi}} = \frac{AE^f}{\ell_{\varphi\psi}}$, for $A = \pi r^2$, radius r and Young's modulus E^f. The second partial derivatives of this energy contribute to the global Hessian matrix at rows and columns corresponding to the displacements of φ and ψ, and the first partial derivatives contribute to the right hand side vector.

2.3 Local Bending Behaviour

For the adjacent node triplet α, ω, β, segment $\alpha\omega\beta$ has bending energy

$$E_{\alpha\omega\beta}^{\text{bend}} = \frac{\kappa_{\alpha\omega\beta}\theta_{\alpha\omega\beta}^2}{2\bar{\ell}_{\alpha\omega\beta}} = \frac{\kappa_{\alpha\omega\beta}\left[(\mathbf{s}_\omega - \mathbf{s}_\alpha) \times (\mathbf{u}_\beta - \mathbf{u}_\omega) + (\mathbf{u}_\omega - \mathbf{u}_\alpha) \times (\mathbf{s}_\beta - \mathbf{s}_\omega)\right]^2}{\ell_{\alpha\omega}^2\ell_{\omega\beta}^2(\ell_{\alpha\omega} + \ell_{\omega\beta})},$$

for a bending constant $\kappa_{\alpha\omega\beta} = \frac{AE^f r^2}{4}$, with $A = \pi r^2$ and Young's modulus E^f. Similar to the stretching case, local contributions to the global Hessian and right hand side vector can be derived from this energy.

2.4 Global Assembly

Combining the displacements $\mathbf{u}_\varphi$ of every node φ, in the vector $\mathbf{U}$,

$$\mathbf{U} = [U_1, U_2, U_3, U_4, ...] = [u_{1_x}, u_{2_x}, ..., u_{1_y}, u_{2_y}, ..., u_{1_z}, u_{2_z}, ...],$$

then a Taylor series expansion of the energy of the system about $\mathbf{U} = 0$ is

$$E(\mathbf{U}) = E_0 + \sum_{i=1}^{N} U_i \frac{\partial E}{\partial U_i}\bigg|_{\mathbf{U}=0} + \frac{1}{2}\sum_{i=1}^{N}\sum_{j=1}^{N} U_i U_j \frac{\partial^2 E}{\partial U_i \partial U_j}\bigg|_{\mathbf{U}=0} + \, ...$$

The resulting system represents the network in mechanical equilibrium with the applied load or perturbation;

$$\sum_j U_j \frac{\partial^2 E}{\partial U_i \partial U_j} = B_i, \quad \text{where } B_i = -\frac{\partial E}{\partial U_i}\bigg|_{\mathbf{U}=0}$$

for $i, j = 1, ..., N$. Denoting the Hessian matrix as H, this gives the global linear matrix system $H\mathbf{U} = \mathbf{B}$ which can be assembled from the local contributions defined in Sects. 2.2 and 2.3.

2.5 Cross-Linking Fibres

In a two dimensional plane, a pair of randomly orientated fibres are trivially cross-linked at their unique point of intersection, if this exists. In three dimensions direct intersections of randomly orientated fibres occur with probability zero for radius $r \to 0^+$. In this case, the minimum separation is considered, and if this is below a given threshold, a short cross-linking fibre is inserted between the two closest points of the original pair of fibres. The cross-linking fibre can either be treated as a stiff and non-rotating inextensible rod, in which case each end point displaces identically, or as an elastic spring, in which case each end point can displace individually. An alternative approach is to constrain fibre orientation to lie along lattice vectors, such that fibres directly intersect. This lattice-based approach can achieve mechanical rigidity at lower fibre counts due to the higher coordination (connectivity) number, z_c, and has been used successfully to predict mechanical properties comparable to randomly orientated fibre networks seen in real-world materials [2]. We have made some preliminary lattice-based experiments (discussed in Sect. 4.3) before moving to randomly orientated networks.

3   Numerical Model

3.1   Linear System Structure (in 2D)

By grouping the unknown displacements in $\mathbf{U}$ by coordinate direction, the block structure of H can be written as

$$H = \begin{bmatrix} H_{xx} & H_{xy} \\ H_{xy}^T & H_{yy} \end{bmatrix}, \quad \text{where } H_{\varphi\psi} = \frac{\partial^2 E}{\partial u_{\varphi_i} \partial u_{\psi_j}}, \tag{2}$$

for $i, j = 1, ..., N$.

The sparsity of H is determined from the local connectivity of each fibre with the other fibres in the network. An initial estimate of the sparsity pattern can be obtained from the adjacency matrix of the internal nodes of a network, and is repeated for each sub-block $H_{\varphi\psi}$ of the global matrix H. To exploit the sparsity of H, we store the non-zero values in compressed sparse row (CSR) format. The matrix H is symmetric, as are each of the diagonal sub-blocks, but each off-diagonal sub-block is not necessarily symmetric.

3.2   Iterative Solution Strategy

Our system is symmetric but may not be positive definite, hence we choose the MINRES iterative method from the family of Krylov subspace methods. These methods perform best when the system is preconditioned to cluster the matrix eigenvalues. Exploiting the block structure seen in (2), we take the diagonal blocks

$$P_{Db} = \begin{bmatrix} H_{xx} & O \\ O & H_{yy} \end{bmatrix}, \tag{3}$$

as the first choice of preconditioner. A similar preconditioner can be defined when extending to three dimensions. Additionally we can also consider a simple asymmetric preconditioner and a symmetric preconditioner with a Schur complement block

$$P_{Ab} = \begin{bmatrix} H_{xx} & H_{xy} \\ O & H_{yy} \end{bmatrix}, \quad P_S = \begin{bmatrix} H_{xx} & O \\ O & S \end{bmatrix}, \tag{4}$$

where $S = H_{yy} - H_{xy}^T H_{xx}^{-1} H_{xy}$. GMRES was used with P_{Ab}, since this preconditioner does not preserve the symmetry of H.

4   Results

4.1   Performance Comparison in 2D

To evaluate the performance of the different choices of preconditioner, we measured the number of iterations required to converge to a relative tolerance of 10^{-3}, using the MINRES and GMRES solvers in MATLAB. The rate of convergence was measured for increasing system size by increasing the number of fibres, N_f, with fixed length $\ell = 0.25$ and radius $r = 0.01$. For each interval of

N_f, the results were averaged over 10 reproducibly-seeded randomly generated networks. Examples of individual network generations with varying N_f can be seen in Fig. 1.

As can be seen from Table 1, P_{Db} shows a consistent performance as N_f increases, with good evidence that we would expect to see the iteration number converge if N_f were to increase further. P_S also performs well, and P_{Ab} consistently shows the lowest iteration count for increasing N_f, but the additional cost in building the P_S preconditioner, and the additional cost per iteration of GMRES versus MINRES leaves P_{Db} as potentially the preferred choice. The diagonally scaled preconditioner D performs the worst, with large fluctuations in the iteration count and a large standard error.

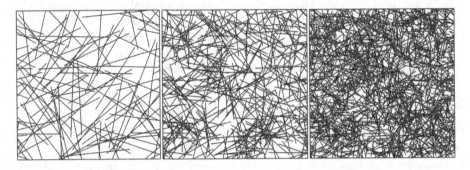

Fig. 1. Examples of generated 2D random fibre networks increasing in area with a fixed fibre density. The number of fibres, N_f, is 100 *(left)*, 400 and 1600 *(right)*, and fibre length, ℓ, and radius, r, are fixed at 0.25 and 0.01 respectively, but r is not to scale. Dangling ends have not been removed.

Table 1. Number of iterations required to converge for different preconditioners and an increasing number of fibres, N_f, of length 0.25 and radius 0.01. Standard error was calculated from a sample of 10 networks.

N_f	100	200	400	800
D	300.2 ± 54.0	480.8 ± 67.3	570.6 ± 77.3	410.9 ± 103.0
P_{Db}	8.0 ± 0.8	8.5 ± 0.5	7.7 ± 0.7	7.2 ± 0.6
P_S	12.3 ± 1.1	14.6 ± 1.2	13.8 ± 1.3	14.1 ± 0.9
P_{Ab}	5.3 ± 0.2	5.4 ± 0.2	5.4 ± 0.2	5.3 ± 0.2

As a further indication of performance we also collected estimates for the condition numbers of the preconditioned system and for H. To do this we consider a similar range of N_f as previously and provided the default seed to the `condest` function in MATLAB for 10 network generations at each interval. From the results in Table 2 we find additional evidence for the poor performance of D, with condition number estimates consistently higher than H without any pre-conditioning. P_S reduces the condition number of H by roughly an order of magnitude, and P_{Db} and P_{Ab} demonstrate the best performance, showing similar estimates to each other with the symmetric having slightly lower values on average for larger N_f.

Table 2. Estimated condition numbers, κ, of different preconditioners applied to H, for varied number of fibres, N_f.

N_f	100	200	400	800	1600
$\kappa(H)$	3.74e6	7.58e6	1.81e7	4.34e7	9.72e7
$\kappa(D^{-1}H)$	8.51e6	2.41e7	7.09e7	1.98e8	4.54e8
$\kappa(P_{Db}^{-1}H)$	7.32e4	1.47e5	4.85e5	1.60e6	5.12e6
$\kappa(P_S^{-1}H)$	1.39e5	4.97e5	1.18e6	4.22e6	-
$\kappa(P_{Ab}^{-1}H)$	6.98e4	1.96e5	5.99e5	2.11e6	-

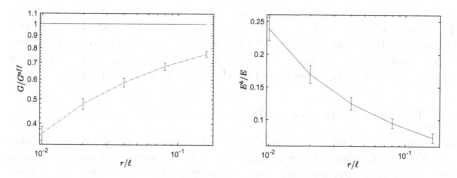

Fig. 2. r/ℓ, for $\ell = 0.25$ and $N_f = 100$ in a 0.5×0.5 plane against G/G^{aff} *(left)* and against E^b/E *(right)*. Nodes within a distance of 10^{-3} were merged into a single node prior to solution. The line $G/G^{\mathrm{aff}} = 1$ *(left)* corresponding to affine response is also shown.

In addition to analysing the performance of the preconditioners, we also investigated the mechanical properties of the networks. In particular for $N_f = 100$, with $\ell = 0.25$, we varied r and measured the shear modulus, G, and ratio of bending energy over total energy, E^b/E. From Fig. 2 *(right)* we can see a smooth crossover from a regime with comparable stretching and bending energies for small r, to a stretching dominated regime for large r. Figure 2 *(left)* demonstrates a smooth transition between affine (i.e. uniform deformation field) dominated behaviour at large r to non-affine behaviour for small r. This coincides well with the known result that stretching-dominated networks become close to affine [4].

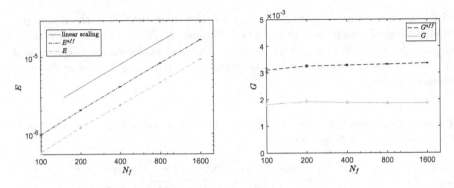

Fig. 3. Calculated values of the energy *(left)* and shear modulus *(right)* verified against the affine predictions for an increasing number of fibres with fixed length 0.25 and radius 0.01. The solid line is added to verify linear scaling.

4.2   Validation

To validate our model results, we verified that the calculated energy, E, and shear modulus, G, were bounded by the affine predictions E^{aff} and G^{aff} respectively. To do this, we increased N_f for $\ell = 0.25$ and $r = 0.01$, and plotted E and G alongside their corresponding affine predictions. Figure 3 demonstrates that E *(left)* and G *(right)* lie below the limits provided by the affine predictions. We can also see that E increases linearly with N_f, as we would expect.

4.3   Preliminary Results in 3D

In a step towards solving randomly orientated three dimensional networks built from the Mikado approach, we were able to obtain preliminary results by solving networks with predetermined direct intersections between fibres. These have irregularity introduced by overlaying different lattice based components together to form a rigid structure. More specifically we take 25 vertical fibres extending the height of a $1 \times 1 \times 1$ cube and insert lattice forming plates through the vertical fibres with varied orientations.

The example seen in Fig. 4 has an average coordination number of $z_c = 4.31$ with fibres of varied length, ℓ, radius $r = 0.01$, and a prescribed but irregular orientation. Given a shear strain $\gamma = 0.05$, the network is fixed in the plane $y = 0$, and sheared in the $y = 1$ plane in the positive x direction. Figure 4 *(left)* shows the structure prior to any shear, Fig. 4 *(centre)* shows the structure displaced after shearing, where the displacements are calculated using a direct solver. Figure 4 *(right)* shows the structure with every node displaced affinely. The total energy in this case is $E_{\text{aff}}^{\text{total}} = 5.0e^{-7}$. Comparing *(centre)* and *(right)*, we can see regions of the network in *(centre)* where bending is more favourable than affine displacement. The total energy here is $E^{\text{total}} = 1.4e^{-7}$, where bending energy $E^{\text{bend}} = 1.3e^{-7}$ is the main contribution.

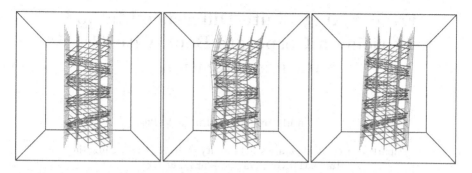

Fig. 4. Visualisations of one of the lattice based 3D networks, for the undisplaced *(left)*, displaced *(centre)* and affinely displaced *(right)* generations of the network where $\gamma = 0.05$.

5   Continued Work

Although lattice-based modelling can provide valuable insight into the mechanical properties of random fibre networks, truly representing real-world materials is still an issue. This will require a model with genuinely random orientation, using one of the two approaches for minimum distance calculation discussed in Sect. 2.5.

The size of the linear system becomes significant in 3D and parallel computing will be essential to model realistic volumes of material. We intend to use PETSc [1] to develop scalable tools for solving our systems, employing the block preconditioned iterative methods demonstrated in Sect. 4.1.

References

1. Balay, S., et al.: PETSc Web page (2018). http://www.mcs.anl.gov/petsc
2. Broedersz, C., Mao, X., Lubensky, T., MacKintosh, F.: Criticality and isostaticity in fibre networks. Nat. Phys. **7**(12), 983–988 (2011)
3. Head, D., MacKintosh, F., Levine, A.: Nonuniversality of elastic exponents in random bond-bending networks. Phys. Rev. E **68**(2), 25101 (2003)
4. Head, D., Levine, A., MacKintosh, F.: Deformation of cross-linked semiflexible polymer networks. Phys. Rev. Lett. **91**(10), 108102 (2003)
5. MacKintosh, F.: Elasticity and dynamics of cytoskeletal filaments and their networks. In: Soft Condensed Matter Physics in Molecular and Cell Biology, pp. 139–145. Taylor & Francis (2006)
6. Storm, C., Pastore, J., MacKintosh, F., Lubensky, T.C., Janmey, P.: Nonlinear elasticity in biological gels. Nature **435**(7039), 191 (2005)
7. Wilhelm, J., Frey, E.: Elasticity of stiff polymer networks. Phys. Rev. Lett. **91**(10), 108103 (2003)

Front Fixing Finite Difference Method for Pricing a Corporate Bond with Credit Rating Migration

Juri Kandilarov$^{(\boxtimes)}$ and Lubin Vulkov

Department of Mathematics, University of Ruse, Ruse, Bulgaria
{ukandilarov,lvalkov}@uni-ruse.bg

Abstract. A front fixing finite difference method for pricing a corporate bond with credit rating migration is developed. Two algorithms are proposed: the first one is of a predictor-corrector type while the second one is a Newton-like method. Comparison numerical experiments show the efficiency and effectiveness of the numerical algorithms.

Keywords: Corporate bond-pricing model · Credit-rating migration
Free boundary problem · Finite difference scheme
Predictor-corrector method · Newton method

1   Introduction

Option pricing and prediction of the variation of options value is a key factor for optimized capitals management. The Black-Sholes model of option pricing is considered to be a major tool of financial engineering. A financial derivative is called an American option if the payoff can take place before the expiration date $T \in [0, \infty)$. American option problem is formulated as parabolic complimentary problem or as *one-phase free boundary value problem*. On this base many methods are developed for pricing of American options, see e.g. [1–3,5–9,13].

In the last years, the modeling of credit rating migration becomes increasingly important in the world of finance. The most credit risk research in academic studies is focused on default, see e.g. the references in [4,10]. In this paper is proposed a new free boundary model in which the rating migration boundary depends on the proportion of the debt and the value of the firms. The pricing corporate bond with credit rating migration problem can be posed as a *two-phase free boundary value problem*. Let ϕ_H and ϕ_L be the functions of the firm value S and time t. Then they satisfy the following PDEs in their regions, [4,11,12]:

$$\frac{\partial \phi_H}{\partial t} + \frac{1}{2}\sigma_H^2 S^2 \frac{\partial^2 \phi_H}{\partial S^2} + rS\frac{\partial \phi_H}{\partial S} - r\phi_H = 0, \qquad S > \frac{1}{\gamma}\phi_H, \ 0 \leq t < T, \quad (1)$$

$$\frac{\partial \phi_L}{\partial t} + \frac{1}{2}\sigma_L^2 S^2 \frac{\partial^2 \phi_L}{\partial S^2} + rS\frac{\partial \phi_H}{\partial S} - r\phi_L = 0, \qquad 0 < S < \frac{1}{\gamma}\phi_L, \ 0 \leq t < T, \quad (2)$$

ⓒ Springer Nature Switzerland AG 2019
G. Nikolov et al. (Eds.): NMA 2018, LNCS 11189, pp. 416–423, 2019.
https://doi.org/10.1007/978-3-030-10692-8_47

with the terminal condition at the maturity time T

$$\phi_H(S,T) = \phi_L(S,T) = \min\{S, F\}. \tag{3}$$

Here σ_L and σ_H $(0 < \sigma_L < \sigma_H)$ represent the volatilities of the firm under the high and low credit grades respectively, r is the risk free interest rate, $0 < \gamma < 1$ represents the threshold proportion of the debt and the value of the firm's rating and F is the face value of the zero-coupon bond.

The value of the bond is constant when it passes the rating threshold, i.e.

$$\phi_H = \phi_L \quad \text{on the rating } migration\, boundary\, S = s(t), t \in (0, T). \tag{4}$$

Also, if one constructs a risk free portfolio π by longing a bond and shorting $\triangle$ amount asset value S, i.e., $\pi_t = \phi_t - \triangle_t S_t$ and such that $d\pi_t = r\pi_t$, this portfolio is also continuous when it passes the rating migration boundary:

$$\pi_H = \pi_L, \tag{5}$$

or by (4) on the rating migration boundary

$$\triangle_H = \triangle_L. \tag{6}$$

By the Black-Scholes theory, (see e.g. [10]), it is equivalent to

$$\frac{\partial \phi_H}{\partial S} = \frac{\partial \phi_L}{\partial S} \quad \text{on the rating migration boundary.} \tag{7}$$

In this paper a front fixing finite difference method for solving of the formulated free boundary value problem is proposed. In the next section the front fixing transformation is introduced and a finite difference approximations for the fixed boundary problem are derived. In Sect. 3 two algorithms are developed: the first one is of a predictor-corrector type while the second one is a Newton-like method. Comparison numerical experiments which show the efficiency and effectiveness of the numerical algorithms are presented in Sect. 4. Finally, some conclusions and future work are commented.

2 Front Fixing Transformation and Approximation

After changing of the variable $x = \frac{S}{s(t)}$, i.e. $S = xs(t)$ and renaming $t := T - t$ (*the time to maturity*), then

$$\varphi(x, t) = \phi(S, t) = \phi(xs(t), t)$$

and from the Eqs. (1), (2) we derive for $0 < t \le T$

$$\frac{\partial \varphi_H}{\partial t} - \frac{1}{2}\sigma_H^2 x^2 \frac{\partial^2 \varphi_H}{\partial x^2} - x\left(r - \frac{\dot{s}(t)}{s(t)}\right)\frac{\partial \varphi_H}{\partial x} + r\varphi_H = 0, \quad 1 < x < \infty, \tag{8}$$

$$\frac{\partial \varphi_L}{\partial t} - \frac{1}{2}\sigma_L^2 x^2 \frac{\partial^2 \varphi_L}{\partial x^2} - x\left(r - \frac{\dot{s}(t)}{s(t)}\right)\frac{\partial \varphi_L}{\partial x} + r\varphi_L = 0, \quad 0 \le x < 1. \tag{9}$$

The terminal condition (3) now is an initial condition

$$\varphi(x,0) = \min(xF/\gamma, F), \qquad 0 \le x < \infty. \tag{10}$$

Across the free boundary $s(t)$ $(x \to 1)$ the following conditions are fulfilled

$$\phi_H(s(t)+,t) = \phi_L(s(t)-,t) = \gamma s(t), \quad \frac{\partial \phi_H}{\partial x}(s(t)+,t) = \frac{\partial \phi_L}{\partial x}(s(t)-,t),$$

from where we have

$$\varphi_H(1,t) = \varphi_L(1,t) = \gamma s(t), \tag{11}$$

$$\frac{\partial \varphi_H}{\partial x}(1,t) = \frac{\partial \varphi_L}{\partial x}(1,t). \tag{12}$$

In order to solve the problem (8)–(12) numerically, we introduce X as a large value of x, where we impose the boundary condition

$$\varphi(X,t) = F. \tag{13}$$

Next, for given positive integers I, k, M and $N = kI$ we define the meshes:

$$\overline{\omega}_h^L = \{0\} \cup \{1\} \cup \omega_h^L, \quad \omega_h^L = \{x_i = ih^L, \ i = 1,\ldots,I-1, \ h^L = \frac{1}{I}\}$$

$$\overline{\omega}_h^H = \{1\} \cup \{X\} \cup \omega_h^H, \quad \omega_h^H = \{x_{I+i} = 1+ih^H, \ i = 1,\ldots,N-I-1, \ h^H = \frac{X-1}{N-I}\}$$

$$\overline{\omega}_\tau = \{0\} \cup \omega_\tau, \quad \omega_\tau = \{t^j = j\tau, \ j = 1,\ldots,M, \ \tau = \frac{T}{M}\}.$$

Our goal is to propose a finite difference method for computing $\varphi_i^j \approx \varphi(x_i, t^j)$ for $(x_i, t^j) \in \omega_h \times \omega_\tau$, $\omega_h = \omega_h^L \cup \{1\} \cup \omega_h^H$ and associated free boundary position $s^j \approx s(t^j)$ for $t^j \in \omega_\tau$. First, from the continuity condition (4) we have $\varphi_{H,I}^j = \varphi_{L,I}^j$ and hence we will omit the subindices H and L.

An implicit-upwind finite difference approximation of (8), (9) has the form

$$\frac{\varphi_i^{j+1} - \varphi_i^j}{\tau} - \frac{1}{2}\sigma_i^2 x_i^2 \frac{\varphi_{i-1}^{j+1} - 2\varphi_i^{j+1} + \varphi_{i+1}^{j+1}}{h_i^2}$$

$$- x_i \left(r - \frac{s^{j+1} - s^j}{\tau s^{j+1}} \right) \frac{\varphi_{i+1}^{j+1} - \varphi_i^{j+1}}{h_i} + r\varphi_i^{j+1} = 0 \tag{14}$$

for $i = 1,\ldots,I-1,I+1,\ldots,N-1$, $j = 0,\ldots,M-1$, where

$$\sigma_i = \begin{cases} \sigma_L \text{ for } i = 1,2,\ldots,I-1, \\ \sigma_H \text{ for } i = I+1,\ldots,N-1, \end{cases} \qquad h_i = \begin{cases} h^L \text{ for } i = 1,2,\ldots,I-1, \\ h^H \text{ for } i = I+1,\ldots,N-1. \end{cases}$$

From the terminal (initial) condition we have

$$\varphi_i^0 = \min(x_i F/\gamma, F) \qquad i = 0,\ldots N. \tag{15}$$

The boundary conditions imply that

$$\varphi_0^j = 0, \qquad \varphi_N^j = F, \qquad j = 0, \dots, M. \qquad (16)$$

From condition (12) we obtain

$$\frac{\varphi_{I+1}^{j+1} - \varphi_I^{j+1}}{h^H} = \frac{\varphi_I^{j+1} - \varphi_{I-1}^{j+1}}{h^L} \quad \text{for } j = 0, 1, \dots, M - 1. \qquad (17)$$

Finally, from (11) we get

$$\varphi_I^{j+1} = \gamma s^{j+1}. \qquad (18)$$

Assuming that the discrete solution φ_i^j, $i = 1, \dots, N - 1$ and s^j are known, on the new $(j + 1)$-th time layer the canonical form of the FDS is as follows:

$$a_i^{j+1} \varphi_{i-1}^{j+1} + c_i^{j+1} \varphi_i^{j+1} + b_i^{j+1} \varphi_{i+1}^{j+1} = d_i^{j+1}, \quad i = 1, \dots, N - 1 \qquad (19)$$

$$\begin{aligned}
a_i^{j+1} &= -\tfrac{\tau}{2h_i^2} \sigma_i^2 x_i^2, \, d_i^{j+1} = \varphi_i^j, & i \neq I, \\
c_i^{j+1} &= 1 + \tfrac{\tau}{h_i^2} \sigma_i^2 x_i^2 + \tfrac{\tau}{h_i} x_i \left(r - \tfrac{s^{j+1} - s^j}{\tau s^{j+1}} \right) + r\tau, & i \neq I, \\
b_i^{j+1} &= -\tfrac{\tau}{h_i^2} \sigma_i^2 x_i^2 - \tfrac{\tau}{h_i} x_i \left(r - \tfrac{s^{j+1} - s^j}{\tau s^{j+1}} \right), & i \neq I, \\
a_I^{j+1} &= -\tfrac{1}{h^L}, \quad b_I^{j+1} = -\tfrac{1}{h^H}, \quad c_I^{j+1} = \tfrac{1}{h^L} + \tfrac{1}{h^H}, \, d_I^{j+1} = 0.
\end{aligned}$$

The initial condition for the free boundary is $s^0 = F/\gamma$.

In this way we obtain for the unknowns φ_i^{j+1}, $i = 1, \dots, N - 1$ and s^{j+1} a nonlinear system of N algebraic equation.

3   Numerical Algorithms

In order to solve the nonlinear system of algebraic equations we developed the following two algorithms.

Algorithm 1. This algorithm is based on the *predictor-corrector* scheme and consists in the following steps (see also [13] for the case of pricing American Put options and [7] for the case of Asian options):

Step 1. *Predictor.* Let the solution and the free boundary position on the time level t^j be known. Instead of the implicit scheme (14) we make use of its explicit variant (with respect to φ) and combining with the second order central difference for the convective term for $i = I - 1$ and $i = I + 1$ we derive

$$\frac{\varphi_{I-1}^{j+1} - \varphi_{I-1}^j}{\tau} - \frac{1}{2}\sigma_{I-1}^2 x_{I-1}^2 \frac{\varphi_{I-2}^j - 2\varphi_{I-1}^j + \varphi_I^j}{h_{I-1}^2}$$

$$- x_{I-1} \left(r - \frac{s^{j+1} - s^j}{\tau s^{j+1}} \right) \frac{\varphi_I^j - \varphi_{I-2}^j}{2h_{I-1}} + r\varphi_{I-1}^j = 0 \qquad (20)$$

$$\frac{\varphi_{I+1}^{j+1} - \varphi_{I+1}^j}{\tau} - \frac{1}{2}\sigma_{I+1}^2 x_{I+1}^2 \frac{\varphi_I^j - 2\varphi_{I+1}^j + \varphi_{I+2}^j}{h_{I+1}^2}$$

$$- x_{I+1} \left(r - \frac{s^{j+1} - s^j}{\tau s^{j+1}} \right) \frac{\varphi_{I+2}^j - \varphi_I^j}{2h_{I+1}} + r\varphi_{I+1}^j = 0. \qquad (21)$$

We find from (20) and (21) φ_{I+1}^{j+1} and φ_{I-1}^{j+1} and put them into (17), from where we find an expression for φ_I^{j+1} as a function of s^{j+1}. Finally, putting φ_I^{j+1} in (18) we obtain a quadratic equation for the unknown position s^{j+1}. It has roots with difference signs and the positive one forms the predictor value of the moving boundary, denoted by $\widetilde{s}^{j+1}$.

Step 2. *Corrector.* From (19) for $i = 1, ..., N - 1$ and replacing s^{j+1} by the predicted value $\widetilde{s}^{j+1}$ we obtain the linear system for the unknowns φ_i^{j+1}, $i = 1, ..., N - 1$. Solving this system, we find the numerical solution on the new time layer. Then we correct the value of the free boundary using (18), $s^{j+1} = \varphi_I^{j+1}/\gamma$.

Algorithm 2. We now describe an algorithm based on the *Newton method*. This method was applied for an American Call option pricing model in [6].

Step 1. We eliminate the known boundary values $\varphi_0^{j+1} = 0$ and $\varphi_N^{j+1} = F$ from (19) for $i = 1$ and $i = N - 1$. Taking into account (18) and (19) we obtain a nonlinear system for N unknowns: φ_i^{j+1}, $i = 1, 2, ..., N - 1$ and s^{j+1}. We denote by $\overset{l}{\mathbf{Y}}$ the vector of these N unknowns at the l-th iteration.

Step 2. We have to solve the equation $\overset{l}{\mathbf{F}} = 0$, with $\overset{l}{\mathbf{F}} = (\overset{l}{\mathbf{F}_1}\overset{l}{\mathbf{F}_2})^T$ where $\overset{l}{\mathbf{F}_i}$, $i = 1, 2$ corresponds to Eqs. (19) and (18), respectively. To this end, we apply the Newton method in the following form:

$$\overset{l}{\mathbf{J}} (\overset{l+1}{\mathbf{Y}} - \overset{l}{\mathbf{Y}}) = - \overset{l}{\mathbf{F}}, \tag{22}$$

with the Jacobi matrix defined by: $\overset{l}{\mathbf{J}} = (\overset{l}{\mathbf{J}_{ij}})_{i,j=1,2}$ where

$$\overset{l}{\mathbf{J}_{11}} = \begin{pmatrix} c_1^{j+1} & b_1^{j+1} & & & \\ a_2^{j+1} & c_2^{j+1} & b_2^{j+1} & & \\ & \ddots & \ddots & \ddots & \\ & & a_{N-2}^{j+1} & c_{N-2}^{j+1} & b_{N-2}^{j+1} \\ & & & a_{N-1}^{j+1} & c_{N-1}^{j+1} \end{pmatrix} \quad \overset{l}{\mathbf{J}_{12}} = \begin{pmatrix} \frac{\partial b_1^{j+1}}{\partial s^{j+1}} y_2^{j+1} \\ \frac{\partial a_2^{j+1}}{\partial s^{j+1}} y_1^{j+1} + \frac{\partial b_2^{j+1}}{\partial s^{j+1}} y_3^{j+1} \\ \vdots \\ \frac{\partial a_{N-2}^{j+1}}{\partial s^{j+1}} y_{N-3}^{j+1} + \frac{\partial b_{N-2}^{j+1}}{\partial s^{j+1}} y_{N-1}^{j+1} \\ \frac{\partial a_{N-1}^{j+1}}{\partial s^{j+1}} y_{N-2}^{j+1} + \frac{\partial b_{N-1}^{j+1}}{\partial s^{j+1}} F \end{pmatrix}$$

$\overset{l}{\mathbf{J}_{21}} = (0, ..., 0, -1/\gamma, 0, ..., 0)$ and $\overset{l}{\mathbf{J}_{22}} = 1$. The position of the nonzero element of $\overset{l}{\mathbf{J}_{21}}$ is I. Similarly $\overset{l}{\mathbf{Y}} = \left(\overset{l}{\mathbf{Y}_{11}}\overset{l}{\mathbf{Y}_{12}}\right)^T$, $\overset{l}{\mathbf{Y}_{11}} = \left(\varphi_1^{j+1}, ..., \varphi_{N-1}^{j+1}\right)$, $\overset{l}{\mathbf{Y}_{12}} = s^{j+1}$.

The iteration process is repeated until the condition $\max |(\overset{l+1}{\mathbf{Y}} - \overset{l}{\mathbf{Y}})| < tol$ is fulfilled.

Step 3. The solution on the $(j+1)$-th time layer is taken as an initial iteration for the next time layer.

Table 1. Mesh-refinement analysis of the predictor-corrector method for Example 1.

N	M	T = 5			T = 10			T = 50		
		s(T)	Differ.	Rate	s(T)	Differ.	Rate	s(T)	Differ.	Rate
50	10	0.92395	-	-	0.73946	-	-	0.36212	-	-
100	20	0.93099	7.0389e−03	-	0.74066	1.1970e−03	-	0.35137	1.0747e−02	-
200	40	0.93442	3.4354e−03	1.035	0.74109	4.3336e−04	1.466	0.34514	6.2309e−03	0.786
400	80	0.93612	1.7022e−03	1.013	0.74129	1.9421e−04	1.158	0.34180	3.3421e−03	0.899
800	160	0.93697	8.4741e−04	1.006	0.74138	9.2651e−05	1.068	0.34007	1.7241e−03	0.955

Table 2. Mesh-refinement analysis of the Newton method for Example 1.

N	M	T = 5			T = 10			T = 50		
		s(T)	Differ.	Rate	s(T)	Differ.	Rate	s(T)	Differ.	Rate
50	10	0.93098	-	-	0.73074	-	-	0.3348	-	-
100	20	0.93433	3.3500e−03	-	0.73658	5.84e−03	-	0.33680	2.00e−03	-
200	40	0.93606	1.7300e−03	0.953	0.73195	2.57e−03	1.184	0.33762	8.20e−04	1.286
400	80	0.93693	8.7000e−03	0.992	0.74034	1.19e−03	1.111	0.33799	3.70e−04	1.148
800	160	0.93737	4.4000e−04	0.984	0.74091	5.70e−04	1.062	0.33816	1.70e−04	1.122

4   Numerical Experiments

Example 1. We consider problem (1)–(7) with parameter values $r = 0.05$, $\sigma_L = 0.3$, $\sigma_H = 0.2$, $F = 1$, $\gamma = 0.8$ and $T = 5$, see [4].

From practical point of view it is enough to take $k = 5$, i.e. $X = 5$. Since there exists no analytical solution to the proposed free boundary problem, we use the mesh refinement analysis with doubling the mesh size h. Since the approximation on time and space is a first order, in the numerical experiments we keep the proportion N/M to be constant $N/M = k$. We denote the absolute value of the difference between the final position of the free boundary obtained on two consecutive meshes and the rate of convergence, respectively by:

$$s_{diff}^M = |s^M - s^{M/2}|, \qquad rate = log_2(s_{diff}^M / s_{diff}^{2M}).$$

In Table 1 we give the results for the free boundary position $s(t)$ at different times $t = 5,\ 10,\ 50$, obtained by the predictor-corrector scheme. Also, the difference in maximum norm between two consecutive values and the rate of convergence are presented. The results show first order of accuracy for the moving boundary. The results obtained by the Newton method are presented in Table 2, where the tolerance value is $tol = 1.0e - 06$. Again the first order of the method and the position of the free boundary are confirmed. In Table 3 we make a CPU time analysis of the predictor-corrector method versus Newton method. It is clearly seen that the predictor-corrector method is faster than the Newton method and for bigger values of the mesh parameters N and M the Newton method uses approximately eight times bigger CPU time in seconds. Also the average number of iterations for the Newton method are presented and it is obvious that the method converges after three iterations for bigger N and M.

Table 3. CPU time analysis of the predictor-corrector method versus Newton method for Example 1 and the averaged number of iterations for the Newton method.

N	50	100	200	400	800
M	10	20	40	80	160
Pr.-Corr. (in sec.)	0.03126	0.03318	0.05640	0.26914	2.08979
Newton (in sec.)	0.07682	0.17689	0.48847	1.95180	15.42786
aver. numb. iter.	4.100	4.000	4.000	3.100	3.0125

In Fig. 1a) the free boundary position $s(t)$, obtained for $N = 100$, $M = 20$: solid line - by the predictor-corrector method; circles - by the Newton method; dashed line - theoretical upper bound, proposed in [4]. It is clearly seen the coincidence of the results from both numerical algorithms to the same free boundary. Also, the decreasing of the free boundary is numerically confirmed. In Fig. 1b) the 3D plot of the function $\varphi(x,t)$ for $T = 5$, $N = 100$, $M = 20$ in the new coordinate system is presented.

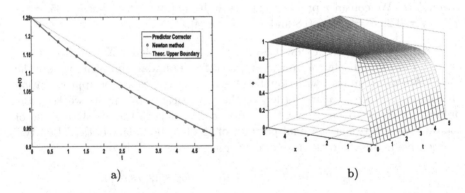

a)   b)

Fig. 1. (a) The free boundary position for $s(t)$, obtained for $N = 100$, $M = 20$: solid line - by the predictor-corrector method; circles - by the Newton method; dashed line - theoretical upper bound; (b) The solution φ in new coordinates (x, t).

5   Conclusion

In this work we construct front fixing finite difference schemes for pricing a corporate bond with credit rating migration. We exploit two algorithms: one of a predictor-corrector type and the other of Newton-like method, which also work well for time-dependent volatility models $\sigma = \sigma(t)$. We plan to extend our numerical methods to 2D models [11].

Acknowledgements. This work was partially supported by the Project 2018-FNSE-03 of the University of Ruse and by the Bulgarian National Fund of Science under the Project DN 12/4-2017.

References

1. Chernogorova, T., Koleva, M., Valkov, R.: A two-grid penalty method for American options. Comp. Appl. Math. (2018). https://doi.org/10.1007/s40314-017-0457-6
2. Company, R., Egorova, V.N., Jodar, L.: Solving American option pricing models by the front fixing method: numerical analysis and computing. Abstr. Appl. Anal. **2014** (2014). Article ID 146745
3. Gyulov, T., Valkov, R.: American option pricing problem transformed on finite interval. Intern. J. of Comp. Math. **93**, 821–836 (2016)
4. Hu, B., Liang, J., Wu, Y.: A free boundary problem for corporate bond with credit rating migration. J. Math. Anal. Appl. **428**(2), 896–909 (2015)
5. Jiang, L.: Modeling and Methods for Option Pricing. World Scientific, Singapore (2005)
6. Kandilarov, J.D., Valkov, R.L.: A numerical approach for the American call option pricing model. In: Dimov, I., Dimova, S., Kolkovska, N. (eds.) NMA 2010. LNCS, vol. 6046, pp. 453–460. Springer, Heidelberg (2011). https://doi.org/10.1007/978-3-642-18466-6_54
7. Kandilarov, J.D., Ševčovič, D.: Comparison of two numerical methods for computation of American type of the floating strike Asian option. In: Lirkov, I., Margenov, S., Waśniewski, J. (eds.) LSSC 2011. LNCS, vol. 7116, pp. 558–565. Springer, Heidelberg (2012). https://doi.org/10.1007/978-3-642-29843-1_63
8. Koleva, M.N., Valkov, R.L.: Modified barrier penalization method for pricing American options. In: Ehrhardt, M., Günther, M., ter Maten, E.J.W. (eds.) Novel Methods in Computational Finance. MI, vol. 25, pp. 215–226. Springer, Cham (2017). https://doi.org/10.1007/978-3-319-61282-9_11
9. Kwok, J.: Mathematical Models of Financial Derivatives. Springer, Heidelberg (1998). https://doi.org/10.1007/978-3-540-68688-0
10. Leland, H.: Corporate debt value, bond covenants, and optimal capital structure. J. Finance **49**(4), 1213–1252 (1994)
11. Liang, J., Chen, X., Wu, Y., Yin, H.-M.: On a corporate bond pricing model with credit rating migration risks and stochastic interest rate. Quant. Financ. Econ. **1**(3), 300–319 (2017)
12. Liang, J., Zhao, Y.J.: Utility indifference valuation of corporate bond with credit rating migration by structure approach. Econ. Model. **54**, 339–346 (2016)
13. Zhu, S.-P., Zang, J.: A new predictor-corrector scheme for valuing American puts. Appl. Math. Comput. **2017**, 4439–4452 (2011)

Positivity Preserving Numerical Method for Optimal Portfolio in a Power Utility Two-Dimensional Regime-Switching Model

Miglena N. Koleva[(✉)] and Lubin G. Vulkov

University of Ruse, 8 Studentska str., 7017 Ruse, Bulgaria
{mkoleva,lvalkov}@uni-ruse.bg

Abstract. We consider a two-dimensional regime switching model with power utility function. The problem is a system of parabolic partial differential equations with non-linear gradient terms and weakly coupled by non-linear exponential terms. We establish lower bounds for the solutions and then we construct an adequate finite difference method, preserving the qualitative properties of the exact solution. Finally, we present and discuss numerical results.

1 Introduction

Regime-switching models allow to capture the dynamics of the price, risk etc., when the economy switches between different states. In this work, we consider the regime-switching problem, derived in [12], with power utility function $U(y) = y^\gamma/\gamma$, $\gamma < 1$, $\gamma \neq 0$, $y \in \mathbb{R}$. Let $s = (s_1, s_2, \dots, s_d) \subset \mathbb{R}^d$ be a vector of stock prices $s_i \in [0, \infty)$, $i = 1, 2, \dots, d$, $\overline{s} := \mathrm{diag}(s)$ and $t \in [0, T]$ is a time variable. The value functions $V^k(t, s, y)$, $k = 1, 2, \dots m$ are defined by $V^k = (ye^{c^k})^\gamma/\gamma$, where $c^k(t, s)$ are solutions of the following system of parabolic equations

$$c_t^k + \frac{1}{2}\mathrm{tr}(\overline{s}\Sigma_k\Sigma_k^T\overline{s}c_{ss}^k) + \left(\mu^k + \frac{\gamma}{1-\gamma}(\mu^k - r\mathbf{1})\right)\overline{s}c_s^k + \frac{1}{\gamma}\sum_{j=1}^m(e^{-\gamma(c^k-c^j)} - 1)\lambda^{kj}$$

$$+ \frac{1}{2(1-\gamma)}z_k^2 + \frac{\gamma}{2(1-\gamma)}\|\overline{s}c_s^k\Sigma_k\|^2 + r = 0, \tag{1}$$

$$c^k(T, s) = c_0(s) \geq 0.$$

Here r is the risk-free interest rate, $\mu^k = \mu^k(t, s) : [0, T] \times B \subset \mathbb{R}^d \to \mathbb{R}^d$ are drift coefficients, $\Sigma_k\Sigma_k^T$ are positive definite, locally Lipschitz and bounded, $\Sigma_k = \Sigma_k(t, s) : [0, T] \times B \subset \mathbb{R}^d \to \mathbb{R}^{d \times d}$ are nonsingular for all (t, s), $\Sigma_k^{-1}\mu^k$ and $\lambda^{kj} : [0, T] \times B \to [0, \infty)$, $\lambda^{kj} \in C_b^1([0, T] \times B)$ are bounded on $\Omega_T = [0, T] \times B$, $B = (0, \infty)^d$ for all $k, j = 1, \dots, m$, $\mathbf{1}$ is the d-dimensional unit column vector, $\|\cdot\|$ is Euclidean norm, c_s^k is the gradient with respect to the vector s, c_{ss}^k is the Hessian matrix with entries $c_{s_i s_j}^k$, $i, j = 1, 2, \dots, d$ and functions z_k^2 are given by

$$z_k^2(t, s) := (\mu^k(t, s) - r\mathbf{1})^T \left(\Sigma_k(t, s)\Sigma_k^T(t, s)\right)^{-1} (\mu^k(t, s) - r\mathbf{1}).$$

G. Nikolov et al. (Eds.): NMA 2018, LNCS 11189, pp. 424–432, 2019.
https://doi.org/10.1007/978-3-030-10692-8_48

Available in the literature results for existence and uniqueness of the solution for parabolic systems, see e.g. [10], exclude the models with exponential non-linearity. In this work, assuming existence and uniqueness of classical solution, we establish minimum principle for the problem (1).

In our previous papers [3,4] were considered the one-dimensional case of (1). The focus in the present work is the numerical investigation of the two-dimensional case of the system (1). We will construct numerical method that preserves the qualitative properties of the exact solution. The results for higher-dimensional case can be generalized in a similar way as we did in [6] for the exponential utility regime-switching model.

The remaining part of the paper is organized as follows. In the next section, we formulate the model problem and establish minimum principle, Theorem 1. Numerical method and its properties are presented in Sect. 3 (see Theorems 2, 3). In Sect. 4 we discuss numerical results and finally we give concluding remarks.

2   The Differential Problem

Consider the two-dimensional case of the system (1), namely

$$
c_t^k + \frac{1}{2}(\sigma_{11}^k s_1^2 c_{s_1 s_1}^k + 2\sigma_{12}^k s_1 s_2 c_{s_1 s_2}^k + \sigma_{22}^k s_2^2 c_{s_2 s_2}^k) + \left(\mu_1^k + \frac{\gamma}{1-\gamma}(\mu_1^k - r)\right) s_1 c_{s_1}^k
$$

$$
+ \left(\mu_2^k + \frac{\gamma}{1-\gamma}(\mu_2^k - r)\right) s_2 c_{s_2}^k + \frac{1}{\gamma}\sum_{j=1}^m (e^{-\gamma(c^k - c^j)} - 1)\lambda^{kj} + \frac{1}{2(1-\gamma)} z_k^2 \tag{2}
$$

$$
+ \frac{\gamma}{2(1-\gamma)}\left((\widetilde{\sigma}_{11}^k s_1 c_{s_1} + \widetilde{\sigma}_{21}^k s_2 c_{s_2})^2 + (\widetilde{\sigma}_{12}^k s_1 c_{s_1} + \widetilde{\sigma}_{22}^k s_2 c_{s_2})^2\right) + r = 0,
$$

$$
c^k(T, s_1, s_2) = c_0(s), \quad k = 1, 2, \ldots, m.
$$

where $\Sigma_k = \{\widetilde{\sigma}_{il}^k\}_{i,l=1}^{2,2}$, $\Sigma_k \Sigma_k^T = \{\sigma_{il}^k\}_{i,l=1}^{2,2}$ and $\sigma_{11}^k = (\widetilde{\sigma}_{11}^k)^2 + (\widetilde{\sigma}_{12}^k)^2$, $\sigma_{22}^k = (\widetilde{\sigma}_{22}^k)^2 + (\widetilde{\sigma}_{21}^k)^2$, $\sigma_{12}^k = \widetilde{\sigma}_{11}^k \widetilde{\sigma}_{21}^k + \widetilde{\sigma}_{22}^k \widetilde{\sigma}_{12}^k$, $\mu_i^k = \mu^k(t, s_i)$.

Note that for $s_1 = 0$ and/or $s_2 = 0$, from (2) we obtain natural boundary conditions. Applying the variable change $s_i^* = 1/s_i$ in the equations in (2), we get a similar to (2) system. Letting $s_i^* = 0$ (i.e. $s_i \to \infty$), $i = 1, 2$, we obtain natural boundary conditions. Returning to the original variable s_i, we deduce that at infinity we may consider similar conditions as for $s_i = 0$.

Further, in (2) we apply logarithmic change of the space variables $x_i = \ln s_i$, $i = 1, 2$ and invert the time $\tau = T - t$. As a result the space region $s \in [0, \infty) \times [0, \infty)$ transforms to the infinite domain $x \in (-\infty, \infty) \times (-\infty, \infty)$. In order to construct numerical method, the unbounded spatial domain is truncated by large enough finite region $\overline{D} = [L_1^-, L_1^+] \times [L_2^-, L_2^+]$, where $L_i^- < 0$ and $L_i^+ > 0$. Taking into account the above consideration, we impose natural boundary conditions on the remote boundaries. Let us define the function δ_i:

$$
\delta_i = \delta(x_i) = \begin{cases} 0, & \text{if } x_i = L_i^{\pm}, \\ 1, & \text{otherwise}, \end{cases} \qquad \delta_i^2 = \delta_i.
$$

The resulting initial-boundary value problem, defined in $\overline{Q}_T = [0,T] \times \overline{D}$ is

$$c_\tau^k - \frac{1}{2}(\delta_1 \sigma_{11}^k c_{x_1 x_1}^k + 2\delta_1 \delta_2 \sigma_{12}^k c_{x_1 x_2}^k + \delta_2 \sigma_{22}^k c_{x_2 x_2}^k)$$

$$- \delta_1 \left(\mu_1^k + \frac{\gamma}{1-\gamma}(\mu_1^k - r) - \frac{1}{2}\sigma_{11}^k \right) c_{x_1}^k - \delta_2 \left(\mu_2^k + \frac{\gamma}{1-\gamma}(\mu_2^k - r) - \frac{1}{2}\sigma_{22}^k \right) c_{x_2}^k$$

$$- \frac{\gamma}{2(1-\gamma)} \left(\delta_1 \sigma_{11}^k (c_{x_1}^k)^2 + 2\delta_1 \delta_2 \sigma_{12}^k c_{x_1}^k c_{x_2}^k + \delta_2 \sigma_{22}^k (c_{x_2}^k)^2 \right) \tag{3}$$

$$- \frac{1}{\gamma} \sum_{j=1}^m (e^{-\gamma(c^k - c^j)} - 1)\lambda^{kj} - \frac{1}{2(1-\gamma)} z_k^2 - r = 0, \quad (\tau, x) \in Q_T = (0,T] \times D,$$

$$c^k(0, x) = c_0^k(x), \quad x = (x_1, x_2) \in \overline{D}, \quad k = 1, 2, \ldots, m.$$

Note that $\delta_1 \sigma_{11}^k (c_{x_1}^k)^2 + 2\delta_1 \delta_2 \sigma_{12}^k c_{x_1}^k c_{x_2}^k + \delta_2 \sigma_{22}^k (c_{x_2}^k)^2 = (\delta_1 \tilde{\sigma}_{11}^k c_{x_1} + \delta_1 \delta_2 \tilde{\sigma}_{21}^k c_{x_2})^2 + (\delta_1 \delta_2 \tilde{\sigma}_{12}^k c_{x_1} + \delta_2 \tilde{\sigma}_{22}^k c_{x_2})^2 \geq 0$.

For the system (3) we establish minimum principle.

Theorem 1. *Let* $\mathbf{m} = \min_k m_k$, $m_k \leq z_k^2(\tau, x)$, $(\tau, x) \in \overline{Q}_T$ *and suppose that the functions* $c^k \in C(\overline{Q}_T) \cap C^{1,3}(Q_T)$ *satisfy the problem (3) in* Q_T. *If* $c_0^k(x) \geq 0$, $k = 1, 2, \ldots, m$, *then*

$$c^k(\tau, x) \geq \left(\frac{\mathbf{m}}{2(1-\gamma)} + r \right) \tau, \quad k = 1, 2, \ldots, m.$$

Proof (outline). We follow the same line of considerations as in [4].

3   Numerical Method

We define an uniform space mesh $\overline{\omega}_h = \overline{\omega}_{h_1} \times \overline{\omega}_{h_2}$ with mesh step sizes h_i, $i = 1, 2$

$$\overline{\omega}_{h_i} = \{x_{i,j_i} : x_{i,j_i} = L_i^- + (j_i - 1)h_i, \ j_i = 1, \ldots, N_i, \ h_i = (L_i^+ - L_i^-)/(N_i - 1)\}$$

and a non-uniform mesh ω_τ in time with increments $\triangle\tau^n$, i.e. $\tau^{n+1} = \tau^n + \triangle\tau^n$, $n = 0, 1, \ldots, N_\tau$. The numerical solution at point $(\tau^n, x_{1j_1}, x_{2j_2})$ is denoted by $C_{n,j_1,j_2}^k := C^k(\tau^n, x_{1j_1}, x_{2j_2})$. We will exploit also the following notations

$$C_{j_1,j_2}^k := C_{n,j_1,j_2}^k, \quad \widehat{C}_{j_1,j_2}^k := C_{n+1,j_1,j_2}^k, \quad C_{t_{j_1,j_2}}^k = \frac{\widehat{C}_{j_1,j_2}^k - C_{j_1,j_2}^k}{\triangle\tau^n},$$

$$C_{e_i+q}^k = \begin{cases} C_{j_1+q,j_2}^k, & i = 1, \\ C_{j_1,j_2+q}^k, & i = 2, \end{cases} \quad C_{\overline{x}_i e_i}^k = \frac{C_{e_i}^k - C_{e_i-1}^k}{h}, \quad C_{x_i e_i}^k = C_{\overline{x}_i e_i+1}^k,$$

$$C_{\dot{x}_i}^k = \frac{1}{2}[C_{\overline{x}_i}^k + C_{x_i}^k], \quad C_{x_p x_q}^k = (C_{x_p}^k)_{x_q}, \quad C_{\overline{x}_p x_q}^k = (C_{\overline{x}_p}^k)_{x_q}, \quad C_{x_p \overline{x}_q}^k = (C_{x_p}^k)_{\overline{x}_q},$$

$$(C^k)_{x_p x_q}^+ = \frac{1}{2}[C_{x_p x_q}^k + C_{\overline{x}_p \overline{x}_q}^k], \quad (C^k)_{x_p x_q}^- = \frac{1}{2}[C_{\overline{x}_p x_q}^k + C_{x_p \overline{x}_q}^k].$$

In order to approximate the first spatial derivative in (3), we use also the following conservative discretization

$$\xi^k(x_{1j_1}, x_{2j_2}) \frac{\partial C^k}{\partial x_i}(x_{1j_1}, x_{2j_2}) \simeq \xi^k_{e_i} \frac{C^k_{e_i+1/2} - C^k_{e_i-1/2}}{h_i} =: \xi^k_{e_i} C^k_{\tilde{x}_i e_i}. \qquad (4)$$

Next, for $C^k_{e_i \pm 1/2}$ in (4), we apply van Leer flux limiter [2,5,8,9] in each spatial direction, namely

$$\Phi(\theta^k) = \frac{|\theta^k| + \theta^k}{1 + |\theta^k|}, \quad \text{where} \quad \theta^k_{e_i+1/2} = \frac{C^k_{x_i e_i}}{C^k_{\tilde{x}_i e_i}}, \qquad (5)$$

$\Phi(\theta^k)$ is Lipschitz continuous, continuously differentiable for all $\theta^k \neq 0$, and

$$\Phi(\theta^k) = 0, \quad \text{if} \quad \theta^k \leq 0 \quad \text{and} \quad \Phi(\theta^k) \leq 2 \min\{1, \theta^k\}. \qquad (6)$$

Following [2] the numerical flux $C^k_{e_i+1/2}$ is approximated in a non-linear way

$$C^k_{e_i+1/2} = \begin{cases} C^k_{e_i} + \frac{1}{2}\Phi(\theta^k_{e_i+1/2})(C^k_{e_i} - C^k_{e_i-1}), & \xi^k_{e_i} \leq 0, \\ C^k_{e_i+1} + \frac{1}{2}\Phi((\theta^k_{e_i+3/2})^{-1})(C^k_{e_i+1} - C^k_{e_i+2}), & \xi^k_{ei} > 0. \end{cases} \qquad (7)$$

and the flux $C^k_{e_i-1/2}$, corresponding to (7) is defined by shifting the indexes j_i.

Using the symmetry property of the flux limiter $\Phi(\theta) = \theta\Phi(\theta^{-1})$ [7] and (5), we approximate $\xi^k C^k_{\tilde{x}_i}$, at each grid node $(\tau, x_{1j_1}, x_{1j_2})$, applying (4), (7) in dependence of the sign of $\xi_{e_i} = (\xi_{e_i})^+ - (\xi_{e_i})^-$, $\xi^+_{e_i} = \max\{0, \xi_{e_i}\}$ and $\xi^-_{e_i} = \min\{0, -\xi_{e_i}\}$:

$$\xi^k_{e_i} \frac{\partial C^k}{\partial x_i} \simeq (\xi^k_{e_i})^+(\Lambda^k_i)^+ C^k_{x_i} - (\xi^k_{e_i})^-(\Lambda^k_i)^- C^k_{\tilde{x}_i}, \quad i = \{1, 2\},$$

$$(\Lambda^k_i)^+ = 1 + \frac{1}{2}\Phi((\theta^k_{e_i+1/2})^{-1}) - \frac{1}{2}\Phi(\theta^k_{e_i+3/2}), \qquad (8)$$

$$(\Lambda^k_i)^- = 1 + \frac{1}{2}\Phi(\theta^k_{e_i+1/2}) - \frac{1}{2}\Phi((\theta^k_{e_i-1/2})^{-1}),$$

where in view of (5), (6) we have $0 \leq (\Lambda^k_i)^- \leq 2$ and $0 \leq (\Lambda^k_i)^+ \leq 2$.

For computing the gradient ratio in (5) at points near to the boundaries, namely $j_i = \{2, N_i - 1\}$, we need the values of $C^k_{j_1, j_2}$ at the outer grid nodes, i.e for $j_i = 0$ or $j_i = N_i + 1$. Then, the second-order extrapolation formulas [11] will be used.

Applying different approximation for the mixed derivative, depending on the sign of σ^k_{12}, and (8) for the discretization of the first derivative, where in the the gradient term we use either (8) or central finite difference depending on the sign of γ, we construct implicit-explicit discretization of (3). Using the notation $A_i^k = \mu_i^k + \frac{\gamma}{1-\gamma}(\mu_i^k - r) - \frac{1}{2}(\widehat{\sigma}_{ii}^k)^2$, at each grid node of $\overline{\omega}_h \times \omega_\tau$, we have

$$C_t^k - \frac{1}{2}(\delta_1\widehat{\sigma}_{11}^k\widehat{C}_{\overline{x}_1 x_1}^k + 2\delta_1\delta_2(\widehat{\sigma}_{12}^k)^+(\widehat{C}^k)_{x_1 x_2}^+ - \delta_1\delta_2(\widehat{\sigma}_{12}^k)^-(\widehat{C}^k)_{x_1 x_2}^- + \delta_2\widehat{\sigma}_{22}^k\widehat{C}_{\overline{x}_2 x_2}^k)$$

$$= \delta_1[(A_1^k)^+(\Lambda_1^k)^+C_{x_1}^k - (A_1^k)^-(\Lambda_1^k)^-C_{\overline{x}_1}^k] + \delta_2[(A_2^k)^+(\Lambda_2^k)^+C_{x_2}^k - (A_2^k)^-(\Lambda_2^k)^-C_{\overline{x}_2}^k]$$

$$+ \frac{\gamma^+}{2(1-\gamma^+)}\left(\delta_1\sigma_{11}^k(c_{\dot{x}_1}^k)^2 + 2\delta_1\delta_2\sigma_{12}^kc_{\dot{x}_1}^kc_{\dot{x}_2}^k + \delta_2\sigma_{22}^k(c_{\dot{x}_2}^k)^2\right)$$

$$- \frac{\gamma^-\delta_1\sigma_{11}^k}{2(1+\gamma^-)}\left((C_{\dot{x}_1}^k)^+(\Lambda_1^k)^-C_{\overline{x}_1}^k - (C_{\dot{x}_1}^k)^-(\Lambda_1^k)^+C_{x_1}^k\right)$$

$$- \frac{\gamma^-\delta_2\sigma_{22}^k}{2(1+\gamma^-)}\left((C_{\dot{x}_2}^k)^+(\Lambda_2^k)^-C_{\overline{x}_2}^k - (C_{\dot{x}_2}^k)^-(\Lambda_2^k)^+C_{x_2}^k\right)$$

$$- \frac{\gamma^-\delta_1\delta_2\sigma_{12}^k}{2(1+\gamma^-)}\left((C_{\dot{x}_1}^k)^+(\Lambda_2^k)^-C_{\overline{x}_2}^k - (C_{\dot{x}_1}^k)^-(\Lambda_2^k)^+C_{x_2}^k + (C_{\dot{x}_2}^k)^+(\Lambda_1^k)^-C_{\overline{x}_1}^k\right.$$

$$\left. - (C_{\dot{x}_2}^k)^-(\Lambda_1^k)^+C_{x_1}^k\right) + \frac{1}{\gamma}\sum_{j=1}^{m}(e^{-\gamma(C^k-C^j)}-1)\lambda^{kj} + \frac{1}{2(1-\gamma)}\widehat{z}_k^2 + r,$$

$$C^k(0, x_1, x_2) = C_0^k(x_1, x_2).$$

The numerical scheme (9) is constructed so that we can guarantee that the discrete solution will have the same lower bound as the exact solution. For further investigations it is convenient to represent the unknown solution in vector form, reordering C_{j_1,j_2}^k and the corresponding equations row by row. Thus the J-th $(J = 1, 2, \ldots, N_1N_2)$ finite difference equation corresponding to (j_1, j_2) has the following relation $J = j_1 + (j_2 - 1)N_1$, $j_1 = 1, 2, \ldots, N_1$, $j_2 = 1, 2, \ldots, N_2$. Therefore, the unknowns C_{j_1,j_2}^k according to the ordering, are expressed as column-vector $C^k = [C_1^k, C_2^k, \ldots, C_J^k, \ldots, C_{N_1N_2}^k]^T$ and (9) can be written in an equivalent matrix-vector form (EMVF).

Let denote by $\|\cdot\|_h$ the maximal discrete norm, $\circ$ is a point-wise (Hadamard) matrix product and β^k is N_1N_2-dimensional vector with entries

$$\beta_J^k = \begin{cases} 0, \text{ if } C_J^k = \alpha, \quad \alpha = \left(\frac{m}{2(1-\gamma)}+r\right)\tau, \\ 1, \text{ otherwise.} \end{cases}$$

Theorem 2. *Suppose that the conditions of Theorem 1 are fulfilled and $\gamma > 0$. Then, if*

$$\frac{\delta_2|\widehat{\sigma}_{12}^k|}{\widehat{\sigma}_{11}^k} \leq \frac{h_2}{h_1} \leq \frac{\widehat{\sigma}_{22}^k}{\delta_1|\widehat{\sigma}_{12}^k|} \quad \text{for all} \quad J = 1, 2, \ldots, N_1N_2, \tag{10}$$

$$\triangle\tau^n \leq \frac{\gamma h_1 h_2}{2\gamma(h_2\|A_1^k \circ \beta^k\|_h + h_1\|A_2^k \circ \beta^k\|_h) + h_1 h_2\|\mathcal{P}^{kj}\|_h}, \tag{11}$$

$$\mathcal{P}_J^{kj} = \begin{cases} \frac{1}{C_J^k-\alpha}\left(\sum_{j=1}^{m}(1 - e^{-\gamma(C_J^k-C_J^j)^+})\lambda_J^{kj}\right), & C_J^k > \alpha, \\ 0, & \text{otherwise,} \end{cases} \quad J = 1, \ldots, N_1N_2,$$

for the numerical solution of (9) at each time level we have

$$C_J^k \geq \left(\frac{\mathrm{m}}{2(1-\gamma)} + r \right) \tau, \quad J = 1, \ldots, N_1 N_2.$$

Proof (outline). Substitute $U_J^k = C_J^k - \alpha$ in the EMVF of (9). For the resulting problem we prove that the coefficient matrix is an M-matrix if the restriction (10) is satisfied. Then, we show that the right-hand side is non-negative, if (11) is fulfilled. Thus, we deduce that $U_J^k \geq 0$, $J = 1, 2, \ldots, N_1 N_2$ [1].

The condition (10) is not restrictive with respect to the financial modeling [12].
In the same manner we establish the following result.

Theorem 3. *Let $\gamma < 0$. If the restriction (10) and conditions of Theorem 1 are fulfilled and*

$$\triangle \tau^n \leq \frac{\gamma^-(1+\gamma^-)h_1 h_2}{2\gamma^-(1+\gamma^-)\mathcal{A}^k + h_1 h_2 ((\gamma^-)^2 \mathcal{Q}^k + (1+\gamma^-)\|\mathcal{N}^{kj}\|)}, \quad (12)$$

$$\mathcal{N}_J^{kj} = \begin{cases} \frac{1}{C_J^k - \alpha} \left(\sum_{j=1}^{m} (e^{\gamma^-(C_J^k - C_J^j)^+} - 1)\lambda_J^{kj} \right), & C_J^k > \alpha, \\ 0, & otherwise, \end{cases} \quad J = 1, \ldots, N_1 N_2,$$

$$\mathcal{A}^k = h_2 \|A_1^k \circ \beta^k\| + h_1 \|A_2^k \circ \beta^k\|,$$

$$\mathcal{Q}^k = \|\beta^k \circ C_{x_1^\circ}^k\|(\|\sigma_{11}^k\| + \|\sigma_{12}^k\|) + \|\beta^k \circ C_{x_2^\circ}^k\|(\|\sigma_{22}^k\| + \|\sigma_{12}^k\|),$$

then, for the numerical solution of (9) at each time level, we have

$$C_J^k \geq \left(\frac{\mathrm{m}}{2(1-\gamma)} + r \right) \tau, \quad J = 1, 2, \ldots, N_1 N_2.$$

4   Numerical Results

We will verify the *order of convergence* of the solution, obtained by (9) and the statements of Theorems 2 and 3. For the first test problem (TP1) we chose the model parameters, similar to the Example 1.1 in [12]: $m = 2$, $\gamma = 0.1$, $\sigma_{11}^1 = \sigma_{22}^1 = 0.04$, $\sigma_{12}^1 = 0.004$, $\sigma_{11}^2 = \sigma_{22}^2 = 0.09$, $\sigma_{12}^2 = 0.054$, $r = 0.005$, $\lambda^{12} = 0.01(\sqrt{s_1 s_2})$, $\lambda^{21} = 0.03(\sqrt[3]{s_1 s_2})$, $\mu_i^k = 0.015$, $k, i = 1, 2$ and $c_0^1(s) \equiv c_0^2(s) \equiv 0$. For the second test problem (TP2) we set $\gamma = -3$ and $\sigma_{12} = -0.004$. Mesh parameters are $h = h_1 = h_2$ (the condition (10) is satisfied), $N = N_1 = N_2$, $L_i^{\pm} = \pm \ln(100)$, $T = 0.5$.

For the convergence test we set fixed time step $\triangle \tau = h^2$. Let C_h^k be the solution of (9), computed on the mesh with space step size h. Thus, we define the error $\mathcal{E}_h^k = \|C_{2h}^k - C_h^k\|_h$. On Fig. 1 we plot errors for TP1 and TP2 at final time T, computed by space meshes with different step sizes for both solutions

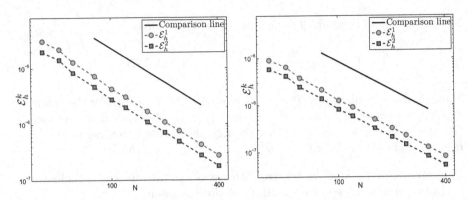

Fig. 1. $\mathcal{E}_h^k$ vs. N for TP1 (*left*) and TP2 (*right*)

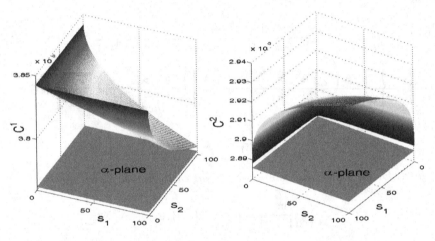

Fig. 2. C^1 (*left*), C^2 (*right*) for TP1 and α-plane

C^1 and C^2 vs. the number N of the space grid nodes in each spatial direction. We also plot comparison line, which correspond to a second order convergence rate. We observe second order convergence in space of C^1 and C^2 and as the ratio $\triangle \tau = h^2$ is fixed, we may conclude that the order of convergence in time is not less than one.

Next, we compute the solution of (9) with *variable time step*, satisfying the equalities in (11) for TP1 and (12) for TP2 with $c_0^1(s) \equiv 8.10^{-5}$, $c_0^2(s) \equiv 0$. On Figs. 2 and 3 we plot the solutions at final time $\tau = T$ in the original (s_1, s_2) variables for TP1 and TP2, respectively. For comparison, on the same graphs we depict also α - plane. We observe that at each grid node $C^k \geq \alpha$.

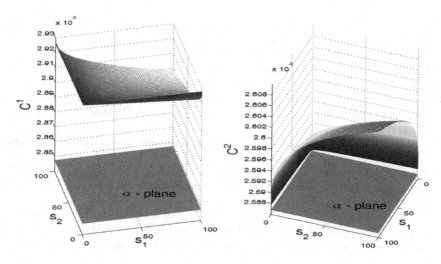

Fig. 3. C^1 (*left*), C^2 (*right*) for TP2 and α-plane

5   Conclusions

The main results of this paper are summarized as follows. We establish lower a-priory positive bounds for the differential problem solution. We construct implicit-explicit finite difference method preserving the qualitative properties of the exact solution.

Numerical results illustrate the order of convergence of the numerical solution - second order in space and first in time, and confirm theoretical statements.

Acknowledgements. This research is supported by the Bulgarian National Science Fund under Project DN 12/4 "Advanced analytical and numerical methods for non-linear differential equations with applications in finance and environmental pollution", 2017.

References

1. Faragó, I., Horváth, R.: Discrete maximum principle and adequate discretizations of linear parabolic problems. SIAM J. Sci. Comp. **28**, 2313–2336 (2006)
2. Gerisch, A., Griffiths, D.F., Weiner, R., Chaplain, M.A.J.: A positive splitting method for mixed hyperbolic - parabolic systems. Num. Meth. for PDEs **17**(2), 152–168 (2001)
3. Gyulov, B., Koleva, M.N., Vulkov, L.G.: Numerical approach to optimal portfolio in a power utility regime-switching model. AIP CP **1910**, 030002 (2017)
4. Gyulov, T.B., Koleva, M.N., Vulkov L.G.: Efficient finite difference method for optimal portfolio in a power utility regime-switching model. Int. J. Comp. Math., (2018). https://doi.org/10.1080/00207160.2018.1474207
5. Hundsdorfer, W., Verwer, J.: Numerical Solution of Time-Dependent Advection-Diffusion-Reaction Equations. Springer Series in Computational Mathematics, vol. 33. Springer, Heidelberg (2003). https://doi.org/10.1007/978-3-662-09017-6

6. Koleva, M.N. Vulkov, L.G.: Numerical method for optimal portfolio in an exponential utility regime-switching model. Int. J. Comp. Math. (2018). https://doi.org/10.1080/00207160.2018.1440289
7. Kusmin, D., Turek, S.: High-resolution FEM-TVD schemes based on a fully multidimensional flux limiter. J. Comput. Phys. **198**(1), 131–158 (2004)
8. van Leer, B.: Towards the ultimate conservative difference scheme II. monotonicity and conservation combined in a second order scheme. J. Comput. Phys. **14**, 361–370 (1974)
9. LeVeque, R.J.: Numerical Methods for Conservation Laws. Birkhäuser, Basel (1992)
10. Pao, C.V.: Nonlinear Parabolic and Elliptic Equations. Plenum Press, NY (1992)
11. Samarskii, A.A.: The Theory of Difference Schemes. Marcel Dekker Inc., New York (2001)
12. Valdez, A.R.L., Vargiolu, T.: Optimal portfolio in a regime-switching model. In: Dalang, R.C., Dozzi, M., Russo, F. (Eds.) Proceedings of the Ascona 2011 Seminar on Stochastic Analysis, Random Fields and Applications, pp. 435–449 (2013)

Effect of Ionic Strength on the Electro-Dipping Force

Galina Lyutskanova–Zhekova[1,2]([✉]) and Krassimir Danov[3]

[1] Institute of Mathematics and Informatics,
Bulgarian Academy of Sciences, Sofia, Bulgaria
g.zhekova@math.bas.bg
[2] Faculty of Mathematics and Informatics, Sofia University, Sofia, Bulgaria
[3] Faculty of Chemistry and Pharmacy, Sofia University, Sofia, Bulgaria
kd@lcpe.uni-sofia.bg

Abstract. The calculation of electro-dipping force, acting on a dielectric particle, attached to the boundary between water and nonpolar fluid, is important for the characterization of the surface charge density of micron-size objects and their three-phase contact angles [1]. The problem was solved semi-analytically, using the Mahler–Fox transformation in the simplified case of one phase with infinite dielectric permittivity [4]. We generalize this approach, taking into consideration the finite dielectric permittivity of the polar phase. We propose a numerical method for calculating the distribution of the electrostatic potential in all phases and the respective values of the dimensionless electro-dipping force. The expression for the weak singularity parameter at the three-phase contact line is analytically derived. In all studied cases, it is weaker than that in the model case [2]. The obtained results show that: (i) the electrostatic potential distribution is close to that in the model case for micron-size particles, large values of the ionic strength and dielectric constant of the polar phase; (ii) the force, arising from the electrostatic field in the polar phase, cannot be neglected for small (nano-size) particles and low ionic strengths.

Keywords: Electrostatic potential distribution · Laplace equations
Complex numerical domains and boundary conditions
Toroidal coordinates

1 Introduction

The interactions between electrically charged colloidal particles, adsorbed at an oil–water interface, depend on the magnitude of the surface charge density and the three-phase contact angle. The prediction of the properties of dielectric particles is of crucial importance for the characterization of a particle monolayer,

The work of Galina Lyutskanova–Zhekova has been partially supported by the Sofia University "St. Kl. Ohridski" under contract No. 80-10-139/25.04.2018.

G. Nikolov et al. (Eds.): NMA 2018, LNCS 11189, pp. 433–440, 2019.
https://doi.org/10.1007/978-3-030-10692-8_49

formation of particle-stabilized emulsions and colloidosomes [1]. Experimentally, it has been established that the electrostatic repulsion is due to the presence of charges at the particle–nonpolar phase boundary [2,3].

The problem was solved semi-analytically, using the Mahler-Fox transformation in the simplified case of water phase with infinite dielectric permittivity [4]. The effect of an external electric field, applied to the particle, was discussed in [5]. Our aim in the present study is to analyze the effect of water phase with a finite value of the dielectric constant and to calculate the distribution of the electrostatic potentials in all phases. We solve the Laplace equations in complex physical domains, using appropriate toroidal coordinates. The developed numerical scheme, which is of second order with respect to space and numerical time, allows fast and precise calculations.

2   Mathematical Formulation of the Problem

Let a spherical charged dielectric particle of radius R and dielectric constant ε_{p} is attached to the interface between nonpolar (oil, air) and polar (water) phases with dielectric constants ε_{n} and ε_{w}, respectively (Fig. 1a). The particle position is determined by the central angle α (three-phase contact angle) and therefore the radius of the three-phase contact line is $r_{\mathrm{c}} = R \sin \alpha$. The electric field intensities, $\mathbf{E}_j$ $(j = \mathrm{n, p, w})$, are induced by surface charges of constant surface charge density σ_{pn}, located at the particle–nonpolar phase boundary, S_{pn}. The electric fields in the dielectric phases, occupying volumes V_j $(j = \mathrm{n, p, w})$, obey the equations $\nabla \cdot \mathbf{E}_j = 0$ and the corresponding electrostatic potentials have the form $\mathbf{E}_j = -\nabla \varphi_j$, where ∇ is the gradient operator. Thus, the potentials can be modelled as solutions of the Laplace equations in the volumes, V_j. At the boundaries between water and dielectric phases, S_{pw} and S_{nw}, there are no adsorbed charges, therefore $\varepsilon_{\mathrm{w}} \mathbf{n} \cdot \nabla \varphi_{\mathrm{w}} = \varepsilon_{\mathrm{p}} \mathbf{n} \cdot \nabla \varphi_{\mathrm{p}}$ at S_{pw} and $\varepsilon_{\mathrm{w}} \mathbf{n} \cdot \nabla \varphi_{\mathrm{w}} = \varepsilon_{\mathrm{n}} \mathbf{n} \cdot \nabla \varphi_{\mathrm{n}}$ at S_{nw} hold true, where $\mathbf{n}$ is the outer unit normal vector from the particle surface (Fig. 1a). At the charged part of the particle surface, S_{pn}, we apply the boundary condition $\varepsilon_0 \varepsilon_{\mathrm{p}} \mathbf{n} \cdot \nabla \varphi_{\mathrm{p}} - \varepsilon_0 \varepsilon_{\mathrm{n}} \mathbf{n} \cdot \nabla \varphi_{\mathrm{n}} = \sigma_{\mathrm{pn}}$, where ε_0 is the dielectric permittivity in vacuum [2,4]. The tangential boundary conditions state that all potentials are continuous functions at the dividing boundaries.

For numerical calculations, it is convenient to reformulate the problem in a dimensionless form. We use the following dimensionless electrostatic potentials, Φ_{p}, Φ_{n}, and Φ_{w}, and ratios between the dielectric constants, $\varepsilon_{\mathrm{pn}}$ and $\varepsilon_{\mathrm{wn}}$:

$$\Phi_j = \frac{\varphi_j \varepsilon_0 \varepsilon_{\mathrm{n}}}{r_{\mathrm{c}} \sigma_{\mathrm{pn}}} \; (j = \mathrm{p, n, w}), \quad \varepsilon_{\mathrm{pn}} = \frac{\varepsilon_{\mathrm{p}}}{\varepsilon_{\mathrm{n}}}, \quad \varepsilon_{\mathrm{wn}} = \frac{\varepsilon_{\mathrm{w}}}{\varepsilon_{\mathrm{n}}}. \tag{1}$$

The cylindrical coordinate system $Or\theta z$ with axis of revolution Oz defines the radial, vertical and polar coordinates, r, z and θ, respectively (Fig. 1a). The complex geometry of the domains is transformed into rectangles (Fig. 1b) by introducing modified toroidal coordinates t and s analogous to those in [2]:

$$\frac{r}{r_{\mathrm{c}}} = \frac{1-t^2}{h}, \quad \frac{z}{r_{\mathrm{c}}} = \frac{2t \sin s}{h}, \quad h(t,s) = 1 + t^2 - 2t \cos s. \tag{2}$$

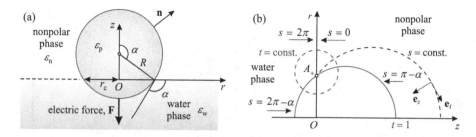

Fig. 1. (a) Sketch of a particle at the interface between nonpolar and water phases. (b) Toroidal coordinate system (t, s), in which the physical domains are transformed into rectangles.

For the curvilinear orthogonal coordinate system, we denote the unit vectors of the local basis by $\mathbf{e}_t$ and $\mathbf{e}_s$. Let us remark that the introduced t and s are similar to the classical toroidal coordinates—we transform the original first coordinate to be in the interval $[0, 1]$. In the new variables, the positions of the interfaces are: $s = 0$ and $s = 2\pi$ from both sides of S_{nw}; $s = \pi - \alpha$ at S_{pn}; $s = 2\pi - \alpha$ at S_{pw}. The axis of revolution corresponds to $t = 1$ and the three-phase contact line—to the pole, A_+, where $t = 0$. The Lamé coefficients, h_t, h_s, and h_θ, of the toroidal coordinate system are calculated from the following relationships:

$$h_t = \frac{2}{h} r_c, \quad h_s = \frac{2t}{h} r_c, \quad h_\theta = r = \frac{1 - t^2}{h} r_c. \qquad (3)$$

Using the general formulae for the Laplace operator and directional derivatives in orthogonal coordinates [2], we get

$$\frac{h}{t(1 - t^2)} \frac{\partial}{\partial t} \left[\frac{t(1 - t^2)}{h} \frac{\partial \Phi_j}{\partial t} \right] + \frac{h}{t^2} \frac{\partial}{\partial s} \left(\frac{1}{h} \frac{\partial \Phi_j}{\partial s} \right) = 0 \text{ in } V_j \ (j = \text{n,p,w}), \qquad (4)$$

$$\mathbf{n} \cdot \nabla u = -\frac{h}{2tr_c} \frac{\partial u}{\partial s} \text{ at } S_{pn}, \ \mathbf{n} \cdot \nabla u = \frac{h}{2tr_c} \frac{\partial u}{\partial s} \text{ at } S_{pw}. \qquad (5)$$

From the latter, we obtain the dimensionless formulation of the considered problem for the electrostatic potentials: (i) in the volumes—equation (4); (ii) at the dividing physical boundaries:

$$\Phi_p|_{s=2\pi-\alpha} = \Phi_w|_{s=2\pi-\alpha}, \ \varepsilon_{pn} \frac{\partial \Phi_p}{\partial s} \bigg|_{s=2\pi-\alpha} = \varepsilon_{wn} \frac{\partial \Phi_w}{\partial s} \bigg|_{s=2\pi-\alpha}, \qquad (6)$$

$$\Phi_n|_{s=0} = \Phi_w|_{s=2\pi}, \ \frac{\partial \Phi_n}{\partial s} \bigg|_{s=0} = \varepsilon_{wn} \frac{\partial \Phi_w}{\partial s} \bigg|_{s=2\pi}, \qquad (7)$$

$$\Phi_n = \Phi_p, \ \frac{\partial \Phi_n}{\partial s} - \varepsilon_{pn} \frac{\partial \Phi_p}{\partial s} = \frac{2t}{1 + t^2 + 2t \cos \alpha} \text{ for } s = \pi - \alpha; \qquad (8)$$

(iii) at the axis of revolution, $t = 1$, and at the three-phase contact line, $t = 0$:

$$\frac{\partial \Phi_j}{\partial t} = 0 \text{ for } t = 1 \ (j = \text{n,p,w}), \ \Phi_n = \Phi_p = \Phi_w = 0 \text{ for } t = 0, \qquad (9)$$

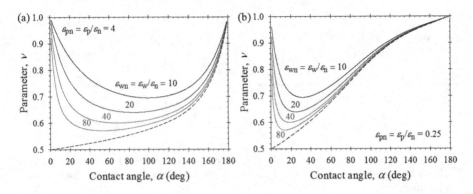

Fig. 2. Dependence of singularity parameter ν on the contact angle α and ratio ε_{wn}: (a) $\varepsilon_{pn} = 4$; (b) $\varepsilon_{pn} = 0.25$. Dashed lines show the model case, studied in [4].

where the electrostatic potentials at the pole A_+ are defined to be equal to zero.

It is shown in [4] that the particulate problem has a weak singularity of the electric field intensity in the close vicinity of the contact line. The leading order solutions of Eq. (4) for $t \to 0$ are $\Phi_j = t^\nu (A_j \sin(\nu s) + B_j \cos(\nu s))$, where A_j and B_j ($j =$ n, p, w) are unknown constants and $0.5 < \nu < 1$. The substitution of these solutions into the boundary conditions (6)–(8) leads to a homogeneous system of six linear equations for A_j and B_j. This system has a nontrivial solution, when its determinant is equal to zero. Thus, we arrive to the following equation for the singularity parameter ν:

$$
\frac{2\varepsilon_{pn}(1 - \varepsilon_{wn})^2}{(1 + \varepsilon_{pn})(1 + \varepsilon_{wn})(\varepsilon_{pn} + \varepsilon_{wn})} - \sin^2(\nu\pi) =
$$
$$
\frac{(1 - \varepsilon_{pn})(1 - \varepsilon_{wn})}{(1 + \varepsilon_{pn})(1 + \varepsilon_{wn})} \cos^2(\nu\alpha) + \frac{(1 - \varepsilon_{wn})(\varepsilon_{pn} - \varepsilon_{wn})}{(1 + \varepsilon_{wn})(\varepsilon_{pn} + \varepsilon_{wn})} \cos^2[\nu(\pi - \alpha)]. \quad (10)
$$

The model case, studied in [4], follows from (10) when $\varepsilon_{wn} \to \infty$ (dashed lines in Fig. 2). The solutions of Eq. (10) for the singularity parameter, ν, for two different ratios ε_{pn} are shown in Fig. 2. The following conclusions can be drawn from it. In all cases, the singularity is weaker than that in the model case. The values of ν increase with the decrease of the ratio between the dielectric constants of water and nonpolar phase, ε_{wn}. This effect is more pronounced for larger values of ε_{wn} and more hydrophilic particles.

3   Numerical Method

In order to find a numerical solution of the problem, we apply the alternating direction implicit method (ADIM), introducing numerical time τ. The continuous function $f(\tau, t, s)$ describes the electrostatic potentials, ordered as follows:

Φ_{n}, Φ_{p}, and Φ_{w}. In the volumes, $f(\tau, t, s)$ obeys an equation of the form (4):

$$\frac{\partial f}{\partial \tau} = \mathbf{T}[f] + \mathbf{S}[f],$$

$$\mathbf{T}[f] = \frac{h}{t(1 - t^2)} \frac{\partial}{\partial t} \left[\frac{t(1 - t^2)}{h} \frac{\partial f}{\partial t} \right], \ \mathbf{S}[f] = \frac{h}{t^2} \frac{\partial}{\partial s} \left(\frac{1}{h} \frac{\partial f}{\partial s} \right), \quad (11)$$

where the operators $\mathbf{T}$ and $\mathbf{S}$ act in directions t and s, respectively. For the discretization of (11), we introduce an uniform mesh with time step δ_τ and space steps δ_n, δ_p, δ_w by dividing each numerical domain $\tilde{V}_j$ ($j = n, p, w$) into squares of length δ_j. We denote the solution at a given moment τ by subscript "0" and at the moment $\tau + 2\delta_\tau$—by subscript "2". Using the Crank–Nicolson method in the ADIM scheme, we obtain the following second order numerical model:

$$(\mathbf{U} - \delta_\tau \mathbf{S})(\mathbf{U} - \delta_\tau \mathbf{T})[f_2 - f_0] = 2\delta_\tau \mathbf{T}[f_0] + 2\delta_\tau \mathbf{S}[f_0] + O(\delta_\tau^3), \quad (12)$$

where $\mathbf{U}$ is the unit operator. Firstly, we solve numerically an equation for a function $g(t, s)$ in the s-direction, and subsequently—an equation in the t-direction

$$(\mathbf{U} - \delta_\tau \mathbf{S})[g] = 2\delta_\tau \mathbf{T}[f_0] + 2\delta_\tau \mathbf{S}[f_0] \text{ and } (\mathbf{U} - \delta_\tau \mathbf{T})[f_2 - f_0] = g \quad (13)$$

as the derivatives in $\mathbf{T}$ and $\mathbf{S}$ are approximated with second-order central differences with respect to t and s.

The main problem in the ADIM arises from the complexity of the boundary conditions, applied to the function g. In order to have a second order precision with respect to t and s, we modify the boundary conditions, assuming the validity of the Laplace equations in the close vicinity of the dividing surfaces and the axis of revolution [6]. We use the following approximation:

$$u''(x) = \frac{-7u(x) + 8u(x \pm \delta_x) - u(x \pm 2\delta_x)}{2\delta_x^2} \mp \frac{3}{\delta_x} u'(x) + O(\delta_x^2), \quad (14)$$

which relates the second and the first derivative of a given function. For example, the limit of the operator $\mathbf{T}$ at the axis of revolution, taking into account the boundary conditions (9) and equation (11), yields

$$\lim_{t \to 1} \mathbf{T}[u] = 2 \left. \frac{\partial^2 u}{\partial t^2} \right|_{t=1} = \frac{-7u(1) + 8u(1 - \delta_t) - u(1 - 2\delta_t)}{2\delta_t^2}. \quad (15)$$

Therefore, we replace the boundary conditions (9) with Eq. (13), in which the operator $\mathbf{T}$ is given by the difference form (15). The result is a modified boundary condition in relaxed form.

From Eqs. (11) and (14), the finite difference representations of the operator $\mathbf{S}$ in the close vicinity of the boundary $s = \pi - \alpha$ are

$$t^2 a_{\mathrm{n}} \mathbf{S}[u] = \frac{-7u(s) + 8u(s - \delta_{\mathrm{n}}) - u(s - 2\delta_{\mathrm{n}})}{2\delta_{\mathrm{n}}^2} a_{\mathrm{n}} + u'(s - 0), \quad (16)$$

$$t^2 a_{\mathrm{p}} \mathbf{S}[u] = \frac{-7u(s) + 8u(s + \delta_{\mathrm{p}}) - u(s + 2\delta_{\mathrm{p}})}{2\delta_{\mathrm{p}}^2} a_{\mathrm{p}} - \varepsilon_{\mathrm{pn}} u'(s + 0), \quad (17)$$

where the functions $a_n(t)$ and $a_p(t)$ are defined with

$$a_n(t) = \delta_n \left(3 - \frac{\delta_n}{h} \frac{\partial h}{\partial s} \right)^{-1} , \quad a_p(t) = \varepsilon_{pn}\delta_p \left(3 + \frac{\delta_p}{h} \frac{\partial h}{\partial s} \right)^{-1} \quad \text{for } s = \pi - \alpha. \quad (18)$$

The sum of Eqs. (16) and (17) along with the boundary conditions (8) leads to the final expression for the finite difference form of the operator $\mathbf{S}$ at the boundary S_{pn}:

$$\mathbf{S}[u] = \frac{-7u(s) + 8u(s - \delta_n) - u(s - 2\delta_n)}{2\delta_n^2 t^2 (a_n + a_p)} a_n$$
$$+ \frac{-7u(s) + 8u(s + \delta_p) - u(s + 2\delta_p)}{2\delta_p^2 t^2 (a_n + a_p)} a_p + \frac{2}{ht(a_n + a_p)} \quad \text{for } s = \pi - \alpha. \quad (19)$$

Thus, the boundary conditions (8) are replaced by Eq. (13), in which the operator $\mathbf{S}$ is given by definition (19). Following an analogous procedure, we derive the respective expression for the difference form of $\mathbf{S}$ at the boundary $s = 2\pi - \alpha$ and on both sides of S_{nw} ($s = 0$ and $s = 2\pi$).

Because of the boundary conditions, the considered algorithm reduces the matrix of the linear algebraic system in the t-direction to a five-diagonal matrix. The respective matrix of the system in the s-direction is again with five non-zero diagonals, but because of the periodicity of the solution at S_{nw} its first row contains two more elements at the end and the last row contains two more elements at the beginning. We implemented a direct elimination numerical method in order to solve the system.

4   Results and Discussion

To achieve good precision of the numerical calculations, we discretize each numerical domain $\tilde{V}_j$ ($j = $ n, p, w) by introducing a 100×100 uniform mesh (see Sect. 3). The time step is chosen to be equal to the minimum of δ_n, δ_p and δ_w. The illustrative figures (Figs. 3b and 4b) correspond to experimental system parameters [2] $\varepsilon_{pn} = 2$ and $\varepsilon_{wn} = 40$. If the oil phase has a larger dielectric constant (for example, castor oil with $\varepsilon_n = 4.54$), then the system parameters are $\varepsilon_{pn} = 0.874$ and $\varepsilon_{wn} = 17.2$ (Figs. 3a and 4a, respectively). The mathematical formulation of the problem assumes that the electrostatic potential at the three-phase contact line is equal to zero (see Eq. (9)) and the intensity of the electric field at large distances from the charged particle is equal to zero. Thus, the calculated potential at infinity has a non-zero value. It is subtracted from the calculated Φ_j in order to obtain the physical electrostatic potential.

Figure 3 shows the distribution of the physical values of Φ_j in the numerical domains for three-phase contact angle $\alpha = 90°$. The considerably higher dimensionless potentials are well illustrated for larger values of the dielectric constant of the nonpolar phase. At the coordinate lines, $s = 0$ (S_{nw}) and $s = 3\pi/2$ (S_{pw}), the electrostatic potentials are considerably lower than those at coordinate line

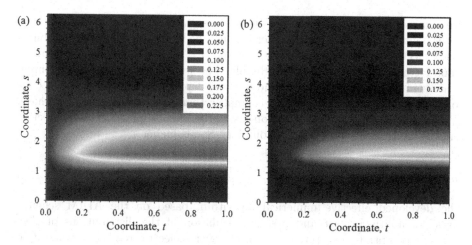

Fig. 3. Distribution of the electrostatic potentials in numerical domains for contact angle $\alpha = 90°$: (a) $\varepsilon_{pn} = 0.874$ and $\varepsilon_{wn} = 17.2$; (b) $\varepsilon_{pn} = 2$ and $\varepsilon_{wn} = 40$.

$s = \pi/2$ (S_{pn}). As it can be expected, the maxima of the electrostatic potentials are at the cross-section of the particle–nonpolar interface and the axis of revolution. The dielectric constant of water is so high that the water phase suppresses the penetration of the electric field in polar phase.

The calculations in [2,4] are performed, assuming zero values of the potentials at the boundaries of the polar fluid. The magnitude of the electro-dipping force decreases if the electrostatic potentials of these boundaries are different than zero. Figure 4 shows the distribution of the surface potentials along the boundaries (solid lines correspond to S_{pn}; dashed lines to S_{pw}; dot-dashed lines — to S_{nw}). The increase of the three-phase contact angle (more hydrophobic particles) leads to higher potentials because of the more charges adsorbed at the particle–nonpolar fluid interface. The effect of the water phase becomes more

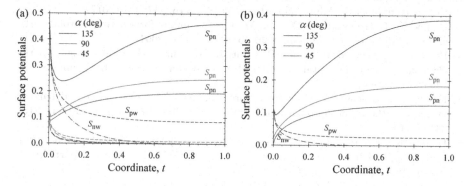

Fig. 4. Distribution of potentials along the interfaces for different values of contact angle α: (a) $\varepsilon_{pn} = 0.874$ and $\varepsilon_{wn} = 17.2$; (b) $\varepsilon_{pn} = 2$ and $\varepsilon_{wn} = 40$.

pronounced. It is important to note that the surface potentials at the particle–water boundary are different from zero. Thus, the boundary S_{pw} also contributes to the electro-dipping force. For $\alpha = 45°$ and $\alpha = 90°$ this contribution is small, while for $\alpha = 135°$ — it is not negligible. The weak singularities at the three-phase contact line $(t = 0)$ correspond to those depicted in Fig. 2. If the dielectric constant of the particle phase, ε_{p}, is smaller than that of the nonpolar phase, ε_{n}, then the electric field penetration inside the particle phase is more effective and the electrostatic potential at boundary S_{pw} is higher.

5   Conclusions

The developed effective numerical algorithm, based on the ADIM scheme, gives possibility for fast and precise calculation of the distributions of electrostatic potentials, generated from a charged dielectric particle, attached to the nonpolar–water interface. For faster calculations, the complex numerical domains are transformed into rectangles, using appropriate toroidal coordinates. The resulting cyclic five diagonal systems of linear equations in the respective directions of ADIM are solved, using a direct elimination method.

The numerical results show the effect of the three-phase contact angle and the dielectric properties of the phases on the induced electric fields and the magnitude of the electro-dipping force. Generally, the decrease of the ratios of the dielectric constants of the particle and nonpolar phase, $\varepsilon_p/\varepsilon_{\mathrm{n}}$, and that of water and nonpolar phase, $\varepsilon_{\mathrm{w}}/\varepsilon_{\mathrm{n}}$, leads to the more pronounced penetration of electric field and higher surface potentials at the particle–water and nonpolar fluid–water boundaries. The magnitude of the potentials (electro-dipping force) is larger for more hydrophobic particles. The calculations generalize what is known from literature [2,4] and give a more precise description of the problem.

References

1. Lotito, V., Zambelli, T.: Approaches to self-assembly of colloidal monolayers: a guide to nanotechnologists. Adv. Colloid Interface Sci. **246**, 217–274 (2017)
2. Danov, K., Kralchevsky, P., Boneva, M.: Electrodipping force acting on solid particles at a fluid interface. Langmuir **20**(15), 6139–6151 (2004)
3. Horozov, T., Aveyard, R., Binks, P., Clint, J.: Structure and stability of silica particle monolayers at horizontal and vertical octane-water interfaces. Langmuir **21**(16), 7405–7412 (2005)
4. Danov, K., Kralchevsky, P.: Electric forces induced by a charged colloid particle attached to the water-nonpolar fluid interface. J. Colloid Interface Sci. **298**(1), 213–231 (2006)
5. Danov, K., Kralchevsky, P.: Forces acting on dielectric colloidal spheres at a water/nonpolar fluid interface in an external electric field. 2. charged particles. J. Colloid Interface Sci. **405**, 269–277 (2013)
6. Wang, Y., Wang, T.: A compact ADI method and its extrapolation for time fractional sub-diffusion equations with nonhomogeneous Neumann boundary conditions. Comput. Math. Appl. **75**(3), 721–739 (2018)

Accuracy of Different Machine Learning Type Methodologies for EEG Classification by Diagnosis

Andrius Vytautas Misiukas Misiūnas[1]([⊠]), Tadas Meškauskas[1],
and Rūta Samaitienė[2,3]

[1] Institute of Computer Science, Faculty of Mathematics and Informatics,
Vilnius University, Didlaukio 47, 08303 Vilnius, Lithuania
andrius.misiukas@mif.vu.lt
[2] Children's Hospital, Affiliate of Vilnius University Hospital Santaros Klinikos,
Santariškių 7, 08406 Vilnius, Lithuania
[3] Clinic of Children's Diseases, Faculty of Medicine, Vilnius University,
Santariškių 4, 08406 Vilnius, Lithuania

Abstract. Electroencephalogram (EEG) classification accuracy of different automatic algorithms (including their setup) is discussed. Two patient groups, characterized by visually similar (to neurologists) EEG rolandic spikes, are under classification. The first group consists of patients with benign focal childhood epilepsy. Patients with structural focal epilepsy define the second group. We analyzed 94 EEGs (with known diagnosis) obtained from Children's Hospital, Affiliate of Vilnius University Hospital Santaros Klinikos.

The EEGs are preprocessed by applying these steps: (i) spike detection; (ii) extraction of spike parameters. After preprocessing of EEGs we gather parameters of detected spikes into lists of equal length N_{spikes}.

The classification algorithms are trained employing one set of patients (containing patients from both groups) and tested on another non-overlapping set of patients (also from both groups). This prevents artificial accuracy inflation due to overfitting.

We compared eight machine learning type classifiers: (1) random forest, (2) decision tree, (3) extremely randomized tree, (4) adaptive boosting (AdaBoost), (5) artificial neural network (ANN), (6) supported vector machine (SVM), (7) linear discriminant analysis (LDA), (8) logistic regression. To estimate quality of classifiers we discuss a set of metrics. The results are following: (I) as expected, for all examined algorithms, the accuracy tends to grow (when N_{spikes} increases), saturating at some asymptotic value; (II) ANN has prevailed as best classifier.

Impact of: (a) different training strategies and (b) spike detection errors on classification accuracy is also discussed.

Novelty and originality of this study comes not only from classifying different types of epilepsy, but also from employed computational methodology (involving parameters of EEG spikes and machine learning type classifier), as well as comparing different methodologies of such type, based on their accuracy and other classifier metrics.

Keywords: Machine learning · EEG · Epilepsy · EEG spikes

© Springer Nature Switzerland AG 2019
G. Nikolov et al. (Eds.): NMA 2018, LNCS 11189, pp. 441–448, 2019.
https://doi.org/10.1007/978-3-030-10692-8_50

1 Introduction

Manual analysis of electroencephalograms (EEGs) is a very difficult and time consuming process, thus a lot of automatic algorithms dedicated to help neurologists with this analysis are proposed, *e.g.* [4,13]. EEG classification analysis may aim at: classifying healthy *vs* ill patient EEGs (*e.g.* ictal and non-ictal EEGs [5], patients diagnosed with epilepsy and healthy [11]), patients with addiction (*e.g.* from alcohol) and addiction-free [1]. Various machine learning (ML) algorithms are employed in solving ever increasing variety of EEG analysis related problems. The choice of classification algorithm is not the only important choice to make. Algorithm parameters and training strategies, EEG preprocessing are significant too.

In this study problem of *classification by diagnosis* is tackled (different from most other works, dealing with healthy *vs* ill type tasks). Previously (in [9]) we proposed the three-stage (EEG spikes detection, evaluation of spikes characteristics, classification) algorithm, based on artificial neural network (ANN) for this problem, as well as some analysis of ANN and SVM classifiers. In this paper we analyze more ML type methods for the same classification problem.

Details and in-depth analysis of data preprocessing (including EEG spikes detection), see *e.g.* [7,8], are out of scope of this article. Instead, here we try to answer the following questions: which classification algorithm is most accurate and most robust (to low amount of spikes in one EEG, false positive and false negative detections of EEG spikes). By discussing these questions (see Sects. 4 and 5), we challenge ANN based method against other known classifiers: random forest, decision tree, extremely randomized tree [3], adaptive boosting (AdaBoost) [2], supported vector machine (SVM), linear discriminant analysis (LDA) and logistic regression.

In numerical experiments Python 3.6.4 programming language was employed as well as `scikit-learn` library (for implementations of ML algorithms and their performance metrics, see Sect. 4.1).

2 Data

EEGs analyzed in this study were provided by Children's Hospital, Affiliate of Vilnius University Hospital Santaros Klinikos. EEG recordings span over a 2010 – 2017 period, only 3 – 17 year old patients (all with known exact diagnosis) were included.

94 EEGs (of 86 different patients) were processed – divided into two groups, characterised by visually similar (to neurologists) EEG spikes:

- **RE Group:** Benign childhood epilepsy (rolandic epilepsy) with centrotemporal spikes (62 EEGs or about 66% of all EEGs), 35 of them boys, ≈56.4%. 75% of all detected EEG spikes were in this group.
- **CP Group:** Structural focal epilepsy patients with cerebral palsy, dysplastic brain lesion, gliosis *etc.* (32 EEGs, or about 34% of all EEGs), 18 of them boys, ≈56.3%. 25% of all detected EEG spikes were in this group.

It should be pointed out that only 30% of all EEGs were manually cleaned by neurologist from artifacts (*e.g.* patient movement) – thus decreasing error of spikes detection (see Sect. 3.1). Rest of the data were processed uncleaned. Cleaned data were used for classification algorithm training while uncleaned data – for algorithm validation (except for testing robustness of the algorithm, by including uncleaned data in the training data set, see Sect. 4.2).

3   Preprocessing of Data

In this section we briefly discuss the steps and algorithms that are performed on the data before it is been tried to classify. Our algorithm consists of three basic steps:

1. EEG spike detection by morphological filter, discussed in [6,7,10], also see Sect. 3.1;
2. Extraction of EEG parameters, introduced and discussed in [8], also see Sect. 3.2;
3. EEG parameter validation, for details see [8], also see Sect. 3.2.

3.1   EEG Spike Detection by Morphological Filter

The first step of our algorithm is EEG spike detection by morphological filter based algorithm. The morphological filter is defined using close-opening and open-closing operations which can be defined using morphological erosion and dilation. For exact definitions please see [6,7].

The idea of morphological filter is to filter out known normal brain activity (brain rhythms, patient movements, *etc.*) leaving abnormal brain activity. Then a detection limit is calculated, all maxima (higher than the detection limit) in filtered signal are treated as potential spikes. However this generates some false positive detections, this problem is addressed in the third step of this algorithm.

3.2   EEG Spike Parameter Extraction and Validation

After EEG spikes are detected, some parameters of these spikes are extracted. These parameters include upslope, downslope, width and baseline at half maximum. Our earlier investigations show that upslope and downslope are the most significant for the classification. In this work upslope and downslope value pairs for each spike are employed for classification.

These values are validated against range of known medically possible values of these parameters. This helps us to decrease false positive detections [8] (however some false positive detections remain). All values that do not meet these criteria are excluded from further analysis.

4   Comparison of Various Classification Methodologies by Performance

A strict rule of keeping each patient in either testing or training data set was kept during this work. The main reason this was done is that each patients EEG spikes are similar to one another and mixing patients in both training and testing samples would result in artificially inflated accuracy values since it would result in same effect as training and testing the classifiers on same data.

Since all EEGs have different amount of spikes and most classification algorithms can be trained and used with fixed number of inputs, EEGs were cut into non-overlapping lists of length N_{spikes}. This allowed us to treat each list as virtual EEG thus making our data accessible to algorithms and allowing us to measure the performance metrics of algorithms more reliably. For most experiments $N_{spikes} = 100$, except for experiment to test accuracy *vs* length of spike lists (see Sect. 4.3). As mentioned in Sect. 2 we had limited amount of artifact-free data as well. All the clean data were used to train or fit algorithms and rest of the data – to test except for testing robustness of the classifiers (see Sect. 4.2).

In this section we are going to explore quality and robustness of different machine learning type classifiers (presented in the Introduction) in respect to various ML classifier metrics (see Sect. 4.1), by falsely detected spikes (see Sect. 4.2) and dependency accuracy spike series length (see Sect. 4.3).

4.1   Performance Metrics of EEG Classification Algorithms

The main aim of this study is to find the ML classification algorithm best suited for classifying EEGs obtained from patients from RE and CP groups. In order to achieve this task, some quantifiable parameters of algorithm performance are needed. The most obvious metric for this task is accuracy, which is sum of true positives and true negatives divided over all detections. This metric is very useful in detecting poorly performing algorithms.

After measuring the accuracy, LDA and logistic regression algorithms were excluded from further analysis due to their accuracy being poor: 53% and 59% respectively. Multiple supported vector machine (SVM) classifier configurations were tested as well. SVM classifiers with linear and quadratic kernels performed consistently with worse accuracy than SVM with cubic kernel thus were removed from further analysis.

While accuracy is a great tool for finding out some poorly performing algorithms, it does not show all of them. For that reason some true positive rate (TPR) and true negative rate (TNR) analysis was done. Although SVM with both RBF and sigmoid kernels were performing with good accuracy of 75%, they were classifying all the data as RE group. The accuracy was achieved purely due to our data set being biased towards RE group (see Sect. 2). Due to this reason these algorithms were excluded from further analysis.

Random forest, decision tree, extremely randomized trees, AdaBoost and ANN presented comparable results for both groups and thus were analyzed further. Table 1 presents the commonly used performance metrics [12] for algorithms tested. These tests were performed to evaluate overall quality of the discussed classifiers. In order to minimize statistical error caused by oscillations (*e.g.* see Fig. 1) of these metrics we first apply linear fitting (for large $30 \leqslant N_{spikes} \leqslant 100$) and then take numerical value of the fitted trend.

Table 1. Performance metrics [12] for algorithms selected group of algorithms with $N_{spikes} = 100$. Ideal classifier column represents metric values for theoretical ideal classifier.

Score/ Algorithm	Random forest	Decision tree	Extremely randomised tree	AdaBoost	ANN	SVM N=3	Ideal classifier
Accuracy	0.78	0.76	0.80	0.81	0.75	0.69	**1.00**
TNR	0.79	0.76	0.83	0.90	0.79	0.79	**1.00**
TPR	0.74	0.77	0.71	0.52	0.74	0.48	**1.00**
F1 score	0.76	0.76	0.75	0.64	0.78	0.57	**1.00**
ROC AUC	0.53	0.49	0.56	0.69	0.64	0.49	**1.00**
Cohen kappa	0.06	−0.01	0.12	0.38	0.28	0.26	**1.00**
Hamming loss	0.48	0.52	0.45	0.32	0.37	0.38	**0.00**
Jaccard simmilarity score	0.52	0.48	0.55	0.68	0.63	0.62	**1.00**
Log loss	16.72	17.92	15.55	11.00	12.93	12.96	**0.00**
Matthews correlation coefficient	0.07	−0.01	0.15	0.42	0.38	0.28	**1.00**
Recall score	0.78	0.76	0.81	0.84	0.78	0.69	**1.00**
Zero one loss	0.48	0.52	0.45	0.32	0.25	0.38	**0.00**

AdaBoost seems to be the best algorithm by most metrics presented in Table 1, except a couple key ones: TPR and F_1 score. This is due to the fact that AdaBoost classifies RE (dominant group) correctly 90% of the time and CP group – only about 52% of the time, SVM with cubic kernel suffers from the same problem. Despite of good performance of AdaBoost across all other metrics, this algorithm is not suited for the task at hand – detecting more rare CP group cases in the pool of CP and RE group data. However AdaBoost could be explored further for potential use in ensemble (voting) type of classifier. This leads to discussion that some classifier quality metrics can be misleading in this case (see Sect. 5).

Table 1 shows some more interesting results. Although random forest, decision tree and extremely randomized trees show both high TPR and TNR, their

ROC AUC, Cohen kappa and Matthews correlation coefficient are poor. This is probably due to the reason that these metrics are designed to take into account chance of classifying a record correctly by guessing, therefore these metrics suggest that these algorithms are getting the correct answer by guessing it. Extremely randomized tree suffers less from this problem, its Cohen kappa and Matthews correlation coefficient scores are still poor. This means that these algorithms are less suited for EEG classification than ANN and are excluded from further analysis.

This leaves us with ANN, SVM (with cubic kernel) and AdaBoost classifiers. Of these three, the ANN classifier is better considering all metrics, thus it is recommended to be used for automatic classification by diagnosis.

4.2 Accuracy Dependency on Falsely Detected Spikes in Training Set

In this section we test robustness of algorithm to both small (by percentage) training sample and falsely positively or falsely negatively detected spikes in the training data set. It should be noted that quality of classification depends not only on choice of classification algorithm, but also on: (a) employed training strategy (percentage of training data vs testing data); (b) spike detection errors (during EEG preprocessing, falsely positively or falsely negatively detected spikes).

As expected, more false detections exist when analyzing not cleaned EEGs (see Sect. 2). Our results show that all testing classifiers are sensitive to inclusion of not cleaned data to some extent, however ANN is most sensitive to inclusion of large amounts of artifacts. Similar results are observed when only clean data is used for training, but the percentage of the training sample size is reduced. This result is expected since our analysis (see Sect. 4.1) show that ANN is the most intelligent classifier while other classifiers are guessing the answer by using uneven distribution across groups of our data sample (see Sect. 2). However this is not an issue if large and clean enough data sample is available for the training.

4.3 Accuracy Dependency on Length of Spikes Lists

It is difficult to tell the diagnosis from a single or a few spikes and their parameters. In order to make more accurate attempt at classification more spike parameters are needed. Thus we divided spikes found in all EEGs into lists (see Sect. 4) containing up to 100 spikes ($1 \leqslant N_{spikes} \leqslant 100$) and examined the dependency between number of spikes in series and accuracy of classification for all classification algorithms employed in this study (see Fig. 1).

For most classifiers (except SVM with cubic kernel) accuracy increases as N_{Spikes} grows, for higher N_{Spikes} all classification algorithms reach saturation levels. For this reason length of 100 spike parameter series was used in all other experiments in this work. Accuracy of SVM with $N_{spikes} < 30$ is not presented due to very long calculation times.

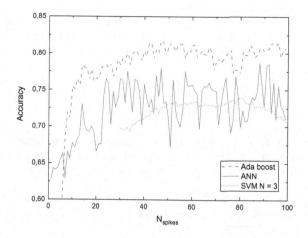

Fig. 1. Accuracy dependency on spike count.

5   Results and Discussion

In this paper we compared quality of different machine learning (ML) type methodologies for EEG classification by diagnosis (see Sect. 4.1 and Table 1).

The analysis of ML classification algorithms present an important point about the metrics (see Table 1) themselves. If unevenly distributed across groups data set is analyzed (see Sect. 2), and classifier (*e.g.* AdaBoost) is biased towards larger group, most metrics still score high, except for F_1 score and either TPR or TNR. Another important point is that if data set is unevenly distributed across groups – it is much easier to guess the correct answer. Random forest, decision tree and extremely randomized tree seems to be doing that as shown by Cohen kappa and Matthews correlation coefficient. Other metrics (ROC AUC, Hamming loss, Jaccard similarity score, log loss, recall score and zero one loss) do not present any new information compared to plain accuracy. Furthermore one could consider these metrics misleading (*e.g.* with AdaBoost) for binary classification problem with unevenly distributed across groups data set.

For algorithm selection to solve classification problems similar to one presented in this work, we recommend evaluation of F_1 score, TPR, TNR and either Cohen kappa or Matthews correlation coefficient metrics. For purposes of selecting candidate algorithms for further analysis accuracy could be used since low accuracy always means poor performance, while high accuracy does not universally suggest good overall performance of algorithm.

Considering all available data we determined that ANN currently is best tested algorithm for classifying spike series of CP and RE groups.

Robustness of ANN classifier (see Sect. 4.2) is not an issue if correct training strategy is selected (clean data employed in training) and big enough clean data sample is available. Following that conclusion we employed all clean data available in the training data sample.

These results also raise some possibilities for future work: improving spike detection accuracy, using ensemble (voting) or statistical boosting for ANN classifier in order to achieve higher classification accuracy with higher both TPR and TNR.

References

1. Byvatov, E., Fechner, U., Sadowski, J., Schneider, G.: Comparison of support vector machine and artificial neural network systems for drug/nondrug classification. J. Chem. Inf. Comput. Sci. **43**(6), 1882–1889 (2003). https://doi.org/10.1021/ci0341161
2. Freund, Y., Schapire, R.E.: A decision-theoretic generalization of on-line learning and an application to boosting. J. Comput. Syst. Sci. **55**(1), 119–139 (1997). https://doi.org/10.1006/jcss.1997.1504
3. Geurts, P., Ernst, D., Wehenkel, L.: Extremely randomized trees. Mach. Learn. **63**(1), 3–42 (2006). https://doi.org/10.1007/s10994-006-6226-1
4. Halford, J.J.: Computerized epileptiform transient detection in the scalp electroencephalogram: obstacles to progress and the example of computerized ECG interpretation. Clin. Neurophysiol. **120**(11), 1909–1915 (2009). https://doi.org/10.1016/j.clinph.2009.08.007
5. Joshi, V., Pachori, R.B., Vijesh, A.: Classification of ictal and seizure-free EEG signals using fractional linear prediction. Biomed. Signal Process. Control **9**, 1–5 (2014). https://doi.org/10.1016/j.bspc.2013.08.006
6. Juozapavičius, A., Bacevičius, G., Bugelskis, D., Samaitienė, R.: EEG analysis - automatic spike detection. Nonlinear Anal. Model. Control **16**(4), 375–386 (2011)
7. Misiukas Misiūnas, A.V., Meškauskas, T., Juozapavičius, A.: On the implementation and improvement of automatic EEG spike detection algorithm. Proc. Lith. Math. Soc. **56**(Ser. A), 60–65 (2015)
8. Misiukas Misiūnas, A.V., Meškauskas, T., Samaitienė, R.: Derivative parameters of electroencephalograms and their measurement methods. Proc. Lith. Math. Soc. **57**(Ser. A), 47–52 (2016)
9. Misiukas Misiūnas, A.V., Meškauskas, T., Samaitienė, R.: Algorithm for automatic EEG classification according to the epilepsy type: benign focal childhood epilepsy and structural focal epilepsy. Biomed. Signal Process. Control **48**, 118–127 (2019). https://doi.org/10.1016/j.bspc.2018.10.006
10. Nishida, S., Nakamura, M., Ikeda, A., Shibasaki, H.: Signal separation of background EEG and spike by using morphological filter. IFAC Proc. Vol. 14th World Congr. IFAC **32**(2), 4301–4306 (1999)
11. Patnaik, L.M., Manyam, O.K.: Epileptic EEG detection using neural networks and post-classification. Comput. Methods Programs Biomed. **91**(2), 100–109 (2008). https://doi.org/10.1016/j.cmpb.2008.02.005
12. Sammut, C., Webb, G.I.: Encyclopedia of Machine Learning and Data Mining. Springer, Boston (2017)
13. Wilson, S.B., Emerson, R.: Spike detection: a review and comparison of algorithms. Clin. Neurophysiol. **113**(12), 1873–1881 (2002). https://doi.org/10.1016/S1388-2457(02)00297-3

Alternate Overlapping Schwarz Method for Singularly Perturbed Semilinear Convection-Diffusion Problems

S. Chandra Sekhara Rao[✉] and Varsha Srivastava

Department of Mathematics, Indian Institute of Technology Delhi,
Hauz Khas, New Delhi 110 016, India
scsr@maths.iitd.ac.in, varsha.iitd@gmail.com

Abstract. We propose an alternate overlapping Schwarz method to solve the singularly perturbed semilinear convection-diffusion problems. The method decomposes the original domain into two overlapping subdomains. One, the outside boundary layer subdomain, and the other, inside boundary layer subdomain. On the outside boundary layer subdomain, a combination of the second order compact difference scheme and central difference scheme with uniform mesh is considered; while on the inside boundary layer subdomain a central difference scheme with a special piecewise-uniform mesh is considered. The convergence analysis is given and the method is shown to have almost second order parameter-uniform convergence. Numerical experiments are presented to demonstrate the efficiency of the method.

Keywords: Alternate overlapping Schwarz method
Central difference scheme · Compact difference scheme
Semilinear convection-diffusion problems
Singular perturbation problems

1 Introduction

Consider the singularly perturbed semilinear convection-diffusion problem

$$Tu := -\varepsilon u'' + au' + f(x, u) = 0, \qquad x \in \Omega = (0, 1) \tag{1a}$$

$$u(0) = p_0, \qquad u(1) = p_1, \tag{1b}$$

where $0 < \varepsilon < 1$ is a small positive perturbation parameter. a and f are sufficiently smooth functions, satisfying the following assumptions:

$$a(x) \geq \alpha > 0, \tag{2}$$

$$\beta \leq f_u(x, u) \leq \delta, \quad \forall \, (x, u) \in (0, 1) \times \mathbb{R}, \tag{3}$$

where β and δ are some positive constants. Under the above assumptions, there exists a unique solution to the problem (1) ([6]). The solution has exponential boundary layer of width $O\left(\varepsilon \ln(\frac{1}{\varepsilon})\right)$ at $x = 1$ ([8]). Singular perturbation

© Springer Nature Switzerland AG 2019
G. Nikolov et al. (Eds.): NMA 2018, LNCS 11189, pp. 449–457, 2019.
https://doi.org/10.1007/978-3-030-10692-8_51

problems arise frequently in many physical phenomena and chemical processes. Model problems of the form (1) are the semi-conductor device simulations based on the drift-diffusion modelling ([15]), modelling in mathematical biology ([10]), fluid dynamics, heat conduction ([1]) and counterflow flames modelling ([11]). Linearized version of problem (1) was investigated in [3,17]. Farrel et. al. [4] considered upwind scheme on piecewise uniform meshes for singularly perturbed semilinear elliptic problems and showed almost first order parameter uniform convergence. An almost second order parameter uniform convergent numerical methods for singularly perturbed semilinear reaction-diffusion problems using the B-spline collocation method and exponential spline difference method were considered in [14] and [12] respectively. For the solution of semilinear self adjoint and non-self adjoint singularly perturbed problems, iterative domain decomposition algorithms based on Schwarz type alternating methods were studied in [2,5,9,13] and the references therein. In the present paper, we design and analyze an alternate overlapping Schwarz method for semilinear convection-diffusion problems using the HODIE technique of [7]. The method decomposes the given domain into two overlapping subdomains. On the outside boundary layer subdomain, we use a combination of the compact difference scheme and central difference scheme, which depends on the relationship between the mesh width and the perturbation parameter; while on the inside boundary layer subdomain, we use a combination of the schemes at the second mesh point and a central difference scheme at the other $(N-1)$ mesh points with uniform mesh spacing.

This paper is arranged as follows. In Sect. 2, the continuous overlapping Schwarz method and the bounds on the derivatives of the solution, and the bounds on the regular and the singular components of the solution are given. Discretization of the domain and a discrete alternate overlapping Schwarz method are given in Sect. 3. Almost second order parameter uniform convergence of the method is proved in Sect. 4. Numerical experiments are presented in Sect. 5 and finally conclusions are given in Sect. 6.

Notations. Throughout the paper, we use C with or without a subscript to denote a generic positive constant independent of the perturbation parameter ε, iteration parameter k and discretization parameter N. We consider the maximum norm and denote it by $||.||_D$, where D is a closed and bounded subset of $\overline{\Omega}$. For a real valued function $f \in C(D)$, we define $||f||_D = \max_{x \in D} |f(x)|$. The analogous discrete maximum norm on the mesh D^N is denoted by $||.||_{D^N}$. We drop D from the notation if $D = \overline{\Omega}$. If $g, z_s \in C(\overline{\Omega})$, then $g_j = g(x_j)$, $z_{s,j} = z_s(x_j)$. The analogous discrete maximum norm on the mesh $\overline{\Omega}^N$ is denoted by $||.||_{\overline{\Omega}^N}$.

2   Properties of the Exact Solution

Continuous Overlapping Schwarz Method. The continuous overlapping Schwarz method finds the solution u of (1) by generating a sequence of approximations $u^{[k]}$ which converge to the exact solution u as $k \to \infty$. First, we decompose the domain $\Omega = (0,1)$ into two overlapping subdomains $\Omega_l = (0, \xi^+)$ and

$\Omega_r = (\xi^-, 1)$, where $0 < \xi^- < \xi^+ < 1$. Assume $\Omega_p = (c, d)$, $p = l, r$. The iterative process is defined as follows

$$u^{[0]}(x) \equiv u_0(x), \qquad 0 < x < 1, \qquad u^{[0]}(0) = u(0), \qquad u^{[0]}(1) = u(1).$$

For each $k \geq 1$, the iterates $u^{[k]}$ are defined by

$$u^{[k]}(x) = \begin{cases} u_l^{[k]}(x), \; x \in \overline{\Omega}_l \setminus \overline{\Omega}_r, \\ u_r^{[k]}(x), \; x \in \overline{\Omega}_r, \end{cases} \tag{4}$$

where for each $k \geq 1$, $u_p^{[k]}$, $p = l, r$ are defined by

$$\begin{aligned} Tu_l^{[k]} &= 0 \text{ in } \Omega_l, & u_l^{[k]}(0) &= u(0), & u_l^{[k]}(\xi^+) &= u_r^{[k-1]}(\xi^+), \\ Tu_r^{[k]} &= 0 \text{ in } \Omega_r, & u_r^{[k]}(\xi^-) &= u_l^{[k]}(\xi^-), & u_r^{[k]}(1) &= u(1). \end{aligned}$$

In the case of semilinear problems, the stability is essentially obtained by inverse-monotonicity properties of the classes of linear boundary value problems [6].

From (1a), we have $-\varepsilon u'' + au' + \int\limits_{s=0}^{1} f_u(x, su)ds \, u = -f(x, 0)$. Define the linear operator L as $Lu := -\varepsilon u'' + au' + \int\limits_{s=0}^{1} f_u(x, su)ds \, u$, $x \in \Omega$, $u \in C^2(\Omega)$, where $\int\limits_{0}^{1} f_u(x, su)ds \geq \beta > 0$. Observe that $L(\pm u) = \mp f(x, 0)$.

Lemma 1 (*Maximum Principle*). *Assume that* $u \in C^2(\overline{\Omega})$, *satisfying* $u(0) \geq 0$ *and* $u(1) \geq 0$. *Then* $Lu \geq 0$ *for* $x \in \Omega$ *implies that* $u \geq 0$ *for* $x \in \overline{\Omega}$.

The following parameter-uniform stability result is a straightforward consequence of the maximum principle.

Lemma 2 (*Stability Estimate*). *Let* u *be the exact solution of (1). Then*

$$||u||_{\overline{\Omega}} \leq \frac{1}{\alpha}||f(\cdot, 0)||_{\overline{\Omega}} + \max\{|p_0|, |p_1|\}.$$

Lemma 3. *Let* u *be the exact solution of (1). Then*

$$||u^{(n)}(x)||_{\overline{\Omega}} \leq C \left(1 + \varepsilon^{-n} \exp\left(\frac{-\alpha(1-x)}{\varepsilon}\right)\right), \; 0 \leq n \leq 4.$$

Proof. Follows from the similar approach used in [18].

To derive the parameter-uniform convergence of the numerical method, a decomposition of the exact solution u, into a regular component v and a singular component w is required. That is, u = v + w. Then $Tu = Tv + {}_vTw$, where $Tv := -\varepsilon v'' + av' + f(x, v)$ and ${}_vTw := -\varepsilon w'' + aw' + f(x, v+w) - f(x, v)$. Note that ${}_vT$ is a new operator.

Lemma 4. *For all $x \in \overline{\Omega}$, the solution u of (1) has the decomposition as $u = v + w$, where the regular component v satisfies*

$$|v^{(k)}(x)| \leq C(1 + \varepsilon^{4-k}), \qquad 0 \leq k \leq 4, \tag{5}$$

and for all $x \in \overline{\Omega}$, the singular component w satisfies

$$|w^{(k)}(x)| \leq C\varepsilon^{-k} \exp(\frac{-\alpha(1-x)}{\varepsilon}), \qquad 0 \leq k \leq 4. \tag{6}$$

Proof. Follows from the similar approach used in [16].

3   Discrete Problem

The transition parameter σ is defined as $\sigma := \min\left\{\frac{1}{3}, \sigma_0 \varepsilon \ln N\right\}$, in which the constant σ_0 will be chosen later in Sect. 5. A uniform mesh $\overline{\Omega}_l^N = \{x_j = j\frac{(1-\sigma)}{N}\}_{j=0}^N$ is placed on the subdomain $\overline{\Omega}_l$ with mesh spacing $h_{l,j} = H_l = \frac{(1-\sigma)}{N}$ for $1 \leq j \leq N$. Note that $x_N = \xi^+$. A special type of piecewise-uniform mesh $\overline{\Omega}_r^N$ is constructed on the subdomain $\overline{\Omega}_r$. The mesh points on the subdomain $\overline{\Omega}_r^N$ are given by $x_0 = \xi^- = \frac{N-1}{N}(1-\sigma)$, $x_j = (1-\sigma) + (j-1)\frac{\sigma}{N}$, $1 \leq j \leq N+1$ with mesh spacing $h_{r,1} = \frac{(1-\sigma)}{N}$ and $h_{r,j} = H_r = \frac{\sigma}{N}$ for $2 \leq j \leq N+1$. Define $\overline{\Omega}^N := \overline{\Omega}_r^N \cup (\overline{\Omega}_l^N \setminus \overline{\Omega}_r^N)$.

On each subdomains $\overline{\Omega}_p^N, p = l, r$, the corresponding discretization is :
$[T_p^N U_p]_j = 0$, where
$[T_p^N U_p]_j := r_{p,j}^- U_{p,j-1} + r_{p,j}^c U_{p,j} + r_{p,j}^+ U_{p,j+1} + q_{p,j}^- f(x_{j-1}, U_{p,j-1}) + q_{p,j}^c f(x_j, U_{p,j})$
and the coefficients $r_{p,j}^*$, $* = -, c, +$ are given by

$$r_{p,j}^- = \frac{-2\varepsilon - a_j h_{p,j+1} + q_{p,j}^-[-(2h_{p,j} + h_{p,j+1})a_{j-1} + a_j h_{p,j+1}]}{h_{p,j}(h_{p,j} + h_{p,j+1})},$$

$$r_{p,j}^+ = \frac{-2\varepsilon + a_j h_{p,j} - q_{p,j}^- h_{p,j}(a_j + a_{j-1})}{h_{p,j+1}(h_{p,j} + h_{p,j+1})}, r_{p,j}^c = -r_{p,j}^- - r_{p,j}^+, \quad q_{p,j}^c = 1 - q_{p,j}^-.$$

Here the coefficients are determined so that the scheme is exact for the polynomials up to degree two and satisfies the normalization condition. The coefficients $q_{p,j}^-$ are free parameters and their values are given later in (8).

The sequence of discrete alternate overlapping Schwarz iterates are defined by
$U^{[0]}(x_j) \equiv u_0(x_j)$, $0 < x_j < 1$, $U^{[0]}(0) = u(0)$, $U^{[0]}(1) = u(1)$.
For $k \geq 1$,

$$U^{[k]}(x_j) = \begin{cases} U_l^{[k]}(x_j), x_j \in \overline{\Omega}_l^N \setminus \overline{\Omega}_r^N, \\ U_r^{[k]}(x_j), x_j \in \overline{\Omega}_r^N, \end{cases}$$

where for all $k \geq 1$, $U_p^{[k]}, p = l, r$ satisfy

$$T_l^N U_l^{[k]} = 0 \text{ in } \Omega_l^N, \quad U_l^{[k]}(0) = u(0), \qquad U_l^{[k]}(\xi^+) = U_r^{[k-1]}(\xi^+),$$
$$T_r^N U_r^{[k]} = 0 \text{ in } \Omega_r^N, \quad U_r^{[k]}(\xi^-) = U_l^{[k]}(\xi^-), \qquad U_r^{[k]}(1) = u(1).$$

4   Error Analysis

The error of the scheme for each $x_j \in \Omega_p^N, p = l, r$ is given by

$$T_p^N u_j - T u_j = T_p^N u_j - T_p^N U_{p,j}^{[k]}$$

$$= r_{p,j}^- \zeta_{p,j-1}^{[k]} + r_{p,j}^c \zeta_{p,j}^{[k]} + r_{p,j}^+ \zeta_{p,j+1}^{[k]} + Q_p \left(\int_0^1 f_u(x_j, U_{p,j}^{[k]} + s\,\zeta_{p,j}^{[k]})ds \right) \zeta_{p,j}^{[k]}$$

$$= [U_p^{[k]} L_p^N \zeta_p^{[k]}]_j$$

where the error function $\zeta_p^{[k]}$ is defined as $\zeta_p^{[k]} := u - U_p^{[k]}$ and the linear operator is defined by $\quad [^y L_p^N z]_j := r_{p,j}^- z_{j-1} + r_{p,j}^c z_j + r_{p,j}^+ z_{j+1} + Q_p \left(\int_0^1 f_u(x_j, y_j + s z_j) ds \right) z_j,$

with $[Q_p g]_j := q_{p,j}^- g_{j-1} + q_{p,j}^c g_j.$

Next is to prove that the matrix associated with $U_p^{[k]} L_p^N$, $p = l, r$ is an M-matrix and the scheme is uniformly stable.

Define $R_{p,j}^- := r_{p,j}^- + q_{p,j}^- f_u$, $R_{p,j}^c := r_{p,j}^c + q_{p,j}^c f_u$, and $R_{p,j}^+ := r_{p,j}^+.$

Lemma 5. *Let N_0 be the smallest positive integer such that $\dfrac{\sigma_0 ||a||}{2} < \dfrac{N_0}{\ln N_0}$, $\dfrac{(||a'|| + \delta)}{N_0} < \alpha$ hold. Also, assume that $||a|| h_{p,j} \geq 2\varepsilon$, $p = l, r$, and $q_{p,j}^-$ is chosen as $q_{p,j}^- \geq \dfrac{a_j}{(a_j + a_{j-1})}$. Then the matrix associated with $U_p^{[k]} L_p^N$ becomes an M-matrix, if there exist positive constants C_1 and C_2 such that for each $x_j \in \Omega_p^N$, $p = l, r$*

$$R_{p,j}^- < 0, \quad R_{p,j}^+ < 0, \quad C_1 \leq R_{p,j}^- + R_{p,j}^c + R_{p,j}^+ \leq C_2. \tag{7}$$

Moreover, the scheme is uniformly stable in the maximum norm, if $h_{p,j+1} R_{p,j}^+ - h_{p,j} R_{p,j}^- \geq C > 0$, where C is a positive constant.

For $x_j \in \Omega_p^N$, $p = l, r$, $q_{p,j}^-$, $p = l, r$ is chosen as

$$q_{p,j}^- = \begin{cases} \dfrac{a_j}{(a_j + a_{j-1})}, & ||a|| h_{p,j} \geq 2\varepsilon \\ \dfrac{(h_{p,j} - h_{p,j+1})}{3 h_{p,j}}, & ||a|| h_{p,j} < 2\varepsilon. \end{cases} \tag{8}$$

For $k \geq 1$, the discrete alternate overlapping Schwarz iterate $U^{[k]}(x_j)$ is decomposed as $\quad U^{[k]}(x_j) = V^{[k]}(x_j) + W^{[k]}(x_j), \forall x_j \in \overline{\Omega}^N$, where

$$V^{[k]}(x_j) := \begin{cases} V_l^{[k]}(x_j), x_j \in \overline{\Omega}_l^N \setminus \overline{\Omega}_r^N, \\ V_r^{[k]}(x_j), x_j \in \overline{\Omega}_r^N, \end{cases} \text{ and } W^{[k]}(x_j) := \begin{cases} W_l^{[k]}(x_j), x_j \in \overline{\Omega}_l^N \setminus \overline{\Omega}_r^N, \\ W_r^{[k]}(x_j), x_j \in \overline{\Omega}_r^N. \end{cases}$$

Here for all $k \geq 1$, $V_{p,j}^{[k]}$, $p = l, r$ are the solutions of

$$T_l^N V_l^{[k]} = 0 \text{ in } \Omega_l^N, \qquad V_l^{[k]}(0) = v(0), \qquad V_l^{[k]}(\xi^+) = V_r^{[k-1]}(\xi^+),$$
$$T_r^N V_r^{[k]} = 0 \text{ in } \Omega_r^N, \qquad V_r^{[k]}(\xi^-) = V_l^{[k]}(\xi^-), \qquad V_r^{[k]}(1) = v(1).$$

Likewise, for all $k \geq 1$, $W_{p,j}^{[k]}$, $p = l, r$ are the solution of

$$_{V_l^{[k]}} T_l^N W_l^{[k]} = 0 \text{ in } \Omega_l^N, \quad W_l^{[k]}(0) = w(0), \quad W_l^{[k]}(\xi^+) = W_r^{[k-1]}(\xi^+),$$
$$_{V_r^{[k]}} T_r^N W_r^{[k]} = 0 \text{ in } \Omega_r^N, \quad W_r^{[k]}(\xi^-) = W_l^{[k]}(\xi^-), \quad W_r^{[k]}(1) = w(1).$$

Note that $_{V_p^{[k]}} T_p^N$, $p = l, r$ are the discrete analogue of the operator $_v T$ in their respective subdomains. The nodal error estimate in the regular component of the solution is given in the following lemma.

Lemma 6. *Let v and $V^{[k]}$ denote the regular components of u and $U^{[k]}$ respectively. Then, for all $k \geq 1$, $||v - V^{[k]}||_{\overline{\Omega}^N} \leq C_1 \lambda^{-k} + C_2 \sigma_0^2 N^{-2} \ln^2 N$, where $\lambda = (1 + \frac{\alpha(1-\sigma)}{\varepsilon N})$.*

To find an error estimate for the singular component in $\overline{\Omega}_p^N$, $p = l, r$, we use the mesh functions $\phi_{p,j}(\gamma)$, $p = l, r$ for some positive constant γ defined

by $\phi_{p,j}(\gamma) := \prod_{k=j+1}^{\Lambda} (1 + \frac{\gamma h_{p,k}}{\varepsilon})^{-1}$, $0 \leq j \leq \Lambda - 1$, where $\phi_{p,\Lambda}(\gamma) = 1$; with

$\Lambda = N$ for $\overline{\Omega}_l^N$ and $\Lambda = N + 1$ for $\overline{\Omega}_r^N$.

Lemma 7. *Suppose that all the assumptions of Lemma 5 hold, then for $x_j \in \Omega_p^N$, $p = l, r$ there exists a constant $C(\gamma)$, such that $_{U_p^{[k]}} L_p^N \phi_{p,j}(\gamma) \geq \frac{C(\gamma)}{\max\{\varepsilon, h_{p,j}\}} \phi_{p,j}(\gamma)$, where $\gamma \leq \alpha/2$ and α is same as defined in (2).*

The nodal error estimate in the singular component of the solution is given in the following lemma.

Lemma 8. *Let w and $W^{[k]}$ be the singular components of u and $U^{[k]}$ respectively. Then for all $k \geq 1$, $||w - W^{[k]}||_{\overline{\Omega}^N} \leq C \left(N^{-\gamma\sigma_0} + \sigma_0^2 N^{-2} \ln^2 N \right)$, where $\gamma \leq \alpha/2$.*

Now, we state and prove the main result of the Section, using the Lemmas 6 and 8.

Theorem 1. *Let u be the exact solution of the problem (1) and let $U^{[k]}$ be the kth iterate of the discrete alternate overlapping Schwarz method. If $\gamma \leq \alpha/2$ then for any $N \geq N_0$, $||(u - U^{[k]})||_{\overline{\Omega}^N} \leq C_1 \lambda^{-k} + C_2 \left(N^{-\gamma\sigma_0} + \sigma_0^2 N^{-2} \ln^2 N \right)$, where $\lambda = (1 + \frac{\alpha(1-\sigma)}{\varepsilon N})$.*

Proof. The triangle inequality gives
$$||u - U^{[k]}|| \leq ||v - V^{[k]}|| + ||w - W^{[k]}|| \leq C_1 \lambda^{-k} + C_2 \left(N^{-\gamma\sigma_0} + \sigma_0^2 N^{-2} \ln^2 N \right),$$
where $\lambda = \left(1 + \frac{\alpha(1-\sigma)}{\varepsilon N}\right)$; and for almost second order uniformly convergent method, we need to take $\gamma\sigma_0 \geq 2$.

5   Numerical Experiments

Example 1. Consider the following singularly perturbed semilinear convection-diffusion problem $-\varepsilon u'' + (1+x)u' + \exp(u) = 0$, $x \in (0,1)$, $u(0) = 0$, $u(1) = 0$.

To solve the corresponding discrete nonlinear systems, the Newton's method is used with the initial approximation $w^{[0]} = (u_0(x_0), u_0(x_1), \ldots, u_0(x_N))^T$, where $u_0(x)$ is the solution of the reduced problem. The stopping criterion is $\|w^{[k]} - w^{[k-1]}\| < 10^{-12}$. For each N and ε, it takes only 2 iterations to satisfy the stopping criterion to get the discrete solution. The stopping criterion for the present high order Schwarz iteration is $\|U^{[k+1]} - U^{[k]}\|_{\overline{\Omega}^N} \le 10^{-8}$. We omit the superscript k on the final iterate and write simply $\tilde{U}$. As the exact solution is not known for the test problem, a variant of the double mesh principle is used to calculate the maximum pointwise errors for different values of ε and N using $E_\varepsilon^N := \|U^N - U^{2N}\|_{\overline{\Omega}^N}$ and the parameter-uniform errors by $E^N := \max_\varepsilon E_\varepsilon^N$. We calculate the parameter-uniform numerical order of convergence by $\rho^N := \dfrac{\ln E^N - \ln E^{2N}}{\ln(2 \ln N) - \ln(\ln(2N))}$. For different values of ε and N, Table 1 represent the maximum pointwise errors E_ε^N, rates of convergence ρ_ε^N,

Table 1. Maximum pointwise-errors E_ε^N, E^N, numerical rates of convergence ρ_ε^N, and ρ^N for Example 1 with $\sigma_0 = 4$.

$\varepsilon = 2^{-j}$	$N = 2^5$	$N = 2^6$	$N = 2^7$	$N = 2^8$	$N = 2^9$	$N = 2^{10}$	
$j = 4$	1.20E−03	2.98E−04	7.43E−05	1.86E−05	4.80E−06	1.49E−06	
	2.73	2.58	2.47	2.35	1.99		ρ_ε^N
8	1.01E−02	3.39E−03	1.13E−03	3.61E−04	1.14E−04	3.52E−05	
	2.13	2.04	2.04	2.00	2.00		ρ_ε^N
12	1.01E−02	3.40E−03	1.12E−03	3.65E−04	1.15E−04	3.56E−05	
	2.13	2.06	2.00	2.01	1.99		ρ_ε^N
16	1.01E−02	3.40E−03	1.13E−03	3.65E−04	1.15E−04	3.55E−05	
	2.13	2.04	2.02	2.01	2.00		ρ_ε^N
20	.	.	.	.	.	.	
							ρ_ε^N
24	.	.	.	.	.	.	
							ρ_ε^N
28	.	.	.	.	.	.	
							ρ_ε^N
32	1.01E−02	3.40E−03	1.13E−03	3.65E−04	1.15E−04	3.55E−05	
	2.13	2.04	2.02	2.01	2.00		ρ_ε^N
E^N	1.01E−02	3.40E−03	1.13E−03	3.65E−04	1.15E−04	3.56E−05	
	2.13	2.04	2.02	2.01	2.00		ρ^N

Table 2. Number of alternate Schwarz iterations required to satisfy the stopping criterion for Example 1 with $\sigma_0 = 4$.

$\varepsilon = 2^{-j}$	$N = 2^5$	$N = 2^6$	$N = 2^7$	$N = 2^8$	$N = 2^9$	$N = 2^{10}$
$j = 4$	31	59	112	213	405	769
8	8	10	13	8	20	40
12	5	5	6	7	8	10
16	4	4	4	4	5	5
20	3	3	3	3	4	4
24	3	3	3	3	3	3
28	3	3	3	3	3	3
32	2	2	2	3	3	3

parameter uniform errors E^N and parameter uniform rates of convergence ρ^N, of the discrete alternate overlapping Schwarz method for the Example 1. The iteration counts for different values of ε and N for Example 1 is given in Table 2. In all the computations the value of σ_0 is chosen as 4.

6   Conclusions

In this work, we proposed and analyzed an alternate overlapping Schwarz method for numerical solution of singularly perturbed semilinear convection-diffusion problems. The stability of both the continuous and discrete semilinear problems, is proved by inverse-monotonicity properties of the classes of linear boundary value problems. The error analysis is done using truncation error and barrier function approach. The present scheme is almost second order parameter uniform convergent. From Table 1, we see that the numerical results are in agreement with the theoretical results. From Table 2, we observe that the number of iterations are less for small values of ε.

Acknowledgements. The authors gratefully acknowledge the valuable comments and suggestions from the anonymous referees. The research work of the second author is supported by Council of Scientific and Industrial Research, India.

References

1. Allen III, M., Herrera, I., Pinder, G.: Numerical Modelling in Science and Engineering. Wiley-Interscience, New York (1988)
2. Boglaev, I.: The solution of a singularly perturbed convection-diffusion problem by an iterative domain decomposition method. Numer. Algorithms **31**, 27–46 (2002)
3. Clavero, C., Gracia, J.L., Lisbona, F.: High order methods on Shishkin meshes for singular perturbation problems of convection-diffusion type. Numer. Algorithms **22**, 73–97 (1999)

4. Farrel, P.A., Miller, J.J.H., O'Riordan, E., Shishkin, G.I.: A uniformly convergent finite difference scheme for a singularly perturbed semilinear equations. SIAM J. Numer. Anal. **33**(3), 1135–1149 (1996)
5. Garbey, M.: A Schwarz alternating procedure for singular perturbation problems. SIAM J. Sci. Comput. **17**, 1175–1201 (1996)
6. Lorenz, J.: Stability and monotonicity properties of stiff quasilinear boundary problems. Univ. u Novom Sadu Zb. rad. Prir.-Mat. Fak. Ser. Mat. **12**, 151–175 (1982)
7. Lynch, R.E., Rice, J.R.: A high-order difference method for differential equations. Math. Comput. **34**(150), 333–372 (1980)
8. O'Malley Jr., R.E.: Singular Perturbation Methods for Ordinary Differential Equations. Springer, New York (1991). https://doi.org/10.1007/978-1-4612-0977-5
9. Mathew, T.P.: Uniform convergence of the Schwarz alternating method for solving singularly perturbed advection-diffusion equations. SIAM J. Numer. Anal. **35**, 1663–1683 (1998)
10. Murray, J.D.: Lectures on Nonlinear Differential Equation Models in Biology. Clarendon Press, Oxford (1977)
11. Pouly, L., Pousin, J.: A spray combustion problem. Math. Models Methods Appl. Sci. **2**, 237–249 (1994)
12. Rao, S.C.S., Kumar, M.: Parameter-uniformly convergent exponential spline difference scheme for singularly perturbed semilinear reaction-diffusion problems. Nonlinear Anal. **71**, e1579–e1588 (2009)
13. Rao, S.C.S., Kumar, S.: Robust high order convergence of a overlapping Schwarz method for singularly perturbed semilinear reaction-diffusion problem. J. Comput. Math. **31**, 509–521 (2013)
14. Rao, S.C.S., Kumar, S., Kumar, M.: A parameter-uniform B-spline collocation method for singularly perturbed semilinear reaction-diffusion problems. J. Optim. Theory Appl. **146**, 795–809 (2010)
15. van Roosbroeck, W.V.: Theory of flows of electrons and holes in germanium and other semiconductors. Bell Syst. Tech. J. **29**, 560–607 (1950)
16. Shishkin, G.I., Shishkina, L.P.: The Richardson extrapolation technique for quasilinear parabolic singularly perturbed convection-diffusion equations. J. Phys. Conf. Ser. **55**, 203–213 (2006)
17. Stynes, M., Tobiska, L.: A finite difference analysis of a streamline diffusion method on a Shishkin mesh. Numer. Algorithms **18**, 337–360 (1998)
18. Vulanovic, R.: A uniform numerical method for quasilinear singular perturbation problems without turning points. Computing **41**, 97–106 (1989)

On the Calculation of Electromagnetic Fields in Closed Waveguides with Inhomogeneous Filling

Dmitry V. Divakov[ID], Mikhail D. Malykh[(✉)][ID], Leonid A. Sevastianov[ID], and Anastasia A. Tiutiunnik[ID]

Department of Applied Probability and Informatics,
Peoples' Friendship University of Russia (RUDN University),
6 Miklukho-Maklaya Street, Moscow 117198, Russian Federation
`malykh_md@rudn.university`

Abstract. We consider a waveguide having the constant cross-section S with ideally conducting walls. We assume that the filling of waveguide does not change along its axis and is described by the piecewise continuous functions ε and μ defined on the waveguide cross section. We show that it is possible to make a substitution, which allows dealing only with continuous functions.

Instead of discontinuous cross components of the electromagnetic field $\boldsymbol{E}$ and $\boldsymbol{H}$ we propose to use four potentials. Generalizing the Tikhonov-Samarskii theorem, we have proved that any field in the waveguide allows such representation, if we consider the potentials as elements of respective Sobolev spaces.

If ε and μ are piecewise constant functions, then in terms of four potentials the Maxwell equations are reduced to a pair of independent equations. This fact means that a few dielectric waveguides placed between ideally conducting walls can be described by a scalar boundary problem. This statement offers a new approach to the investigation of spectral properties of waveguides. First, we can prove the completeness of the system of the normal waves in closed waveguides using standard functional spaces. Second, we can propose a new technique for calculating the normal waves using standard finite elements. Results of the numerical experiments using FEA software FreeFem++ are presented.

Keywords: Waveguide · Maxwell equations · Sobolev spaces
Normal modes

1 Introduction

In a hollow waveguide it is possible to introduce two scalar potentials, using which the Maxwell equations are reduced to a pair of uncoupled wave equations,

The publication has been prepared with the support of the "RUDN University Program 5-100" and funded by RFBR according to the research projects No. 18-07-00567 and No. 18-51-18005.

G. Nikolov et al. (Eds.): NMA 2018, LNCS 11189, pp. 458–465, 2019.
https://doi.org/10.1007/978-3-030-10692-8_52

as it was proved in the classical papers by Tikhonov and Samarskii [1]. The most important consequence from the Tikhonov and Samarskii theorem is the completeness of the system of normal waves in a hollow waveguide, according to which any wave propagating through the waveguide can be presented as a superposition of transverse electric and transverse magnetic waves (TE and TM waves) [2]. In 1990s this consequence extremely important for substantiating the partial radiation conditions and the incomplete Galerkin method [3] was generalized for the case of a waveguide, in which the filling varies over the transverse section, but is constant along the waveguide axis [4]. As a result, the theorem of field representation using potentials became shadowed by its consequence.

Due to this circumstance, the computational complexity of the spectral problems for hollow waveguides and for the waveguides filled with inhomogeneous matter differs in principle. In the first case, the resulting problems are scalar, and one can use the well-developed methods, equally applicable to acoustics and quantum mechanics. In the case of a waveguide filled with inhomogeneous matter, one has to solve numerically the full vector electrodynamic problem. Such problems possess the zero eigenvalue of infinite multiplicity, due to which their numerical solution requires nontrivial procedures hard for computer implementation, e.g., the method of mixed finite elements [5,6].

We should also note that for the problems of radiophysics the case of piecewise constant filling is of particular interest, since a waveguide with smoothly changing filling can be practically fabricated only by using thin homogeneous layers with finite but small difference of ε and μ between the adjacent layers. At the junction between different layers the transverse components of the vector fields $\boldsymbol{E}$ and $\boldsymbol{H}$ have discontinuities, which lead to additional difficulties in their approximation by continuous finite elements.

In the present paper we return to the problem of presenting an arbitrary electromagnetic field in a waveguide with piecewise constant filling in its classical formulation. Usually, as in the case of a hollow waveguide, the introduction of potentials allows integrating some of the Maxwell equations and reducing the number of desired functions. It is well known that in a waveguide filled by inhomogeneous medium this is not the case. However, we believe that the main advantage of introducing potentials is dealing with continuous potentials instead of discontinuous field components. From this point of view, the introduction of potentials can be considered as a change of variables providing a transition from discontinuous functions to continuous ones.

2   Notations

In this paper the subject of study is a closed waveguide having the constant cross section S with piecewise constant distribution of ε and μ filling invariable along the waveguide axis. The line of filling discontinuity will be denoted by Γ. Let the axis Oz of Cartesian coordinates be directed along the waveguide axis and assume for brevity that

$$\boldsymbol{A}_\perp = (A_x, A_y, 0)^T \quad \text{and} \quad \nabla = (\partial_x, \partial_y, 0)^T, \quad \nabla' = (-\partial_y, \partial_x, 0)^T.$$

The electromagnetic field in the closed waveguide $S \times Z \times T$ with the filling ε, μ is described by the vector fields $\boldsymbol{E}, \boldsymbol{H}$ with the components defined in $(S - \Gamma) \times Z \times T$ under the condition that the contraction of $\boldsymbol{E}, \boldsymbol{H}$ and their partial derivatives in z and t to the section S for all values of z and t represents piecewise smooth functions satisfying Maxwell equations

$$\begin{cases} \operatorname{curl} \boldsymbol{E} = -\dfrac{\mu}{c} \partial_t \boldsymbol{H}, & \operatorname{div} \varepsilon \boldsymbol{E} = 0, \\ \operatorname{curl} \boldsymbol{H} = +\dfrac{\varepsilon}{c} \partial_t \boldsymbol{E}, & \operatorname{div} \mu \boldsymbol{H} = 0 \end{cases} \tag{1}$$

in the waveguide $S \times Z \times T$, ideal conductivity conditions of waveguide walls

$$\boldsymbol{E} \times \boldsymbol{n} = 0, \quad \boldsymbol{H} \cdot \boldsymbol{n} = 0 \tag{2}$$

at the regular points of the boundary $\partial S \times Z \times T$, matching conditions

$$\begin{cases} [\boldsymbol{E} \times \boldsymbol{n}] = \mathbf{0}, & [\varepsilon \boldsymbol{E} \cdot \boldsymbol{n}] = 0 \\ [\boldsymbol{H} \times \boldsymbol{n}] = \mathbf{0}, & [\mu \boldsymbol{H} \cdot \boldsymbol{n}] = 0 \end{cases} \tag{3}$$

at regular points of the filling discontinuity boundary $\Gamma \times Z \times T$.

3   Helmholtz Decomposition

Let us define the relation between the fields and potentials as

$$\boldsymbol{E}_\perp = \nabla u_e + \frac{1}{\varepsilon} \nabla' v_e, \quad \boldsymbol{H}_\perp = \nabla v_h + \frac{1}{\mu} \nabla' u_h. \tag{4}$$

Each of these formulae is a 2D analogue of Helmholtz decomposition, well known in the elasticity theory.

Note 1. In electrodynamics, for the field $\boldsymbol{H}_\perp$ such potentials arose in the proof of completeness of the system of normal modes as an auxiliary construction [7]. The four potentials were introduced in our papers [8,9] for smooth filling without the coefficients $\frac{1}{\varepsilon}$ and $\frac{1}{\mu}$, important only in the case of discontinuity.

Theorem 1. *For any electromagnetic field* $\boldsymbol{E}, \boldsymbol{H}$ *in the waveguide, one can find such functions* u_e, u_h *of the variables* z, t *taking the values in the Sobolev space* $\overset{o}{W}{}_2^1(S)$ *and such functions* v_e, v_h *of the variables* z, t *taking the values in the Sobolev space* $W_2^1(S)$, *that the equality* (4) *is valid. The above representation is unique up to additive constants.*

The Theorem 1 means that under the change of variables $\boldsymbol{E}, \boldsymbol{H}$ by two potentials and two components E_z, H_z using the Eq. (4) no solutions of Maxwell equations are lost. The conditions

$$u_e, u_h, E_z \in \overset{o}{W}{}_2^1(S) \quad \text{and} \quad v_e, v_h, H_z \in W_2^1(S)$$

replace the conditions at the filling discontinuity, as well as the boundary conditions. Since the potentials are elements of Sobolev spaces, it is natural to consider the Maxwell equations in the weak form [10]. In the case of arbitrary ε and μ these equations are cumbersome and not presented here.

4   Splitting of the System of Maxwell Equations in Waveguides with Piecewise Constant Filling

Now let us dwell on the case of particular interest for practice, when the filling of the waveguide is piecewise constant. According to the Theorem 1, the electromagnetic field $\boldsymbol{E}, \boldsymbol{H}$ in such waveguide can be presented in the form (4). From Maxwell equations it follows that the potentials u_e, u_h and E_z are elements of $\overset{0}{W}{}^1_2(S)$, coupled by the equations

$$
\begin{cases}
\displaystyle\iint_S \varepsilon(\nabla u, \nabla u_e)\,dxdy = \partial_z \iint_S \varepsilon u E_z\,dxdy, \\[4mm]
\displaystyle\iint_S \frac{c}{\mu}(\nabla u, \nabla u_h)\,dxdy = -\partial_t \iint_S \varepsilon u E_z\,dxdy,
\end{cases}
\tag{5}
$$

for any u from $C_0^\infty(S)$, where $E_z = \partial_z u_e + \partial_t u_h$, the potentials v_e, v_h and H_z are elements of $W^1_2(S)$, coupled by the equations

$$
\begin{cases}
\displaystyle\iint_S \frac{c}{\varepsilon}(\nabla v, \nabla v_e)\,dxdy = \partial_t \iint_S \mu v H_z\,dxdy, \\[4mm]
\displaystyle\iint_S \mu(\nabla v, \nabla v_h)\,dxdy = \partial_z \iint_S \mu v H_z\,dxdy,
\end{cases}
\tag{6}
$$

for any v from $C^\infty(S)$, where $H_z = \partial_z v_h - \partial_t v_e$.

The Eqs. (5) and (6) can be used also for waveguide field tailoring. If u_e, u_h and E_z from $\overset{0}{W}{}^1_2(S)$ satisfy the Eq. (5), and v_e, v_h and H_z from $W^1_2(S)$ satisfy Eq. (6), then the field $\boldsymbol{E}, \boldsymbol{H}$, calculated using Eq. (4), satisfies Maxwell equations in the generalized sense. Moreover, if this field has continuous partial derivatives of the first order in all variables everywhere except the filling discontinuity points, and the filling discontinuities are of the first kind, then this field satisfies Maxwell equations (1) in the continuity domain, the matching conditions (3) at the discontinuities, and the conditions of ideal conductivity at the boundary (2).

Since the system of Maxwell equations has been separated into two independent systems, the electromagnetic field $\boldsymbol{E}, \boldsymbol{H}$ in the waveguide with the filling described by piecewise constant functions ε and μ is a superposition of TE and TM fields.

5   Monochromatic Fields in Waveguides with Piecewise Filling

Let us apply the developed theory to the case of monochromatic fields, when the time dependence is described by the factor $e^{-i\omega t}$.

The monochromatic TM field is described by the potentials

$$
u_e = \tilde{u}_e e^{-i\omega t}, \quad u_h = \tilde{u}_h e^{-i\omega t},
$$

that satisfy the equations

$$\begin{cases} \iint\limits_{S} \varepsilon(\nabla u, \nabla \tilde{u}_e)dxdy = \partial_z^2 \iint\limits_{S} \varepsilon u \tilde{u}_e dxdy - ik\partial_z \iint\limits_{S} \varepsilon u \tilde{u}_h dxdy, \\ \iint\limits_{S} \frac{1}{\mu}(\nabla u, \nabla \tilde{u}_h)dxdy = ik\partial_z \iint\limits_{S} \varepsilon u \tilde{u}_e dxdy + k^2 \iint\limits_{S} \varepsilon u \tilde{u}_h dxdy, \end{cases} \tag{7}$$

for any $u \in \overset{0}{W}{}_2^1(S)$, here $k = \omega/c$. Let us rewrite this system of equations in the operator form using the standard technique of the theory of Sobolev spaces [11]. The symmetric bilinear form

$$\iint\limits_{S} (\nabla u, \nabla \tilde{u}) k(x,y)dxdy$$

for any piecewise smooth k is bounded in the norm W_2^1, therefore, such bounded self-adjoint operator A_k exists that this form equals $(u, A_k \tilde{u})$. The symmetric bilinear form

$$\iint\limits_{S} u \tilde{u} k(x,y)dxdy$$

for any piecewise smooth k is completely continuous in the norm W_2^1, so that such bounded self-adjoint operator B_k exists that this form equals $(u, B_k \tilde{u})$. Therefore, the system (7) can be rewritten as

$$\begin{cases} A_\varepsilon \tilde{u}_e = \partial_z^2 B_\varepsilon \tilde{u}_e - ik\partial_z B_\varepsilon \tilde{u}_h, \\ A_{\frac{1}{\mu}} \tilde{u}_h = ik\partial_z B_\varepsilon \tilde{u}_e + k^2 B_\varepsilon \tilde{u}_h. \end{cases} \tag{8}$$

Assume that the frequency $\omega = kc$ of the considered field differs from the special frequencies (magnetic cutoff frequencies) at which the operator $A_{\frac{1}{\mu}} - \omega^2 B_\varepsilon$. Excluding $\tilde{u}_h$ from this system, we get

$$A_\varepsilon \tilde{u}_e = \partial_z^2 \left(B_\varepsilon + k^2 B_\varepsilon (A_{\frac{1}{\mu}} - k^2 B_\varepsilon)^{-1} B_\varepsilon \right) \tilde{u}_e. \tag{9}$$

Therefore, the TM field can be described as the solution of Eq. (9), which in the case of piecewise constant filling plays the same role as the Helmholtz equation in the case of hollow waveguides.

For linear differential equations, the coefficients of which are compact operators, one can always write a general solution by means of a system of root vectors of the appropriate polynomial operator bundle [12]. In particular, one can present the general solution of the Eq. (9) as a sum of TM fields having the form

$$\boldsymbol{E}(x,y)e^{i\gamma z - i\omega t}, \quad \boldsymbol{H}(x,y)e^{i\gamma z - i\omega t},$$

each of them being a generalized solution of Maxwell equations in the waveguide. Here the parameter γ can take purely real and purely imaginary values.

The object resulting from this consideration is well known in the waveguide theory. The nontrivial field E, H in the waveguide $S \times Z \times T$, depending on z and t as $e^{i\gamma z - i\omega t}$, where ω, γ are constants, is referred to as a normal mode of the waveguide, ω being the frequency and γ being the wavenumber. Complex values of γ are possible.

The consideration of TE modes is quite similar.

Theorem 2. *Let the filling of the waveguide be described by piecewise continuous functions ε and μ. For the frequencies different from the cutoff ones, any monochromatic electromagnetic field E, H in the waveguide can be presented as a superposition of normal modes of the waveguide, the corresponding series for the fields E, H converging in norm $L^2(S)$.*

Note 2. The completeness of the system of normal modes for the waveguides with composite filling was established by Delitsyn [4]. In his papers, the initial system of Maxwell equations was reduced to a new one, which could be written in compact, but, unfortunately, non-self-adjoint operators. The theory developed by Keldysh [12] allowed proving the completeness of the system of eigenvectors and adjoined vectors. However, for non-self-adjoint operators the completeness is not identical to basisness; the latter was established using this method only for waveguides having circular cross section [13]. In the proposed version of the theory of normal modes, based on the technique of four potentials, the search for the normal modes is reduced to a self-adjoint problem, which removes multiple delicate issues, including the question of basisness.

The Theorem 2 reduces the calculations in waveguide problems to the determination of normal modes. For example, the TM mode is an eigenfunction of the problem

$$\begin{cases} A_\varepsilon \tilde{u}_e = -\gamma^2 B_\varepsilon \tilde{u}_e + k\gamma B_\varepsilon \tilde{u}_h, \\ A_{\frac{1}{\mu}} \tilde{u}_h = -k\gamma B_\varepsilon \tilde{u}_e + k^2 B_\varepsilon \tilde{u}_h. \end{cases} \tag{10}$$

All the points of the $k\gamma$-plane where this problem has nontrivial solution form the dispersive curve of the waveguide.

To solve this problem it is natural to use the truncation method: we will use finite element spaces instead of Sobolev spaces and change the operators $A_\varepsilon, A_{\frac{1}{\mu}}$ and B_ε to the sparse matrices, generated by the same bilinear forms.

For calculation of these matrices and further manipulations with block-sparse matrices we use free FEA software FreeFem++ [14]. Our FreeFem++ program for the construction of waveguide dispersion curves can work with any waveguide cross-sections if the boundaries can be described parametrically with the help of elementary functions and with any piecewise-constant filling described with the help of algebraic inequalities.

Example 1. On the Fig. 1 we can see the dispersive curve of the waveguide with the cross-section

$$S = \{0 < x < 1\} \times \{0 < y < 1\},$$

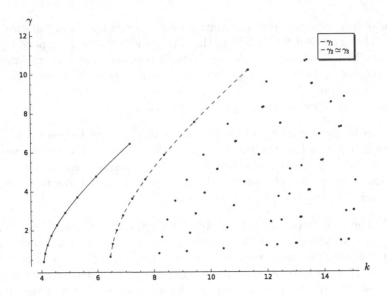

Fig. 1. The dispersive curve for the Example 1.

the piecewise-constant filling is

$$\varepsilon = \begin{cases} 1.2, & (x-0.5)^2 + (y-0.3)^2 < 0.5 \\ 1, & \text{otherwise} \end{cases}, \quad \mu = 1. \tag{11}$$

We use the mesh with 2120 triangles and solve the eigenvalue problem at several values of β (the step $\Delta\beta = 0.1$).

To check the convergence we made the series of the numerical experiments for the hollow waveguide with a box cross-section.

6   Conclusion

In the present paper, the main theorems of the hollow waveguide theory are generalized over the case of waveguides with piecewise constant filling. Similar to the case of hollow waveguides, an arbitrary field can be presented as a sum of TE and TM fields. In the case of a hollow waveguide, the monochromatic TE and TM fields satisfy Helmholtz equation. Instead, for a filled waveguide the following equation arises

$$Au = \partial_z^2 Ku, \tag{12}$$

where A and K are bounded self-adjoint operators, A determines strictly positive defined quadratic form, and K is a completely continuous operator. This equation is by no means more complicated than the Helmholtz equation, so that the proposed method allows applying the numerical methods developed for scalar waveguides to the vector model. Presented results of the numerical experiments made in FEA software FreeFem++ confirm this statement.

References

1. Samarskii, A.A., Tikhonov, A.N.: About representation of the field in a waveguide in the form of the sum of fields TE and TM. J. Teor. Phys. **18**(7), 959–970 (1948). (in Russian)
2. Chew, W.C.: Lectures on theory of microwave and optical waveguides (2012). http://wcchew.ece.illinois.edu
3. Sveshnikov, A.G.: A substantiation of a method for computing the propagation of electromagnetic oscillations in irregular waveguides. U.S.S.R. Comput. Math. Math. Phys. **3**(2), 413–429 (1963)
4. Delitsyn, A.L.: On the completeness of the system of eigenvectors of electromagnetic waveguides. Comput. Math. Math. Phys. **51**(10), 1771–1776 (2011)
5. Delitsyn, A.L.: Application of the finite element method to the calculation of modes of dielectric waveguides. Comput. Math. Math. Phys. **39**(2), 298–304 (1999)
6. Lezar, E., Davidson, D.B.: Electromagnetic waveguide analysis. In: Automated Solution of Differential Equations by the Finite Element Method. The FEniCS Project, pp. 629–643 (2011)
7. Delitsyn, A.L.: An approach to the completeness of normal waves in a waveguide with magnetodielectric filling. Differ. Equ. **36**(5), 695–700 (2000)
8. Malykh, M.D., Sevastianov, L.A., Tiutiunnik, A.A., Nikolaev, N.E.: On the representation of electromagnetic fields in closed waveguides using four scalar potentials. J. Electromagn. Waves Appl. **32**(7), 886–898 (2017)
9. Malykh, M.D., Sevastianov, L.A., Tiutiunnik, A.A., Nikolaev, N.E.: Diffraction of electromagnetic waves on a waveguide joint. In: EPJ Web of Conferences, vol. 173, p. 02014 (2018)
10. Duvaut, G., Lions, J.-L.: Les inéquations en mécanique et en physique. Dunod, Paris (1972)
11. Stummel, F.: Rand- und Eigenwertaufgaben in Sobolewschen Räumen. LNM, vol. 102. Springer, Heidelberg (1969). https://doi.org/10.1007/BFb0059060
12. Keldysh, M.V.: On the completeness of the eigenfunctions of some classes of nonselfadjoint linear operators. Russ. Math. Surv. **26**(4), 15–44 (1971)
13. Bogolyubov, A.N., Delitsyn, A.L., Malykh, M.D.: On the root vectors of a cylindrical waveguide. Comput. Math. Math. Phys. **41**(1), 121–124 (2001)
14. Hecht, F.: New development in FreeFem++. J. Numer. Math. **20**(3–4), 251–265 (2012). 65Y15

Analysis of Hierarchical Compression Parallel Solver for BEM Problems on Intel Xeon CPUs

Dimitar Slavchev[✉] and Svetozar Margenov

Institute of Information and Communication Technologies
at the Bulgarian Academy of Sciences, Sofia, Bulgaria
dimitargslavchev@parallel.bas.bg

Abstract. We compare the performance of traditional Gaussian elimination with a solver utilizing hierarchical compression of the matrix. The test problems are obtained by Boundary Element Method (BEM) simulation of laminar flow around airfoils. The most computationally expensive part of the BEM algorithm is to solve the arising system of linear algebraic equations. The related dense matrix can be compressed using a Hierarchically Semi-Separable (HSS) representation. This significantly lowers the computational complexity of the solution method, thus allowing faster overall execution.

The performance of STRUMPACK library implementation of HSS and the MKL direct solver is compared on Intel Xeon architecture. At the end, we examine the accuracy of the HSS approximation using the (exact) results of Gaussian elimination as a reference solution.

1 Introduction

This work is motivated by the recent development of heterogeneous high performance computing (HPC) architectures. Solving systems of linear algebraic equations with dense matrices is among the most computationally intensive numerical linear algebra problems. This is the topic of the article. The focus is on the comparative performance analysis of a solution method based on hierarchical compression of a class of test matrices obtained by BEM simulation of laminar flows around airfoils.

The traditional Gaussian elimination has computational complexity $O(n^3)$, where n is the number of unknowns (degrees of freedom). The methods based on hierarchical compression have nearly optimal complexity, i.e. $O(r^2 n)$, where r is the maximum rank of off-diagonal blocks of the matrix. Typically r is much smaller than n. For some problems it is either a constant or it grows slowly like $O(\ln n)$.

The STRUMPACK (STRUctured Matrices PACKage) software package implements HSS compression to solve structured matrices (both sparse and dense). This is done by exploiting the structure of such matrices in order to lower the complexity of the overall problem.

© Springer Nature Switzerland AG 2019
G. Nikolov et al. (Eds.): NMA 2018, LNCS 11189, pp. 466–473, 2019.
https://doi.org/10.1007/978-3-030-10692-8_53

The contribution of our study includes both, sequential performance analysis and parallel scalability analysis. The organization of this article is as follows. A brief overview of the HSS implementation in STRUMPACK is given in Sect. 2. The presented numerical results are analyzed in Sect. 3 ending with a brief summary in Sect. 4.

2   HSS Method

A summary of Hierarchically Semi-Separable (HSS) matrices is available in [1], while the parallel algorithm implemented in the STRUMPACK package is described in [2]. A more involved theoretical analysis of HSS matrices can be found in [3].

The HSS framework developed in STRUMPACK consists of:

1. Compression into HSS form using random sampling.
2. Solving linear systems using ULV-like factorization.
3. Computing HSS matrix-vector products.

The HSS compression is the most important part of the HSS framework. Once the matrix is compressed we can efficiently solve the system of equations.

2.1   HSS Representation

The HSS compression uses a *cluster tree* that defines a hierarchical partitioning of the dense matrix A. This decomposition can be performed for any matrix, but it has a practical value mostly when the off-diagonal blocks of A have a *low-rank*. An $n \times n$ matrix A have an HSS form if:

1. The off-diagonal blocks of a 2×2 partitioning of A are low-rank:

$$A = \begin{bmatrix} A_{1,1} & A_{1,2} \\ A_{2,1} & A_{2,2} \end{bmatrix} = \begin{bmatrix} D_1 & U_1^{\text{big}} B_{1,2} V_2^{\text{big}*} \\ U_2^{\text{big}} B_{2,1} V_1^{\text{big}*} & D_2 \end{bmatrix}$$

U_i, $B_{i,j}$ and V_i^* are called *generators*. The D_i matrices are the diagonal blocks of A. If the off-diagonal blocks are "low rank" than the U_i matrices will be "tall and skinny", $B_{i,j}$ matrices will be "small and square" and the V_i^* matrices will be "short and wide". Their aspect ratio depends on the rank.
2. Recursively we can repartition the diagonal blocks D_1 and D_2 and so on. After second level of recursion and renumbering we get:

$$A = \begin{bmatrix} \begin{bmatrix} D_1 & U_1^{\text{big}} B_{1,2} V_2^{\text{big}*} \\ U_2^{\text{big}} B_{2,1} V_1^{\text{big}*} & D_2 \end{bmatrix} & U_3^{\text{big}} B_{3,6} V_6^{\text{big}*} \\ U_6^{\text{big}} B_{6,3} V_3^{\text{big}*} & \begin{bmatrix} D_4 & U_4^{\text{big}} B_{4,5} V_5^{\text{big}*} \\ U_5^{\text{big}} B_{5,4} V_4^{\text{big}*} & D_5 \end{bmatrix} \end{bmatrix}$$

3. The following recurrence relation holds true for the generators appearing on two consecutive levels of recursion:

$$U_3^{\text{big}} = \begin{bmatrix} U_1^{\text{big}} & 0 \\ 0 & U_2^{\text{big}} \end{bmatrix} U_3 \text{ and } V_3^{\text{big}} = \begin{bmatrix} V_1^{\text{big}} & 0 \\ 0 & V_2^{\text{big}} \end{bmatrix} V_3.$$

Then:

$$A = \begin{bmatrix} \begin{bmatrix} D_1 & U_1^{\text{big}} B_{1,2} V_2^{\text{big}*} \\ U_2^{\text{big}} B_{2,1} V_1^{\text{big}*} & D_2 \end{bmatrix} & \begin{bmatrix} U_1^{\text{big}} & 0 \\ 0 & U_2^{\text{big}} \end{bmatrix} U_3 B_{3,6} V_6^* \begin{bmatrix} V_4^{\text{big}*} & 0 \\ 0 & V_5^{\text{big}*} \end{bmatrix} \\ \begin{bmatrix} U_4^{\text{big}} & 0 \\ 0 & U_5^{\text{big}} \end{bmatrix} U_6 B_{6,3} V_3^* \begin{bmatrix} V_1^{\text{big}*} & 0 \\ 0 & V_2^{\text{big}*} \end{bmatrix} & \begin{bmatrix} D_4 & U_4^{\text{big}} B_{4,5} V_5^{\text{big}*} \\ U_5^{\text{big}} B_{5,4} V_4^{\text{big}*} & D_5 \end{bmatrix} \end{bmatrix}$$

The HSS representation of an $n \times n$ matrix A relies on the recursive clustering of the index set $\{1, ..., n\}$. This partitioning is referred to as *HSS tree*. Each node τ is associated with a subset I_τ of $\{1, ..., n\}$. The root node is associated with $\{1, ..., n\}$, and every non-leaf node τ with its children ν_1 and ν_2.

$$I_\tau = I_{\nu_1} \cup I_{\nu_2}$$

The HSS representation of A follows the tree structure:

- For leaf node τ, the diagonal blocks $D_\tau = A(I_\tau, I_\tau)$ are uncompressed.
- For each non leaf node τ, with children ν_1 and ν_2, the corresponding off-diagonal blocks A_{ν_1, ν_2} and A_{ν_2, ν_1} are represented by

$$A_{\nu_1, \nu_2} = U_{\nu_1}^{\text{big}} B_{\nu_1, \nu_2} V_{\nu_2}^{\text{big}*} \tag{1}$$

Furthermore, the following hierarchical relation holds:

$$U_\tau^{\text{big}} = \begin{bmatrix} U_{\nu_1}^{\text{big}} & 0 \\ 0 & U_{\nu_2}^{\text{big}} \end{bmatrix} \text{ and } V_\tau^{\text{big}} = \begin{bmatrix} V_{\nu_1}^{\text{big}} & 0 \\ 0 & V_{\nu_2}^{\text{big}} \end{bmatrix}.$$

We never have to calculate or store explicitly the "big" matrices at non-leaf nodes. We can always construct them from their children, grandchildren and so on until we go down to the leaf nodes. The resulting from the above example is shown on Fig. 1.

The importance of ordering the rows and columns of A has to be noticed. In some particular cases, the matrices coming from real life problems are generated in an order that preserves the low-rank property.

In general, for a given dense matrix A, Eq. (1) does not hold exactly. A proper threshold ε is introduced to calculate the generators. Typically a large threshold will result in higher compression (i.e. the HSS form of the matrix will take less space and will be faster to compute) but lower accuracy.

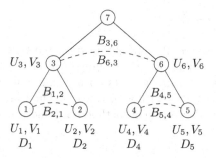

Fig. 1. 3-level HSS tree.

2.2 HSS Compression with Random Sampling

The compression algorithm implemented in STRUMPACK uses randomized sampling, which is implemented by multiplying the matrix with a set of random vectors. The method is introduced by Martinsson [1]. The main advantage of this approach is that it doesn't need to access the entire matrix A, but only parts of it.

If a fast sparse matrix-vector product is utilized, the complexity of the HSS compression is $O(r^2 n)$, where n is the size of the matrix and r is the maximum rank of the off-diagonal blocks, found during the (approximate) compression. For matrices with proper structural properties, r is much smaller than n. A typical behaviour of r for matrices arising form BEM discretization is $O(\ln n)$.

2.3 ULV-like Factorization and Solution

Once (approximately) compressed into HSS form the matrix A can be factorized using ULV factorization [4] (a special form of LU factorization). This factorization uses orthogonal transformations to transform the problem of eliminating n unknowns into the problem of eliminating $O(r)$ unknowns. The remaining unknowns are then eliminated using LU factorization.

The factorization used in STRUMPACK does not use orthogonal transformation, and instead uses the special structure of the HSS generators therefore it is referred to as "ULV-like" factorization (Fig. 2). The complexity is $O(r^2 n)$ [4,5].

After the ULV-like factorization, the linear system is solved using triangular solution in two passes. The complexity is $O(rn)$ [4,5].

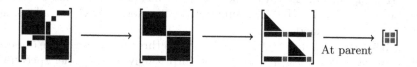

Fig. 2. Illustration of the *ULV-like* factorization process.

3 Numerical Results

The presented numerical results are obtained on the HPC cluster AVITOHOL of the Institute of Information and Communication Technologies, Bulgarian Academy of Sciences. We run the tests on a single node with two Intel Xeon E5-2650v2 8C 2.6 GHz CPUs with 8 cores each. The examined test problem is based on applying boundary element method for numerical simulation of laminar flow around airfoils [6].

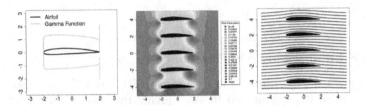

Fig. 3. Values of the γ function on the central airfoil, velocity field around the wing profiles (center), streamlines (right)

In a nutshell we simulate the laminar flow around five airfoils as shown on Fig. 3. After solving the arising system of linear equations we obtain the values of the γ function for each boundary element on the five airfoils (the solution for the middle one is shown on the left). From there we can also calculate the velocity field around the airfoils as well as the streamlines.

The compression tolerance ε, required by the HSS algorithm, is associated with the relative and absolute thresholds ε_{rel} and ε_{abs}. STRUMPACK allows both to be specified by the user. The compression process stops when one of them is reached or the algorithm determines that the matrix is singular. The latter happens when the tolerance is too large and, consequently, STRUMPACK is unable to get a proper approximate compression of the matrix.

We present numerical tests for several different settings of ε_{rel} (the default value of 10^{-2} as well as 10^{-4}, 10^{-8} and 10^{-12}). For the absolute threshold ε_{abs}, the default value of 10^{-8} is used.

The times of the sequential tests are presented on Fig. 4 (left). The asymptotic behavior of $O(n^3)$ is clearly seen for the MKL implementation of the Gaussian elimination. We see also the nearly optimal complexity of the STRUMPACK of the HSS algorithm, as well as the impact of increasing the rank r with the decrease of ε_{rel}. STRUMPACK significantly outperforms MKL for all settings of ε_{rel} in sequential mode.

The best parallel times are observed when 16 threads are used. The obtained results are plotted on Fig. 4 (right). Not surprisingly, the parallel speedup of Gaussian solver from MKL is better than the STRUMPACK ones. The second important conclusion concerns the overall performance. For larger n, the observed advantage of STRUMPACK is restricted to the case of relatively larger

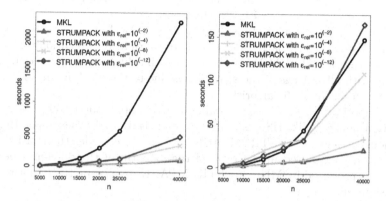

Fig. 4. Performance of STRUMPACK and MKL: sequential test (left), and parallel scalability tests using 16 cores (right)

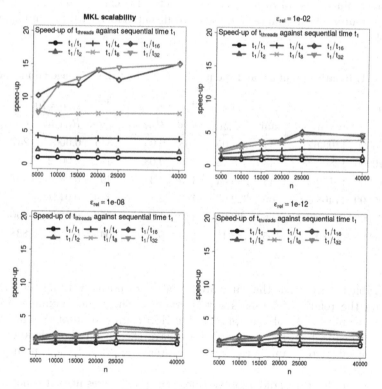

Fig. 5. Parallel speedup of MKL (up, left), STRUMPACK with $\varepsilon_{rel} = 10^{-2}$(up, right), 10^{-8}(down, left), 10^{-12}(down, right),

$\varepsilon_{rel} = 10^{-2}, 10^{-4}$ and $\varepsilon_{rel} = 10^{-8}$. For the highest accuracy tested $\varepsilon_{rel} = 10^{-2}$ the results are comparable with the MKL performance.

The parallel speedup is presented in Fig. 5. MKL shows almost optimal speedup. Using 16 threads gives us almost 16 times faster execution times. This is not the case for STRUMPACK though, which achieves only a fraction of this speedup with all settings of the ε_{rel}. This is due to the much more complicated recursive structure of the algorithm.

An important question when working with an approximate compression (factorization) method, like HSS, is how accurate it is. In order to examine the used threshold settings of the STRUMPACK package we consider the relative accuracy $R_{relative}$ expressed in the ratio of the ℓ_2 norm of the difference between the results produced by MKL and STRUMPACK divided by the ℓ_2 norm of the MKL solution.

$$R_{relative} = \frac{\left| x^{Gauss} - x^{HSS} \right|_{l_2}}{\left| x^{Gauss} \right|_{l_2}} = \frac{\sqrt{\sum_{i=1}^{n} (x_i^{Gauss} - x_i^{HSS})^2}}{\sqrt{\sum_{i=1}^{n} (x_i^{Gauss})^2}}$$

The MKL solution is taken as reference/exact. This choice is ruled by the fact that the computed solution as well as the resultant velocity field seem physically plausible.

Table 1. Relative accuracy and the maximum rank of the off-diagonal matrices

n	$\varepsilon_{rel} = 10^{-2}$		$\varepsilon_{rel} = 10^{-4}$		$\varepsilon_{rel} = 10^{-8}$		$\varepsilon_{rel} = 10^{-12}$	
	$R_{relative}$	rank	$R_{relative}$	rank	$R_{relative}$	rank	$R_{relative}$	rank
5000	1703.62	34	3.05	163	0.05	345	0.0001	237
10000	12820.38	43	8.48	159	0.08	452	0.0012	269
15000	10150.31	43	235.54	149	0.36	494	0.0066	307
20000	17408.79	65	322.98	160	2.8	549	0.0131	415
25000	2489700.59	66	37.05	150	1.09	566	0.0089	490
40000	227805	105	188.1	213	0.56	645	0.052	838

In Table 1 we present the calculated relative ℓ_2 norms varying the problem sizes and the relative tolerances tested. We also show the maximum rank of the off-diagonal blocks of the approximate HSS compression of A. The rank is directly dependent on the values that we have chosen for the tolerances. The higher the calculated rank is the better accuracy we get but the time needed is also increased.

Decreasing the threshold improves the accuracy. The results with the smallest relative threshold $\varepsilon_{rel} = 10^{-12}$ show the best accuracy. It should be noted that for a given set of thresholds, the accuracy of the results degrades as n grows. With the default threshold of $\varepsilon_{rel} = 10^{-2}$ the relative error is particularly high.

4   Concluding Remarks

Performance analysis of the STRUMPACK package for solving dense systems of linear algebraic equations arising from the use of Boundary Element Method is presented. A Hierarchical Semi-Separable (HSS) based method is tested on Intel Xeon E5-2650v2 8C 2.6 GHz CPUs and the parallel performance is measured against using a direct method. Several different tolerances for the HSS compression are examined in order to evaluate the accuracy of the method.

The HSS based method works significantly faster in the sequential mode but it's not as parallel efficient as the direct method from MKL. The accuracy of the method is sensitive with respect to threshold parameters which must be fine tuned for a given size of the problem in order to achieve acceptable results.

Acknowledgments. The partial support by the Bulgarian NSF Grant DN 12/2 is acknowledged. The firs author is also supported trough Bulgarian Academy of Sciences Program for support of Ph.D. Students.   ·

We acknowledge the opportunity to run the numerical tests on the HPC cluster AVITOHOL [7] of the Institute of Information and Communication Technologies, Bulgarian Academy of Sciences.

References

1. Martinsson, P.G.: A fast randomized algorithm for computing a hierarchically semiseparable representation of a matrix. SIAM J. Matrix Anal. Appl. **32**(4), 1251–1274 (2011)
2. Rouet, F.-H., Li, X.S., Ghysels, P., Napov, A.: A distributed-memory package for dense hierarchically semi-separable matrix computations using randomization. ACM Trans. Math. Softw. **42**(4), 27:1–27:35 (2016)
3. Xia, J., Chandrasekaran, S., Gu, M., Li, X.S.: Fast algorithms for hierarchically semiseparable matrices. Numer. Linear Algebr. Appl. **17**(6), 953–976 (2010)
4. Chandrasekaran, S., Gu, M., Lyons, W.: A fast adaptive solver for hierarchically semiseparable representations. CALCOLO **42**(3), 171–185 (2005)
5. Xia, J.: Randomized sparse direct solvers. SIAM J. Matrix Anal. Appl. **34**(1), 197–227 (2013)
6. Slavchev, D., Margenov, S.: Performance analysis of Intel Xeon Phi MICs and Intel Xeon CPUs for solving dense systems of linear algebraic equations: case study of boundary element method for flow around airfoils. In: Georgiev, K., Todorov, M., Georgiev, I. (eds.) Advanced Computing in Industrial Mathematics. SCI, vol. 793, pp. 369–381. Springer, Cham (2019). https://doi.org/10.1007/978-3-319-97277-0_30
7. Avitohol supercomputer. http://www.hpc.acad.bg/system-1/

Application of WRF-CMAQ Model System for Analysis of Sulfur and Nitrogen Deposition over Bulgaria

Dimiter Syrakov, Emilia Georgieva$^{(\boxtimes)}$ (ID), Maria Prodanova, Elena Hristova, Ilian Gospodinov, Kiril Slavov, and Blagorodka Veleva

National Institute of Meteorology and Hydrology, Bulgarian Academy of Sciences, 66, Tsarigradsko Shose blvd., 1784 Sofia, Bulgaria
{dimiter.syrakov,emilia.georgieva,maria.prodanova,elena.hristova, ilian.gospodinov,kiril.slavov,blagorodka.veleva}@meteo.bg
http://www.meteo.bg

Abstract. The advanced air quality modelling system WRF-CMAQ is applied to estimate the spatial distribution of sulfur and nitrogen wet deposition on seasonal basis for 2016 and 2017. The numerical system is set-up for nested domains, from European scale (d1-81 km resolution) to country level (d3-9 km resolution) to account for transport and chemistry processes taking place over broad range of scales and impacting the deposition at given location. A precipitation bias adjustment approach is applied to all grid nodes of domain d3 in order to reduce effects of precipitation overestimation by the model. The effect of the bias adjustment on the seasonal deposition pattern is discussed. The approach leads to 25% decrease in annual wet depositions for the country.

Keywords: Depositions · Modelling · Precipitation chemistry

1 Introduction

Atmospheric deposition is part of complex air pollution and environmental processes that link emissions of pollutants, their chemical transformation and sinks, and their effects on the earth's surface. While the deposition of air pollutants is seen as a natural cleansing process of the atmosphere, and thus beneficial for the air quality, the uptake of deposited substances has been investigated for the adverse effects on terrestrial and aquatic ecosystems (acidification and eutrophication).

Nowadays, with the sharp reductions of sulfur oxides (SO_x) emissions in Europe, nitrogen oxides and ammonia (NO_x and NH_3) are main acidifying compounds in Western and Northern Europe, although emissions of SO_x have higher acidifying potential and are still contributing to acidification [1]. The atmospheric deposition occurs through "wet" and "dry" mechanisms, and the sum of both defines the total deposition. The wet deposition is mainly due to precipitation, the dry deposition - to gravitational settlement, diffusion and turbulent

© Springer Nature Switzerland AG 2019
G. Nikolov et al. (Eds.): NMA 2018, LNCS 11189, pp. 474–482, 2019.
https://doi.org/10.1007/978-3-030-10692-8_54

transfer processes. To monitor the wet deposition, precipitation chemistry networks have been established worldwide, e.g. EMEP in Europe [2]. Monitoring of dry deposition is still challenging, with sparse data for limited periods. Evaluation of deposition based on observations has lower spatial resolution with significant interpolation errors over larger areas [3].

Chemical transport models (CTM) based on numerical modelling of atmospheric transport and chemical transformation processes are recognized as powerful tools to fill in gaps in monitoring data. Their use is driven both by advances in high performing computer technology and availability of open source codes. Numerical models for deposition studies have been applied over the last decade for many regions of the world, [4–6].

In Bulgaria observations on precipitation chemistry are available only for short term periods and in few locations only [7,8]. Numerical simulations have been carried out recently at the National Institute of Meteorology and Hydrology (NIMH) for wet, dry and total sulfur (S) and nitrogen (N) depositions in Bulgaria for the years 2016 and 2017 [9]. The approach of precipitation bias adjustment (PBA) for wet depositions following [4] was tested for single monitoring sites.

This study is an extension of the previous work, with a methodology for PBA not only at single sites, but for the whole territory of the country. The main objective is to investigate the spatio-temporal distribution of these new, "adjusted", S and N wet depositions and to check model performance based on comparison to wet depositions data for sulfur, reduced nitrogen and oxidized nitrogen for 2017 in Sofia.

2   The Modelling System

The Bulgarian Chemical Weather Forecasting System (BgCwFS), operationally running at NIMH since 2012, is applied for the estimation of monthly accumulated depositions of different pollutants.

2.1   Overview and Set-Up

BgCwFS has four main computational modules [10]:

- Meteorological model – the Weather Research and Forecasting Model (WRF v.3.6.1) for 3D fields of winds, temperature, precipitation etc. [11];
- CTM of Eulerian type – the Community Multi-scale Air Quality model, CMAQ v.4.6 - for chemistry transformations, transport and deposition of pollutants [12];
- Interface for linking meteorological data to CTM – the Meteorology-Chemistry Interface Processor; MCIP v.3.6
- Emission module – The Sparse Matrix Operator Kernel Emissions Modelling System (SMOKE v.2.4) - used partly for calculating biogenic emissions and for merging area sources (AS), large point sources (LPS), and biogenic emission files.

As deposition fluxes are the end result of complex atmospheric and chemistry processes taking place over broad range of spatial scales, the nesting approach is applied. Five nested modelling domains are defined: Europe (d1) with grid resolution of 81 km, Balkan Peninsula (d2) - 27 km, Bulgaria (d3) - 9 km, Sofia region (d4) - 3 km, and Sofia city (d5) - 1 km. The vertical structure of the atmosphere is represented by 14 σ-levels with 8 levels in the Planetary Boundary Layer (PBL). The initial and boundary conditions for WRF are provided by the National Centers for Environmental Prediction Global Forecast System (NCEP GFS) data with space resolution of $1° \times 1°$ and temporal resolution of 6 h. Initial conditions for CMAQ are part of previous day calculations. A predefined set of vertical concentration profiles is used as chemical boundary condition for the European domain (d1), all other domains receive their boundary conditions from the previous one in the hierarchy.

Extensively tested parameterization schemes have been selected in the model set-up – the Yonsei University scheme for the PBL, the WSM6 scheme for the microphysics, the Noah Land-Surface model. The chemical mechanism in CMAQ v.4.6 is the 4th generation module "cb4_ae4_aq".

Results for domain d3 ("Bulgaria") are used in the analysis.

2.2 Emissions

The emissions are based on the TNO inventory for 2009 [13], for Bulgaria national emission inventories for 2010 are used. The primary emission source for S deposition is SO_2 coming from combustion of fossil fuels in thermal power plants (TPP) and from industrial processes (refineries). N deposition is due to emissions of NO_x from transport, combustion processes and to NH_3 emissions from agriculture and livestock manure. These gases have also significant effect on the formation of secondary aerosols – sulfates (SO_4^{2-}), nitrate (NO_3^-) and ammonium (NH_4^+), which may account for a substantial part of the total S (N) depositions.

Sulfur emissions are prevailing in SE Europe, while higher emissions of nitrogen oxides and ammonia are noted in the northern part of Central Europe (Fig. 1). The largest LPS of SO_2 in Bulgaria is the coal fired TPPs "Maritza East" in the south-eastern part of the country, other coal operating TPPs are located to the west and north of the country. The nitrogen emissions relevant to deposition in Bulgaria are mainly in the Lower Danube Plain (agricultural regions).

2.3 Deposition Calculations

The atmospheric deposition outputs of CMAQ have been archived in separate files containing hourly values for 29 species, gas and aerosol particles. The S deposition is estimated as the sum of the depositions of sulfate (SO_4^{2-}) and sulfur dioxide (SO_2). The oxidized nitrogen N_{oxi} deposition flux includes nitrate (NO_3), nitrogen oxide (NO), and nitrogen dioxide (NO_2). The reduced nitrogen N_{red} is the sum of ammonia (NH_3) and ammonium (NH_4^+). The sum of N_{oxi}

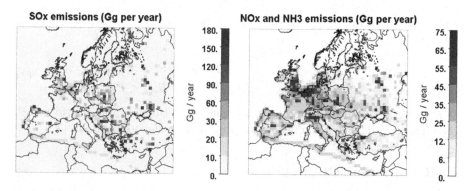

Fig. 1. Emissions (Gg per year) of SO_x (left) and (NH_3 plus NO_x) (right) for 2010 in domain d1

and N_{red} is denoted as N deposition, while the sum of wet and dry is denoted as total deposition.

3 Precipitation Bias Adjustment (PBA) for Wet Depositions

Previous studies on model evaluation for wet deposition in Bulgaria [9,14] indicated that BgCwFS overestimates the precipitation amounts. The application of PBA [4] as post-processing to model wet depositions at single observational sites has shown positive effect on seasonal and annual depositions.

The model wet depositions WD_{mod} were linearly corrected by the ratio of the observed PR_{obs} to estimated precipitation PR_{mod}, following [4]:

$$WD_{mod}^{adj} = WD_{mod} \cdot (PR_{obs}/PR_{mod}) \qquad (1)$$

The extension of the approach to all grid nodes in d3 requires the construction of a reference, "observed" precipitation field. The objective analysis of precipitation is a challenging task due to the heterogeneity in the spatial distribution. A precipitation analysis method, developed at the Forecast Center at NIMH, consist in combination of data from the operational numerical weather prediction model ALADIN [15] and from observations using Cressman [16] analysis. The precipitation model field is used as a first guess, further corrected for the difference between forecasted and observed precipitation amount at a given point. The weight function depends on the distance between the grid points and the observation point, as well as on the difference in their elevation. The current version of the precipitation analysis system has horizontal resolution of about 1 km and uses topographic data from the U.S. Geological Survey with resolution 30 arc-seconds. This system is used to produce diurnal, monthly and seasonal maps of accumulated precipitations in an operational way.

Here, the monthly accumulated precipitation, obtained by the above analysis method, is noted as gridded "observed" precipitation and is used for corrections of model wet depositions in d3. At each grid point the correction factor RAT is calculated:

$$RAT = (PR_{obs}/PR_{mod}) \quad \begin{cases} = 1, & \text{for } PR_{obs} \geq PR_{mod}; \\ \neq 1, & \text{for } PR_{obs} < PR_{mod} \end{cases} \qquad (2)$$

RAT is then used to scale the model monthly wet depositions at grid nodes, where monthly model precipitations are higher than the observed ones.

4   Results and Discussion

4.1   Precipitation

Figure 2a shows average values of gridded precipitation in domain d3 for different seasons in 2016 and 2017 as estimated by the model and by the observations. The normalized mean bias (NMB) has absolute values between 2.5% and 25.4%. In summer (JAS) the underestimation is about 16%, probably due to sub-scale events typical for this season. In spring (AMJ) and autumn (OND) the model overestimation is about 16%. On annual basis, the overestimation is 8.4% (2017) and 1.2% (2016). Figure 2b shows the scatterplot of gridded precipitations for April 2016 as an example for model overestimation in the range (100–150 mm). This overestimation is observed mainly in the mountain areas.

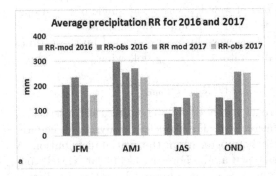 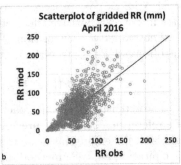

Fig. 2. (a) Precipitation (mm) – Model and observed for different seasons in 2016 and 2017 (b) Scatterplot of gridded precipitations (mm) for April 2016

The spatial distribution of simulated and observed precipitations, and the corresponding correction factor RAT (Fig. 3 for April 2016 as example) show overestimation mainly in the mountainous regions and in the NE part of the country.

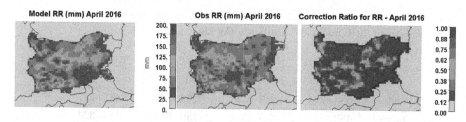

Fig. 3. Monthly precipitation (mm) – model (left), observed (center), RAT (right), April 2016

4.2 Simulated S and N Wet Depositions with PBA

Figure 4 shows the spatial distribution of accumulated S-WD and N-WD (kg.km^{-2}) for the period from January to March 2017 (winter) obtained without PBA and with PBA (adjusted), along with the model and observed precipitation. The extreme SE part of the country has both higher precipitation and wet depositions. This area includes nature parks and depositions might impact adversely the ecosystems. More significant emission sources are located outside of the area, indicating that the simulated deposition is due to regional and long-range transport processes. The maps of the adjusted deposition fields indicate also areas with higher depositions but with not so high precipitation – the areas around the TPP "Maritza East" in southern Bulgaria and the NW part of the country.

The seasonal and annual S-WD and N-WD without PBA (mod) and with PBA (mod-a) are summarized in Table 1. The values represent the average of the depositions for the grid nodes located in Bulgaria. On annual basis, the adjustment leads to 28% decrease for S-WD and to 25% decrease for N-WD.

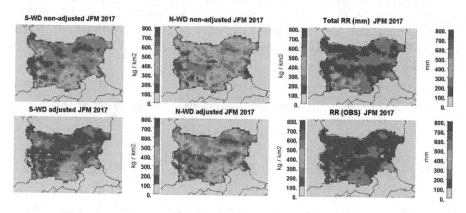

Fig. 4. S-WD (left) and N-WD (center) for winter 2017: without PBA (top) and with PBA (bottom). Accumulated precipitation (mm) (right): model (top), observed (bottom).

Table 1. Simulated seasonal and annual S-WD and N-WD, without PBA (mod) and PBA (mod-a) for 2016 and 2017. Units are (kg.km^{-2} period^{-1}.)

	S-WD 2016		N-WD 2016		S-WD 2017		N-WD 2017	
	mod	mod-a	mod	mod-a	mod	mod-a	mod	mod-a
Winter	333	224	401	328	262	201	361	277
Spring	482	319	377	301	486	383	413	324
Summer	190	163	125	107	250	210	187	158
Autumn	238	177	302	224	337	177	403	224
Annual	**1243**	**883**	**1205**	**960**	**1335**	**971**	**1364**	**983**

The smallest decrease is obtained for summer (14–16%), the greatest decrease (48%) for S-WD in autumn 2017.

4.3 Comparison to Observed Deposition in Sofia

Figure 5 shows results for the wet depositions of S, N_{red} and N_{oxi} for Sofia in 2017. Daily precipitation samples were collected at the NIMH site (42.655N, 23.384E, 586 m a.s.l.) with an automatic wet only precipitation sampler WADOS. 78 precipitation samples were collected and analyzed for acidity-pH, electrical conductivity-EC, main anions and cations, and some elements. The depositions were estimated by multiplying the measured concentration of SO_4^{2-}, NO_3^-, NH_4^+ by the observed precipitation amount and summing for the relative periods. The model values are two – without PBA (mod) and with PBA (adj). The nearest grid node is used for the interpolation of model values to the site of the sampler.

The model simulates correctly the prevalence of S over N depositions for Sofia. N_{oxi}-WD represents the main part of the N-WD. On annual basis NMB for S-WD decreases from 43% (overestimation) to −12% (slight underestimation). NMB for N_{red}-WD is respectively 27% and −26%, NMB for N_{oxi}-WD varies from 35% to −15%.

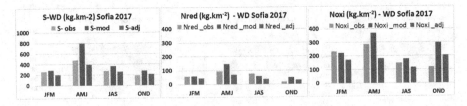

Fig. 5. S-WD (left), N_{red}-WD (center) and N_{oxi}-WD (right) for Sofia on seasonal basis in 2017. First bar – observations, middle bar – model; third bar – model with PBA. Units are (kg.km^{-2}).

5   Conclusion

The Bulgarian Chemical Weather Forecasting System has been set up for calculations of deposition fluxes using nested domains from European to country and city level. The focus in the analysis here is on the effect of the precipitation bias adjustment on the spatial distribution of sulfur and nitrogen wet depositions, as well as on mean gridded values for different seasons in 2016 and 2017. The precipitation bias adjustment approach [4] has been applied to all grid points of the modelling domain. The reference ("observed") precipitation field is obtained by an objective analysis technique, combining observations and data from the analysis of an operational weather forecasting model at NIMH.

The precipitation bias adjusted wet depositions are lower than the original model depositions – on annual basis by about 25%, with higher impact in spring and autumn – up to 48% for sulfur in autumn. The spatial distribution of the adjusted depositions has similar pattern to the original one, but provides better evidence on areas with elevated depositions where the precipitation amounts are not very high. For example the south eastern part of the country is characterized by higher sulfur wet depositions and northern Bulgaria has more nitrogen than sulfur wet depositions in winter.

The results indicate possible beneficial effect of applying precipitation bias adjustment as post-processor to model depositions on seasonal or annual basis. The techniques adopted represent a step towards a methodology for long-term deposition calculations in Bulgaria on regular basis.

Acknowledgements. This study was performed with the financial support from the Bulgarian National Science Fund trough contract N. DN-04/4-15.12.2016. We acknowledge TNO for providing emission data, US EPA and US NCEP for providing free-of-charge air quality models and meteorological data.

References

1. EEA, 2017: Air Quality in Europe-2017 Report, European Environmental Agency, Copenhagen (2017). ISBN 978-92-9213-921-6, Publication Office of EC-Luxembourg, 80 pp (2017)
2. Tørseth, K., et al.: Introduction to the European Monitoring and Evaluation Programme (EMEP) and observed atmospheric composition change during 1972–2009. Atmos. Chem. Phys. **12**, 5447–5481 (2012). https://doi.org/10.5194/acp-12-5447-2012
3. Im, U., et al.: Atmospheric deposition of nitrogen and sulfur over southern Europe with focus on the Mediterranean and the Black Sea. Atmos. Environ. **81**, 660–670 (2013). https://doi.org/10.1016/j.atmosenv.2013.09.048
4. Appel, K.W., et al.: A multi-resolution assessment of the Community Multiscale Air Quality (CMAQ) model v4.7 wet deposition estimates for 2002–2006. Geosci. Model Dev. **4**, 357–371 (2011). https://doi.org/10.5194/gmd-4-357-2011

5. Vivanco, M.G., Theobald, M.R., García-Gómez, H., Garrido, J.L., Prank, M., Galmarini, S.: Modelled deposition of nitrogen and sulfur in Europe estimated by 14 air quality model-systems: evaluation, effects of changes in emissions and implications for habitat protection, Atmos. Chem. Phys. Discuss. (2018). https://doi.org/10.5194/acp-2018-104

6. Ge, B.Z., Wang, Z.F., Xu, X.B., Wu, J.B., Yu, X.L., Li, J.: Wet deposition of acidifying substances in different regions of China and the rest of East Asia: modeling with updated NAQPMS. Environ. Pollut. **187**, 10–21 (2014). https://doi.org/10.1016/j.envpol.2013.12.014

7. Iordanova, L.: Local and advective characteristics of the precipitations' chemical composition in Sofia, Bulgaria. Compt. Rend. Acad. Bulg. Sci. **63**(2), 295–302 (2010)

8. Hristova, E.: Chemical composition of precipitation in urban area. Bul. J. Meteo Hydro. **22**(1–2), 41–49 (2017)

9. Syrakov, D., Prodanova, M., Georgieva, E., Hristova, E.: Applying WRF-CMAQ models for assessment of sulfur and nitrogen deposition in Bulgaria for years 2016 and 2017. Intern. J. Env. Pollut. (submitted in April 2018)

10. Syrakov, D., et al.: A multi-domain operational chemical weather forecast system. In: Lirkov, I., et al. (Eds.) LSSC 2013. LNCS, vol. 8353, pp. 413–420. Springer, Heidelberg (2014)

11. Skamarock, W.C., Klemp, J.B.: A time-split non-hydrostatic atmospheric model. J. Comput. Phys. **227**, 3465–3485 (2008). https://doi.org/10.1016/j.jcp.2007.01.037

12. Byun, D., Schere, K.L.: Review of the governing equations, computational algorithms and other components of the Models-3 Community Multiscale Air Quality (CMAQ) modeling system. Appl. Mech. Rev. **59**, 51–77 (2006). https://doi.org/10.1115/1.2128636

13. Kuenen, J.J.P., Visschedijk, A.J.H., Jozwicka, M., Denier van der Gon, H.A.C.: TNO-MACC_II emission inventory; a multi-year (2003–2009) consistent high-resolution European emission inventory for air quality modelling. Atmos. Chem. Phys. **14**, 10963–10976 (2014). https://doi.org/10.5194/acp-14-10963-2014

14. Georgieva, E., Hristova, E., Syrakov, D., Prodanova, M., Batchvarova, E.: Preliminary evaluation of CMAQ modelled wet deposition of sulphur and nitrogen over Bulgaria. In: Proceedings of the 18th International Conference on HARMO18, 9–12 October 2017, Bologna, Italy, pp. 51–55 (2017)

15. ALADIN home page http://www.cnrm.meteo.fr/aladin/

16. Cressman, G.P.: An operational objective analysis system. Mon. Weather Rev. **87**, 367–374 (1959)

Digital Approximations
for Pseudo-Differential Equations
and Error Estimates

Vladimir B. Vasilyev$^{(\boxtimes)}$ (iD)

Belgorod National Research University, Belgorod 308007, Russia
vbv57@inbox.ru

Abstract. We consider a discrete version of pseudo-differential operators and equations in appropriate discrete functional spaces. Using a special factorization for an elliptic symbol we obtain certain results on a solvability for such discrete equations and give a comparison between discrete and continuous solutions.

Keywords: Digital pseudo-differential operator
Periodic wave factorization · Solvability · Approximation rate

1 Introduction

This paper is devoted to the approximation of the solution of a very wide class of integro-differential equations, namely pseudo-differential equations [1,2]. A pseudo-differential operator A in a domain D of Euclidean space $\mathbf{R}^m$ is defined by its symbol $A(x, \xi)$, i.e. function defined in $D \times \mathbf{R}^m$, by the following formula

$$(Au)(x) = \int\limits_{D} \int\limits_{\mathbf{R}^m} A(x, \xi) e^{i(x-y)\cdot \xi} u(y) d\xi dy, \quad x \in D.$$

To construct a good approximate solution for the equation

$$(Au)(x) = v(x), \quad x \in D, \tag{1}$$

we use some discrete-periodic constructions. First we start from simplest equation (1) in which the symbol $A(x, \xi)$ does not depend on a spatial variable x. Thus, for the Eq. (1) with symbol $A(\xi)$ we construct a discrete operator A_d with the symbol $A_d(\xi)$. Such discrete operator is defined for functions $u_d(\tilde{x})$ of a discrete variable $\tilde{x} \in h\mathbf{Z}^m, h > 0$, and acts in certain discrete functional spaces. Then we replace the Eq. (1) by its discrete analogue

$$(A_d u_d)(\tilde{x}) = v_d(\tilde{x}), \quad \tilde{x} \in D_d, \tag{2}$$

where $D_d = D \cap h\mathbf{Z}^m$.

Supported by the State contract of the Russian Ministry of Education and Science (contract No. 1.7311.2017/8.9).

G. Nikolov et al. (Eds.): NMA 2018, LNCS 11189, pp. 483–490, 2019.
https://doi.org/10.1007/978-3-030-10692-8_55

Further we study solvability of the equation (2) for some canonical domains D using the discrete Fourier transform F_d

$$(F_d u_d)(\xi) = \sum_{\tilde{x} \in h\mathbf{Z}^m} e^{-i\tilde{x}\cdot\xi} u_d(\tilde{x})h^m, \quad \xi \in h^{-1}[-\pi,\pi]^m,$$

and its inverse, and give some comparison estimates for solutions of the equations (1) and (2) in some discrete functional spaces.

If $u_d(\tilde{x}), \tilde{x} \in h\mathbf{Z}^m$, is a function of a discrete variable then we say "discrete function". For such discrete functions one can define the discrete Fourier transform

$$(F_d u_d)(\xi) \equiv \tilde{u}_d(\xi) = \sum_{\tilde{x} \in h\mathbf{Z}^m} e^{-i\tilde{x}\cdot\xi} u_d(\tilde{x})h^m, \quad \xi \in \hbar\mathbf{T}^m,$$

if the latter series converges. The obtained function $\tilde{u}_d(\xi)$ is periodic in $\mathbf{R}^m$ with basic cube of periods $\hbar\mathbf{T}^m$. Such discrete Fourier transform preserves all key properties of the integral Fourier transform, particularly the inverse discrete Fourier transform is given by the formula

$$(F_d^{-1}\tilde{u}_d)(\tilde{x}) = \frac{1}{(2\pi)^m} \int_{\hbar\mathbf{T}^m} e^{i\tilde{x}\cdot\xi} \tilde{u}_d(\xi)d\xi, \quad \tilde{x} \in h\mathbf{Z}^m.$$

2   Digital Pseudo-differential Operators

Let $A_d(\xi)$ be a periodic function in $\mathbf{R}^m$ with basic cube of periods $h^{-1}[-\pi,\pi]^m$ so that

$$c_1(1+|\zeta_h^2|)^{\frac{\alpha}{2}} \leq |A_d(\xi)| \leq c_2(1+|\zeta_h^2|)^{\frac{\alpha}{2}}, \tag{3}$$

where $\zeta_h^2 = h^{-2}\sum_{k=1}^{m}(e^{-ih\xi_k}-1)^2$, and the positive constants c_1, c_2 do not depend on h.

Let $D \subset \mathbf{R}^m$ be a domain (finite or infinite). We will consider the functions $u_d(\tilde{x})$ defined in $D_d \equiv D \cap h\mathbf{Z}^m, h > 0$, and introduce the following operator

$$(A_d u_d)(\tilde{x}) = \sum_{\tilde{y} \in h\mathbf{Z}^m} \int_{\hbar\mathbf{T}^m} A_d(\xi)u_d(\tilde{y})e^{i(\tilde{x}-\tilde{y})\cdot\xi}h^m d\xi, \quad \tilde{x} \in D_d,$$

where $\hbar \equiv h^{-1}, \mathbf{T}^m \equiv [-\pi,\pi]^m$.

Definition 1. *The operator A_d is called a discrete pseudo-differential operator or shortly h-operator. The periodic function $A_d(\xi)$ is called its $\hbar$-symbol.*

Let us remind that a symbol (operator) is called elliptic if

$$ess \inf_{\xi \in \hbar\mathbf{R}^m} |A_d(\xi)| > 0,$$

and obviously all symbols under consideration are elliptic.

3   Equations in a Cone

Definition 2. *By definition the space $H^s(h\mathbf{Z}^m)$ is a closure of the discrete Schwartz space $S(h\mathbf{Z}^m)$ with respect to the norm*

$$||u_d||_s = \left(\int\limits_{\hbar\mathbf{T}^m} (1 + |\zeta_h^2|)^s |\tilde{u}_d(\xi)|^2 d\xi \right)^{1/2}. \tag{4}$$

Definition 3. *The space $H^s(D_d)$ consists of discrete functions from $H^s(h\mathbf{Z}^m)$ with supports in $\overline{D_d}$. The norm in the space $H^s(D_d)$ is induced by the norm of the space $H^s(h\mathbf{Z}^m)$. The space $H_0^s(D_d)$ consists of discrete functions (distributions from $S'(\mathbf{R}^m)$) u_d with supports in D_d, additionally these discrete functions must admit a continuation ℓ onto $H^s(h\mathbf{Z}^m)$. A norm in the space $H_0^s(D_d)$ is given by the formula*

$$||u_d||_s^+ = \inf ||\ell u_d||_s,$$

where the infimum is taken over all continuations ℓ.

The solvability of the equation (2) in such spaces for the half-space $D = \mathbf{R}_+^m$ was studied earlier [5–9]. Below we will consider briefly a more complicated case than a half-space.

3.1   Tube Domains and Periodic Bochner Kernel

Let D be a convex cone that does not include a whole straight line, and $\overset{*}{D}$ be a conjugate cone for D, i.e.,

$$\overset{*}{D} = \{x \in \mathbf{R}^m : x \cdot y > 0, \ y \in D\}.$$

Let $T(\overset{*}{D}) \subset \mathbf{C}^m$ be a set of the type $\hbar\mathbf{T}^m + i \overset{*}{D}$. For $\hbar\mathbf{T}^m \equiv \mathbf{R}^m (h \to 0)$ such a domain of multidimensional complex space is called a radial tube domain over the cone $\overset{*}{D}$ [3, 4, 13]. We introduce the function which is called periodic Bochner kernel

$$B_d(z) = \sum_{\tilde{x} \in D_d} e^{i\tilde{x} \cdot z} h^m, \quad z = \xi + i\tau, \quad \xi \in \hbar\mathbf{T}^m, \quad \tau \in \overset{*}{D},$$

and define the operator

$$(B_d u)(\xi) = \lim_{\tau \to 0} \int\limits_{\hbar\mathbf{T}^m} B_d(z - \eta) u_d(\eta) d\eta.$$

This operator is roughly speaking a conical analogue of the periodic Hilbert transform [5, 6].

3.2   The Periodic Wave Factorization

To describe solvability conditions for the equation (2) we introduce the following concept.

Definition 4. *The periodic wave factorization for an elliptic symbol $A_d(\xi)$ is called its representation in the form*

$$A_d(\xi) = A_{d,\neq}(\xi)A_{d,=}(\xi),$$

where the factors $A_{d,\neq}(\xi), A_{d,=}(\xi)$ admit an analytical continuation into domains $T(\overset{*}{D}), T(-\overset{*}{D})$ respectively and satisfy the estimates

$$|A^{\pm 1}_{d,\neq}(\xi)| \le c_1(1 + |\hat{\zeta}^2|)^{\pm \frac{\ae}{2}}, \quad |A^{\pm 1}_{d,=}(\xi)| \le c_2(1 + |\hat{\zeta}^2|)^{\pm \frac{\alpha - \ae}{2}},$$

with constants c_1, c_2 non-depending on h,

$$\hat{\zeta}^2 \equiv \hbar^2 \left(\sum_{k=1}^{m} (e^{-ih(\xi_k + \tau_k)} - 1)^2 \right), \quad \xi \in \hbar\mathbf{T}^m, \tau \in \pm \overset{*}{D}.$$

The number $\ae \in \mathbf{R}$ is called an index of the periodic wave factorization.

3.3   A Solvability

Theorem 1. *If the elliptic symbol $\tilde{A}_d(\xi)$ admits periodic wave factorization with the index $\ae$ so that $|\ae - s| < 1/2$ then the operator $A_d : H^s(D_d) \to H^{s-\alpha}(D_d)$ is invertible and a solution of the equation (2) for arbitrary right-hand side $v_d \in H_0^{s-\alpha}(D_d)$ in Fourier images is given by the formula*

$$\tilde{u}_d(\xi) = A^{-1}_{d,\neq}(\xi)B_d(A^{-1}_{d,=}(\xi)\widetilde{\ell v_d}(\xi)), \tag{5}$$

where ℓv_d is an arbitrary continuation of v_d into $H^s(h\mathbf{Z}^m)$.

Proof. Let ℓv_d be an arbitrary continuation of v_d on the whole $h\mathbf{Z}^m$ so that $\ell v_d \in H^{s-\alpha}(h\mathbf{Z}^m)$. Let

$$w_d(\tilde{x}) = (\ell v_d)(\tilde{x}) - (A_d u_d)(\tilde{x})$$

and rewrite

$$(A_d u_d)(\tilde{x}) + w_d(\tilde{x}) = (\ell v_d)(\tilde{x}).$$

Further applying the discrete Fourier transform F_d and using the periodic factorization we write

$$A_{d,\neq}(\xi)\tilde{u}_d(\xi) + A^{-1}_{d,=}(\xi)\tilde{w}_d(\xi) = A^{-1}_{d,=}(\xi)\widetilde{\ell v_d}(\xi).$$

According to considerations from [10–12] we have $A_{d,\neq}(\xi)\tilde{u}_d(\xi) \in \tilde{H}^{s-\ae}(h\mathbf{Z}^m)$, $\tilde{A}^{-1}_{d,=}(\xi)\tilde{w}_d(\xi) \in \tilde{H}^{s-\alpha+\alpha-\ae}(h\mathbf{Z}^m)$ and analogously $A^{-1}_{d,=}(\xi)\widetilde{\ell v_d}(\xi) \in \tilde{H}^{s-\ae}(h\mathbf{Z}^m)$.

Moreover, really $A_{d,\neq}(\xi)\tilde{u}_d(\xi) \in \widetilde{H}^{s-\infty}(D_d)$ in view of a holomorphy property, and accurate considerations with supports of $A_{d,=}(\xi)$ and $\tilde{w}_d(\xi)$ show that in fact $A_{d,-}^{-1}(\xi)\tilde{w}_d(\xi) \in \widetilde{H}^{s-\infty}(h\mathbf{Z}^m \setminus D_d)$.

Thus we obtain a variant of a jump problem for the space $\widetilde{H}^{s-\infty}(h\mathbf{Z}^m)$ which was considered in [12] and according to this result we have

$$\tilde{A}_{d,\neq}(\xi)\tilde{u}_d(\xi) = B_d(\tilde{A}_{d,=}^{-1}(\xi)\widetilde{\ell v_d}(\xi))$$

or finally

$$\tilde{u}_d(\xi) = \tilde{A}_{d,\neq}^{-1}(\xi)B_d(\tilde{A}_{d,=}^{-1}(\xi)\widetilde{\ell v_d}(\xi)).$$

This finishes the proof.   △

4   Approximation Rate

4.1   Test Operators and the Periodic Wave Factorization

It is natural there is a question if we have such factorization for some cases. Here we consider simple variant of a pseudo-differential operator, namely the Calderon–Zygmund operator. Such operators satisfy the condition (3) with $\alpha = 0$ so we can consider these operators as linear bounded operators in the space $H^0(h\mathbf{Z}^m) = L^2(h\mathbf{Z}^m)$ (see also [5,8]). We will give here certain sufficient conditions for an existence of the periodic wave factorization for an elliptic symbol.

Theorem 2. *Let an elliptic symbol $A_d(\xi) \in C(\hbar\mathbf{T}^m)$ be such that*

$$\text{supp } F_d^{-1}(\ln A_d(\xi)) \subset D_d \cup (-D_d), \tag{6}$$

Then the symbol $A_d(\xi)$ admits a periodic wave factorization with vanishing index.

Proof. If we start from equality

$$A_d(\xi) = A_{\neq}(\xi) \cdot A_{=}(\xi)$$

then taking a logarithm we obtain

$$\ln A_d(\xi) = \ln A_{\neq}(\xi) + \ln A_{=}(\xi)$$

and we have a special kind of a jump problem [10–12]. Let us note that $\ln A_d(\xi)$ is a univalent function because the domain is a simply connected.

Further, we will denote by $A_1(\hbar\mathbf{T}^m)$ a subspace of the space $L_2(\hbar\mathbf{T}^m)$ consisting of functions which admit a holomorphic bounded continuation into $T(\overset{*}{D})$. So evidently we consider the possibility of decomposition of the function $\ln A_d(\xi)$ into two summands one of which belongs to the space $A_1(\hbar\mathbf{T}^m)$ and the second one belongs to the space $A_2(\hbar\mathbf{T}^m)$ (bounded holomorphic continuations into $T(-\overset{*}{D})$). Let us denote

$$F^{-1}(\ln \tilde{A}_d(\xi)) \equiv v(x).$$

If $supp\ v \subset D_d \cup (-D_d)$ then we have the unique representation

$$v = \chi_+ v + \chi_- v$$

where $\chi_\pm$ is an indicator of the discrete set $\pm D_d$.

Further passing to the Fourier transform and potentiating we obtain the required factorization. $\triangle$

Remark 1. The condition (6) is not necessary but we have no algorithm for constructing a periodic wave factorization. For $D = \mathbf{R}_+^m$ a such algorithm always exists (see [6,7]).

4.2   A Comparison

For a comparison we need to introduce the special projector $Q_h : \mathbf{R}^m \to h\mathbf{Z}^m$ which is defined for smooth functions at least. For such a function v we take its Fourier transform $Fv \equiv \tilde{v}$, then we take a restriction of $\tilde{v}$ to the $\hbar\mathbf{T}^m$ and periodically continue it to the whole $\mathbf{R}^m$. Finally we take its inverse discrete Fourier transform F_d^{-1} and denote the result by $Q_h v$. Obviously this is a discrete function defined in $h\mathbf{Z}^m$.

If we impose strong enough restrictions on the right-hand side and the factorization elements then one can give a comparison between discrete and continuous solutions.

Let $S(\mathbf{R}^m)$ be the Schwartz space of infinitely differentiable rapidly decreasing at infinity functions, and P_h be a restriction operator on $h\mathbf{Z}^m$, i.e. for $u \in S(\mathbf{R}^m)$

$$(P_h u)(x) = \begin{cases} u(\tilde{x}), & x = \tilde{x} \in h\mathbf{Z}^m; \\ 0, & x \notin h\mathbf{Z}^m. \end{cases}$$

Lemma 1. *For* $u \in S(\mathbf{R}^m), \forall \beta > 0$, *we have*

$$|(P_h u)(\tilde{x}) - (Q_h u)(\tilde{x})| \leq Ch^\beta, \quad \forall \tilde{x} \in h\mathbf{Z}^m,$$

where the constant C *depends on* u *only.*

Proof. Indeed, we need to compare two Fourier transforms. By definition

$$(P_h u)(\tilde{x}) = \frac{1}{(2\pi)^m} \int\limits_{\mathbf{R}^m} e^{i\tilde{x}\cdot\xi} \tilde{u}(\xi)d\xi,$$

and respectively

$$(Q_h u)(\tilde{x}) = \frac{1}{(2\pi)^m} \int\limits_{\hbar\mathbf{T}^m} e^{i\tilde{x}\cdot\xi} \tilde{u}(\xi)d\xi,$$

thus this difference is given by the integral

$$(P_h u)(\tilde{x}) - (Q_h u)(\tilde{x}) = \frac{1}{(2\pi)^m} \int\limits_{\mathbf{R}^m \setminus \hbar\mathbf{T}^m} e^{i\tilde{x}\cdot\xi} \tilde{u}(\xi)d\xi.$$

The conclusion of Lemma 1 follows from the invariance of the Schwartz class $S(\mathbf{R}^m)$ with respect to the Fourier transform and the simple estimate

$$|\tilde{u}(\xi)| \le C_u|\xi|^{-\gamma}$$

$\forall \gamma > 0.$ △

A non-periodic analogue of the operator B_d is the following [3, 4, 13]

$$(B\tilde{u})(\xi) = \lim_{\tau \to 0+} \int\limits_{\mathbf{R}^m} B(x - y + i\tau)\tilde{u}(y)dy,$$

where $B(z)$ is the Bochner kernel

$$B(z) = \int\limits_{D} e^{iy \cdot z}dy, \quad z = x + i\tau, \quad x \in \mathbf{R}^m, \tau \in \pm \overset{*}{D}.$$

Lemma 2. *If $u \in S(\mathbf{R}^m)$ then the following estimate*

$$|(F^{-1}B\tilde{u})(\tilde{x}) - (F_d^{-1}B_d\widetilde{Q_h u})(\tilde{x})| \le Ch^\beta, \quad \tilde{x} \in D_d,$$

holds for $\forall \beta > 0$, and the constant C depends on u only.

Proof. Here we need the description and comparison for two projectors related to the Hilbert transform, both standard and periodic. Let us denote by $\chi(x)$ an indicator of the cone D and $\chi_d(\tilde{x})$ an indicator of the discrete cone D_d. Then according to structural properties of two mentioned transforms we have the following equalities

$$F^{-1}B\tilde{u} = \chi \cdot u, \quad F_d^{-1}B_d\widetilde{Q_h u} = \chi_d \cdot (Q_h u).$$

Further one can apply the Lemma 1.   △

4.3   Continuous and Discrete Solutions

Starting from Lemma 2 and Theorem 1 we are able to compare discrete and continuous solutions in a cone. Below we give this comparison under the conditions of Theorem 2 for the existence of a unique solution. We suppose in this section that initial symbol $A(\xi)$ admits the wave factorization with respect to the cone D [4]

$$A(\xi) = A_{\ne}(\xi) \cdot A_{=}(\xi),$$

and $A_d(\xi)$ is a restriction of $A(\xi)$ on $\hbar\mathbf{T}^m$ which is periodically extended to the whole $\mathbf{R}^m$.

Theorem 3. *If the symbol $A(\xi)$ is infinitely differentiable in $\mathbf{R}^m$ with the factors $A_{\ne}(\xi), A_{=}(\xi)$, u is the unique solution of the equation (1) with support in D, u_d is a solution of the equation (2) then for $v \in S(\mathbf{R}^m)$ we have the following error estimate*

$$|u(\tilde{x}) - u_d(\tilde{x})| \le Ch^\beta, \quad \forall \tilde{x} \in h\mathbf{Z}_+^m,$$

for arbitrary $\beta > 0$.

To refine this theorem we will show how to choose a right-hand side for solving the Eq. (2). The solution of the equation (1) in Fourier images has the form [4]

$$\tilde{u}(\xi) = A_{\neq}^{-1}(\xi)BA_{=}^{-1}(\xi)\widetilde{\ell v}(\xi),$$

where ℓv is an arbitrary continuation of v from D onto the whole $\mathbf{R}^m$ in the corresponding functional space. Since the right-hand side in the Eq. (2) is defined in D_d only then one needs to choose $Q_h(\ell v)$ instead of ℓv_d to obtain the required estimate.

References

1. Taylor, M.E.: Pseudodifferential Operators. Princeton University Press, Princeton (1981)
2. Eskin, G.: Boundary Value Problems for Elliptic Pseudodifferential Equations. AMS, Providence (1981)
3. Bochner, M., Martin, W.T.: Several Complex Variables. Princeton University Press, Princeton (1948)
4. Vasil'ev, V.B.: Wave Factorization of Elliptic Symbols: Theory and Applications. Introduction to the Theory of Boundary Value Problems in Non-smooth Domains. Kluwer academic Publishers, Dordrecht (2000)
5. Vasilyev, A.V., Vasilyev, V.B.: Discrete singular operators and equations in a half-space. Azerb. J. Math. 3(1), 84–93 (2013)
6. Vasil'ev, A.V., Vasil'ev, V.B.: Periodic Riemann problem and discrete convolution equations. Differ. Equ. 51(5), 652–660 (2015)
7. Vasil'ev, A.V., Vasil'ev, V.B.: On the solvability of certain discrete equations and related estimates of discrete operators. Doklady Math. 92(2), 585–589 (2015)
8. Vasilyev, A.V., Vasilyev, V.B.: Discrete singular integrals in a half-space. In: Current Trends in Analysis and its Applications. Trends in Mathematics, pp. 663–670. Birkhäuser, Basel (2015)
9. Vasilyev, A.V., Vasilyev, V.B.: On a digital approximation for pseudo-differential operators. Proc. Appl. Math. Mech. 17(1), 763–764 (2017)
10. Vasilyev, V.B.: Discrete equations and periodic wave factorization. In: AIP Conference Proceedings, vol. 1759, pp. 0200126-1–0200126-5 (2016)
11. Vasilyev, V.B.: The periodic Cauchy kernel, the periodic Bochner kernel, and discrete pseudo-differential operators. In: AIP Conference Proceedings, vol. 1863, pp. 140014-1–140014-4 (2017)
12. Vasilyev, V.B.: Discrete pseudo-differential operators and boundary value problems in a half-space and a cone. Lobachevskii J. Math. 39(2), 289–296 (2018)
13. Vladimirov, V.S.: Methods of the Theory of Functions of Many Complex Variables. Dover Publications, Mineola (2007)

Some Difference Algorithms for Nonlinear Klein-Gordon Equations

Asuman Zeytinoglu[1] and Murat Sari[2(✉)]

[1] Suleyman Demirel University, Isparta, Turkey
asumanzeytinoglu@sdu.edu.tr
[2] Yildiz Technical University, Istanbul, Turkey
sarim@yildiz.edu.tr

Abstract. In this study, sixth and eighth-order finite difference schemes combined with a third-order strong stability preserving Runge-Kutta (SSP-RK3) method are employed to cope with the nonlinear Klein-Gordon equation, which is one of the important mathematical models in quantum mechanics, without any linearization or transformation. Various numerical experiments are examined to verify the applicability and efficiency of the proposed schemes. The results indicate that the corresponding schemes are seen to be reliable and effectively applicable. Another salient feature of these algorithms is that they achieve high-order accuracy with relatively less number of grid points. Therefore, these schemes are realized to be a good option in dealing with similar processes represented by partial differential equations.

Keywords: Klein-Gordon equation · Nonlinear processes
High-order finite difference scheme
Strong stability preserving Runge-Kutta

1 Introduction

Several physical phenomena of quantum mechanics, electricity, plasma physics, fluid dynamics, propagation of waves and many other physical processes like these are described by partial differential equations within their range of validity. Since the partial differential equations have become a very useful tool for describing these natural models, a wide variety of physically significant problems modeled by nonlinear partial differential equations has been the focus of extensive studies. One of the most important mathematical models is the nonlinear Klein-Gordon equation in quantum field theory. The equation is the basic evolution equation in relativistic field theory [1] and is known as one of the nonlinear wave equations. It arises in nonlinear optics, plasma physics and quantum field theory. This equation has attracted much attention in studying condensed matter physics, in investigating the interaction of solitons in collisionless plasma, the recurrence of initial states [2]. Due to these reasons, the Klein-Gordon equations have been the focus of a great deal of studies in that either proofs of

© Springer Nature Switzerland AG 2019
G. Nikolov et al. (Eds.): NMA 2018, LNCS 11189, pp. 491–498, 2019.
https://doi.org/10.1007/978-3-030-10692-8_56

the existence of solutions or seeking analytical and numerical solutions were considered. There are many researchers who used various approaches [3–14] to produce the solutions of the Klein-Gordon equation. Although the equation has been extensively studied, it is still a problem of continuing interest because different physical phenomena are modeled by the Klein-Gordon equation. It is desirable to use higher-order numerical schemes to obtain significantly accurate and numerically more economic responses of many problems. As pointed out in [15], the high-order schemes are not only accurate and effective but also can give rise to satisfactory results with less number of grid points. The current paper, therefore, explores the utility of two high order schemes named sixth and eighth-order finite difference (FD6 and FD8) to solve different forms of the non-linear Klein-Gordon equations. These methods are applied for discretizing spatial derivatives. Then, the third-order strong stability preserving Runge-Kutta (SSP-RK3) method is accepted to deal with the temporal derivation of the processes. The computed results are compared with available results in the literature and thus the accuracy of them have been shown in terms of some error norms. This paper is arranged as follows. In Sect. 2, the model equation called the general form of one-dimensional Klein-Gordon equation is introduced. In Sect. 3, FD6 and FD8 schemes for spatial discretization, and the SSP-RK3 approximation for the time integration are presented. In order to demonstrate the applicability and efficiency of the proposed methods, numerical experiments are conducted in Sect. 4. Finally, conclusions are given that briefly summarize the results in Sect. 5.

2   The Model Equation

The following Klein-Gordon equation

$$u_{tt} + \alpha u_{xx} + \beta u + \gamma u^k = f(x,t), \qquad a \leq x \leq b, \qquad t \geq 0 \qquad (1)$$

with the initial and boundary conditions

$$u(x,0) = g_1(x), \quad u_t(x,0) = g_2(x), \quad u(a,t) = g_3(a,t), \quad u(b,t) = g_4(b,t) \quad (2)$$

where α, β and γ are known constants, $f(x,t), g_1, g_2, g_3, g_4$ are known functions, and unknown function $u(x,t)$ are considered in this paper. In order to assess the applicability and accuracy of the high order finite difference methods for considered problems, different forms of the nonlinear Klein-Gordon equation mentioned above are taken into consideration.

3   Discretizations

3.1   Spatial Variation via High-Order FD Schemes

Spatial derivatives are evaluated by various orders of finite difference schemes. At first, the domain of the problem $[a, b]$ is divided into N subintervals as follows

$$a = x_1 < x_2 < ... < x_N < x_{N+1} = b, \qquad h = \Delta x = x_{i+1} - x_i$$

for $i = 1, 2, ..., N$. Thus, the numerical solution of u is denoted by u_i^n at grid point (x_i, t^n). To discretize the term u_{xx} in Eq. (1), the FD6 and FD8 schemes are derived for the second order derivatives. The second order spatial derivative u_i'' at point i can be approximated by the following $(R+L)$-order finite difference scheme using $(R + L + 1)$-point stencil

$$u_i'' = \frac{1}{h^2} \sum_{j=-L}^{R} a_{j+L} u_{i+j}, \qquad 1 \le i \le N + 1. \tag{3}$$

In the above formulae, R and L indicate the number of grid points in the right and left hand sides for the taken stencil, respectively. At internal points, R and L are equal while they are different for the boundary nodes. The coefficients a_j are unknown constants which need to be known at each point i. To determine the coefficients a_j, Taylor series expansion is used with 7-point stencil generating the FD6 schemes. After these operations, the computed values of a_j belonging to the FD6 are written into matrix forms as follows:

$$A = \frac{1}{180} \begin{pmatrix} 812 & -3132 & 5265 & -5080 & 2970 & -972 & 137 & 0 & \cdots & \cdots & 0 \\ 137 & -147 & -255 & 470 & -285 & 93 & -13 & 0 & \cdots & \cdots & 0 \\ -13 & 228 & -420 & 200 & 15 & -12 & 2 & 0 & \cdots & \cdots & 0 \\ 2 & -27 & 270 & -490 & 270 & -27 & 2 & 0 & \cdots & \cdots & 0 \\ 0 & 2 & -27 & 270 & -490 & 270 & -27 & 2 & 0 & \cdots & 0 \\ 0 & 0 & 2 & -27 & 270 & -490 & 270 & -27 & 2 & 0 & 0 \\ \vdots & & & & & & & & & & \\ 0 & \cdots & \cdots & 0 & 2 & -12 & 15 & 200 & -420 & 228 & -13 \\ 0 & \cdots & \cdots & 0 & -13 & 93 & -285 & 470 & -255 & -147 & 137 \\ 0 & \cdots & \cdots & 0 & 137 & -972 & 2970 & -5080 & 5265 & -3132 & 812 \end{pmatrix}.$$

Thus, the second order spatial derivative term can be re-written into matrix form as follows:

$$U'' = AU$$

where $U = (u_1, u_2, ..., u_{N+1})^T$. The coefficients a_j for FD8 schemes are determined by using 9-point stencil in a similar way (see [16]).

3.2 Temporal Variation via Third-Order Strong Stability Preserving Runge-Kutta Method

After application of the FD techniques to the related equation, the equation can be reduced into a set of ordinary differential equations in time. Then the governing equation becomes

$$\frac{d^2 u_i}{dt^2} = P u_i \tag{4}$$

where P indicates a spatial linear/nonlinear differential operator. The spatial terms are approximated by the schemes. To solve Eq. (4), the SSP-RK3 is applied. Each spatial derivative on the right hand side of Eq. (4) was computed using the present methods and then the semi-discrete Eq. (4) was solved considering the SSP-RK3 through the following process:

$$u_i^{(1)} = u_i^m + \Delta t P u_i^m,$$

$$u_i^{(2)} = \frac{3}{4} u_i^m + \frac{1}{4} u_i^{(1)} + \frac{1}{4} \Delta t P u_i^{(1)},$$

$$u_i^{m+1} = \frac{1}{3} u_i^m + \frac{2}{3} u_i^{(2)} + \frac{2}{3} \Delta t P u_i^{(2)}.$$

To use these operations, Eq. (4) was re-writen as a system of first-order ordinary differential equations.

It is well-known that several factors such as computer speed, available memory, desired accuracy and stability influence the choice of time integration technique. For solving hyperbolic conservation laws with stable spatial discretizations, Gottlieb et al. [17] developed a high-order SSP time discretization technique. As was pointed out by them, there is no stability criterion for fully discrete methods where P is nonlinear, unlike in the case of linear operators. However, the SSP methods guarantee the stability properties expected of the forward Euler method [17]. Notice that it is possible to use existing stability conditions when the problems are linearized. It can be recognized that the linearization of a model leads to loss of originality of the physical problem.

4   Numerical Experiments

The above approaches are applied to obtain numerical solutions for certain forms of the nonlinear Klein-Gordon equation. Three examples are considered to show the applicability and efficiency of the proposed methods. The numerical computations have been performed using uniform grids. To show the accuracy of the schemes, the following error norms

$$L_\infty = \max |u_i^{exact} - u_i^{numerical}|, \quad RMS = \sqrt{\frac{\sum_{i=1}^{N+1} |u_i^{exact} - u_i^{numerical}|^2}{N+1}}$$

are used via the corresponding exact solutions and the results are compared with available results in the literature.

Example 1. The following nonlinear and nonhomogenous form of the Klein-Gordon equation

$$u_{tt} = u_{xx} - u^2 + 6xt(x^2 - t^2) + x^6 t^6,$$

is taken with the initial conditions $u(x,0) = 0, u_t(x,0) = 0$ where $0 \leq x \leq 1$. The exact solution is given by $u(x,t) = x^3 t^3$. For this example, the RMS and L_∞ error norms of the proposed schemes are calculated up to $t = 6$ and are documented in Table 1 together with the literature results. As seen in Table 1, the results computed with less grid points are relatively more accurate than the literature results given in the table.

Table 1. The RMS and L_∞ errors of the presented methods at various times for Example 1

	FD6 $\Delta t = 0.00001, h = 0.1$		FD8 $\Delta t = 0.00001, h = 0.1$		Ref. [12] $\Delta t = 0.0001, h = 0.05$		Ref. [10] $\Delta t = 0.00005, h = 0.01$		Ref. [6] $\Delta t = 0.0001, h = 0.02$	
t	L_∞	RMS	L_∞	RMS	L_∞	RMS	L_∞	RMS	L_∞	RMS
1	1.50E-05	6.36E-06	1.50E-05	6.36E-06	7.80E-06	4.86E-06	5.87E-04	1.92E-04	1.10E-05	5.47E-06
2	6.00E-05	2.54E-05	6.00E-05	2.54E-05	1.23E-04	7.77E-05	4.66E-03	2.15E-03	1.65E-04	1.15E-04
3	1.35E-04	5.73E-05	1.35E-04	5.73E-05	5.30E-04	2.45E-04	1.51E-02	4.92E-03	5.97E-04	3.24E-04
4	2.40E-04	1.02E-04	2.10E-04	1.03E-04	1.86E-03	9.70E-04	3.42E-02	8.47E-03	1.83E-03	9.77E-04
5	3.50E-04	1.47E-04	4.20E-02	1.80E-02	3.52E-03	1.90E-03	6.32E-02	1.30E-02	3.69E-03	1.90E-03
6	2.85E-02	1.04E-02	—	—	—	—	—	—	—	—

Example 2. Now, the following cubically nonlinear homogenous Klein-Gordon equation

$$u_{tt} = \frac{5}{2}u_{xx} - u - \frac{3}{2}u^3$$

is taken with the interval $0 \le x \le 1$. The initial and boundary conditions are given by

$$u(x,0) = \sqrt{\frac{2}{3}}\tan(\sqrt{\frac{2}{9}}x), \quad u_t(x,0) = \frac{1}{2}\sqrt{\frac{2}{3}}\sqrt{\frac{2}{9}}\sec^2(\sqrt{\frac{2}{9}}x),$$

$$u(0,t) = \sqrt{\frac{2}{3}}\tan(\sqrt{\frac{2}{9}}\frac{t}{2}), \quad u(1,t) = \sqrt{\frac{2}{3}}\tan(\sqrt{\frac{2}{9}}(1+\frac{t}{2}))$$

with the exact solution

$$u(x,t) = \sqrt{\frac{2}{3}}\tan(\sqrt{\frac{2}{9}}(x+\frac{t}{2})).$$

The L_∞ and RMS error norms of the proposed schemes are calculated at different time t and presented in Table 2 together with some available results in the literature. It can be seen from the corresponding table that the results especially produced by the FD6 scheme are in good agreement with the compared results. Furthermore, the number of grid points used in the proposed schemes is less than the number of grid points used in the literature works. On the other hand, the results revealed that the FD8 scheme starts to lose the accuracy earlier than the FD6 scheme, because in higher order methods the overflow can be seen more quickly as compared to lower order methods.

Example 3. As a final example, a quadratically nonlinear, nonhomogenous form of the Klein-Gordon equation is taken by

$$u_{tt} = u_{xx} - \frac{\pi^2}{4}u - u^2 + x^2(\sin\frac{\pi t}{2})^2$$

over $0 \le x \le 1$ with the initial and boundary conditions

$$u(x,0) = 0, \quad u_t(x,0) = \frac{\pi x}{2}, \quad u(0,t) = 0, \quad u(1,t) = \sin\frac{\pi t}{2}$$

extracted from the exact solution $u(x,t) = x\sin\frac{\pi t}{2}$.

Table 2. The RMS and L_∞ errors of the presented methods at various times for Example 2

	FD6 $\Delta t = 0.001, h = 0.1$		FD8 $\Delta t = 0.001, h = 0.1$		Ref. [6] $\Delta t = 0.001, h = 0.01$		Ref. [8] $\Delta t = 0.0001$	
t	L_∞	RMS	L_∞	RMS	L_∞	RMS	L_∞	RMS
1	8.90E-08	4.89E-08	4.89E-07	2.00E-07	4.08E-05	4.06E-06	6.12E-06	3.56E-06
2	2.80E-06	1.61E-06	5.97E-04	3.30E-04	1.58E-04	1.57E-05	2.22E-05	1.35E-05
3	7.08E-05	4.40E-05	1.63E-00	7.33E-01	6.48E-04	6.45E-05	9.13E-05	5.42E-05
4	2.56E-03	1.07E-03	—	—	5.36E-03	5.33E-03	7.79E-04	4.48E-04

	Ref. [10] $\Delta t = 0.00005, h = 0.01$		Ref. [11] $\Delta t = 0.001, h = 0.01$		Ref. [14] $\Delta t = 0.00001, h = 0.00625$	
t	L_∞	RMS	L_∞	RMS	L_∞	RMS
1	1.42E-04	6.62E-05	3.57E-05	2.23E-06	1.36E-07	7.45E-08
2	4.66E-04	1.54E-04	1.32E-04	7.43E-06	9.96E-07	7.15E-07
3	1.95E-03	4.93E-04	4.29E-04	2.18E-05	6.32E-06	4.59E-06
4	2.82E-02	7.15E-03	2.18E-03	8.63E-05	1.99E-04	1.33E-04

Table 3. The RMS and L_∞ errors of the present methods for Example 3

		FD6									
		$t = 0.1$		$t = 1$		$t = 2$		$t = 3$		$t = 4$	
Δt	h	L_∞	RMS	L_∞	RMS	L_∞	RMS	L_∞	RMS	L_∞	RMS
0.001	0.1	3.13E-07	1.49E-07	3.60E-05	2.44E-05	6.71E-05	4.45E-05	1.19E-04	6.88E-05	3.66E-04	2.09E-04
	0.05	3.27E-07	1.60E-07	3.62E-05	2.50E-05	6.78E-05	4.58E-05	1.73E-04	1.08E-04	4.75E-03	1.87E-03
0.0001	0.1	3.16E-08	1.50E-08	3.60E-06	2.44E-06	6.67E-06	4.45E-06	1.15E-05	6.76E-06	3.05E-05	1.75E-05
	0.05	3.30E-08	1.62E-08	3.62E-06	2.50E-06	6.66E-06	4.56E-06	1.33E-05	7.98E-06	1.67E-04	9.21E-05
0.00001	0.1	3.16E-09	1.50E-09	3.60E-07	2.44E-07	6.67E-07	4.45E-07	1.14E-06	6.75E-07	2.99E-06	1.72E-06
	0.05	3.30E-09	1.62E-09	3.62E-07	2.50E-07	6.65E-07	4.56E-07	1.30E-06	7.82E-07	1.61E-05	8.60E-06

		$t = 5$		$t = 6$		$t = 7$		$t = 8$	
Δt	h	L_∞	RMS	L_∞	RMS	L_∞	RMS	L_∞	RMS
0.001	0.1	2.18E-03	1.48E-03	1.20E-02	8.70E-03	1.07E-01	4.89E-02	9.56E-01	5.82E-01
	0.05	1.36E-01	6.04E-02	4.52E-00	2.06E-00	—	—	—	—
0.0001	0.1	2.11E-04	1.26E-04	1.20E-03	8.21E-04	7.37E-03	4.45E-03	5.90E-02	2.75E-02
	0.05	3.65E-03	1.58E-03	7.02E-02	3.54E-02	2.30E-00	1.04E-00	—	—
0.00001	0.1	2.07E-05	1.23E-05	1.21E-04	8.13E-05	7.05E-04	4.46E-04	5.73E-03	2.62E-03
	0.05	3.17E-04	1.47E-04	6.55E-03	2.96E-03	1.86E-01	8.49E-02	4.32E-00	2.11E-00

		FD8									
		$t = 0.1$		$t = 1$		$t = 2$		$t = 3$		$t = 4$	
Δt	h	L_∞	RMS	L_∞	RMS	L_∞	RMS	L_∞	RMS	L_∞	RMS
0.001	0.1	3.15E-07	1.49E-07	3.64E-05	2.44E-05	4.16E-04	1.62E-04	2.99E-02	1.48E-02	5.09E-00	2.29E-00
	0.05	3.27E-07	1.61E-07	5.03E-05	2.82E-05	2.02E-01	6.15E-02	—	—	—	—
0.0001	0.1	3.18E-08	1.50E-08	3.63E-06	2.44E-06	3.87E-05	1.51E-05	2.57E-03	1.19E-03	3.81E-01	1.78E-01
	0.05	3.30E-08	1.62E-08	3.91E-06	2.71E-06	1.29E-02	3.14E-03	—	—	—	—
0.00001	0.1	3.18E-09	1.50E-09	3.63E-07	2.44E-07	3.85E-06	1.50E-06	2.53E-04	1.17E-04	3.69E-02	1.73E-02
	0.05	3.30E-09	1.62E-09	3.80E-07	2.70E-07	1.23E-03	2.97E-04	8.12E-00	1.96E-00	—	—

		$t = 5$		$t = 6$		$t = 7$		$t = 8$	
Δt	h	L_∞	RMS	L_∞	RMS	L_∞	RMS	L_∞	RMS
0.001	0.1	—	—	—	—	—	—	—	—
	0.05	—	—	—	—	—	—	—	—
0.0001	0.1	—	—	—	—	—	—	—	—
	0.05	—	—	—	—	—	—	—	—
0.00001	0.1	3.26E-00	1.52E-00	—	—	—	—	—	—
	0.05	—	—	—	—	—	—	—	—

The error norms are calculated for various values of Δt and h up to times $t = 8$, and are presented in Table 3. It can be said from the table that the accuracy of the suggested schemes are quite high especially for relatively small time values. It is also clear that the error norms decrease as the values of Δt decrease. However, if the parameter h is decreased simultaneously together with the value of Δt, the errors tend to increase in time. The results also revealed that the FD8 scheme starts to lose the accuracy earlier than the FD6 scheme as in the previous example.

The average absolute errors and convergence rate (CR) of the proposed schemes are produced for values of Δt at $t = 2$ in Table 4. The numerical rate of convergence (CR) has been studied to know about the convergency of the schemes. The CR values are calculated using the following formula:

$$CR \approx \frac{\log(E(\Delta t_1)/E(\Delta t_2))}{\log(\Delta t_1/\Delta t_2)},$$

where $E(\Delta t_j)$ is the average absolute error when using the time steps Δt_j.

Table 4. The CR values of the present methods with $h = 0.1$ at $t = 2$ in Example 3

FD6			FD8		
Δt	Average absolute error	CR	Δt	Average absolute error	CR
1/16	3.0584E-01	-	1/16	7.0698E+01	-
1/20	1.0749E-01	4.69	1/20	1.1676E+01	8.07
1/24	3.1030E-02	6.82	1/24	2.4932E+00	8.47

It can be observed from the examples that the produced results are satisfactory until a certain time t, but this may not always be the case for large time values. This is understandable since the proposed methods are explicit. To overcome this drawback, an exponential, implicit or a finite element method [8,11] etc. can be preferred.

5   Conclusions

In this paper, high-order difference schemes combined with the SSP-RK3 have been proposed to efficiently solve the nonlinear Klein-Gordon equations. The validity and accuracy of the numerical models have been verified through the computed results and the literature. The results indicate that the corresponding schemes have been seen to be reliable and easy to use. Furthermore, the results produced with relatively less grid points have been understood to be more accurate than some available results in the literature. It has been revealed that proposed techniques have been realized to be very good alternatives to some existing ones in solving physical problems represented by the Klein-Gordon equations.

References

1. Arodz, H., Hadasz, L.: Lectures on Classical and Quantum Theory of Fields. Springer, London (2010). https://doi.org/10.1007/978-3-642-15624-3
2. Dodd, R.K., Eilbeck, I.C., Gibbon, J.D., Morris, H.C.: Solitons and Nonlinear Wave Equations. Academic, London (1982)
3. Cao, W.M., Guo, B.Y.: Fourier collocation method for solving nonlinear Klein-Gordon equation. J. Comput. Phys. **108**, 296–305 (1993)
4. El-Sayed, S.M.: The decomposition method for studying the Klein-Gordon equation. Chaos Soliton. Fract. **18**, 1025–1030 (2003)
5. Inc, M., Ergut, M., Evans, D.J.: An efficient approach to the Klein-Gordon equation: an application of the decomposition method. Int. J. Simul. Process Model. **2**, 20–24 (2006)
6. Dehghan, M., Shokri, A.: Numerical solution of the nonlinear Klein-Gordon equation using radial basis functions. J. Comput. Appl. Math. **230**, 400–410 (2009)
7. Bratsos, A.G.: On the numerical solution of the Klein-Gordon equation. Numer. Methods Partial Differ. Equ. **25**, 939–951 (2009)
8. Yin, F., Tian, T., Song, J., Zhu, M.: Spectral methods using Legendre wavelets for nonlinear Klein/Sine-Gordon equation. J. Comput. Appl. Math. **275**, 321–334 (2015)
9. Khuri, S.A., Sayfy, A.: A spline collocation approach for the numerical solution of a generalized nonlinear Klein-Gordon equation. Appl. Math. Comput. **216**, 1047–1056 (2010)
10. Li, Q., Ji, Z., Zheng, Z., Liu, H.: Numerical solution of nonlinear Klein-Gordon equation using Lattice Boltzmann method. Appl. Math. **2**, 1479–1485 (2011)
11. Mittal, R.C., Bhatia, R.: Numerical solution of nonlinear system of Klein-Gordon equations by cubic B-spline collocation method. Int. J. Comput. Math. **92**, 2139–2159 (2015)
12. Sarboland, M., Aminataei, A.: Numerical solution of the nonlinear Klein-Gordon equation using multiquadratic quasi-interpolation scheme. Univ. J. Appl. Math. **3**, 40–49 (2015)
13. Guo, P.F., Liew, K.M., Zhu, P.: Numerical solution of nonlinear Klein-Gordon equation using the element-free kp-Ritz method. Appl. Math. Model. **39**, 2917–2928 (2015)
14. Chang, C.W., Liu, C.S.: An implicit Lie-group iterative scheme for solving the nonlinear Klein-Gordon and sine-Gordon equations. Appl. Math. Model. **40**, 1157–1167 (2016)
15. Sankaranarayanan, S., Shankar, N.J., Cheong, H.F.: Three-dimensional finite difference model for transport of conservative pollutants. Ocean Eng. **25**, 425–442 (1998)
16. Sari, M., Gurarslan, G., Zeytinoglu, A.: High-order finite difference schemes for the solution of the generalized Burgers-Fisher equation. Commun. Numer. Methods Eng. **27**, 1296–1308 (2011)
17. Gottlieb, S., Shu, C.W., Tadmor, E.: Strong stability-preserving high-order time discretization methods. SIAM Rev. **43**, 89–112 (2001)

Author Index

Printed in the United States
By Bookmasters